Light as an Ecological Factor: II

Light as an Ecological Factor: II

The 16th Symposium of
The British Ecological Society
26–28 March 1974

edited by
G. C. Evans
R. Bainbridge
and
O. Rackham

Blackwell Scientific Publications
OXFORD LONDON EDINBURGH MELBOURNE

Osney Mead, Oxford,
8 John Street, London, WC1
9 Forrest Road, Edinburgh,
P.O. Box 9, North Balwyn, Victoria, Australia.

ISBN 0 632 00087 2

Distributed in the U.S.A. by
Halsted Press,
a division of
John Wiley & Sons, Inc.,
New York

Printed in Great Britain by
Western Printing Services Ltd
Bristol
and bound by
The Kemp Hall Bindery
Oxford

Contents

Section 7
Demonstrations

Introduction

In this 16th Symposium of the British Ecological Society we develop the topic of the 6th Symposium of ten years ago. When introducing that symposium volume we said '. . . light occupies a special place as an ecological factor because of the number, diversity and importance of the effects which it produces on both plants and animals; and also because of the complexity of the light climate, and the difficulties which have attended attempts to measure and characterize it. During recent years there has been such a wealth of investigation on these different aspects that full justice can obviously not be done to all in a three-day meeting. We have accordingly felt it wise to focus attention on two particular types of community of plants and animals, one terrestrial and one aquatic, with the intention of discussing a wide range of topics of ecological interest in a connected way.'

The timeliness of this previous volume was sufficiently demonstrated by the fact that the first printing was sold out before many of the English reviews appeared: in planning the present volume it has been our hope to maintain interest by carrying the work forward without repetition. Although many topics have been greatly extended, remarkably little of the work previously described has been falsified or superseded during the last decade. We are therefore able to take it as read and use it as a foundation for the present work, which is largely devoted to topics which are new (e.g. visual effects of polarized light) or which previously received only glancing reference (e.g. heating effects of short-wave radiation).

At the same time we have endeavoured to knit the two volumes into a single whole. A few papers describe substantial extensions of previous work, where this is relevant to the themes of the present meeting, and all the other contributors of papers were invited to supply addenda. Six kindly did so, and these appear as Section 6 (Nos 23–28).

Otherwise the principles we adopted in organizing the meeting were essentially the same as those used before. Once again we endeavoured to maintain a balance between interests in plants and animals, and between terrestrial and aquatic communities. For terrestrial situations we leant away from our previous topic of woods and forests towards communities smaller in scale, and for aquatic situations towards fresh water rather than our previous topic of the sea. Once again we arranged 22 papers for a three-day meeting, which we consider near the optimum. The detailed arrangements having previously met with satisfaction, the reading of communications usually lasted about 30 minutes, as before 'depending on the requirements of the author and the place of the topic in the programme; but authors were encouraged, prior to the meeting, to prepare a longer paper for the symposium volume if they felt it desirable, and then to present its principal features in such a way as to stimulate discussion. In this way the author's full material [including the content of a number of demonstrations given in conjunction with papers] has become available to those unable to attend the meeting in person, while the time available for formal and informal discussion was not curtailed. It thus naturally happens that many questions asked in discussion are answered in the full text of the paper. Authors who wished were also able to take account of matters raised in discussion when revising their papers for the press immediately after the meeting. . . . There remained, however, a number of points of interest.' Mr Stephen Head checked through and transcribed the tape recording of the questions and discussions, and with one exception we have included at the end of the relevant papers those points not already covered.

The exception is a plea for uniformity of units, made by Professor J. S. L. Monteith at the end of the session over which he presided. This is of such general importance as to warrant presentation here. Professor Monteith referred to the bewildering range of different units used in *Light as an Ecological Factor I*, resulting from the absence at that time of clearly accepted units in which to express radiant energy and light. The situation had since been altered by the introduction of the Système International d'Unités. SI units had been widely promulgated in Europe and were coming in, if more slowly, in North America also. The system had been advocated and publicized in Britain by the Royal Society, the British Standards Institute, and scientific societies, but although accepted by schools and by many universities it had not yet been generally adopted. He pointed out that in this current Symposium irradiance had been expressed in several different units based on calories, whereas the recommended SI unit for irradiance was joules per square metre per second. This could be simplified to watts per square metre, the unit which had been used, for instance, by Professor Spence. Professor Monteith was prepared to make small departures from strict SI units for reasons of convenience, for instance in going up by powers of 10^3; but in fact the watt per square metre was a very convenient unit for practical use. Values of hundreds of W m^{-2} were commonly encountered in the open, and values of less than 100 W m^{-2} in shade, which was consistent with the precision of the instruments usually used for measuring irradiance. For energy integrated over time the joule per square metre per day was too small a unit: the megajoule per square metre per day was more convenient, since at this latitude it ranged from the order of 2–5 in the open in winter to the order of 20–25 in summer. Professor Monteith finally expressed the hope that there might be more standardization of units in the rest of the Symposium, and especially in the papers as published.

His plea fell on very sympathetic ears, and a number of authors took the opportunity of revision after the meeting to change their units to the SI equivalent. We have welcomed this, while not feeling able to insist on uniformity in all cases. As university teachers we are aware of the great strides which have been made in the use of SI units for science teaching in British schools, so that it now seems that the majority of undergraduates reading natural sciences come up to their university thinking naturally in SI units. This was certainly not the case ten years ago, when our main aim in this field was to encourage the use of absolute heat or energy units and discourage (in cases other than the study of vision) the use of units depending on the spectral sensitivity of the human eye. But we are also aware that agreement on units is not universal, and that roughly two-fifths of the members of the British Ecological Society reside outside the British Isles. Some authorities, including meteorological services, continue to use units based on the calorie, for understandable historical reasons; and there are even curious hybrids like the milliwatt-hour, which seem to us without merit. The very extensive literature provides a third strand to our problem. The *Smithsonian Meteorological Tables* are still in calories—probably the greatest single step forward in this field would be for them to be brought up to date and reissued in SI units: the bulk of the publications following on the International Geophysical Year are in calories: and so on. For many years to come the student of this field will have to learn to use more than one system of units. For all these reasons we have encouraged authors to use SI units, but have not insisted on it.

The usual compromises have been necessary in the use of common, technical and scientific names for plants and animals, although once again it would be merely pedantic to aim at complete uniformity of treatment. In the index we have endeavoured to link all important English names with their scientific equivalents, for the benefit of readers to whom they are not the common names, although even here there are problems, as with the word 'duck'. In general all scientific names are indexed, the exception being when many species of a single genus are all referred to on the same page of the text. Here '*Genus* spp.' has been used to keep down the length of the index while, it is hoped, sacrificing little of its utility.

We know that all those present at the meeting and many who read this volume will deeply regret the loss of Dr Philip Lloyd, who died in a tragic accident in the summer of 1975.

Our thanks are due to the Master and Fellows of St John's College for accommodation, and for providing a generous subvention which made it possible to reduce the charges to students attending the meeting; to the University of Cambridge for the use of the Lady Mitchell Hall; and to Professor Brian who provided facilities for a Conversazione and for the exhibition of demonstrations in the Department of Botany. We should also like to thank the many members of their staffs, members of the Society and others whose help and co-operation contributed so much to the success of the meeting: and in particular the local secretary Mr Stephen Head, whose hard work and talent for organization ensured that all ran smoothly.

In organizing the meeting and preparing this volume for the press, R. Bainbridge was responsible for papers 8–14; G. C. Evans for papers 2–6, 18 and 21; and O. Rackham for papers 1, 7, 15–17, 19, 20 and 22. The references were checked and the Indexes prepared by Mrs P. A. Chapman. Having again worked equally together, we subscribe ourselves as before in alphabetical order, but this time starting with the letter E,

G. CLIFFORD EVANS
OLIVER RACKHAM
RICHARD BAINBRIDGE

1 The measurement of irradiance with particular reference to marine biology

G. P. ARNOLD *Fisheries Laboratory, Lowestoft, Suffolk*

Introduction

Light is a major ecological factor in the sea affecting a very wide range of processes from primary production to fish vision. The measurement of light has posed a problem in all fields of marine photobiology. The basic difficulty has been that physicists, for practical reasons, developed the science of photometry independently of radiometry, defining a primary standard of light and measuring lights in terms of the visual response of a standardized human observer. Unfortunately biologists have, by and large, borrowed photometry and applied it to other organisms without thinking about the underlying physical principles. The result is a literature full of confused quantities and incorrect units, with very little useful data on the primary motivator in all photobiology—radiant power (Tyler 1971). Although the dangers of using photometric units for organisms other than man have been pointed out for many years by a succession of authors there seems to be a great reluctance among photobiologists to make the transition to radiometric units. In part this seems to be due to the erroneous belief that photometric units can always be converted to radiometric ones and vice versa, when in fact there are only two situations when the conversion can be made. The first is complete darkness and the second is when the radiation has the precise spectral distribution numerically specified by the photopic sensitivity curve of the human eye (Fig. 1.1a) (Tyler 1971).

The purpose of this paper is to demonstrate simply and clearly: (1) the necessity of using radiometric measurements for organisms other than man; (2) the relationship between photometric and radiometric quantities; (3) how photometric lamps can be used as standards of spectral power distribution; and (4) the essential features of irradiance meters.

Photobiology

Light is a form of electromagnetic energy and like radio waves, for example,

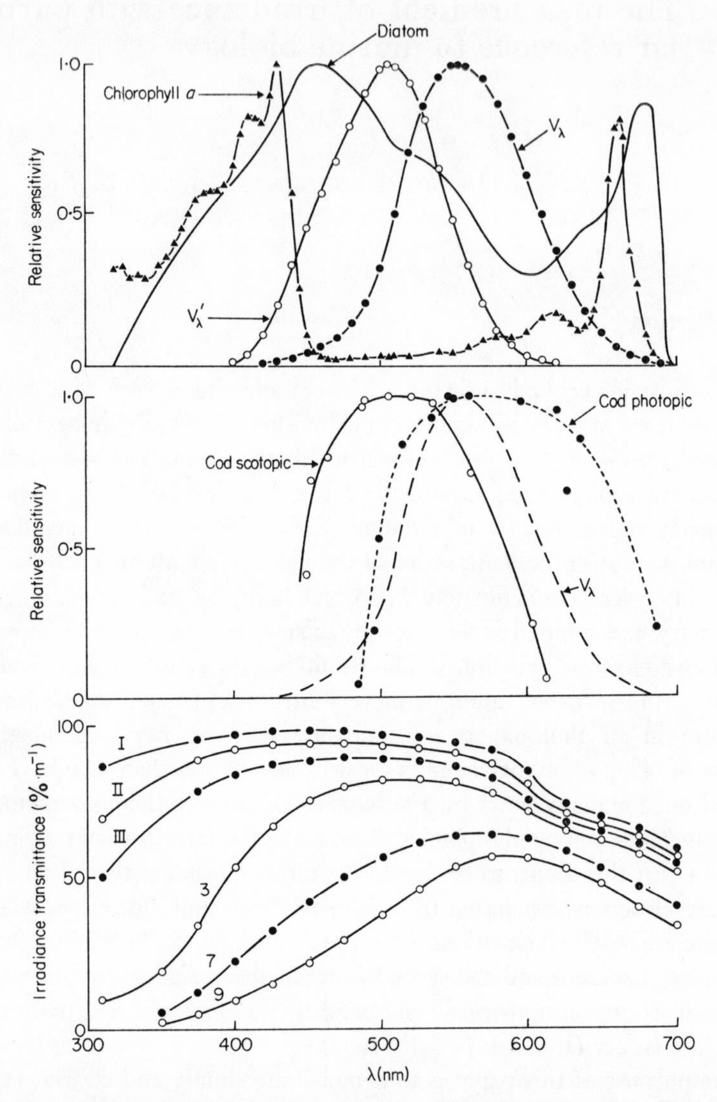

Figure 1.1. (a) Photopic (V_λ) (Jones & Preston 1969) and scotopic (V'_λ) (C.I.E. 1957) spectral sensitivity curves for the human eye, the absorption spectrum of a 90% acetone extract of chlorophyll *a* (Richards 1952), and a typical photosynthetic action spectrum for a diatom and a brown alga (Halldal 1974).
(b) Photopic and scotopic spectral sensitivity curves for the eye of the cod, *Gadus morhua*, determined by electroretinogram (Protasov 1964).
(c) Spectral transmission curves for different types of oceanic (I, II and III) and coastal (3, 7 and 9) water (Jerlov 1968).

conveys radiant power from a source to a receiver. The electromagnetic spectrum extends over 19 orders of magnitude, from cosmic rays with wavelengths as short as 10^{-6} nm to radiowaves with wavelengths up to 10 km. Light occupies less than one order of magnitude within this vast array, extending at the most from 200–1000 nm.

Quantum theory

In its interaction with matter radiation behaves as if it were a stream of particles, known as quanta. Each quantum, or photon, carries a definite amount of energy and has a wave pattern associated with it in space. The relationship between the radiation velocity (c), its frequency (ν) and its wavelength (λ) is given by $c = \nu\lambda$. The energy (ε) of each photon is given by:

$$\varepsilon = h\nu = \frac{hc}{\lambda}$$

where Planck's constant, $h = 6{\cdot}6256 \times 10^{-34}$ J s and $c = 2{\cdot}9979 \times 10^{8}$ m s^{-1} (Chiswell & Grigg 1971). The quantum energy may be expressed either in joules or electron volts (1 eV $= 1{\cdot}602 \times 10^{-19}$ joules) and

$$\varepsilon = \frac{1{\cdot}9863 \times 10^{-16}\ \text{J}}{\lambda}$$

or $\varepsilon = \dfrac{1239{\cdot}8}{\lambda}$ eV, where λ is measured in nm.

The quantum properties of light are of fundamental importance in photobiology. Light must be absorbed by photoreceptors and this takes place by individual processes in each of which 1 quantum of energy is absorbed by 1 molecule. Quanta are indivisible and the reactions in which they take place are all or nothing. When a quantum is taken up by an atom, an electron is transferred to a more distant orbit and the energy required to do this is taken from the photon. Any excess energy given by the photon to the electron is wasted, either by re-radiation, or by degradation to heat. Since the energy of each photon varies inversely with wavelength more energy is wasted at shorter wavelengths, and for a given energy there are more photons available at longer wavelengths. Light thresholds for all photochemical reactions are determined by the quantum efficiency of the reaction. In photosynthesis it is calculated that for every molecule of oxygen liberated in the production of glucose at least 4 but probably not more than 10 quanta are required (Ramsay 1966). In the human eye the minimum number of quanta required for peripheral vision is of the order of 800 s^{-1} at the cornea, corresponding to about 80 s^{-1} absorbed by the rods (Pirenne 1956). The quantum is the basic quantity to be measured in all photobiological studies.

Spectral sensitivity

How an organism utilizes the light quanta incident upon it depends on the spectral sensitivity of its photoreceptors. Our thoughts on light are dominated by our own spectral sensitivity but other organisms have very different responses.

For the human eye the visible spectrum is defined by two spectral sensitivity curves, one for the light-adapted (photopic) eye and one for the dark-adapted (scotopic) eye. Because of variation in sensitivity between individuals and also changes in sensitivity of the individual with age, standard sensitivity curves have been defined. The photopic sensitivity curve, the V_λ function, was adopted by the Commission Internationale de l'Éclairage (CIE) in 1924, and by the Comité International des Poids et Mesures in 1933 (Jones & Preston 1969). The scotopic sensitivity curve, the V'_λ function, was adopted by the CIE in 1951 (CIE 1957). With photopic vision peak sensitivity is at 555 nm and the bandwidth at half the maximum value is 100 nm; with scotopic vision peak sensitivity shifts to 507 nm (Fig. 1.1a). Some people, however, have extended visibility into the ultraviolet or infrared (Arnold *et al.* 1971).

Chlorophyll is essential for photosynthesis and the spectral absorption curve of extracted chlorophyll *a* (Richards 1952) is shown in Fig. 1.1a. It is clearly very different from the V_λ function. It is also slightly different from that of *in vivo* chlorophyll *a* whose red absorption peak occurs further into the red between 670 and 680 nm (Halldal 1974) compared with 663 nm. The photosynthetic action spectrum of green algae closely follows the chlorophyll absorption curve, having a marked minimum at 550 nm so that these plants appear to waste an appreciable fraction of submarine energy (Strickland 1958). Most marine phytoplankton are, however, like the brown algae and contain accessory carotenoid pigments, such as fucoxanthin and peridinin, which enhance their absorption in the region 500–560 nm (Strickland 1958, Halldal 1974). Brown, like green, algae show good agreement between *in vivo* absorption and action spectra and an *in vivo* photosynthetic action spectrum for a typical diatom or brown alga (Halldal 1974) is shown in Fig. 1.1a.

Fish have photopic and scotopic spectral sensitivy curves comparable to those of the human eye but the curves generally have a different shape and a different peak sensitivity (Blaxter 1970). The spectral sensitivity curves of the cod *Gadus morhua* are compared with the V_λ function in Fig. 1.1b.

Photometry and radiometry

The radiation produced by most sources is not monochromatic but either consists of a discrete number of monochromatic lines or else has a continuous spectral distribution with, perhaps, some lines superimposed. In order to describe radiation completely it is necessary therefore to make spectro-radiometric measurements, which give the power distribution as a function of wavelength. The power in the lines is expressed in absolute measure and that in the continuum as power per unit wavelength, or frequency, interval. In many applications, however, the effectiveness of the radiation may be characterized by means of a single quantity, obtained by weighting the spectral distribution curve according to an appropriate function of wavelength and then integrating. The commonest example of this technique is photometry in which the luminous effect of radiation on the human eye is evaluated by weighting the spectral distribution of the radiation by the V_λ function. The single quantity so defined may be determined either by measuring the spectral distribution and then calculating the integral, or by direct measurement, using a radiation detector whose spectral sensitivity varies with wavelength in agreement with the prescribed weighting function (Gillham 1961).

Because, however, photometric units deliberately incorporate the spectral sensitivity characteristics of the average normal young human eye (for the reason see p. 13) it is clearly a breach of intellectual honesty to use them in any other context (Tyler 1973a). In photobiology it is therefore necessary to make radiometric measurements and in the aquatic field this course has been advocated by a number of authors (Clarke & Wertheim 1956, Strickland 1958, Craig 1964, Westlake 1965, Tyler 1971). The force of this argument is clearly seen by comparing the spectral sensitivity curves in Fig. 1.1a and 1.1b and the seawater transmission curves in Fig. 1.1c. Clearly a photometer, whose spectral sensitivity is by definition matched to the V_λ function, will measure neither the total light that is available in the sea, nor the light that is selectively absorbed by the various organisms. It should be noted in Fig. 1.1b that at the half-power points of the V_λ function (510 and 610 nm), the photopic and scotopic sensitivity curves of the cod eye have values of 98% and 100% respectively.

Phytoplankton organisms are much more sensitive to blue and red light than the human eye which is predominantly sensitive to green. Despite this much broader spectral sensitivity it has been widely thought that the measurement of the green fraction of the total ambient light ought to be proportional to the total quanta available for photosynthesis with sufficient accuracy for most purposes. Recent studies have shown that this is not necessarily so

(Tyler 1973a). Tyler calculated total lux (see p. 10 for definition) and total quanta from spectral irradiance data and determined the ratio quanta: lux as a function of depth in ocean water having a variety of chlorophyll concentrations. The ratio varied by a factor of 6·7 for the ocean waters studied and the actual value depended on the depth and the chlorophyll concentration. In the Sargasso Sea the ratio was large and increased with depth; in homogeneous water rich in phytoplankton the ratio was small and decreased with depth. With stratified phytoplankton the ratio of quanta : lux will be erratic in value as a function of depth. Tyler concluded that errors as high as 600–700% can occur if photometric measurements are used to estimate the relevant radiant energy associated with oceanic photosynthesis.

Attempts to study the interaction of radiant power with an organism can undertake to measure one or more of three quantities: the power *available*, the power *absorbed* or the power *utilized* (Tyler 1973b). Most marine photobiology is only at the stage of measuring the power available and a choice must be made between using either a spectroradiometer or an integrating radiometer with a weighted spectral sensitivity. The weighting factor can be matched to the spectral transmission of a particular type of seawater (Clarke & Wertheim 1956), or to the spectral sensitivity of a particular organism. There are, however, inherent difficulties with integrating instruments. The Scientific Committee on Oceanic Research (SCOR) Working Group No. 15 carefully considered the idea of an equivalent detector for measuring photosynthetically useful radiant power in the sea but gave it up as impractical. They recommended that the total available power or quanta be measured instead (UNESCO 1964). Blaxter & Parrish (1965) used a radiometer whose spectral sensitivity was closely matched to that of the dark-adapted herring (*Clupea harengus*) eye. This radiometer was calibrated in metre-candles (= lux) but these were not true photometric units because the spectral sensitivity of the instrument did not match the V_λ function. Such integrated measurements are, however, of limited use even when they are made in radiometric units. Firstly, fish show a Purkinje shift from photopic to scotopic spectral sensitivity with falling irradiance, so that the spectral sensitivity of the radiometer should be altered comparably. Secondly, the spectral sensitivity curve can vary with the type and size of stimulus used to determine it (Northmore & Muntz 1974). For studies of fish behaviour, therefore, different weighting functions would be required for different facets of behaviour. Thirdly, measurements of this type are restricted to the one particular organism. Spectroradiometric measurements are clearly preferable in that they measure the distribution of the available power with wavelength. The results can subsequently be used to calculate any required quantity: the total available radiant power, the radiant power weighted by any chosen spectral function or the spectral transmission properties of the seawater.

Photometric and radiometric units

Logically, photometry is only a branch of radiometry and, theoretically, if the spectral power distribution of a light source is known, it is possible to calculate its photometric properties. In practice, however, the situation is not so simple, and it was for practical reasons that photometry grew up as a separate science. In the 1920's absolute radiometry was unable to produce either the technique or the necessary accuracy for the satisfactory establishment of a unit of light and, as a result, an empirical unit was defined, based on a material but reproducible primary standard. In the continued absence of any absolute radiometric specifications the empirical unit, the *candela*, is necessarily retained in the present SI system of units. If a radiometric specification of the unit of light could be reached, then the candela could be placed on a proper footing, and ideally could become a derived unit based on the unit of power, the watt (Jones & Preston 1969).

The absolute measurement of spectral power distributions, without reference to some standard source having a known distribution, has so far proved very difficult and inaccurate. Comparison with a standard source, generally a black-body radiator at a known colour temperature, is easier and more accurate, and this technique has recently brought photometry and radiometry together. Standard photometric sources can now be used as radiometric standards and, although the empiricism and comparison remain, it is relatively easy for photobiologists to make the transition from photometric to radiometric units. For practical reasons again, however, the calibration of standard sources will continue to be made for some considerable time by conventional 'one-shot' photometric measurements, rather than by spectroradiometry which is much more laborious (Jones & Preston 1969). It is necessary therefore for the photobiologist to understand the basic photometric as well as the radiometric units.

Photometric quantities

The four basic photometric quantities are described in Fig. 1.2 and defined below.

Luminous intensity (I)

The earliest standard sources of light were candles, but they gave place in the 19th century first to specially designed flame standards and then to incandescent standards. In 1909 the United States of America, France and the United Kingdom adopted a common one-candle unit and this standard was maintained in carbon filament lamps. At the fifth session of the Commission Internationale de l'Éclairage (CIE) in 1921, a standard unit—still conserved in filament lamps—named the 'international candle' was adopted.

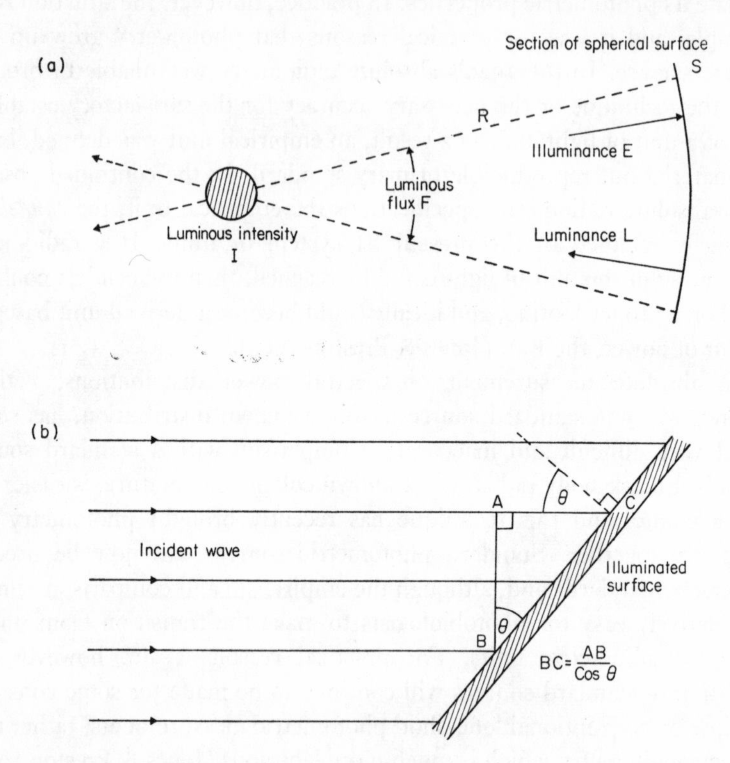

Figure 1.2. (a) The four basic photometric quantities.
(b) The cosine law of illumination.

Subsequently in 1937, with the development of a platinum standard, the Comité Consultatif de Photométrie of the Comité International des Poids et Mesures (CIPM) proposed a 'new candle', which was renamed the *candela* by the CIE in 1948. The final definition of the unit adopted by the General Conference on Weights and Measures 1948 is:

Candela. The unit of luminous intensity. The magnitude of the candela is such that the luminance of a full radiator at the temperature of solidification of platinum is 60 candelas per square centimetre (Jones & Preston 1969). Symbol cd.[1]

[1] It is unfortunately not clear from this definition that intensity is energy per unit solid angle and that 1 candela = 1 lumen steradian^{-1} (see p. 14 and Table 1.1).

The primary standard

The platinum primary standard has been set up and used in the laboratories of eight different countries, and in the United Kingdom two determinations of the candela have been made at the National Physical Laboratory (NPL). The primary standard used at the NPL is a thoria tube (length 47mm, bore 2·5mm, wall thickness c. 0·5mm) closed at the bottom end and standing in a thoria crucible (Fig. 1.3). A machined platinum ingot with an axial hole to

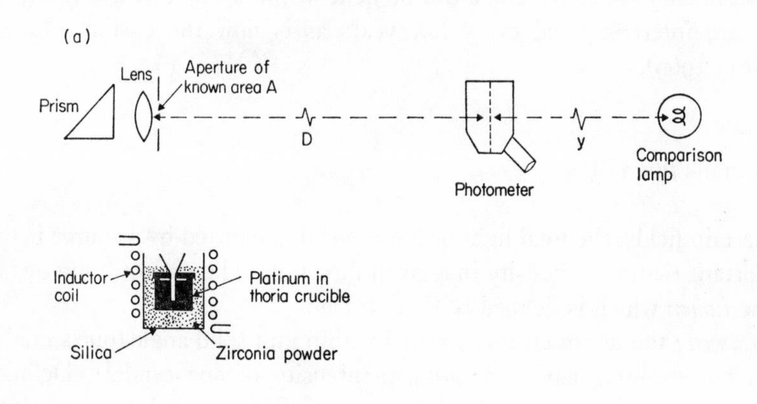

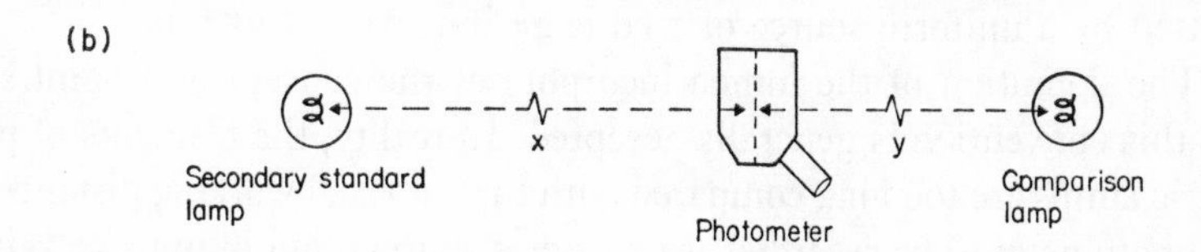

Figure 1.3. Calibration of standard lamps by comparison with the primary standard by the substitution method.
(a) Comparison lamp balanced against primary standard.
(b) Secondary standard lamp balanced against comparison lamp. (Jones & Preston 1969).

accommodate the thoria tube fills the space round the tube, and the tube itself is filled with zirconia powder, so that the effective cavity is well within the region surrounded by the platinum. The whole assembly is contained within a cylindrical silica pot also packed with zirconia powder, and surrounded by the heating coil of a high-frequency induction furnace operated at 1·6 MHz. Above the crucible is a right-angle prism and lens and a fixed circular aperture, which images the radiating cavity onto a photometer (Jones & Preston 1969).

The primary standard is used to calibrate a group of tungsten filament vacuum lamps, which have been set to a colour-match with it. These fundamental secondary standards of luminous intensity are used to monitor the constancy of the primary standard during use, to conserve the unit between successive references to the primary standard and, by exchanges, to compare the values of the candela or the reproducibility of the primary standard as found by the different laboratories. The present total range of uncertainty in the unit derived practically from the primary standard in the different laboratories on various occasions is about 1%. In comparison, the values of the secondary standard lamps can be held within a range of about 0·5%, if they are intercompared every few years as is now the custom (Jones & Preston 1969).

Luminous flux (F)

In certain fields, the total light or luminous flux emitted by a source is more important than the intensity in a given direction. The unit of luminous flux is the *lumen* which is defined (CIE 1957) as:

Lumen: the luminous flux emitted within unit solid angle (one steradian) by a point source having a uniform intensity of one candela. Definition adopted by the 9th General Conference on Weights and Measures 1948. Symbol lm.

There are 4π steradians to a sphere so that the total luminous flux emitted by a uniform source of 1 cd is 4π lumens or 12·57 lm.

The definition of the lumen incorporates the concept of a point source and this convention is generally accepted. In reality, the filaments of photometric lamps are too long compared with their normal working distance from the photometer to be regarded as true point sources, but in practice values of intensity are assigned to them on the assumption that the working distance (see p. 15) will be adhered to within narrow limits (Jones & Preston 1969).

Illuminance (E)

When luminous flux reaches a surface that surface is said to be illuminated.

The *illumination* at a point of a surface is the quotient of the luminous flux incident on an infinitesimal element of surface containing the point under consideration by the area of that element (Walsh 1953, CIE 1957). For quantitative statements the term illuminance is used; the SI unit is the *lux*, which is 1 lumen m^{-2} (Table 1.1).

The two fundamental laws of illumination are the inverse square law and the cosine law. Both follow from the wave theory of radiation if a luminous point source is considered as the source of a system of spherical waves. The area of each wave as it travels out from the point increases as the square of its radius. Since its energy must be regarded as distributed equally over the

Table 1.1. Equivalent photometric and radiometric units

Photometric quantity	Symbol	Photometric unit	Abbreviation	Radiometric quantity	Radiometric unit
Luminous flux	F	Lumen	lm (cd sr)	Radiant power	W
Luminous intensity	I	Candela	cd (lm sr^{-1})	Radiant intensity	W sr^{-1}
Illuminance	E	Lux	lx (lm m^{-2})	Irradiance	W m^{-2}
Luminance	L	Nit	nt (cd m^{-2})	Radiance	W sr^{-1} m^{-2}

surface, the surface density of the energy must vary inversely as the square of the radius of the wave. Thus the quantity of radiant energy received by any area which is normal to the direction of propagation varies inversely as the square of its distance from the source. It is understood that the area is so small in comparison with its distance from the source that it may be considered as part of the spherical wave. Since the rate of propagation of light energy is held to be constant in photometric work, it follows that the luminous flux also varies inversely with the square of the distance of the surface from the source. If a small area S is at a distance D from a source, of which the luminous intensity in the direction of S is I, then the luminous flux incident on S is $F_s = IS/D^2$ and the illuminance $E = F_s/S = I/D^2$, where D is measured in the same units as S. This relationship gives an alternative definition of the unit of illumination as that illumination which is produced at the surface of a sphere of unit radius due to a uniform point source of one candela placed at its centre (p. 131 in Walsh 1953) (Fig. 1.2a). This definition avoids the stipulation necessary in the case of a plane surface that it shall be negligibly small compared with D and normal to the incident light. It follows also from the rectilinear propagation of light that the flux incident on a surface is proportional to cos θ, when θ is the angle between the normal to the surface and the direction of propagation of the wave (Fig. 1.2b). The last equation may thus be expanded to

$$E = \frac{I \cos \theta}{D^2}$$

The two laws of illumination embodied in this equation may be formally stated:

The illumination of an elementary surface due to a point source of light is proportional to the luminous intensity of the source in the direction of that surface, and to the cosine of the angle between this direction and the normal to the surface, and is inversely proportional to the square of the distance between the surface and the source (p. 131 in Walsh 1953).

Luminance (L)

Luminance is the term adopted by the CIE to denote the amount of light emitted from a surface. The term 'brightness' describes the visual sensation experienced on looking at the surface. Luminance is defined as:

The quotient of the luminous intensity in the given direction of an infinitesimal element of the surface containing the point under consideration, by the orthogonally projected area of that element in a plane perpendicular to the given direction (Walsh 1953, CIE 1957.)

The SI unit of luminance is the candela m^{-2}, known as the *nit*.

The light reflected from an ideal surface depends on the illuminance (E) and the reflection factor ρ and $L = E\rho/\pi$. The factor ρ is the ratio of reflected to incident flux and for an ideal surface $\rho = 1{\cdot}0$. In practice values for diffuse surfaces range from $\rho = 0{\cdot}8$ for white paper to $\rho = 0{\cdot}04$ for black cloth (Arnold *et al.* 1971). The ideal surface is a uniform diffuser with a luminance independent of viewpoint and lighting angle, but in practice all surfaces have a certain specular quality, exhibiting different reflection factors under different conditions. The term *luminance factor* (β) denotes the ratio between the luminance of a surface in a particular direction and the luminance of a uniform diffuser under similar conditions of illumination.

Colour temperature and the photometric scale

The primary standard of light emits radiation of black-body spectral type characterized by temperature only. The basic temperature is the thermodynamic temperature, symbol T, the unit of which is the kelvin, symbol K. The kelvin is the fraction 1/273·16 of the thermodynamic temperature of the triple point of water. The International Practical Temperature Scale of 1968 (IPTS-68) distinguishes between the International Practical Kelvin Temperature with the symbol T_{68} and the International Practical Celsius Temperature with the symbol t_{68}. The relation between T_{68} and t_{68} is given by:

$$t_{68} = T_{68} - 273{\cdot}15\text{K}.$$

The freezing point of platinum on the IPTS-68 is 2044·9K; the equivalent Celsius temperature is therefore 1771·75°C (National Physical Laboratory 1969).

Because of the emitting properties of tungsten, the spectral emission of a tungsten filament lamp is very nearly the same as that of a full (black-body) radiator *over the visible spectrum*. It can therefore also be characterized by a temperature known as the colour temperature, which is defined as the temperature of a black-body radiator whose light colour-matches that of the lamp. This temperature is higher than the actual temperature of the filament; Jones (1970a) lists equivalent colour temperatures and true tungsten temperatures.

The fundamental secondary standard lamps are operated at a colour temperature of 2044·9K, so that, when they are calibrated against the primary standard, there is no appreciable difference in colour temperature or spectral distribution of energy. Because general service tungsten lamps operate at much higher colour temperatures (2800–3400K) it is necessary also to maintain standard lamps at these higher temperatures, but still to calibrate them against the secondary standard lamps at 2044·9K. The term photometric scale thus incorporates in the first place the idea of an ascent in colour temperature. However, both the *lux* and the *candela*, from which it is derived, are defined at 2044·9K, and their values would alter with ascending colour temperature unless the intensity of each lamp were compared against the platinum primary standard, using a *standard eye* with a defined spectral sensitivity. This standard eye is the essential feature of the photometric scale and the internationally accepted standard is the V_λ function for the light-adapted human eye (see p. 4).

Radiometric quantities

The four basic photometric quantities are matched by four exactly equivalent radiometric ones, whose units are shown in Table 1.1. Formal definitions can be found in the CIE International Lighting Vocabulary (1957) or in Jerlov (1968), but the units should be self-explanatory. Total radiant power W is given by the equation

$$W = \sum_{\lambda_1}^{\lambda_2} W(\lambda).\Delta\lambda,$$

(Tyler 1973b), and the equivalent photometric quantity by

$$F = K_{max} \sum W(\lambda)V(\lambda).\Delta\lambda \text{ (Walsh 1953)},$$

where $W(\lambda)$ is the radiant power over a narrow band of wavelength $\Delta\lambda$ and K is the luminous efficiency of radiation. The value of K_{max} is calculated as 682 lm W^{-1} (Walsh 1953) and two experimental determinations at the NPL have given values of 685 $\pm$ 4 and 687 $\pm$ 3 lm W^{-1} (Jones & Preston 1969). It is clear from these two equations that the lumen is no more than a specially defined watt, weighted by the average spectral sensitivy of the normal young human eye (Walsh 1953, Tyler 1973b). It can also be seen from Table 1.1 that, although the candela is the basic defined photometric unit, it is in fact only a lm sr^{-1}.

Practical calibration sources

Standardizing laboratories maintain secondary standards of both luminous intensity and luminous flux in the form of either gas-filled, or vacuum, tungsten lamps. The photometric scale at the NPL, for example, is embodied in six groups of lamps, of which three are standards of intensity and three of luminous flux (Jones & Preston 1969). Working laboratories need comparable lamps and these working standards are calibrated in turn against the secondary standards.

Illuminance standards

Whilst projector or car lamps can be used as working standards, specially constructed lamps are available for accurate work. The Wotan Wi 41/G (Osram GmbH, München, Germany) is a straight-wire gas filled uniplanar lamp with a conical bulb. A black opaque coating covers half the surface of the bulb and a window, limiting an angle of radiation of about $\pm$ 5°, is left open in the coating opposite the filament. This prevents light reflected from the filament support frame from travelling in the measuring direction. This lamp is rated at 30V 6A and, when operated at a colour temperature of 2856K, has a luminous intensity of about 250 cd. The General Electric Co. Ltd. (Hirst Research Centre, Wembley, Middx) make a straight-wire uniplanar vacuum lamp suitable for colour temperatures up to 2400K and rated at 100V 75W. At 2400K the intensity is about 62 cd. The National Physical Laboratory supplies miners' lamps for low level photometric calibrations. These are small pear-shaped lamps, with a pearl bulb, rated at 4 V 1A. At a colour temperature of 2000K they have an intensity of about 0·1 cd, and at 2856K an intensity of about 4 cd.

Lamps to be used as working standards are carefully 'aged' for 5% of expected life at their operating voltage (Walsh 1953, Jones & Preston 1969) before calibration at the standardizing laboratory. The voltage and current required to produce the desired colour temperature are then determined and finally the luminous intensity is measured at the same voltage and current. Full details of the lamp calibration are provided by the standardizing laboratory. The accuracy of the luminous intensity of working standards calibrated in this way is ± 1%, but applies only after an initial period of burning of at least 10 min., during which the lamp reaches thermal equilibrium. The intensity of the Wotan Wi 41/G lamp changes less than 1% for 100h of use at 2856K; that of the GEC lamp about 1% for 100h of use at 2400K (R. G. Berry pers. comm.).

Working standards are used to calibrate photometers by the inverse square law method. The basic requirement is an optical bench, which enables the photometer head to be set at a known distance from the lamp filament. Working distances of 1–2 m are commonly used and at these distances the filament is assumed to be a point source (Walsh 1953) (see p. 10). Ideally the distance between filament and photometer head should be measured to an accuracy of 0·05% or 0·5 mm in 1 m, so that the bench scale must be graduated accurately in millimetres. The position of the uniplanar filament of the lamp is determined by sighting it through the clear glass envelope against a plumb line hanging against the bench scale. The position of the photometer is determined in the same way. A very important precaution in photometry is to prevent stray light from reaching the photometer head. Stray light is any light which reaches the instrument other than directly from the lamp and may be due to other sources in the room or to reflections from objects on or near the bench. It is removed by a series of opaque black screens, with holes of various sizes, mounted on the optical bench between the lamp and the photometer. The distances between the screens are adjusted in relation to the sizes of the apertures so that nothing can be seen from the position of the photometer except the lamp itself, with the remainder of the field occupied by black surfaces. As an additional precaution black curtains can be hung along the bench on either side (Walsh 1953).

Accurate photometric measurements necessitate accurate measurement and control of the voltage or current at which the working standard is operated. A stabilized DC power supply is essential. Luminous intensity varies approximately as the fourth power of the voltage and the eighth power of the current; it is natural therefore to specify and control the voltage (Jones & Preston 1969). Setting the current, however, eliminates any errors due to contact resistance in the lamp socket. Further, the rate of change of light output with time is slower, because the filament thins with use. This results in increased wattage and light output for set current, which compensates for progressive blackening of the bulb (R. G. Berry pers. comm.). Whichever quantity is set, the accuracy of the measurement and the electrical stability determines the accuracy of the lamp intensity. If the objective is to set the intensity to 0·5% then either the voltage must be set to 0·1% accuracy or the current to 0·05%.

The three types of lamp described above give values of illuminance ranging from 70 lux down to 0·02 lux at a working distance of 2 m (Table 1.2).

Table 1.2. Photometric properties of three working standard lamps

Lamp	NPL Ref no.	T_c K	Voltage V	Current A	I cd	Illuminance (lux) D = 1·5m	2·0m	2·5m
Wotan (Wi 41/G)	7KA70	2856	32·17	6·010	281·0	124·9	70·3	45·0
GEC (100V 75W)	7KG71	2400	92·6	0·782	63·4	28·2	15·9	10·1
Miners' (4V 1A)	7KE71	2856	3·606	0·9842	3·875	1·8	0·97	0·62
		2000	1·437	0·6573	0·065	0·03	0·02	0·01

For low level photometry the apparatus shown in Fig. 1.4 is used. Light from the lamp is reflected by an aluminium plate, smoked with magnesium

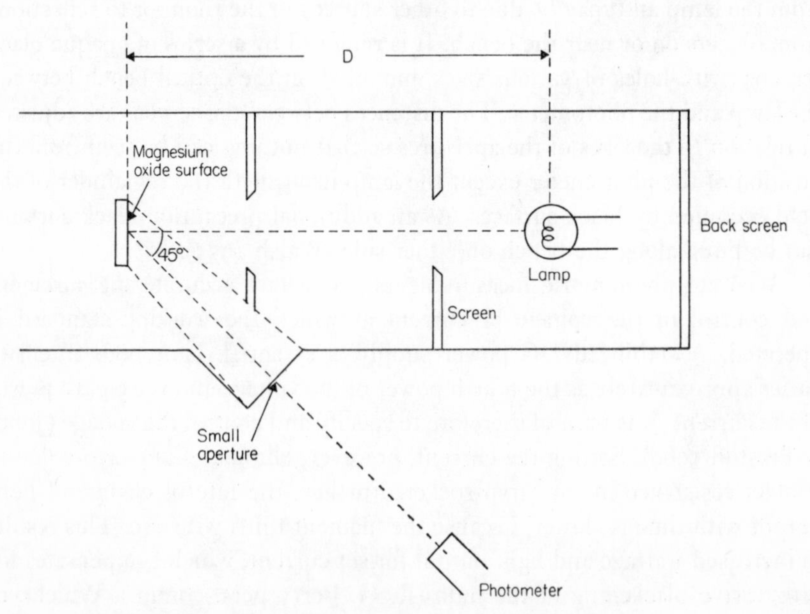

Figure 1.4. Photometric box for calibrations at low levels of illuminance.

oxide, onto a small aperture of known size. The aperture acts as a secondary source of known intensity (Y) calculated from

$$Y = \frac{I}{\pi D^2} \cdot \frac{\pi d^2}{4},$$

where d is the diameter of the aperture in m, and D is usefully about 0·5m (F. A. Garforth pers. comm.). If the magnesium oxide surface is carefully prepared the luminance factor is close to unity throughout the visible spectrum (1·01 ± 0·005) and the total reflection factor (see p. 12) is about 0·98 (National Physical Laboratory 1960). With a lamp of 4 cd intensity, d = 4mm and D = 0·5m, the intensity of the aperture is approximately 6×10^{-5} lux.

Irradiance standards

Photometric lamps of known colour temperature and intensity can be used as standards of spectral power distribution. The NPL has published a set of tables (Jones 1970 a & b) of spectral power distribution, for a wide range of colour temperatures, covering the visible spectrum and also the near ultraviolet and near infra-red regions. The main application of these tables is to produce a known absolute spectral irradiance. The calibrated lamp of intensity I (cd) and colour temperature T_c is set up at a distance D (cm) from a selected area. D is usefully about 2m as in photometric calibrations. The spectral radiant intensity, $P(\lambda)$, of the source at any wavelength λ is obtained by finding the table of Q (tungsten) for the temperature T_c and setting

$$P(\lambda) = I\,Q(\lambda) \quad \text{W sr}^{-1}\,\Delta\lambda^{-1}.$$

The spectral irradiance at the centre of the area is then given by

$$\frac{P(\lambda)\cos\theta}{D^2} \quad \text{W cm}^{-2}\,\Delta\lambda^{-1},$$

where θ is the angle between the normal to the area and the line joining the area and the centre of the lamp.

The NPL tables, which are based on De Vos' values for the emissivity of tungsten, list the absolute spectral radiant intensity per candela, $Q(\lambda)$, for both tungsten and black-body sources in μW sr^{-1} cd^{-1} $\Delta\lambda^{-1}$ for a 10nm wavelength interval. The temperature range covered is 1700–3300 K in 100 K steps, together with two selected values of 2043·2 K and 2855·54 K to represent the platinum point and the CIE illuminant A respectively. $Q(\lambda)$ is given at 10nm intervals over the range 240–830nm and at wider intervals to 2600nm. The tables assume that the source is viewed through a quartz window of constant transmission factor 0·91. In view of the uncertainties associated with photometric calibrations and the deviations of practical lamps from the idealized performance, the accuracy attainable by this method is approximately ± 2% within the visible spectrum and rather worse outside it (Jones 1970b).

The spectral irradiance produced at a distance of 2m by the three lamps, whose photometric characteristics are shown in Table 1.2, are given in Table 1.3, together with the equivalent quantum energy at each wavelength.

Table 1.3. Absolute spectral irradiance produced at 2m distance by three photometric lamps ($\Delta\lambda$ = 20nm).

Wavelength	Quantum energy	Spectral irradiance W m^{-2}		
		Photometric lamp		
nm	joules	Miners' 7KE71 T_c = 2856K	GEC 7KG71 T_c = 2400K	Wotan 7KA70 T_c = 2856K
360	5·518 × 10^{-19}	0·08 × 10^{-4}	0·05 × 10^{-3}	0·06 × 10^{-2}
380	5·227 ,,	0·12 ,,	0·09 ,,	0·09 ,,
400	4·966 ,,	0·19 ,,	0·15 ,,	0·14 ,,
420	4·729 ,,	0·27 ,,	0·24 ,,	0·20 ,,
440	4·514 ,,	0·37 ,,	0·37 ,,	0·27 ,,
460	4·318 ,,	0·49 ,,	0·54 ,,	0·36 ,,
480	4·138 ,,	0·63 ,,	0·75 ,,	0·46 ,,
500	3·973 ,,	0·79 ,,	1·01 ,,	0·57 ,,
520	3·820 ,,	0·96 ,,	1·32 ,,	0·69 ,,
540	3·678 ,,	1·14 ,,	1·68 ,,	0·82 ,,
560	3·547 ,,	1·32 ,,	2·09 ,,	0·96 ,,
580	3·425 ,,	1·51 ,,	2·53 ,,	1·10 ,,
600	3·310 ,,	1·70 ,,	3·01 ,,	1·23 ,,
620	3·204 ,,	1·89 ,,	3·53 ,,	1·37 ,,
640	3·104 ,,	2·08 ,,	4·07 ,,	1·51 ,,
660	3·010 ,,	2·26 ,,	4·64 ,,	1·64 ,,
680	2·921 ,,	2·43 ,,	5·21 ,,	1·76 ,,
700	2·838 ,,	2·59 ,,	5·78 ,,	1·88 ,,
720	2·759 ,,	2·74 ,,	6·36 ,,	1·99 ,,
740	2·684 ,,	2·87 ,,	6·91 ,,	2·08 ,,
760	2·614 ,,	2·98 ,,	7·45 ,,	2·16 ,,
780	2·547 ,,	3·09 ,,	7·97 ,,	2·24 ,,
800	2·483 ,,	3·18 ,,	8·45 ,,	2·31 ,,

The values are given in W m^{-2}, because although the NPL tables list the absolute spectral radiant intensity per candela in μW cm^{-2}, the preferred SI unit is the W m^{-2} (Chiswell & Grigg 1971). Data on spectral irradiance in energy units should therefore be published in this form. Where the source has a continuous spectrum, values of spectral irradiance in W m^{-2} can be converted to quantum units by first converting to J s^{-1} m^{-2} (1W = 1J s^{-1}) and secondly by dividing by the appropriate quantum energy in joules. Thus, for example, a spectral irradiance of $0{\cdot}57 \times 10^{-2}$ W m^{-2} at $\lambda = 500$ nm is equal to $0{\cdot}57 \times 10^{-2}$ J s^{-1} m^{-2}, which divided by $3{\cdot}973 \times 10^{-19}$ joules (Table 1.3) gives a value of $1{\cdot}43 \times 10^{16}$ photons s^{-1} m^{-2}. The quantum values equivalent to the power values of spectral irradiance listed in Table 1.3 are shown in Fig. 1.5.

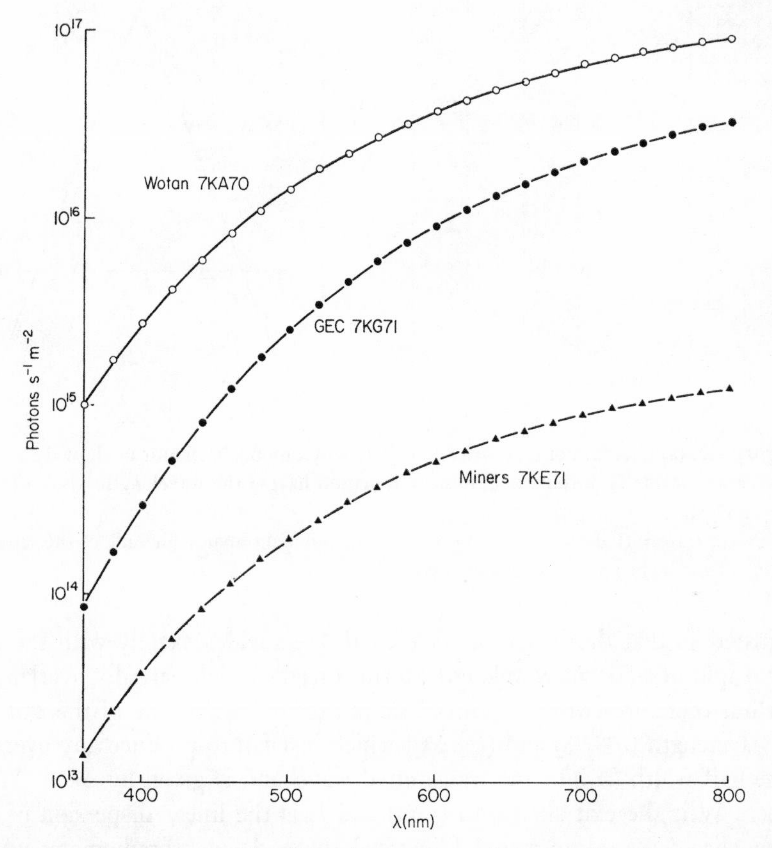

Figure 1.5. **Absolute spectral irradiance in quantum units at 2m from three photometric lamps.**

A spectroradiometer set up with its receiving surface 2 m from one of the lamps in Table 1.2 will receive at any particular wavelength the spectral irradiance shown in Table 1.3 or Fig. 1.5. It will, however, only detect a fraction of this value, the actual amount depending on the band-pass function of its monochromator. White light incident on the narrow rectangular entrance slit of a monochromator is differentially refracted by a prism or a diffraction grating, which can be rotated to produce a series of monochromatic images on a similar narrow exit slit (Fig. 1.6a). The theoretical bandpass function of an instrument with unit magnification and with both slits of identical width, or of one in which the ratio of exit to entrance slit width is the same as the magnification, is shown in Fig. 1.6b. The monochromator is

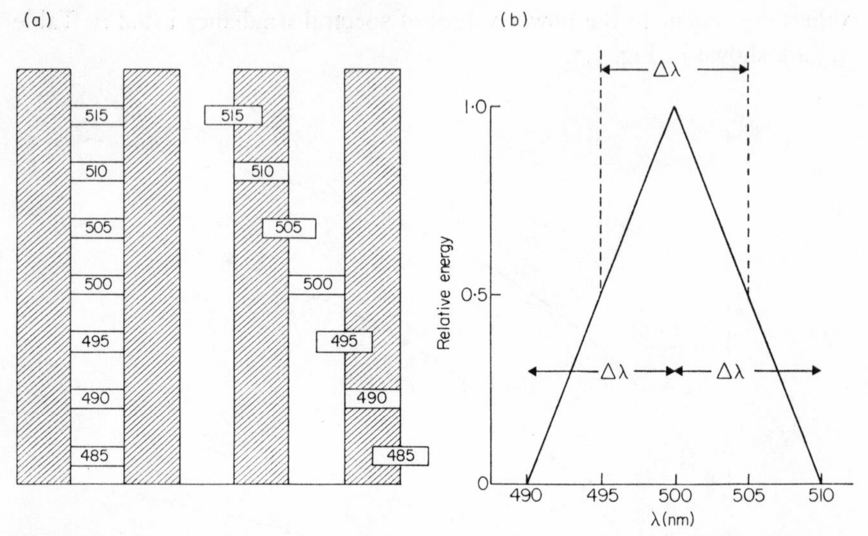

Figure 1.6. (a) Diagram of entrance and exit slits of a monochromator to show dispersion of separate beams of pure monochromatic radiation having the wavelengths shown in nm (Tyler 1973b).
(b) Power transmitted by the exit slit of the same monochromator for each of the separate beams of monochromatic radiant energy.

adjusted so that the image of wavelength λ coincides exactly with the exit slit. Light of adjacent wavelengths forms images which partially overlap the slit but there are two monochromatic images on either side of the exit slit of wavelength $(\lambda+\Delta\lambda)$ and $(\lambda-\Delta\lambda)$, which just fail to produce any overlap. The half width $(\Delta\lambda)$ of the transmitted waveband is given by $\Delta\lambda = W/L$, where W is the exit slit width (mm) and L is the linear dispersion of the monochromator in nm mm^{-1} (Thetford 1968). A spectroradiometer whose monochromator has, for example, a half-power bandpass of 4·8nm will detect 4·8/10 of the spectral irradiance incident upon its receptor, where the spectral irradiance is calculated in 10nm wavelength intervals. In practice, the actual bandpass will deviate from the ideal triangular function and may also vary across the visible spectrum (Tyler & Smith 1966). A precise knowledge of this variation is, however, not required when the instrument is used to compare fairly similar spectral power distributions. The effective bandpass can be estimated by scanning the spectral lines from a mercury-cadmium discharge lamp (O. C. Jones pers. comm.).

Luminance and Radiance Standards

A tungsten strip lamp of known colour temperature can also be used as a standard of fairly high luminance or spectral radiance. The spectral radiance is given by the product of Q(λ) and Z (tungsten), the apparent luminance. The accuracy achieved in establishing radiance standards can be inferior to those for irradiance because temperature gradients of 15K to 150K may occur on practical strip filaments (Jones 1970 a & b).

Irradiance meters

The essential feature of any irradiance meter is an optical collector with a cosine response. Commercially-available laboratory instruments generally incorporate such a collector as do the various types of solarimeter for measuring solar radiation in air (Blackwell 1966). Underwater instruments, however, frequently do not have correctly designed collectors, even though as long ago as 1933 Atkins & Poole (1933) constructed one which closely approximated to the cosine response. Many subsequent authors have not been critical of the properties of their collectors which can be so poor that as a result the instrument does not measure irradiance. Even relative measurements made with poor collectors will contain large variable and unknown systematic errors and if the response of the collector is not stated it is impossible to assign a reliable estimate of systematic error to any published underwater measurements of irradiance (Smith 1969).

The underwater collector which approaches the cosine response most closely was developed for the Scripps Spectroradiometer and was described by Smith (1969). Smith made a critical examination of the collecting properties of this (I) and three other commonly-used underwater collectors (Fig. 1.7) and calculated the systematic error of each design in measuring irradiance (Table 1.4). The errors for collectors I–III, which are made of translucent perspex, are given as a percentage difference between the perfect and actual irradiance values; in all cases the actual values are less than the perfect values. The error is larger for upwelling than downwelling irradiance and also increases slightly with depth; it increases as the radiance distribution becomes more diffuse. The error of collector IV, which is made of frosted glass, is so large that it clearly does not measure irradiance at all and is presented as the ratio perfect response/actual response.

When a translucent diffusing material is submerged in water the refractive index at the boundary of the diffusing material is changed and a larger percentage of the incident radiant power is backscattered in water than in air. This immersion effect results from internal and external changes in interface reflection (Westlake 1965, Smith 1969), and it must be considered whenever measurements above and below the water surface are compared. Since underwater radiometers are almost invariably calibrated against a standard lamp

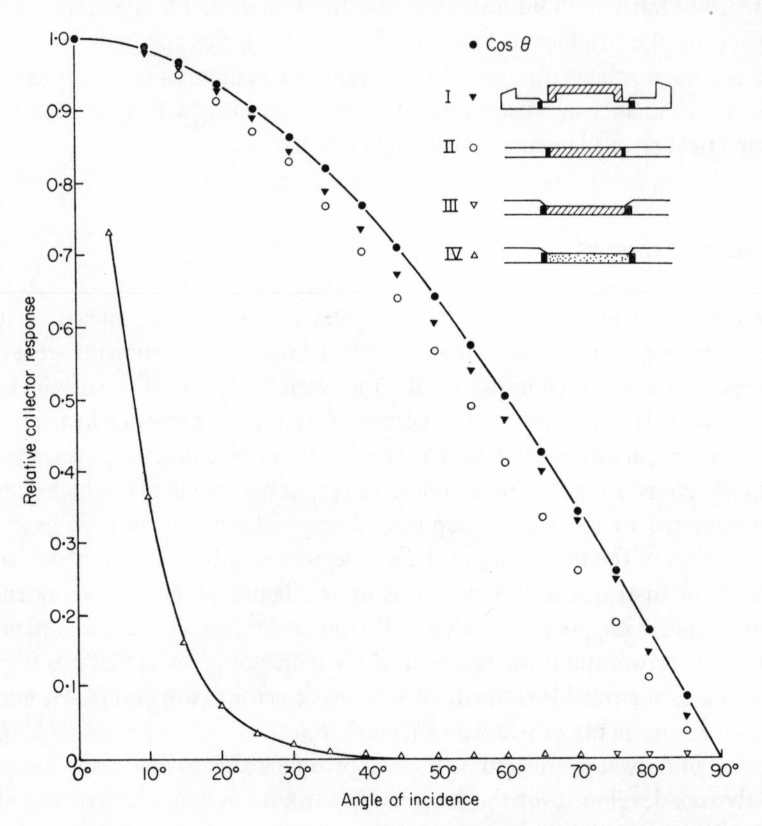

Figure 1.7. Relative response versus angle of incidence for the four representative underwater collectors. The solid curve is a plot of cos θ against θ (Smith 1969).

in air, an immersion-effect correction is needed for all underwater measurements. Smith (1969) described the procedure for determining this correction factor and showed experimentally that the correction factor for the Scripps collector varies as a function of wavelength, being qualitatively correlated with the spectral absorptance of the diffusing plastic. It follows that if underwater irradiance is to be measured accurately by a spectroradiometer, an immersion-effect correction must be made for each wavelength. It follows also that for an integrating radiometer measuring total irradiance between

350 and 700 nm the immersion-effect correction depends on the spectral properties of the water being measured. For a total-irradiance collector the correction factor can change with location and depth by an amount that is half as large as the correction itself. Neglect of the immersion-effect correction will have a diminishing effect with depth on relative measurements of irradiance, but for absolute measurements at any depth the correction must be applied together with another correction for the deviation of the collector from a true cosine response (Smith 1969).

Table 1.4. Systematic error due to the responses of four representative underwater irradiance collectors (Fig. 1.7). A collector having an exact cosine response would have zero systematic error. For collectors I, II and III the percentage error ((perfect response minus actual response)/perfect response) is given; for collector IV the error is given as the ratio (perfect response/actual response). (Smith 1969).

	Depth	Collector			
	m	I	II	III	IV
Downwelling irradiance	4·24	2·3%	5·3%	5·9%	×16
	66·1	2·8%	6·3%	7·4%	×9
Upwelling irradiance	4·24	6·4%	17%	32%	×30
	66·1	8·3%	19%	36%	×35

Thermopiles, which theoretically have a uniform response to equal amounts of energy regardless of wavelength, are generally not sensitive enough for underwater irradiance measurements. Selenium photovoltaic cells, photomultiplier tubes, or photodiodes, which are commonly used instead, all have different spectral responses. Such devices can be simply filtered to give them the V_λ function for measuring illuminance, but their spectral response is a problem in the design of integrating irradiance meters. Recently, however, two integrating radiometers have been developed with an approximately uniform sensitivity to equal numbers of quanta at all wave-lengths. Jerlov & Nygård's (1969) instrument is based on a selenium photovoltaic cell, Prieur's (1970) instrument on a photomultiplier tube. Spectroradiometers are calibrated across the spectrum against a standard source so that the spectral sensitivity of the photodetector presents no real problem.

Acknowledgements

I should like to thank Dr. O. C. Jones and Messrs. R. G. Berry, F. A.

Garforth & J. R. Moore of the National Physical Laboratory for their advice in setting up photometric and radiometric calibration facilities at Lowestoft. Professor J. L. Monteith kindly commented on the manuscript.

References

ARNOLD C.R., ROLLS P.J. & STEWART J.C.J. (1971) *Applied Photography*. 510 pp. Focal Press, London.

ATKINS W.R.G. & POOLE H.H. (1933) The photo-electric measurement of the penetration of light of various wave-lengths into the sea and the physiological bearing of the results. *Philosophical Transactions of the Royal Society B* **222**, 129–64.

BLACKWELL M.J. (1966) Radiation meteorology in relation to field work, pp. 17–39, in *Light as an ecological factor* (edited by R. Bainbridge, G.C. Evans and O. Rackham). Blackwell Scientific Publications, Oxford.

BLAXTER J.H.S. (1970). Light-fishes, pp. 213–85, in *Marine Ecology* (edited by O. Kinne). Wiley-Interscience, London.

BLAXTER J.H.S. & PARRISH B.B. (1965) The importance of light in shoaling, avoidance of nets and vertical migration by herring. *Journal du Conseil* **30**, 40–57.

CHISWELL B. & GRIGG E.C.M. (1971) *SI Units*. John Wiley, Sydney.

COMMISSION INTERNATIONALE DE L'ÉCLAIRAGE (1957) *International Lighting Vocabulary* 1, 1–136.

CLARKE G.L. & WERTHEIM G.K. (1956). Measurements of illumination at great depths and at night in the Atlantic Ocean by means of a new bathyphotometer. *Deep-Sea Research* **3**, 189–205.

CRAIG R.E. (1964) Radiation measurement in photobiology—choice of units. *Photochemistry and Photobiology* **3**, 189–94.

GILLHAM E.J. (1961) Radiometric standards and measurements. *Notes on Applied Science* (23), 1–23.

HALLDAL P. (1974) Light and photosynthesis of different marine algal groups, pp. 345–60, in *Optical Aspects of Oceanography* (edited by N.G. Jerlov & E. Steemann Nielsen). Academic Press, London & New York.

JERLOV N.G. (1968) *Optical Oceanography*. 194pp, Elsevier Publishing Co., Amsterdam.

JERLOV N.G. & NYGÅRD K. (1969) A quanta and energy meter for photosynthetic studies. *Report. Institut for Fysisk Oceanografi, Københavns Universitet* (10), 1–19.

JONES O.C. (1970a) Tables of spectral power distribution. *National Physical Laboratory, Division of Quantum Metrology, Report Qu.* 14.

JONES O.C. (1970b) Standard spectral power distributions. *Journal of Physics D: Applied Physics* **3**, 1967–76.

JONES O.C. & PRESTON J.S. (1969) Photometric standards and the unit of light. *Notes on Applied Science* (24), 1–39.

NATIONAL PHYSICAL LABORATORY (1960) Notes on the preparation of a standard reflecting surface of magnesium oxide. 2pp (mimeo).

NATIONAL PHYSICAL LABORATORY (1969) *The International Practical Temperature Scale of 1968*. 36 pp. H.M.S.O., London.

NORTHMORE D.P.M. & MUNTZ W.R.A. (1974) Effects of stimulus size on spectral sensitivity in a fish (*Scardinius erythrophthalmus*), measured with a classical conditioning paradigm. *Vision Research* **14**, 503–14.

PIRENNE M.H. (1956) Physiological mechanisms of vision and the quantum nature of light. *Biological Reviews* 31, 194–241.

PRIEUR L. (1970) Photomètre marin mesurant un flux de photons (Quanta-mètre). *Cahiers Oceanographiques* 22, 493–9.

PROTASOV V.R. (1964) Some features of the vision of fishes (Nekotorye Osobennosti Zreniia Ryb) pp. 29–48 in D.V. Radakov & V.R. Protasov, *Skorosti Dvizheniia i Nekotorye Osobennosti Zreniia Ryb*. Nauka, Moscow. (Translation 949, Marine Laboratory, Aberdeen.)

RAMSAY J.A. (1966) *The experimental basis of modern biology*. 339 pp. Cambridge University Press.

RICHARDS F.A. (1952) The estimation and characterization of plankton populations by pigment analyses I. The absorption spectra of some pigments occurring in diatoms, dinoflagellates, and brown algae. *Journal of Marine Research* 11, 147–55.

SMITH R.C. (1969) An underwater spectral irradiance collector. *Journal of Marine Research* 27, 341–51.

STRICKLAND J.D.H. (1958) Solar radiation penetrating the ocean. A review of requirements, data and methods of measurement, with particular reference to photosynthetic productivity. *Journal of the Fisheries Research Board of Canada* 15, 453–93.

THETFORD A. (1968) Control of spectral composition and intensity of stimulus, pp. 87–108, in *Techniques of photo-stimulation in biology* (edited by B.H. Crawford, G.W. Granger & R.A. Weale). North-Holland Publishing Co., Amsterdam.

TYLER J.E. (1971) Review of *Marine Ecology* 1 (1970), ed. O. Kinne. *Limnology and Oceanography* 16, 841–2.

TYLER J.E. (1973a) Lux vs. quanta. *Limnology and Oceanography* 18, 810.

TYLER J.E. (1973b) Applied radiometry. *Oceanography & Marine Biology Annual Review* 11, 11–25.

TYLER J.E. & SMITH R.C. (1966) Submersible spectroradiometer. *Journal of the Optical Society of America* 56, 1390–6.

UNESCO (1964) Report of the first meeting of the joint group of experts on photosynthetic radiant energy. *Unesco Technical Papers in Marine Science* (2), 1–5.

WALSH J.W.T. (1953) *Photometry*. 544 pp. Constable & Co. London.

WESTLAKE D.F. (1965) Some problems in the measurement of radiation under water: a review. *Photochemistry and Photobiology* 4, 849–68.

Questions and discussion

In reply to Dr. G. C. Evans, Dr. Arnold said that his receiver could withstand all North Sea depths. An opal-glass cosine collector, sealed with an O-ring, projected from the top of the case. A wavelength-dependent correction for immersion had to be made.

Professor T. H. Waterman pointed out that there was a movement, led by Dr. N. G. Jerlov, for the international standardization of methods of measuring light in the sea.

2 The construction and long term field use of inexpensive aerial and aquatic integrating photometers

D. F. WESTLAKE and F. H. DAWSON *Freshwater Biological Association, River Laboratory, Wareham, Dorset*

Introduction

It was considered that conventional methods of integrating the total incident radiant energy reaching streams in woodlands, and penetrating underwater, were too expensive and would give results that were difficult to interpret in relation to photosynthesis. At vulnerable field sites the expense of the initial purchase may have to be repeated after damage or theft. The proportion of photosynthetically active radiation will vary much more in such situations than at sites above terrestrial vegetation. The integrating photometers we describe are modifications of that of Powell & Heath (1964). The silver voltameter discussed by these authors, but dismissed, is reintroduced and used in conjunction with Macfadyen's technique (1956) which uses acetic acid to reduce the dendroid nature of growth of the silver deposits.

This paper endeavours to meet some of the previous objections to the use of selenium barrier layer photovoltaic cells, in particular their unreliability over long periods which is often assumed.

Apparatus and methods

The local primary standard

The local primary standard is a Kipp solarimeter, calibrated at Kew Observatory and permanently installed at the top of a steel pole on the roof of the laboratory. The output is integrated to give daily totals (Kipp and Zonen solarimeter integrator CCI). The output from the solarimeter head is also fed via the integrator and an impedance matching or buffer amplifier (100 kΩ) to a moving coil chopper-bar chart recorder (Elmes 1002). The normal precautions are taken to maintain the quality of readings at the solarimeter station and there is an automatically operating battery power supply for emergencies.

The Aerial Photometers

The aluminium photometer case contains a barrier layer selenium photocell (Megatron Type B, 36 mm diameter, mounted, and potted in Araldite) retained in position by a friction-fit plastic ring. Stainless steel rods and screws are used to construct holders for the silver rod electrodes of the integrating voltameter, which is directly attached to the photocell (Fig. 2.1a).

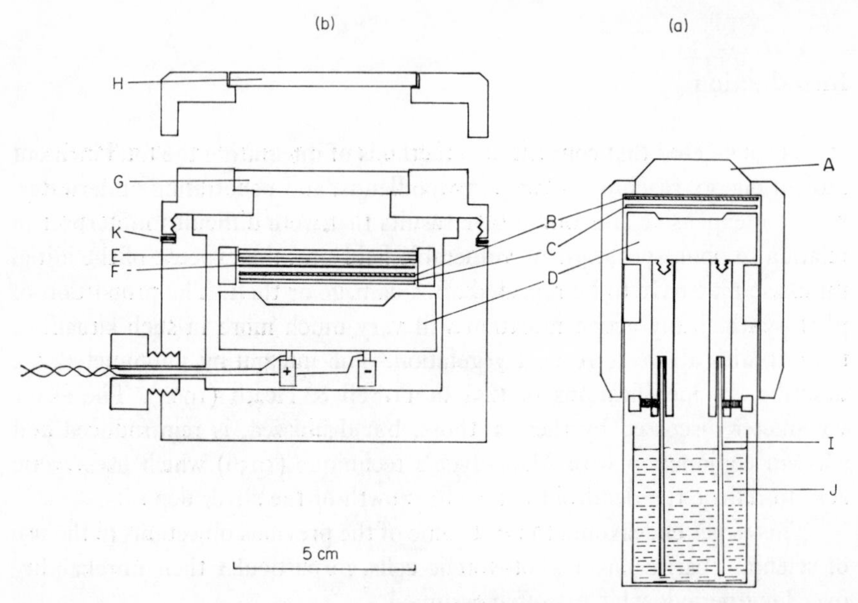

Figure 2.1. The construction of aerial (a) and aquatic (b) photometers. A, Perspex opal. B and C, Steel Blue and Pale Salmon filters. D, Selenium photocell. E, Ultraviolet filter. F, Neutral filter. G, Clear glass. H, Opal glass mounted in removable cap. I, silver rod. J, electrolyte. K, sealing ring. For further details, see text.

A raised and bevelled plastic opal (Perspex acrylic sheet 9·5 mm ($\frac{3}{8}''$, Grade 028), is used to correct the photometer's cosine response. Colour filters, Cinamoid 'Steel Blue' and 'Pale Salmon' (Strand Electric and Engineering Co. Ltd. London) are used to correct the spectral response of the photocells towards equality over the range of photosynthetically active radiation (PhAR, about 400 nm to about 700 nm). If the light reduction achieved by the opal (~90%) and the colour filters (~60%) is insufficient to bring the maximum instantaneous summer irradiance within the linear response range of the photocell (about 96% reduction being necessary) then a layer of woven plastic 'cloth' (Tygan) (~40%) can be used as a further neutral filter.

The photometer was mounted in a rotating clamp for measurement of the angular response of the light collector to a collimated beam of light from a tungsten filament bulb, at 10° intervals (Spence *et al.* 1971). The collection of light was very close to that expected by the cosine law from 0 to 80° from the axis, but beyond this, from 80 to 100°, more light was collected because of the raised nature of the opal. This error was much smaller than the under-response typically obtained at these low angles in the absence of an opal and was therefore not considered serious.

The linearity of the response of the photometer to light was determined by comparison with the solarimeter over the full local range of total irradiance. The maximum sustained irradiance in Southern England is about 860 J m^{-2} s^{-1} with brief increases up to 1070 J m^{-2} s^{-1}. The photometer output was linear in response to these irradiances (Fig. 2.2) when measured with a

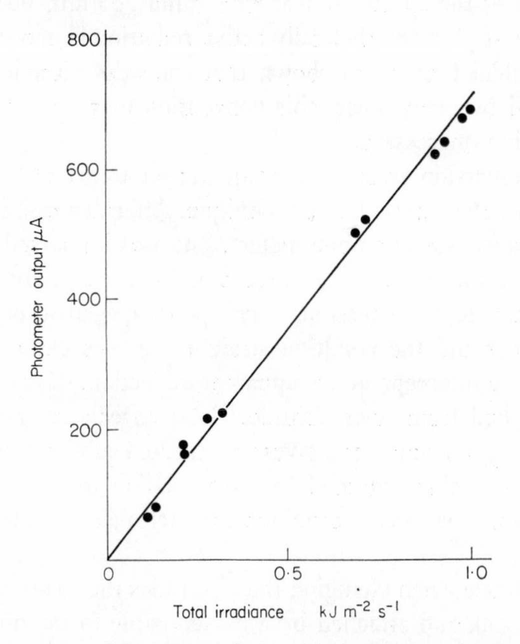

Figure 2.2. A comparison of the output of an aerial photometer with the solarimeter, over the full range of incident total irradiance encountered in the field observations.

Cambridge 'Unipivot' galvanometer, fitted with a range of shunts and of similar low resistance to the voltameters. Recalibrations of each field photometer in turn are made at regular intervals of between four and six weeks, replacing the field photometer with another calibrated photometer (Photometer no. 5, see Fig. 2.3 and Table 2.1).

The Aquatic Photometers

These photometers are constructed in a similar, though larger form to accommodate a 61 mm diameter mounted Megatron B photocell (Fig. 2.1b). A detachable conventional flat glass flashed opal is used instead of a raised acrylic one to provide a smooth, easily cleaned surface less suitable than plastic for the attachment of algae and settlement of detritus. Nevertheless cleaning is necessary every two or three days. The photometer includes the colour filters previously mentioned and a Tygan cloth neutral filter to reduce the incident energy level to the region of linear response of the photocell. An ultraviolet filter (Ilford 805 Q Colourless) was also added to make the cut-off at the blue end of the spectrum sharper around 400 nm, because the conventional range of photosynthetically active radiation is 400–700 nm. However, since Halldal (1967) has shown that the near ultraviolet present in sunlight is used by many algae, this convention may be criticized and the filter is probably unnecessary.

The true immersion error for the aquatic photometers (Westlake 1965) was measured by the extrapolation technique. After determining the matching factor between two dry photometers, one was immersed to a series of depths at 10 cm intervals. After correcting for their relative sensitivities, apparent percentage transmissions were plotted against depth on semi-logarithmic paper and the resulting straight line was extrapolated to zero depth to give an intercept at an apparent reflection. Deducting the true reflection, obtained from solar altitude, cloud cover and graphs published by Anderson (1952), from 100% gives an expected sub-surface value, which is divided by the value obtained by extrapolation to give the immersion correction factor. This was determined for each pair of cells and found to be 1·46.

For convenience when changing the electrodes the silver voltameters are on the stream bank and attached by a buried cable to the photometer cell. One of these aquatic photometers is used on the stream bed and another is supported close by, above the water surface. Another pair are calibrated at the laboratory and are exchanged at monthly intervals.

Integration

The output of the selenium cells is integrated by silver voltameters, silver being transferred from the anode to the cathode proportionally to the number of coulombs of electricity passing. Readily available plastic containers (Weedol tubes), painted matt black on the inside to reduce photo-activated decay of the electrolyte, are used to hold a 5 × 2·5 cm flat-bottomed glass tube containing 20 ml of electrolyte (Fig. 2.1a). This is 25% w/v A.R. grade silver nitrate solution with 0·25% v/v acetic acid. The silver electrodes (76 mm × 1·6 mm (3″ × 0·064″) diameter grade I silver rod) dip about 2·5 cm into the electrolyte. The resistance of these voltameters to current densities of the order of magnitude expected was found to be about 10Ω at 10°C, varying about 3Ω over the normal range of air temperatures. One of the series of photometers was attached to a voltameter using electrical grade copper electrodes of similar size in 20% w/v A.R. grade cupric nitrate solution for comparative studies.

The electrodes are prepared for use by gently filing off irregularities, that is, the silver from the cathode of a previous run. Emery paper is used for polishing as necessary. One of the electrodes is marked with a notch near the upper end to avoid confusion and the electrodes are alternately used as anode and cathode to extend their useful life. Electrodes are discarded when their diameter was reduced to about 1 mm. Before weighing to 0·1 mg the electrodes are thoroughly washed in distilled water and agitated in acetone before air drying. The electrode pairs returning from field cells are washed with distilled water on site before returning to the laboratory, after which they are washed again before weighing.

The accuracy of the silver voltameter was tested by supplying a constant current within the range 5–250 μA for periods of one week to silver voltameters connected in series so that each voltameter received the same current. This current was supplied by an accumulator with a variable resistance in series and was measured using an ammeter permanently connected in series with the voltameter. The apparatus was kept in a constant temperature room, because originally the major variation in output of the accumulator was due to temperature. The electrochemical equivalent of silver found was within one per cent of the theoretical. The repeatability of this result was considered to be obtained from the differences between the members of such a series of voltameters in any particular run, which were within ±0·2%.

It is important to note that using the silver electrodes, a difference of 1 mg could produce a change in photometer calibration factor of about ±10% in the winter but only ±2% in the summer. When copper was used, an error of three times this can be produced because the electrochemical equivalent is only a third of that of silver.

The photometers are calibrated by exposure to daylight near the solarimeter for periods of about a week. At the beginning and ending of the exposure period the reading on the solarimeter integrator is recorded. Regressions of silver transferred on total integrated irradiance measured by the Kipp standard were highly significant ($P \gg 0{\cdot}001$), throughout any one year. Over longer periods the significance decreased slightly as the influence of a drift in sensitivity became more important. The weight of silver deposited by a gram calorie of incident radiation is calculated and called the calibration factor. The reciprocal factors used for the conversion of field determinations are derived from regressions of the calibration factors for each cell against time (see below). Any effects of changes in the ratio of photosynthetic to total irradiance in incident daylight are assumed to be averaged out over long periods. To calculate integrated photosynthetically active irradiance, the Kipp values or the reciprocal factors were multiplied by a standard factor of 0·45.

Results

The Photometers

The photoelectric conversion efficiency as expressed by the calibration factor does decrease with time, but in a regular manner at a rate of between 5%–15% yr^{-1}, for example photometer 5, Fig. 2.3 and Table 2.1. The large F

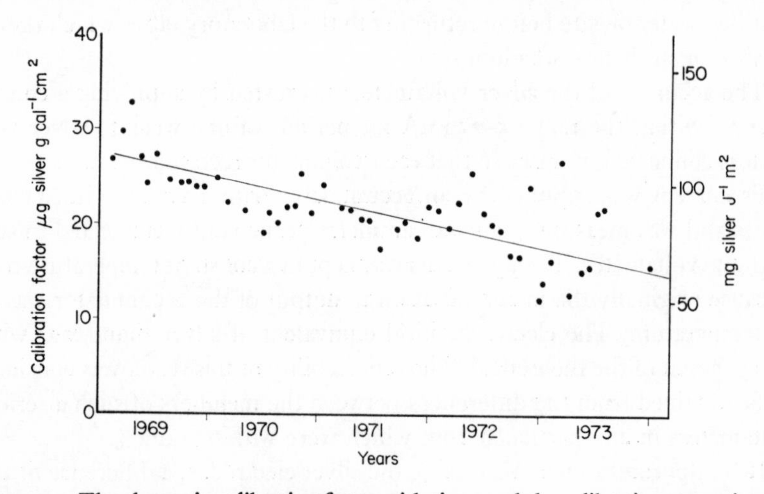

Figure 2.3. The change in calibration factor with time, and the calibration regression, for photometer No. 5. For details of regression see Table 2.1.

ratios obtained show that a large part of the variance is explained by the decrease in sensitivity with time. On theoretical grounds it would be expected in the case of silver electrodes that the gain or loss in weight would be equally suitable for calculation. This is in contrast to copper, as oxides form on both electrodes (Powell & Heath 1964). In general, however, bearing in mind weighing and manipulation errors in the transport, exchange and washing of electrodes for field cells, the mean of the gain and loss of silver, neglecting

Table 2.1. Analysis of the regression of the calibration factors against time for the photometers (calculation to 8 decimal places); μg silver (total gcal)$^{-1}$ cm^{-2} = $a+b$(months).

Photometer No.	Period	No. of Calibrations	Gains			Losses			Means		
			b	*a*	*F* ratio[3]	*b*	*a*	*F* ratio[3]	*b*	*a*	*F* ratio[3]
Silver											
1	1971–73	26							−0·1418	24·91	7·38*
2a[1]	1969	15	−0·2052	28·26	3·35	−0·3453	29·28	12·67**	−0·2752	28·77	9·41**
2b	1971–73	31	−0·1933	25·08	6·51*	−0·1353	23·67	3·89	−0·1643	24·38	8·61**
3	1969–73	51	−0·1651	24·93	11·04**	−0·1627	24·06	9·64**	−0·1639	24·49	11·11**
5	1969–73	53	−0·2170	27·42	25·56***	−0·1701	25·14	14·84***	−0·1935	26·28	21·01***
Copper											
6	1969–73	203	−0·0129	9·96	10·54**	−0·0082	6·44	11·32***	−0·0103	8·18	7·02**
6	1969–73	51[2]							−0·0152	9·11	2·08

[1] The original photocells 2a, and 4, failed in the first year; 2b was a replacement and 4 was not used again.

[2] Every fourth calibration used to make the number comparable with the silver calibrations for 3 & 5.

[3] F ratio probabilities: *, P = 0·05; **, P = 0·01; ***, P = 0·001.

signs, was found to be the most satisfactory value to use (Table 2.1). The weighing of the two electrodes is also helpful as a check for errors.

The variation found in calibration factors is undoubtedly partly due to differences in the spectral energy distribution between dull cloudy and bright periods, which are averaged out by the regressions of the monthly calibrations. Such differences will give rise to real differences in the relations between the total irradiance measured by the solarimeter and the photosynthetically active irradiance measured by the photometers in the field from week to week.

The selenium cells are temperature sensitive and, although all are calibrated and used at ambient temperatures, some differences will undoubtedly exist between unshaded and shaded and above and under water conditions. Temperature will also affect the resistance of the voltameter which will have a small effect on the current from the photocell, but no obvious changes with season are seen in the regressions of calibration factors on time. Frost severe enough to freeze the electrolyte occasionally causes the loss of a week's results.

The Field Sites

The photometers were placed beside the stream under investigation. At Hollybush ('unshaded A') the stream flowed through an open field beside a watercress bed. At Bere Heath the upstream East bank was covered by marsh with only a few scattered willow bushes. Further downstream trees of *Salix viminalis* and *Alnus glutinosa* increased in density and size. The West bank was a marshy meadow at the upstream end of the stretch with scattered bushes increasing downstream, especially opposite the trees on the other bank, until the stream entered a wood of tall mature trees.

Three aerial photometers were situated on the West bank, one in the meadow ('unshaded site B'), one amongst the scattered bushes opposite the smaller trees and one in the wood ('deciduous wood'). The aquatic photometers were also on the West bank, between the unshaded site B and the partial shade site. The grid references for all these sites are given in Table 2.2. Some further details of the Bere Heath stretch, and chalk-stream valleys in general may be found in Westlake *et al.* (1972).

The Field Results

The use of photometers to measure incident photosynthetic irradiance at two unshaded field sites 15 km from the laboratory produced annual means slightly lower than at the local primary standard site but not significantly different from it (Table 2.2). The differences between sites did not exceed 20% in 80–90% of individual weeks and there were no obvious seasonal changes (Fig. 2.4). The unshaded site A photometer was on a post 2 m above the ground near the Hollybush stream mentioned above and may have had some slight distant shading, which was more obvious for the unshaded site B photometer which had some trees to the South-West reducing low-angled light. The week to week differences between sites are to be expected since cloud cover will not be identical.

The partial shade photometer was situated 1 m from the ground on the West bank of the Bere Heath stream which here flowed South-South-Easterly. A line of trees, 10 m high, on the East bank of the stream shaded the photometer. This site received 0·8 of PhAR at the unshaded sites, with little seasonal variation. Presumably changes in leaf cover were matched by changes in shading by permanent features as the maximum solar altitude varied.

The photometer in the deciduous wood was moved 5 m each week along the stream bank within the wood to obtain mean values for the light penetration through the canopy. This is presented as a running mean in Fig. 2.4 to smooth out most of the site-to-site fluctuations. The site ratio varied between 0·06 in late summer and 0·3 to 0·4 in winter, with rapid changes at the times of leaf emergence and leaf fall.

At Bere Heath the three sites, open, partially shaded (0·8 of open) and wooded (0·15 of open), show differences in plant growth in the stream. Much of the open reach is densely covered with *Ranunculus calcareus* in the spring and by *Rorippa nasturtium-aquaticum* in the summer. In the partially shaded reach *Rorippa* is slower to develop and probably never attains the same biomass (per unit area colonized) as in the open reach. The partially shaded reach is deeper, which also restricts the spread of *Rorippa*, and the final result is that the two species are co-dominant in the autumn (Castellano, pers. comm.). There is no difference in benthic algal biomass (on stones) in the spring, but in the summer and autumn the biomass at the unshaded site is over twice that at the partially shaded site (Marker, pers. comm.). Taxonomic differences are also observed. *Ulvella frequens*, *Chamaesiphon* sp. and associated diatoms dominate the summer populations at the partially shaded reach but *Vaucheria* sp. and lime-encrusting organisms such as *Homoeothrix crustacea* are more abundant in the unshaded reach. Angiosperms are almost absent from the wooded reach where there are only dense patches of *Pellia fabbroniana* and some *Spongiella* sp. and *Ephydatia* sp. This is comparable with reaches of the R. Frome deeper than 1 m where the light is reduced to less than 0·3 of incident and *Fontinalis antipyretica* becomes a significant part of the total plant biomass.

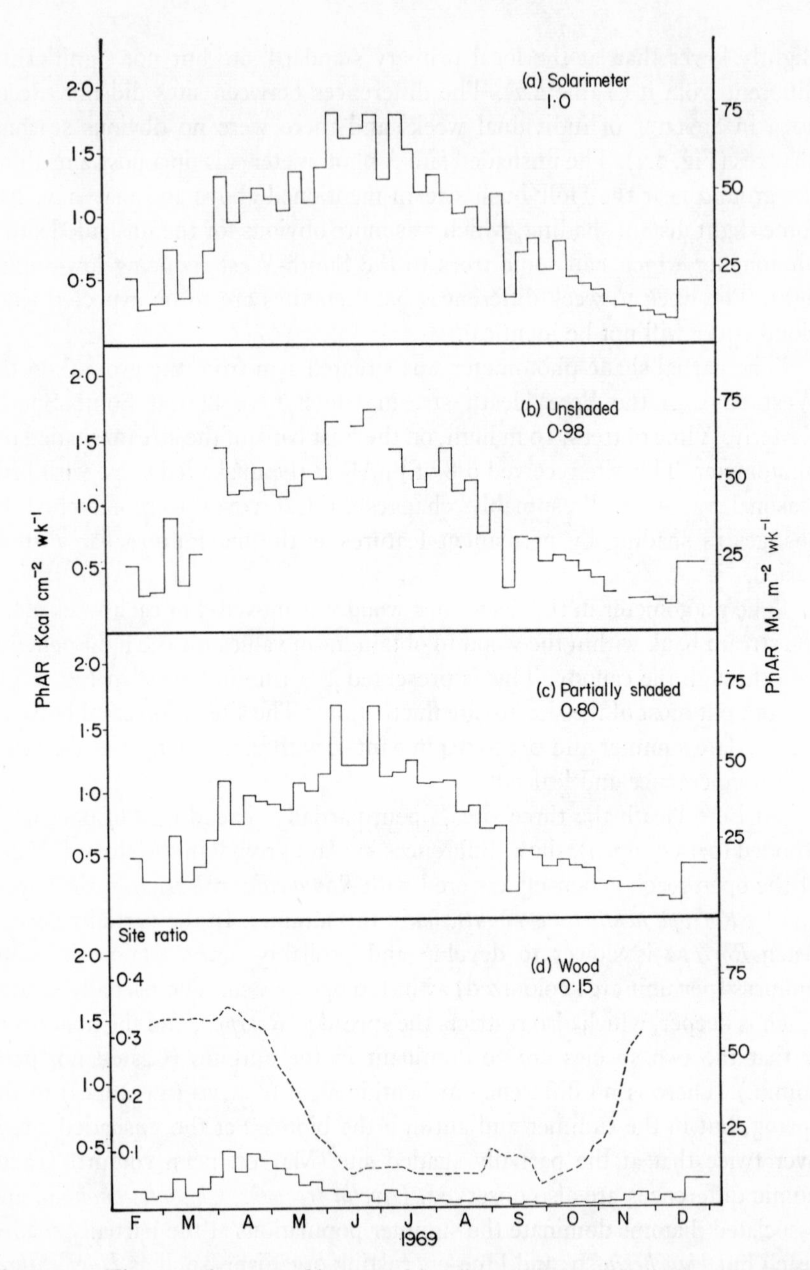

Figure 2.4. The annual course of integrated irradiance (photosynthetically active radiation, PhAR) over periods of a week at field sites compared with the solarimeter in 1969. The last period of the year was two weeks, so the weekly rate would be approximately half the rate shown. Annual mean site ratio (PhAR at site/PhAR at solarimeter) under each title. Fig. 2.4d includes the site ratio (wooded/open) for the wooded site as a continuous line through running means (over 4 week periods).

Table 2.2. Incident photosynthetically active radiation (PhAR) at field sites 1969–73: grid references give site positions, for further details see text. Site ratios[1] and (in brackets) success ratios.[2] n.c., not calculated.

Year	Solarimeter standard (E. Stoke, SY870867) (× 0·45) GJ m^{-2} y^{-1}	Unshaded A Hollybush, SY839956)	Unshaded B (Bere Heath, SY858926)	Partial Shade (Bere Heath, SY858923)	Deciduous Wood (Bere Heath, SY858922)	Surface (Bere Heath, SY858924)	Submerged (Bere Heath, SY858924)
1969	1·69	0·976 (0·902)	—	0·804 (0·994)	0·150 (1·00)	—	—
1970	1·77	0·951 (0·950)	—	0·770 (0·198)	—	—	—
1971	1·84	0·907 (0·985)	0·965 (0·938)	0·776 (0·982)	—	—	—
1972	1·66	0·953 (0·812)	1·033 (0·902)	0·808 (0·871)	—	0·940 (0·960)	0·53 (0·85)[3]
1973	1·78	1·029 (0·988)	0·894 (0·760)	0·839 (0·978)	—	n.c.	n.c.
Mean	1·75	0·96	0·96	0·80	0·15	0·940	0·53

[1] $$\text{Site ratio} = \frac{\text{PhAR measured at site for weeks of successful exposure}}{\text{PhAR from solarimeter for same weeks}}$$

[2] $$\text{Success ratio} = \frac{\text{PhAR measured at site for weeks of successful exposure}}{\text{Annual PhAR received at site estimated for all weeks}}$$

[3] Excluding four anomalous results, discussed in text.

The results from the aquatic photometers were primarily intended to give data on the irradiance available for plant growth on the stream bed at Bere Heath. At the site of the underwater measurements the photosynthetically active irradiance reaching the water surface was reduced to 0·94 of the incident, mainly by tall herbaceous plants on the bank. Over the year between 0·23 and 0·83 of the surface irradiance reached the stream bed (Fig. 2.5).

This excludes the four anomalous values in February and May, when the irradiance on the stream bed appeared to be higher than incident. There is no obvious explanation of these since they do not correspond with changes of weather or photometers, both pairs of photometers were involved in May, the gains and losses agree well and the surface irradiances correspond satisfactorily with measurements at other sites. Furthermore, they cannot be explained by assuming that the surface and underwater electrodes were exchanged, because the actual weighings from the submerged photometer were always less than from the surface photometer and the immersion correction factor produces the actual reversal of magnitudes. Yet this factor is used throughout and there is no reason why it should be different or not required in these particular weeks (rain wetting the surface photometer will not produce the same optical effects as immersion and the depth was never less than 10 cm).

The annual average transmission was 0·56 (excluding these four results) which includes the effects of both surface reflection and attenuation losses. Assuming that the annual average reflection of direct and diffuse radiation from slightly rippled water for all solar altitudes and weathers is about 0·1, the attenuation component corresponds to a transmission of 0·62, or an attenuation coefficient of 1·6 (ε, log base e) for an annual average depth of 0·3 m (range 0·2–0·5 m). This may be compared with attenuation coefficients determined in chalk-streams using conventional selenium photocell vertical transmission meters (Westlake 1966). These are between 0·60 and 1·8, unless the water is exceptionally turbid, but are weighted towards the transmission of green radiation because of the spectral sensitivity of unfiltered photocells. Data in Westlake (1966) for the R. Frome allow a correction factor of 1·3 to be calculated, giving attenuation coefficients for photosynthetically active radiation between 0·78 and 2·3 which straddle the estimated mean integrated value of 1·6. Underestimation of the stream bed irradiance is expected because of the silt, animals and algae that collect on the submerged photometer between cleanings, but the results do not appear to be low. Under shallow (0·2 m), non-turbid (ε, 0·78) conditions transmission would be predicted as about 0·77 (0·83 observed) and under deeper (0·5 m), fairly turbid (ε, 2·3) conditions 0·28 would be predicted (0·23 observed).

The transmission variations shown in Fig. 2.5 will be influenced by depth and turbidity, which at first sight would be expected to be closely related to discharge. However, the correlation between discharge and transmission is poor. Firstly, depth tends to increase in the spring, although discharges are decreasing, because of weed growth in the river, and may decrease when floods wash the plants away. Secondly, suspended solids increase very rapidly in a flood but their concentration falls much faster than the discharge, and an initial small flood may have a concentration of suspended material as high as subsequent larger floods. Thirdly, suspended solids may be affected by cattle or human activity.

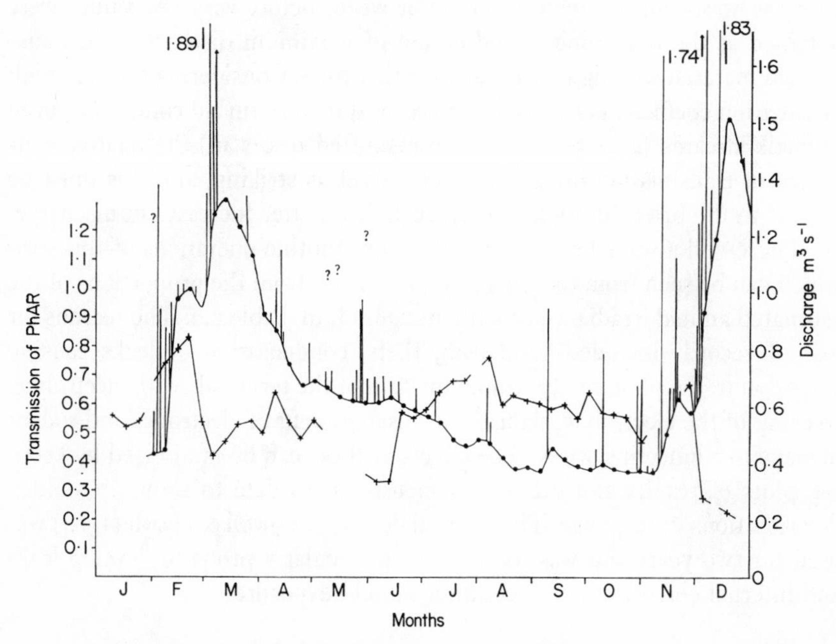

Figure 2.5. The annual course of weekly mean and flood discharges, and transmission of photosynthetically active irradiance to the stream bed (Bere Stream at Bere Heath 1972).
+ Transmission (irradiance at stream bed/irradiance at surface), omitting four anomalous values.
● Weekly mean discharge.
| Discharge at peak of flood.
For details and discussion, see text.

The year illustrated in Fig. 2.5 started with low flows and some weed remained in the river. This was removed by the floods in early February together with much sediment. After these floods the discharge rose considerably as the chalk springs increased their output of clear water and transmission was high in late February during a period of fairly stable discharge and fairly shallow water. A particularly low transmission occurred after another flood and transmission remained fairly low or average during the period of maximum discharge, when suspended solids are generally high (Westlake *et al.* 1972). As the discharge decreased transmission changed little, probably because depth increased with weed growth. In early June transmission was low. There were frequent small floods and suspended solids are often high in May and early June because of work in water-cress beds upstream (Westlake *et al.* 1972). When this period was passed and floods became less frequent, transmission was fairly high or average until a series of floods in late October increased the suspended material. A further fairly high value occurred in November during a period of stable shallow conditions after the wash-out of much of the water weed, before very low values were obtained at the beginning of the period of maximum discharge when suspended material was higher than at any other time. Considering the very high attenuation coefficients that can be measured in very turbid conditions, even in chalk streams ($\varepsilon > 5{\cdot}0$, $< 0{\cdot}07$ transmitted to 0·5 m) the relative constancy of transmission integrated over a week is striking, and this must be related to the brief duration of such turbidities after increases in discharge.

There will always be breakdowns in any routine operations of this sort, which can be seen from the gaps in Fig. 2.4. and from the proportions of the estimated annual irradiance actually measured, in Table 2.2. The reasons for loss of records included vandalism, theft, condensation or leaks causing corrosion in the photometer, contamination of the terminals with electrolyte, freezing of the electrolyte, damage or misalignment of electrodes and rodent damage to connecting wires. The effects of these can be minimized by keeping plots of results and calibration factors up to date to show up sudden deteriorations in response. The original design of Powell & Heath (1964) was used for two years and was found to be particularly prone to develop leaks and internal corrosion under continuous field exposure.

Acknowledgements

Thanks are due to J. Leach and P. Henville for maintaining the photometers and to H. Zschorn and J. Morgan for their construction.

We are indebted to Mr. E. Castellano and Mr. I. Farr for permission to use unpublished data on depth and suspended solids in the discussion of changes in transmission, and to Mr. E. Castellano and Dr. A. F. H. Marker for information on the vegetation at the Bere Heath sites outside the wood.

References

ANDERSON E.R. (1952) Energy-budget studies. Water loss investigations: Vol. 1, Lake Hefner Studies, Tech. Rep. *Circ. U.S. geol. Surv.* **229**, 71–119.

HALLDAL P. (1967) Ultraviolet action spectra in algology. *Photochem. Photobiol.* 6, 445–60.
MACFADYEN A. (1956) The use of a temperature integrator in the study of soil temperature. *Oikos* 7, 56–81.
POWELL M.C. & HEATH O.V.S. (1964) A simple and inexpensive integrating photometer. *J. exp. Bot.* 15, 187–91.
SPENCE D.H.N., CAMPBELL R.M. & CHRYSTAL J. (1971) Spectral intensity in some Scottish freshwater lochs. *Freshwat. Biol.* 1, 321–37.
WESTLAKE D.F. (1965) Some problems in the measurement of radiation under water: a review. *Photochem. Photobiol.* 4, 849–68.
WESTLAKE D.F. (1966) The light climate for plants in rivers. In *Light as an ecological factor* (edited by R. Bainbridge, G.C. Evans & O. Rackham) *Brit. Ecol. Soc. Symp.* No. 6, 99–120. Blackwell Scientific Publications, Oxford.
WESTLAKE D.F., CASEY H., DAWSON F.H., LADLE M., MANN R.H.K. & MARKER A.F.H. (1972) The chalk-stream ecosystem. In *Productivity problems of freshwaters* (edited by Z. Kajak & A. Hillbricht-Ilkowska) *Proc. IBP/UNESCO Symp.* Kazimierz Dolny, May 1970, 615–35. Polish Scientific Publishers, Warszawa-Kraków.

Questions and discussion

In answer to a question from Professor Monteith Dr. Westlake said that he had used the factor of 0·45 throughout to convert the values of irradiance measured by the Kipp solarimeter to photosynthetically active irradiance mainly because one or two of the I.B.P. synthesis volumes had decided to take 0·45 as their standard, which he was inclined to think was unfortunate. Professor Monteith said that when allowance was made for light from blue sky the theoretical factor should be nearer 0·50, and that analysis by Dr. Szeicz of a long series of measurements at Cambridge and Rothamsted had shown this in practice (*this vol.*, pp. 513–19). Dr. Westlake said that in part the fluctuations in calibration factors shown in Fig. 2.3 might be due to systematic differences between very sunny and very cloudy weeks, but that in winter, when the deposit weights were very small, the random errors would be proportionately larger. Dr. Evans said that systematic differences in the conversion factor would be expected under tree canopies, because if there were a dull grey sky and a dense canopy most of the light reaching the ground would come from near the zenith, whereas when the sun is shining in these latitudes most of the light would come from elsewhere, so that there would inevitably be canopy differences between the two conditions.

In answer to a question from Dr. Rackham about the opal diffusers, Dr. Westlake said that in diffuse light the transmission factor was fairly constant over most of the visible spectrum, falling a little approaching 400 nm. In a cosine correction test, the bevelled edge of the diffuser in the aerial photometers tended to give rather too much light at very low angles of incidence.

Dr. Evans pointed out that the spectral transmission of opals can vary with angle of incidence.

Professor Waterman developed the view that the term photometer should be restricted to measurements related to the visual properties of radiation, other measurements being referred to as radiometric. Dr. Westlake defended his usage by pointing to the very large overlap between the responses of plants and the human eye, while the action spectrum of photosynthesis differed widely over the mixture of plant phyla growing in the area of his measurements. No useful purpose would thus be served by attempting a correction for a particular action spectrum, while it is convenient to use figures for the total radiation in the visible region of the spectrum.

3 Phytoplankton production in the Rivers Thames and Kennet at Reading during 1970

T. J. LACK[1] and A. D. BERRIE[2] *Department of Zoology, University of Reading*

Introduction

The primary production and respiration of the phytoplankton in the River Thames and its tributary, the River Kennet, were investigated at Reading from April 1967 to April 1968 (Kowalczewski & Lack 1971). The species composition of the phytoplankton at these sites was studied simultaneously by Lack (1971). These studies demonstrated that the Thames above the confluence had a well developed phytoplankton with a marked seasonal pattern, and the annual net production in that year was 7·979 MJ m^{-2} $year^{-1}$. The Kennet phytoplankton consisted largely of benthic algae in suspension and, due to a higher ratio of suspended solids, respiration exceeded production in the water column giving an annual net production of $-1{\cdot}151$ MJ m^{-2} $year^{-1}$ in that year. These investigations were part of a study of production and energy flow at all trophic levels in the River Thames at Reading which formed part of the British contribution to the International Biological Programme. The overall results showed that the phytoplankton was the main source of primary production in the Thames (Berrie 1972a; Mann *et al.* 1972).

Kowalczewski & Lack (1971) used their data to derive a regression equation relating net production by phytoplankton in the Thames to a combination of solar radiation, chlorophyll concentration and euphotic depth. These latter three variables can be measured quite accurately and easily. If the equation was valid under a wide variety of conditions it would provide a simpler method of calculating net production than the present direct methods. It was clearly desirable to repeat the observations so that the equation could be used to predict net production at the same time as direct measurements were being made.

Present Addresses:

[1] Water Research Centre, Medmenham Laboratory, Medmenham, Marlow, Buckinghamshire.

[2] Freshwater Biological Association, River Laboratory, East Stoke, Wareham, Dorset.

For the overall production and energy flow study, the phytoplankton production in the Thames below the confluence with the Kennet had to be estimated. Kowalczewski & Lack (1971) did this by assuming that the levels of production which they measured in the two rivers above the confluence could be put together in proportion to the relative discharges of the rivers at the confluence on each occasion. Recently there has been increasing interest in the possible consequences of transferring water from one river system to another for supply or regulation purposes and the Thames could be affected in this way in the future. There is little information to indicate what happens to phytoplankton production when water from two rivers with different levels of productivity is mixed. It seemed useful to make direct measurements of phytoplankton production below the confluence at the same time as measurements were made in the two rivers above the confluence.

The present investigation was carried out during 1970 to obtain data for another year for comparison with those obtained in 1967–68; to test the predictive value of the equation for calculating net production indirectly; and to determine whether production below the confluence of the rivers was the same as might be predicted from measurements above the confluence and data on relative discharges.

Study sites and methods

The limnological characteristics of the rivers have been described previously together with details of the two sites above the confluence and of the methods used (Kowalczewski & Lack 1971). This description should be consulted for a fuller account.

Sites

Two sites were studied on the Thames. The first was about 200 m upstream of the confluence, on the north side of the river, and had a maximum depth of 2·4 m. The second was about 1·8 km downstream of the confluence, on the north side of the river, with the same maximum depth. The site studied on the Kennet was about 200 m upstream of the confluence, on the west side of the channel, and had a maximum depth of 1·8 m. The sites were selected to avoid shading during the hours of high solar irradiance.

Methods

The water in both rivers is well mixed and a single sample from just below the surface can be taken as representative of the whole depth and width of the channel. Water samples, usually 1 litre in volume, were taken from each site and suction-filtered through Whatman GF/C filters. The pigments were extracted in 90% aqueous acetone and the optical densities of the extracts were determined at 665 and 750 nm in 4 cm cuvettes. The quantity of algal pigments was calculated according to the formula given by Lorenzen (1967) which permits the separation of the pigments into chlorophyll *a* and its degradation product pheophytin *a*. Algal production and respiration were estimated every second week for most of the year but every fourth week in winter. Values were determined from changes in dissolved oxygen concentration in transparent and opaque bottles suspended horizontally at five depths in the river as described previously, except that on most occasions an extra set of bottles was incubated at 30 cm depth. The production and respiration rates obtained from each depth were weighted, using the morphometric data given in Table 1 of Kowalczewski & Lack (1971), to give an integrated value for the whole water column. These data were plotted to give annual curves and the year's total production and respiration were estimated from the area under the curve.

Integrated daily values of the intensity of total solar radiation were obtained from a Kipp Solarimeter at the Electrical Research Station (University of Reading), 5 km from the river. Transmission of light through the waters of the Thames and Kennet was measured on occasions with a submersible photometer fitted with opal discs. Instantaneous estimates of the volumes of water discharged by the Thames at Caversham Weir (1·2 km upstream) and by the Kennet at its confluence, were provided by the Thames Conservancy.

Results

Algal pigments

The seasonal pattern of variation of chlorophyll concentration (Fig. 3.1) was essentially similar to that found in the earlier series of observations. Both rivers showed a period of high chlorophyll *a* concentration during April, May and early June, rising to a maximum of 196·9 μg l^{-1} in the Thames and 61·8 μg l^{-1} in the Kennet. At the time of the maximum in the Thames there was no detectable pheophytin present and in the Kennet it represented <2% of the total pigment concentration. In both rivers the dominant phytoplankter at the time of the chlorophyll maxima was the centric diatom *Stephanodiscus hantzschii* Grunow, as observed in previous years (Lack 1971). Cell counts at the Water Research Centre, Medmenham Laboratory (21 km downstream), showed that, during the chlorophyll maximum in the Thames, *S. hantzschii* amounted to 72,900 cells ml^{-1} out of a total count of 82,500 cells ml^{-1}.

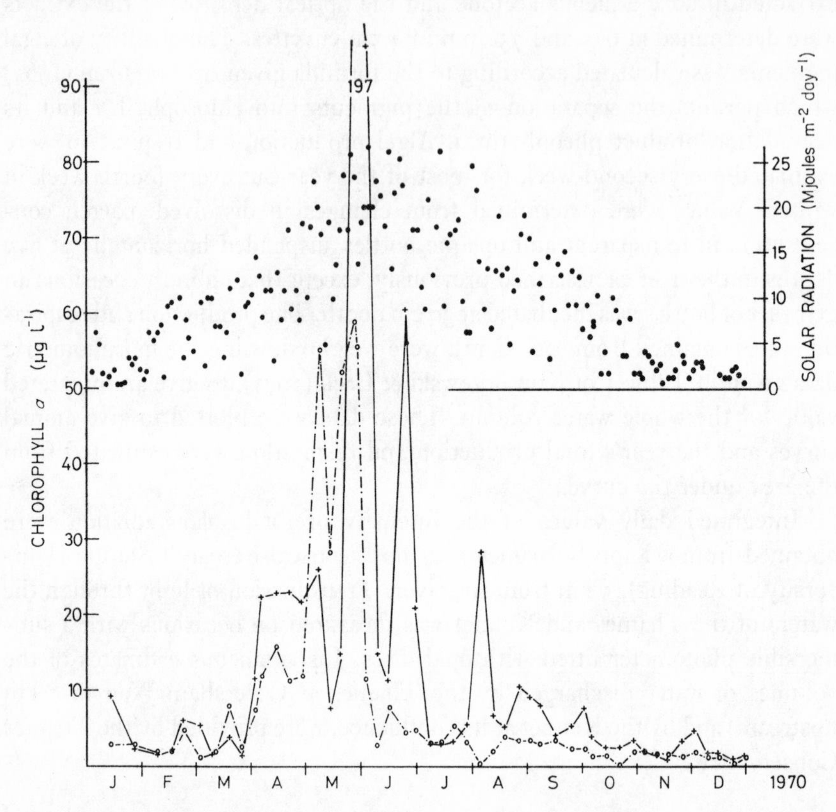

Figure 3.1. Variations in total daily solar radiation (●), and in chlorophyll *a* concentration during the year 1970 in the River Kennet (○ — · — ○) and in the River Thames above Kennetmouth (+ —— +).

The diatom population declined rapidly in the Thames and was followed by a typical midsummer phytoplankton of various Chlorophyceae (Lack 1971). The chlorophyll increased to 60·4 µg l^{-1} and the pheophytin concentration at this time was 5·5 µg l^{-1}. Cell counts at Medmenham showed that, out of a total count of 12,000 cells ml^{-1}, the green algae amounted to 6,700 ml^{-1}, and these were chiefly *Dictyosphaerium pulchellum* Wood. This population also declined rapidly until, by the first week in July, the chlorophyll had fallen to 2·9 µg l^{-1}.

On 4 August the chlorophyll in the Thames rose to 28·3 µg l^{-1} and the pheophytin to 78·5 µg l^{-1}. This situation was very similar to that observed in August 1967. The autumn bloom of centric diatoms appears to have a high concentration of degraded chlorophyll. The total cell count at Medmenham at this time was 37,100 ml^{-1} of which 22,800 ml^{-1} were *S. hantzschii*. One week later the chlorophyll had fallen to 6·7 $\mu g\ l^{-1}$ and it exceeded 10 $\mu g\ l^{-1}$ on only one occasion up to the end of December.

It seems to be characteristic of the Kennet that, once the spring diatom activity has declined, the chlorophyll concentration and cell count remain at comparatively low levels (Lack 1971). From the middle of June the chlorophyll did not exceed 10 $\mu g\ l^{-1}$ and on many occasions was $<5\ \mu g\ l^{-1}$.

Solar radiation

The intensity of solar radiation varied greatly over the period (Fig. 3.1). During January, November and December the integrated daily radiation rarely exceeded 5 MJ m^{-2} day^{-1} and the minimum value was 0·3 MJ m^{-2} day^{-1} on 10 January. Values greater than 20 MJ m^{-2} day^{-1} were recorded in May, June, July and August with a maximum of 27·1 MJ on 21 June.

Discharge

The discharge of the Thames ranged from 9·9 to 117·2 $m^3\ s^{-1}$ with an average value of 34·7 $m^3\ s^{-1}$ (Fig. 3.2). This is lower than the averages of 42·2 and 40·5 $m^3\ s^{-1}$ calculated for 1966 and 1967 respectively. From the beginning of June until the end of October the discharge did not exceed 20 $m^3\ s^{-1}$. This is only half the value that Lack (1971) calculated would sweep away the phytoplankton faster than it could be replaced by recruitment. The reduction of the standing crop in the middle of May (Fig. 3.1) was probably due to an increase in discharge from 28·0 to 41·4 $m^3\ s^{-1}$ which exceeded this critical value. The sudden declines in the chlorophyll concentration following the spring, summer and autumn peaks did not appear to be caused by increased discharge.

The discharge of the Kennet was always less than the Thames ranging from 4·8 to 14·4 $m^3\ s^{-1}$ with an average of 9·2 $m^3\ s^{-1}$ which was considerably lower than in 1966 and 1967. It has been shown previously (Lack 1971) that increases in discharge of the Kennet bring numerous benthic algae into suspension and that these often become dominant over the typically planktonic forms. These discharge effects were not observed to any great extent during 1970. The rapid decreases in chlorophyll concentration in the middle of May and at the beginning of June (Fig. 3.1) did not coincide with any sudden changes of discharge.

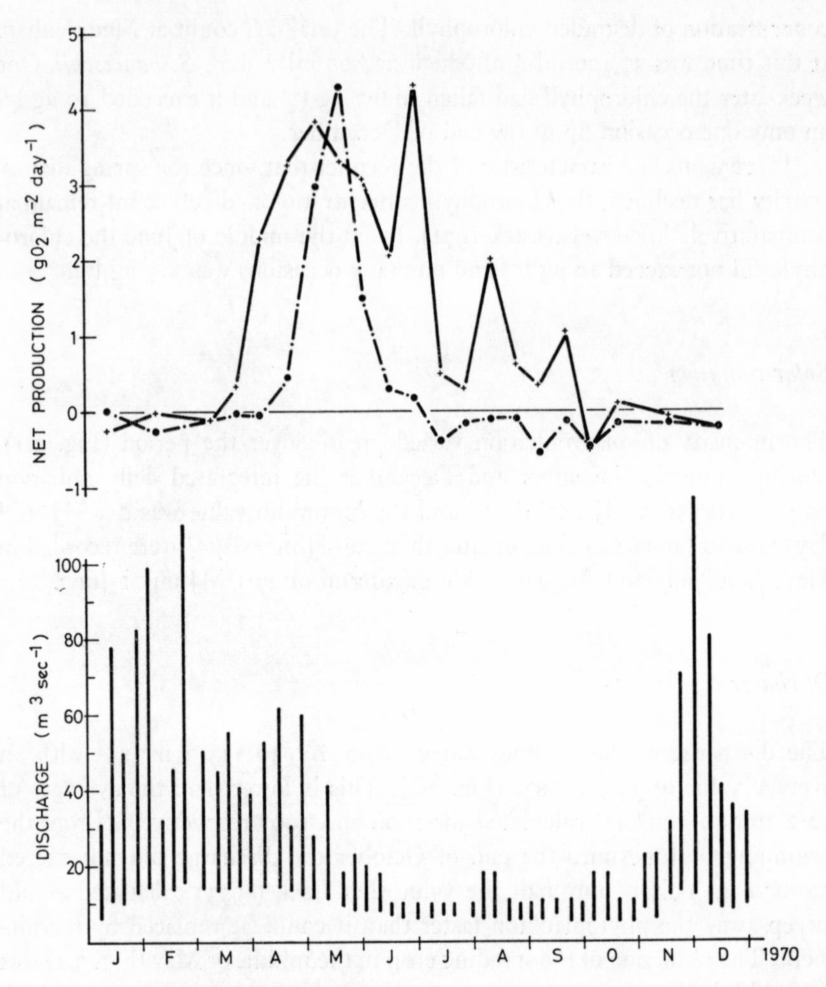

Figure 3.2. Variations in daily net oxygen production in the River Thames above Kennetmouth (+ —— +), and in the River Kennet (● —·— ●), together with the discharges of the two rivers, during 1970. Discharges are instantaneous values on particular days, and the lines join the rate for the River Thames (upper end) and for the River Kennet (lower end).

Net production

Thames

From the beginning of January until 9 March there was no positive net production of oxygen in the Thames (Fig. 3.2). From 23 March onwards the phytoplankton produced a net surplus of oxygen over respiration reaching a peak of 3·83 g O_2 m^{-2} day^{-1} during the spring diatom bloom. Levels of production then declined until the end of June when a secondary peak of activity produced 4·33 g O_2 m^{-2} day^{-1} 5 days after the midsummer peak of green algae had been recorded. Production levels again declined until 11 August when a value of 2·03 g O_2 m^{-2} day^{-1} was measured just after the autumn chlorophyll peak produced by centric diatoms. During the rest of the year only low levels of production were detected and respiration exceeded production in the last 3 months. A minimum production of −0·50 g O_2 m^{-2} day^{-1} was measured on 5 October.

Vertical profiles of net oxygen production in the Thames are shown in Fig. 3.3. Maximum production almost always occurred at a depth of 30 cm. The maximum daily net production was 9·32 mg O_2 l^{-1} day^{-1} on 29 June. On this occasion and several others, net production in bottles just below the surface (3 cm) was lower than that measured at 30 cm, indicating that the algae in the upper bottles were light saturated and further production was being inhibited. On 1 June positive net production was measured only in the upper 60 cm. This rapid decrease with depth suggests that light attenuation was limiting production. The chlorophyll concentration at the start of the exposure period was 95 μg l^{-1} and it is probable that at this level of biomass the phytoplankton had become self-shading.

Total annual net production was calculated as 369 g O_2 m^{-2} $year^{-1}$ which was much lower than the 906 g O_2 m^{-2} $year^{-1}$ calculated for 1967–68.

Kennet

Net oxygen production was considerably lower than in the Thames (Fig. 3.2) and negative values were obtained for most of the year. Positive production was measured on 21 April and levels increased to a maximum of 4·28 g O_2 m^{-2} day^{-1} on 19 May. This value coincided with a very high chlorophyll concentration chiefly due to centric diatoms. Over the next 6 weeks oxygen production decreased to a value of −0·39 g O_2 m^{-2} day^{-1} on 13 July and there was no further positive net production for the remainder of the year. A minimum value of −0·52 g O_2 m^{-2} day^{-1} was measured on 7 September.

Vertical profiles of net oxygen production in the Kennet are shown in Fig. 3.4. Maximum daily rates usually occurred within the top 30 cm of the water column. A maximum production of 13·28 mg O_2 l^{-1} day^{-1} was recorded on 19 May in the bottles 3 cm below the surface. On this occasion and many others, positive net production was measured in the top 60 cm only. Positive production was never recorded below 120 cm throughout the whole period.

The total annual net production was calculated as 81 g O_2 m^{-2} $year^{-1}$

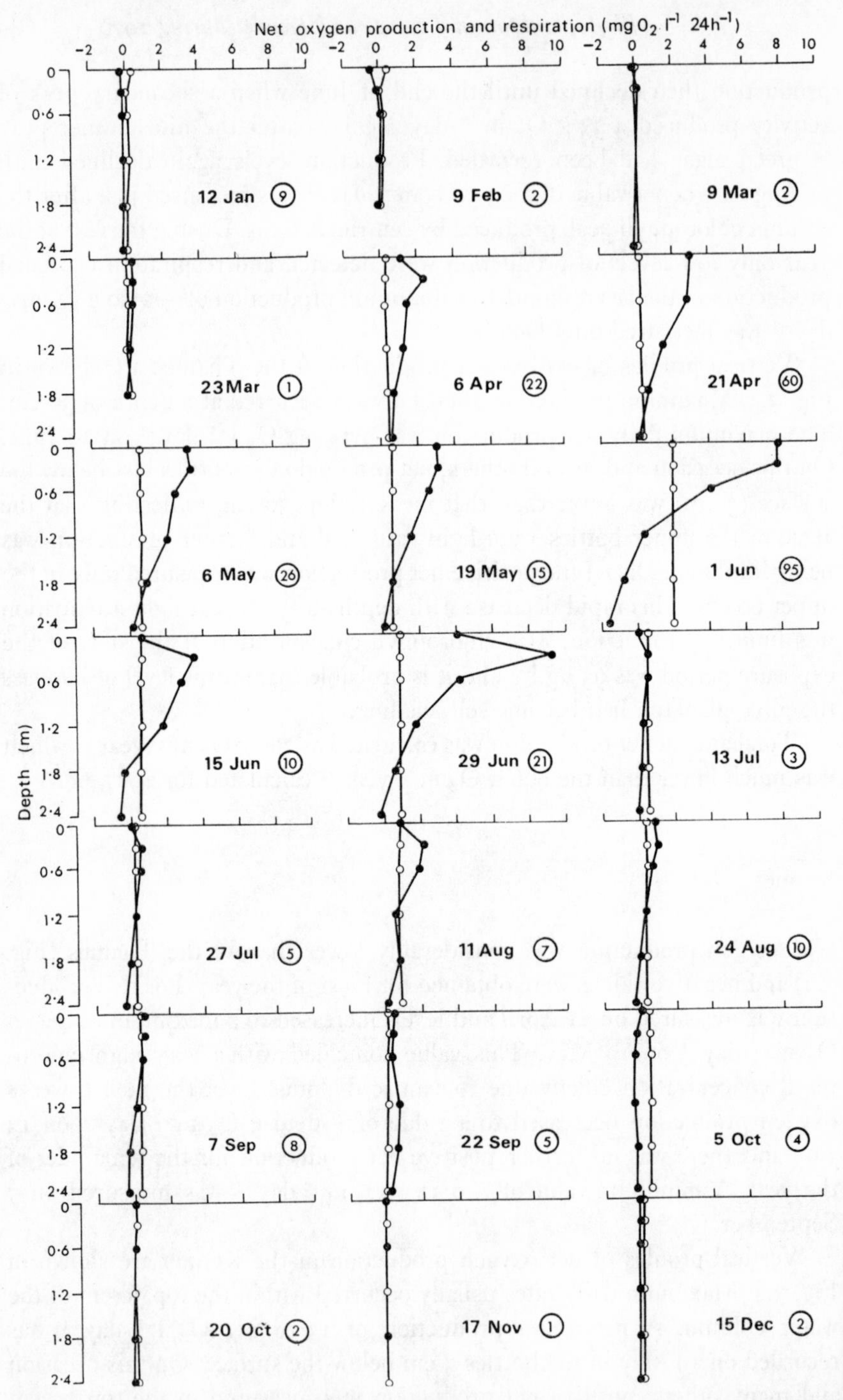

Figure 3.3. Depth profiles of daily oxygen production (●——●) and respiration (○——○) in the Thames above Kennetmouth during the year 1970. Encircled figures represent the chlorophyll concentration (μg l^{-1}) in the river at the beginning of the exposure period.

which was greater than that found in 1967–68 when a value of -57 g O_2 m^{-2} $year^{-1}$ was obtained.

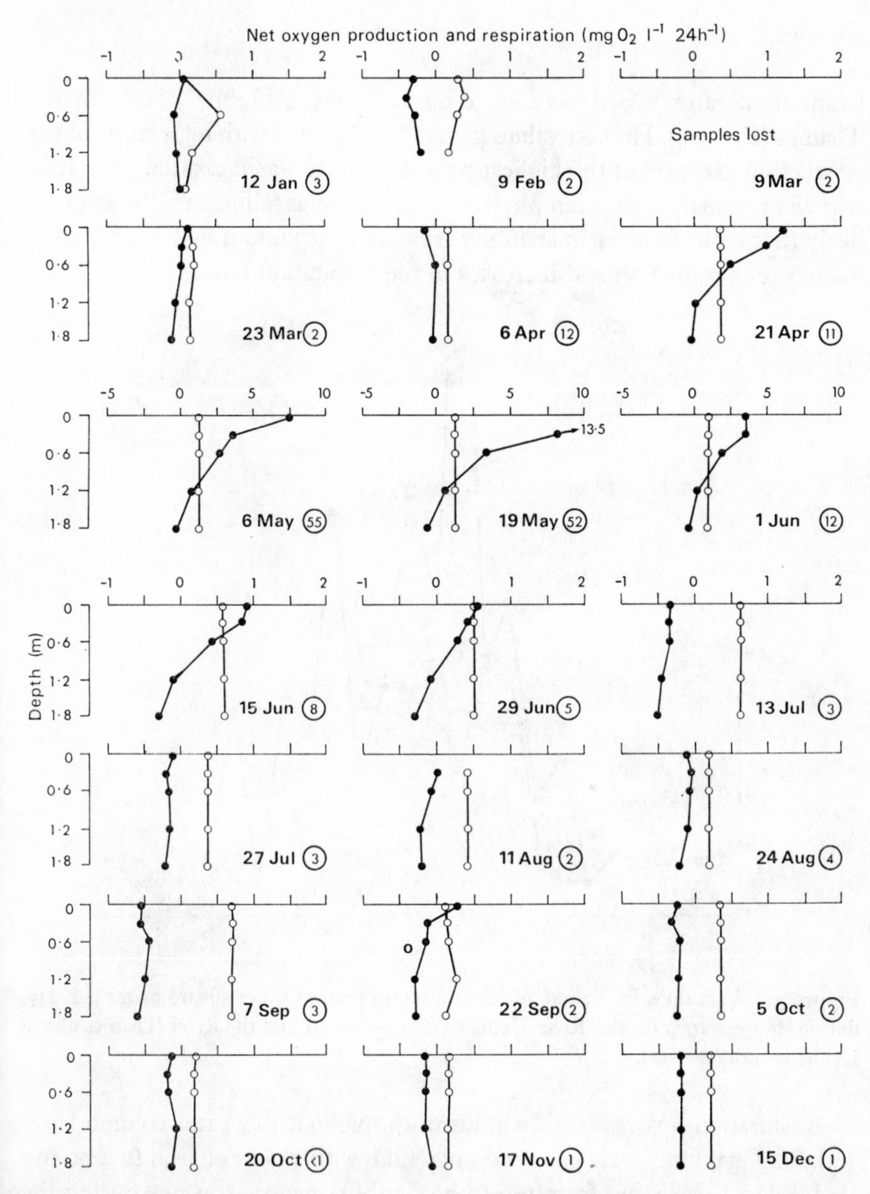

Figure 3.4. Depth profiles of daily oxygen production (● —— ●) and respiration (○ —— ○) in the River Kennet during the year 1970. Encircled figures represent the chlorophyll concentration (μg l^{-1}) in the river at the beginning of the exposure period. Note the scale change for the three profiles dated 6 and 19 May and 1 June: all other profiles to the same scale.

Respiration

Thames

Respiration rates varied between 0·09 and 4·65 g O_2 m^{-2} day^{-1} in the Thames (Fig. 3.5). Highest values generally coincided with high levels of net production except that the highest respiration rate was measured on 1 June when a previously high chlorophyll concentration was falling rapidly. It seems likely that rapid declines in standing crop are concomitant with decreases in net oxygen production and increases in the respiration rate.

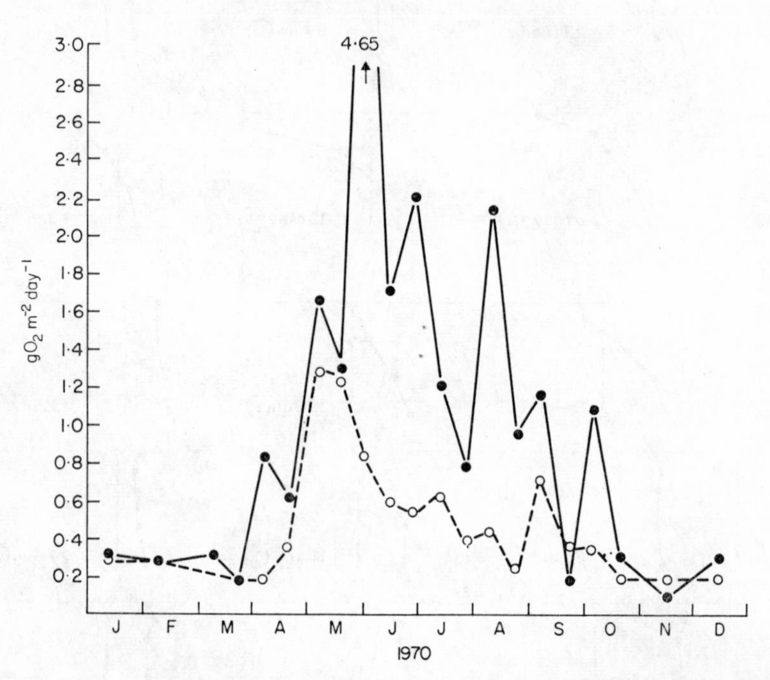

Figure 3.5. Variations in the rate of respiration, in grams O_2 per square metre per day, during the year 1970 for the River Kennet (o — — — o) and the River Thames above Kennetmouth (●—●).

Respiration was reasonably uniform throughout the water column (Fig. 3.3). The maximum rate occurred on 1 June with values of 1·98 to 2·00 mg O_2 l^{-1} day^{-1} occurring from top to bottom. On many occasions during the winter the respiration rate exceeded the rate of oxygen production.

Total annual respiration was calculated as 341 g O_2 m^{-2} $year^{-1}$ which was 27% lower than was determined during 1967–68.

Kennet

Respiration rates in the Kennet were almost always lower than those in the Thames (Fig. 3.5). They ranged from 0·17 to 1·28 g O_2 m^{-2} day^{-1}. The highest value was measured on 6 May at the time of the first peak of the spring bloom (Fig. 3.1). Respiration rates declined gradually after this. Apart from the period of algal activity in the spring, the respiration rate was always $<0{\cdot}8$ g O_2 m^{-2} day^{-1}.

As in the Thames, respiration rates were reasonably uniform throughout the profile (Fig. 3.4). The maximum rate was 1·25 mg O_2 l^{-1} day^{-1} recorded from top to bottom on 6 May. For much of the year respiration exceeded oxygen production at all depths.

The total annual respiration was calculated as 156 g O_2 m^{-2} $year^{-1}$ which was very close to the value of 165 g O_2 m^{-2} $year^{-1}$ found during 1967–68.

Factors influencing net oxygen production in the River Thames

1. *Relationship between net production and solar radiation*

The relationship between net production and total solar radiation was linear during 1967–68 (Fig. 3.6). Although the points were rather scattered a significant regression equation was obtained:

$$P_{\text{net}} = 0{\cdot}59\,I - 2{\cdot}65$$

where P_{net} = net oxygen production (g O_2 m^{-2} day^{-1})
I = total solar radiation (MJ m^{-2} day^{-1})

In 1970 no such relationship was found (Fig. 3.6). There was no correlation between production and solar radiation up to a level of 13 MJ m^{-2} day^{-1}. Between 13 and 18 MJ there was a marked increase in net production but at greater levels of solar radiation the net production was depressed. This effect was so marked that the maximum radiation level of 27·0 MJ m^{-2} day^{-1} coincided with a net production of only 0·48 g O_2 m^{-2} day^{-1}.

2. *Relationship between net production, chlorophyll and euphotic depth*

The product of chlorophyll concentration and euphotic depth showed some relationship with the net production in 1967–68 (Fig. 3.7). The scatter was

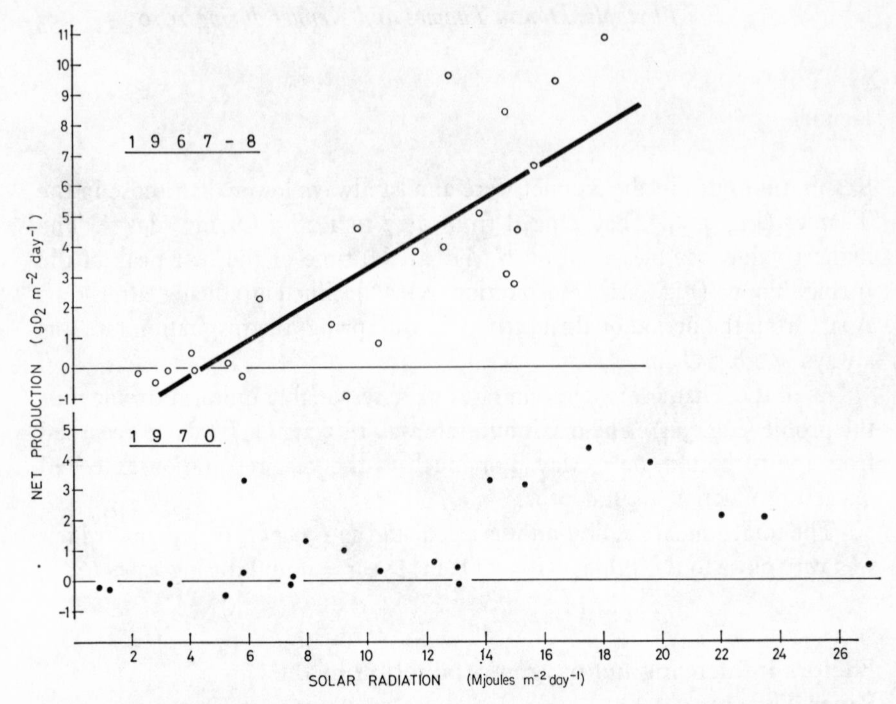

Figure 3.6. Relationships between net oxygen production, in grams O_2 per square metre per day, and total daily solar radiation, in megajoules per square metre per day, for the years 1967–68 (○) and 1970 (●) in the River Thames above Kennetmouth. For discussion, see text.

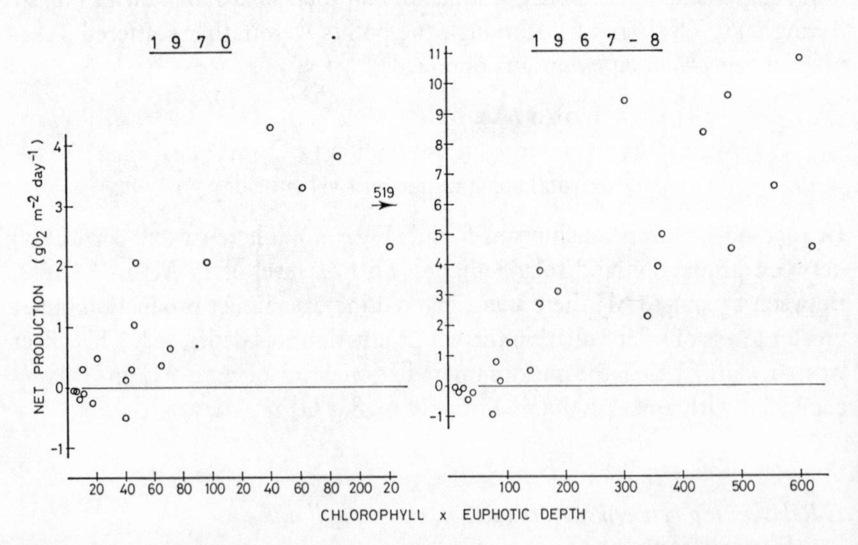

Figure 3.7. Relationships between net oxygen production, in grams O_2 per square metre per day, and the product of the euphotic depth (metres) and the concentration of chlorophyll *a*, in milligrams per cubic metre, during the years 1967–68 and 1970 in the River Thames above Kennetmouth. Note the differences in scale between the two years.

great at the upper levels suggesting that the net production was depressed by self-shading at high chlorophyll concentrations. Measurements made during 1970 fell only in the lower part of the 1967–68 range where the linear relationship had been good (Fig. 3.7) *i.e.* from -1 to 4 g O_2 m^{-2} day^{-1} and up to 200 units on the axis of abscissae. A linear regression on the 1970 data showed that

$$P_{net} = 0{\cdot}019X - 0{\cdot}20$$

where X = chlorophyll concentration (mg m^{-3}) $\times$ euphotic depth (m).

The 1967–68 data points that fall within this same range ($n = 13$) gave a regression equation of:

$$P_{net} = 0{\cdot}020\ X - 0{\cdot}93$$

In both cases the slopes are almost the same indicating that the relationship between P_{net} and chlorophyll $\times$ euphotic depth was similar during both years. The error factor in the 1967–68 data was approximately 4·5 times greater than in 1970.

On some occasions it was not possible to measure the euphotic depth due to equipment failures and use was made of the relationship between euphotic depth and suspended matter in the water (Fig. 3.8). A good inverse relationship was found enabling the euphotic depth to be calculated from the regression equation:

$$Z_{eu} = 5{\cdot}144 - 2{\cdot}4375 \log_{10} Y$$

where Z_{eu} = euphotic depth (m)

$\log_{10} Y$ = logarithm of the amount of material in suspension (mg l^{-1})

3. *Combination of factors*

Derivation of Relationships

Kowalczewski & Lack (1971) reported that net oxygen production (P_{net}) in the Thames could be expressed by a relationship with total solar radiation (I), and with the product of chlorophyll concentration and the euphotic depth (X). Unfortunately the published regression equation was incorrect and should have read:

$$P_{net} = 0{\cdot}323\ I + 0{\cdot}009\ X - 2{\cdot}211$$

The 1970 data were treated similarly and gave the following regression equation:

$$P_{net} = 0{\cdot}070\ I + 0{\cdot}008\ X - 0{\cdot}281$$

Application of the 1967–68 regression equation to the data collected in 1967–68 resulted in a series of points that tended to underestimate the actual net production at its highest levels and to overestimate at intermediate levels (Fig. 3.9a). The measured annual net production in 1967–68 was 906 g O_2 m^{-2} whereas that calculated from the area under the predicted curve was 755 g O_2 m^{-2}.

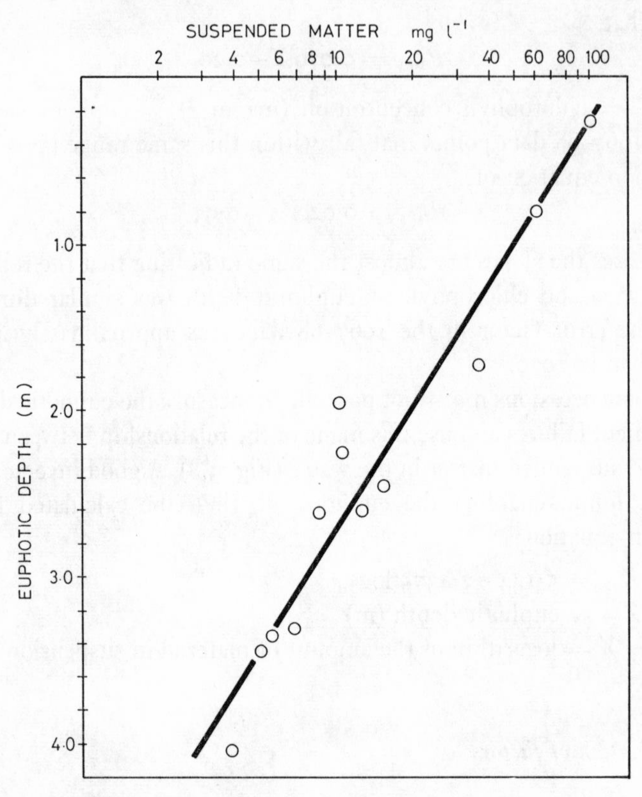

Figure 3.8. The relationship between the euphotic depth (metres) and the material in suspension (milligrams per litre, log. scale) in the River Thames above Kennetmouth during the year 1970.

A similar plot of the 1970 observed and predicted data using the 1970 regression equation showed a similar relationship (Fig. 3.9b). At times the equation underestimated the net production but this was balanced by over-estimates at other times. The measured annual net production in 1970 was 369 g O_2 m^{-2} while the calculated value was 371 g O_2 m^{-2}. Clearly the regression equation for 1970 closely models the actual net production over that one-year period.

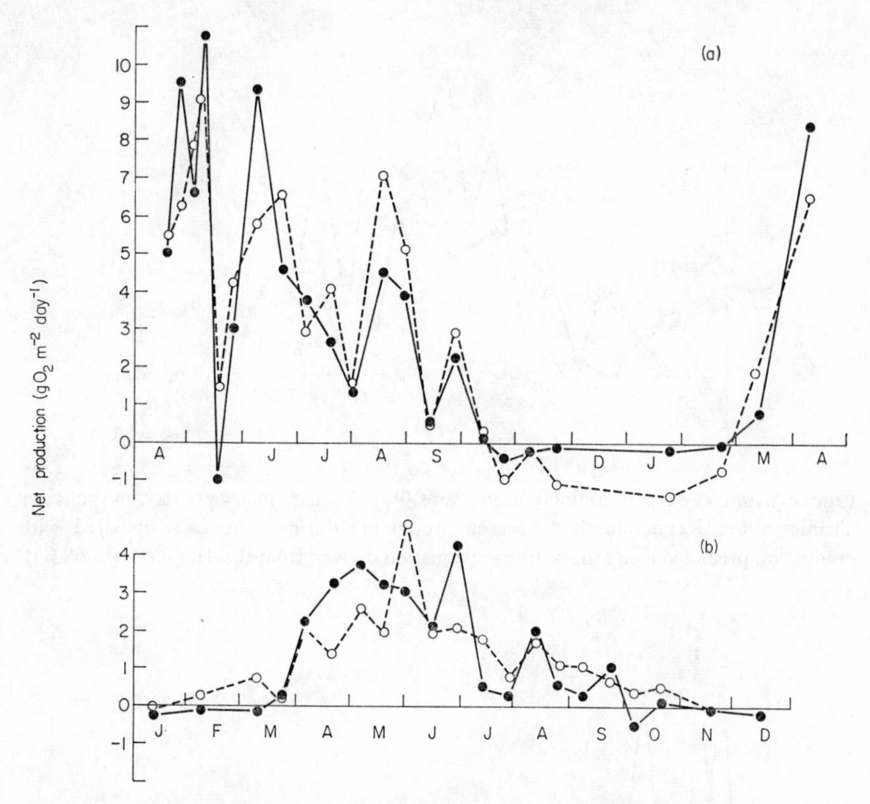

Figure 3.9. Net oxygen production, in grams O_2 per square metre per day, in the River Thames above Kennetmouth. Observed production (●) compared with predicted production (○), (a) for the year 1967–68 and (b) for the year 1970, using in each case the regression equation derived from the observed values for the year concerned.

Predictive Value of Relationships

The real test of these regressions is to see if any one year's equation can predict another year's production. Accordingly, the equation for the 1967–68 data was applied to the 1970 measured data to calculate predicted net productions. These are plotted in Fig. 3.10 and it is clear that the regression equation overestimates the actual net production for most of the year.

A similar application of the 1970 equation to the 1967–68 data is shown in Fig. 3.11. The higher levels of net production are underestimated by the equation but the majority of the predictions are quite close to the measured values. It appears that the 1970 equation can satisfactorily predict the medium and low levels of net production *i.e.* 5 g O_2 m^{-2} day^{-1} or less and during 1967–68 there would have been only six occasions when the observed values were greatly in excess of the predicted.

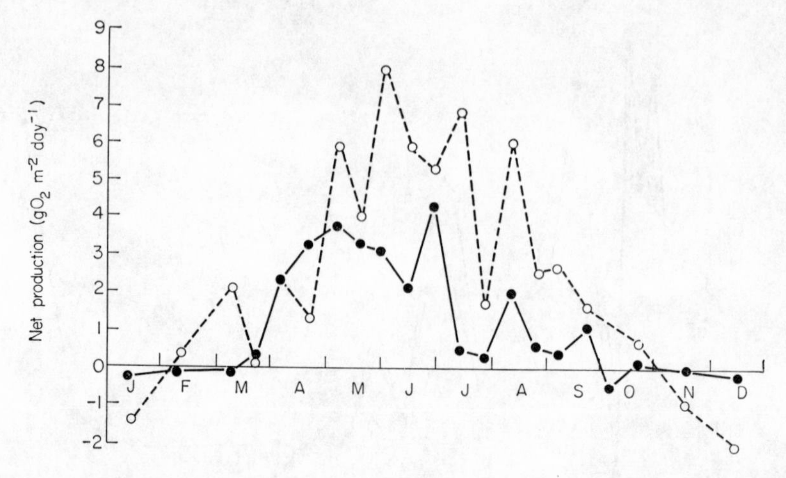

Figure 3.10. Net oxygen production, in grams O_2 per square metre per day, in the River Thames above Kennetmouth. Observed production during 1970 (●) compared with production predicted using the regression equation derived from the data for 1967–68 (○).

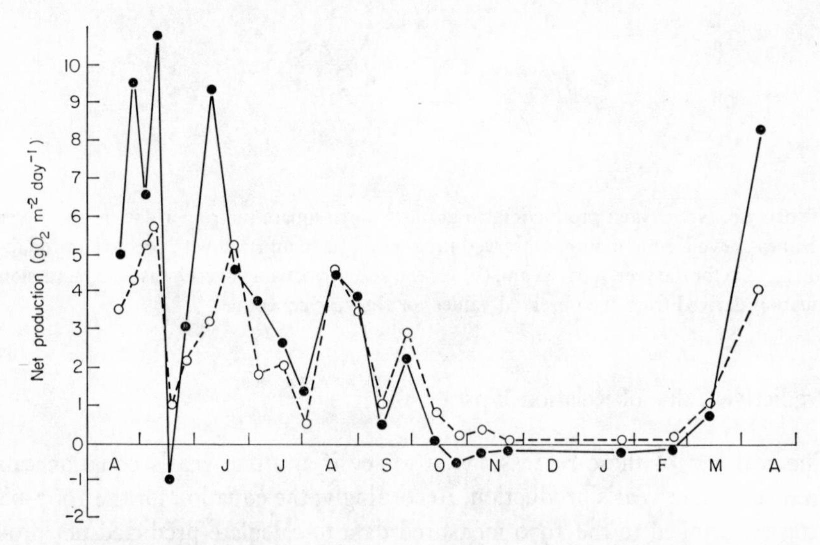

Figure 3.11. Net oxygen production, in grams O_2 per square metre per day, in the River Thames above Kennetmouth. Observed production during 1967–68 (●) compared with production predicted using the regression equation derived from the data for 1970 (○).

A combination of 1967–68 and 1970 data gave a regression equation of:

$$P_{net} = 0{\cdot}115\, I + 0{\cdot}012\, X - 0{\cdot}971$$

The equation has an F value of 56·66 which is significant at the 1% level and a correlation coefficient of 0·86. The relationship between observed and predicted production is shown in Fig. 3.12.

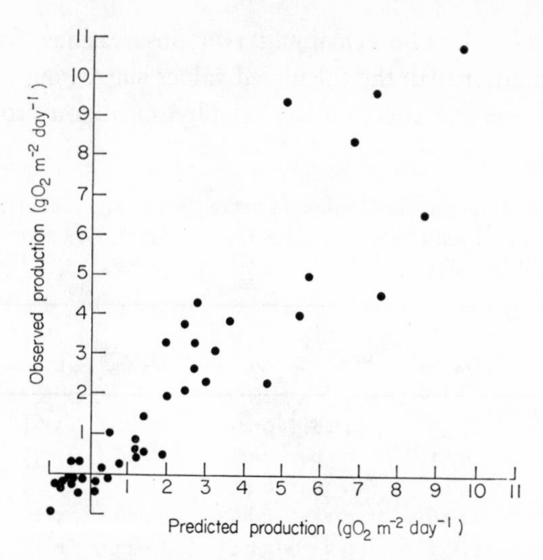

Figure 3.12. Net oxygen production, in grams O_2 per square metre per day, in the River Thames above Kennetmouth. Values of observed production for the periods of observation during 1967–68 and 1970, plotted against production predicted using the regression equation derived from the combined data for the two periods of observation.

Production of River Thames below the Kennet confluence

Kowalczewski & Lack (1971) obtained production values for the Thames below Kennetmouth by considering that reach of the river to be a mixture of waters of two levels of productivity in proportion to the relative discharges on each occasion of the Thames at Caversham and the Kennet at its confluence. The formula used was:

$$P_b = \frac{(P_a \times D_a) + (P_k \times D_k)}{D_a + D_k}$$

where P_b = production in Thames below Kennet confluence
P_a = production in Thames above Kennet confluence
P_k = production in Kennet
D_a = discharge of Thames above Kennet confluence
D_k = discharge of Kennet at confluence

During 1970 measurements of production were carried out in the Thames below the Kennet confluence to test the validity of this assumption. The results are presented in Table 3.1. The calculated values of net production correspond with observed values on only one occasion. On 19 May and 7 September, the measured net production below Kennetmouth exceeded the value calculated from measurements above the confluence. This might suggest that some synergistic effect was occurring when water from the two rivers became mixed. The remaining two observations from below the confluence were lower than the calculated values suggesting that the Kennet was having a depressive effect on the net phytoplankton production of the Thames.

Table 3.1. Calculated and measured values of net oxygen production in the Thames below the Kennet confluence. Production units are g O_2 m^{-2} day^{-1}. Discharge units are m^3 s^{-1}. Symbols are defined in the text.

Date	P_a	D_a	P_k	D_k	D_a+D_k	P_b Calculated	P_b Observed
19/5/70	3·31	17·56	4·28	9·26	26·82	3·64	5·07
15/6/70	2·06	9·91	0·30	7·08	16·99	1·33	1·34
13/7/70	0·48	9·91	−0·39	6·85	16·76	0·12	−0·28
11/8/70	2·03	17·56	−0·11	5·80	23·36	1·50	0·24
7/9/70	0·35	11·33	−0·52	5·10	16·43	0·08	0·41

Discussion

There are obvious difficulties in comparing the 1970 observations with the data from 1967–68. The environmental conditions were different and this was reflected in the variable factors recorded in each year. Although the annual pattern of phytoplankton development in the Thames was similar, with three main peaks in each year, the biomass attained over the two years was quite different. The summer of 1970 was characterized by higher levels of solar radiation combined with lower levels of chlorophyll and it is not surprising that production was affected. The annual net production in 1970 was less than half that attained in 1967–68 and the respiration was correspondingly reduced by 27%. The Kennet, unlike the Thames, had a much higher concentration of chlorophyll during the spring bloom period of 1970. Consequently, a higher level of net production was attained over the year.

During 1970 the lower biomass of phytoplankton in the Thames gave rise to lower levels of production per unit of solar radiation (Fig. 3.6). Figure 3.6 also shows that measurements were made on four days in 1970 when solar radiation was higher than that recorded on any of the days on which production was measured in 1967–68. It was at these high levels of radiation that the effects of light saturation became apparent. The difference in the algal biomass between the two years is also shown in the scales used in the two parts of Fig. 3.7 and causes difficulty in making comparisons between them.

The introduction of an additional set of light and dark bottles at a depth of 30 cm produced interesting results. In the Thames these usually gave the highest level of production and they were important in demonstrating the occurrence of light-saturation and self-shading effects in the upper part of the water column.

Although the differences between the two years make comparisons difficult, the combined data cover a wide range of conditions. They show a fairly high correlation between observed and predicted production, so the regression equation based on these combined data may have greater value for predicting production. It is clear from Fig. 3.12 that, although the correlation may be poor on individual occasions, if the variables are measured on a number of occasions a reasonable prediction should be possible for the whole period.

The observations on production in the Thames below the confluence indicate that it is not satisfactory to calculate the productivity of mixed waters from a knowledge of the phytoplankton production in each source and their relative contributions. Other factors appear to influence the result. This may have significance when considering the possible effects of large-scale transfers of water from one catchment to another.

Light appears to be the most important physical factor determining phytoplankton production in the Thames. Its penetration through the water depends on the amount of material in suspension, which can be quite high (Berrie 1972b), and this determines the euphotic depth. Thus to make a reasonable prediction of net production by the phytoplankton, it is necessary to make use of data on solar radiation, euphotic depth and chlorophyll concentration.

Summary

Primary production by the phytoplankton was estimated from changes in the oxygen concentration in light and dark bottles suspended in the rivers on 21 occasions during the year. The determinations were made just above the confluence of the rivers and, on 5 occasions, additional determinations were made 1·7 km below the confluence. The production estimates are related to data on chlorophyll concentration, solar radiation and euphotic depth and compared with similar data from 1967–68. A regression formula for predicting production from measurements of the other variables is derived and assessed. Production below the confluence of the two rivers could not be predicted satisfactorily from measurements of production in each river above the confluence and of the discharge of each river.

Acknowledgements

This investigation was financed by a grant from the Natural Environment

Research Council. We are grateful to Professor G. Williams for encouragement and research facilities, and to Mrs. Sara White for assistance with the field work. The Thames Conservancy, the Water Research Centre and the Electrical Research Station generously made available unpublished data which are quoted.

References

BERRIE A.D. (1972a) Productivity of the River Thames at Reading. In *Symposium on Conservation and Productivity of Natural Waters*, eds. Edwards R.W. & Garrod D.J. *Symp. zool. Soc. Lond.* 29, 69–86.

BERRIE A.D. (1972b) The occurrence and composition of seston in the River Thames and the role of detritus as an energy source for secondary production in the river. *Mem. Ist. Ital. Idrobiol.* 29 Suppl., 473–83.

KOWALCZEWSKI A. & LACK T.J. (1971) Primary production and respiration of the phytoplankton of the Rivers Thames and Kennet at Reading. *Freshwat. Biol.* 1, 197–212.

LACK T.J. (1971) Quantitative studies on the phytoplankton of the Rivers Thames and Kennet at Reading. *Freshwat. Biol.* 1, 213–24.

LORENZEN C.J. (1967) Determination of chlorophyll and pheo-pigments; spectrophotometric equations. *Limnol. Oceanogr.* 12, 343–6.

MANN K.H., BRITTON R.H., KOWALCZEWSKI A., LACK T.J., MATHEWS C.P. & MCDONALD I. (1972) Productivity and energy flow at all trophic levels in the River Thames, England. In *Productivity problems of freshwaters*, eds. Kajak Z. & Hillbricht-Ilkowska A. Polish Scientific Publishers: Warszawa-Krakow.

Questions and discussion

Dr. Westlake and Dr. Lack discussed problems of interpreting the measurements of community net respiration. Dr. Lack said that non-algal respiration was insignificant at times when net production is highest. Dr. Westlake wondered whether respiration in the community not related to light could explain the discrepancy between the times of year when the plants were truly dominant and times when the plants were perhaps less important than the rest of the community. Dr. Lack agreed that this is a possibility, but thought not, as it had been shown that the zooplankton is almost non-existent in the free water column, while the bacterial respiration of the free water was remarkably low during the time of algal blooms. Dr. Westlake objected that bacteria are in fact more likely to be associated with the algae themselves. Dr. Lack agreed that this was so in lakes but thought it less likely in the Thames where *Stephanodiscus* diatoms form such a high proportion of the phytoplankton. He thought that further study would show that these are more free of bacteria than the green or blue-green algae.

4 Light and the ecology of *Laminaria hyperborea* II

JOANNA M. KAIN (Mrs. N. S. JONES) *Department of Marine Biology, University of Liverpool*
E. A. DREW and B. P. JUPP *Gatty Marine Laboratory, St. Andrews, Scotland*

Introduction

At the previous Symposium, held in 1965, the results of the first few years of work on *Laminaria hyperborea* (Gunn.) Fosl. in relation to light were reported (Kain 1966). It had been shown that this species was dominant on subtidal rock of moderate to severe exposure to wave action down to a variable lower limit, probably determined in some sites by grazing by *Echinus*. It had been found that the light requirements of the early stages of this species and its major competitors in the Laminariales were similar and thus not critical in determining the outcome of competition. In natural populations of *L. hyperborea* plants of a given age were slightly smaller in deeper water and it was concluded that growing conditions deteriorated with depth and thus presumably with decreased irradiance. At a given depth growth of individuals was faster when the community was open than in a closed forest. All the evidence available suggested that factors determining the lower limit of the species in the Isle of Man were acting on the phase of establishment of gametophytes or young sporophytes rather than the ability of sporophytes to grow at more than a minimal rate. Measurements of light in the sea around the Isle of Man, where most of the observations on *L. hyperborea* were made, were confined to selected days when a photocell equipped with blue, green and red filters was lowered through a depth range. Considerable variation in attenuation was found. Likely ranges of irradiance at various depths were calculated, using published data on surface irradiance for the same latitude as the Isle of Man and the attenuation coefficients of the clearest and most turbid water encountered. It seemed from these results that for much of the time the photosynthesis of mature fronds would not be light saturated even in shallow water, thus explaining the decrease in growth rate with depth. It also seemed that in some sites in the Isle of Man the lower limit could be determined by irradiance levels in winter at the time of sporing and thus establishment.

Since 1965 work on this species has continued and considerably more has been accomplished as a result of several workers joining the field.

In previous papers by Kain irradiance has been expressed in μg cal $cm^{-2}s^{-1}$, a unit also used by Lüning (e.g. 1971) and others. In this paper the SI unit of W/m^2 is used and previous data converted using the following factor:

$$1\ \mu g\ cal\ cm^{-2}s^{-1} = 0{\cdot}04185\ W/m^2.$$

Light reaching the habitats

The disadvantage of measuring light from a small boat is the necessary avoidance of rough water and consequent omission of turbulent conditions. The obvious answer to this is to instal photocells in the sea. This was done in Port Erin Bay in 1967 (Kain 1971a). A construction of Dexion angle iron and concrete blocks bore attachment shelves for two Perspex cases containing selenium barrier layer photocells, one exactly 2 m above the other. A cable led to the laboratory and a two-channel recorder. Each cell was covered with a Schott VG9 green filter, a neutral filter of black nylon chiffon and an opal filter. From a Kipp & Zonen solarimeter on a cliff-top a cable led to a further recorder. Twice a year the underwater cells were calibrated against the solarimeter by floating them on a frame just below the sea surface and recording simultaneously. As the cell output deteriorated, a calibration factor was deemed to vary continuously between calibrations. Although cleaned fairly frequently by divers, deposits of diatoms, silt etc. on the opals covering the cells built up considerably at times.

The results on the record charts were treated in two ways. The first was very time consuming but allowed extrapolation to chosen depths (Kain 1971a). For each day simultaneous readings were taken for surface irradiance and the two cells at about five different times and each cell reading expressed as a percentage of the surface (green only), after a correction had been made for dirty opal filters. From predicted tidal levels the depth of water above each cell was calculated for each of these times and the ten readings plotted against depth. A line passing through 100% at the surface was drawn as the best fit to the points. Values for the percentage of surface irradiance were read off for certain depths on the bottom (allowing an effective mean tidal height of 2·5 m). Each value for green irradiance was converted to total visible using a curve (Fig. 4 in Kain 1971a) obtained from numerous readings with three filters. The percentage for each day at each depth, obtained in this way, was combined with the surface irradiance for that day (expressed as mean W/m^2 over 24 hours) to obtain the estimate of irradiance for each day and depth. A further mean was taken of all the daily estimates for each month. The results for seven months were calculated in this way and are shown in Table 4.1.

The second method eliminated the corrections (of doubtful accuracy)

Table 4.1. The estimated mean visible irradiance in W/m² near Port Erin (54°05′ N, 4°46′ W), Isle of Man in 1968. June and December from Kain (1971a). For details, including method of calculation, see text.

	Depth			
Month	0 m	5 m	10 m	15 m
January	1·89	0·30	0·062	0·014
February	14·19	3·29	0·950	0·296
March	14·50	2·49	0·588	0·175
April	29·90	5·61	1·445	0·405
May	43·57	6·32	1·573	0·433
June	60·68	19·04	6·905	2·687
December	2·84	0·72	0·221	0·077

made for dirty opals by confining the days for which calculations were made to within a week or ten days, according to the season, of the opals being cleaned by divers. Instead of plotting and extrapolating (also of doubtful accuracy) for each day, the green irradiance received by each cell was calculated from the area measured off the recorder chart. This was multiplied by the appropriate calibration factor. If each cell were at the surface, covered by its green filter, the reading for the day multiplied by the calibration factor should be equal to the reading from the solarimeter for the day (because this was how the calibration factor was derived). Thus the reading from the cell when underwater, also with a green filter, could be expressed as a percentage of the solarimeter reading to obtain the percentage of the surface value of green irradiance. This was converted to a percentage of surface total visible irradiance as in the other method. The visible irradiance at the depth of the cell was then given by this percentage of the surface visible irradiance (again expressed as mean W/m² over 24 hours) for that day. This gave results for the depths on the bottom of the two cells, which were 3 and 5 m below lowest astronomical tide (LAT), without having to allow for the tidal height. About eight days were selected from each month, the selection depending on opal cleanliness and the full functioning of the system and not being influenced by weather conditions. A mean was produced for each available month in the years 1968 to 1971 inclusive and the results are shown in Table 4.2. The shallower cell was not functioning for the first half of 1970. In mid-1970 the cells were moved to slightly deeper water, 3·6 and 5·6 m below LAT but the results corrected to 3 and 5 m by interpolation.

Table 4.3 shows results from the integrated output of the Kipp & Zonen solarimeter, effectively measuring irradiance above the water surface for each month during the time for which the solarimeter integrator has been in action. Again the values were expressed as mean W/m² for each 24 hours and a further mean taken for each month. There is some deviation from the values interpolated from Kimball's (Sverdrup *et al.* 1942) data for the appropriate latitude, mainly in the summer.

Table 4.2. The estimated mean visible irradiance for each month in W/m² at two depths in Port Erin Bay. For further information and discussion see text.

	3 m				5 m			
	1968	1969	1970	1971	1968	1969	1970	1971
January	0·112				0·054		0·594	
February	6·39				3·87		1·67	
March	6·33			6·57	3·52		6·84	3·08
April				15·3			10·2	5·31
May	16·6	17·2		12·2	9·42	9·49		5·57
June	27·5	28·9		19·5	18·3	18·6		10·9
July	30·9	28·0			23·0	18·8		
August	18·8	20·2	8·49	21·0	13·7	14·2	4·60	12·0
September	4·82		6·74		3·23		4·18	
October	4·90	3·57		4·90	3·29	1·91		2·34
November		2·21	0·833	1·36		1·31	0·310	0·686
December	1·25	1·33	1·88		0·672	0·778	0·799	

Table 4.3. The mean visible surface irradiance for each month measured by a Kipp & Zonen solarimeter and integrator in W/m² at Port Erin. All but two days were included.

	1970	1971	1972	1973	1974
January		10·520	8·011	7·586	9·766
February		19·733	18·549	23·549	18·857
March		38·327	43·300	49·237	45·742
April		66·870	78·246	70·992	74·381
May	86·130	98·977	87·511	87·258	
June	114·011	86·639	96·778	117·364	
July	86·079	111·545	91·414	91·835	
August	80·583	65·021	74·412	74·131	
September	48·909	61·059	54·947	55·164	
October	27·832	31·881	30·212	26·400	
November	13·430	13·951	12·639	12·526	
December	8·595	6·625	5·999	7·128	

An idea of how the water type varies with season can be obtained from Fig. 4.1 which shows the percentage of the visible surface irradiance received at 5 m below LAT on each of the days for which calculations were made from 1968 to 1971, together with monthly means. Clearly there is considerable variation, partly caused by varying tidal heights during daylight hours, particularly in the winter. It is confirmed, as was thought probable from previous results, that light attenuation is generally greater in winter and the most turbid water occurs then. The clearest water occurs in June, July and August, when storms are relatively rare and the spring phytoplankton bloom has died down.

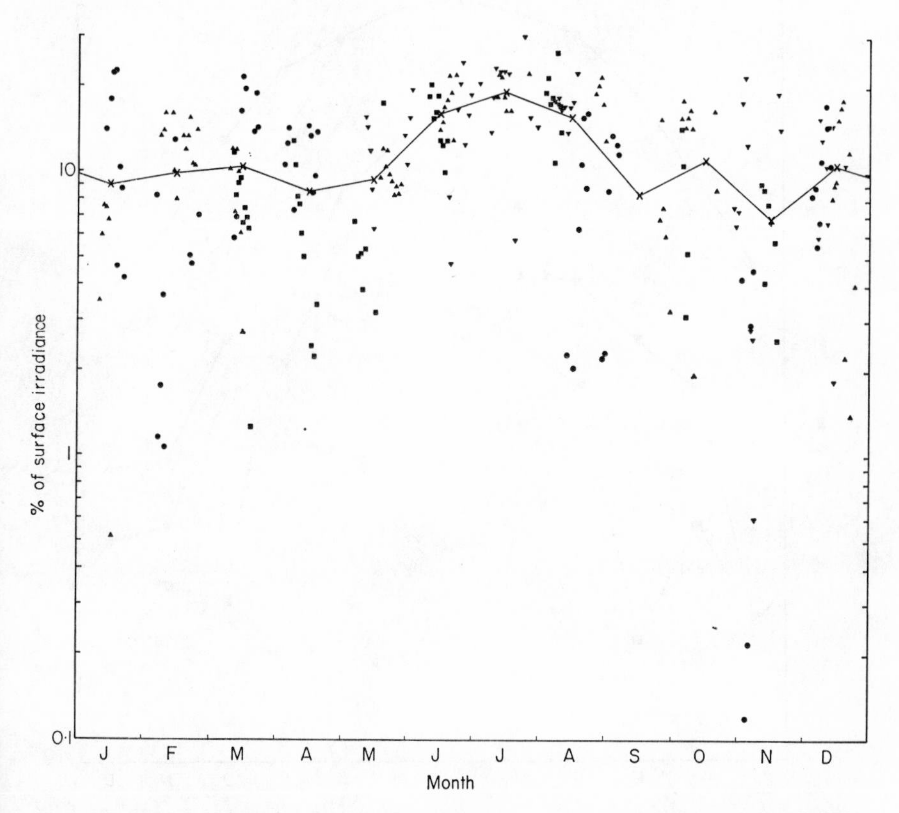

Figure 4.1. The percentage of surface irradiance recorded each day at 5 m below LAT at Port Erin (54°05′N, 4°46′W), Isle of Man. ▲ 1968; ▼ 1969; ● 1970; ■ 1971; × monthly mean.

The actual visible irradiances recorded as monthly means for the years available for surface, 3 and 5 m below LAT are shown in Fig. 4.2. The peak in underwater irradiance is in July rather than June because water clarity for these years was greater in July.

In 1965, after calculating underwater light levels from Kimball's data and the two extreme water types found, it was concluded that water type made more difference than seasonal variation in solar irradiance. In the depth range occupied by *L. hyperborea* results since have shown this to be true only for short periods of time. Of the 259 days for which calculations were made, 3 days gave no record on the charts, while the lowest percentage of the surface irradiance at 5 m below LAT was 0·12 and the highest 30. Thus the combination of water type and tidal effect gave a variation of up to 250-fold. This was even greater than could have been predicted from the previous measure-

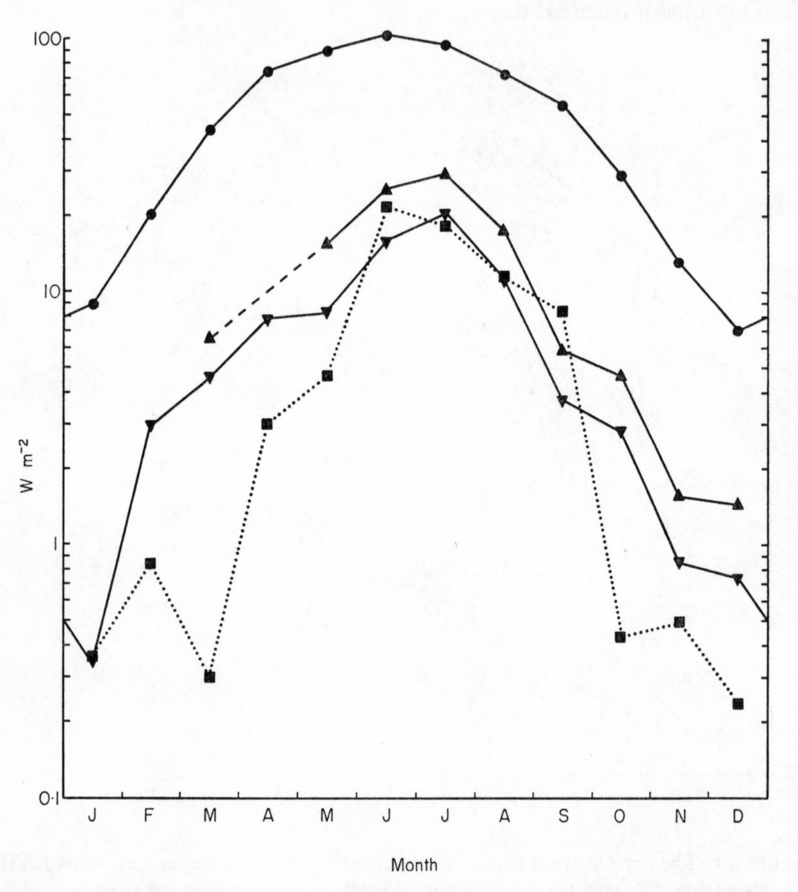

Figure 4.2. Monthly means of recorded visible irradiance. ● surface at Port Erin 1970–74; ▲ 3 m below LAT at Port Erin 1968–71; ▼ 5 m below LAT at Port Erin 1968–71; ■ 2·2 m below LAT at Helgoland 1967 (replotting from Lüning, 1971, using the original data). At Port Erin the means at 3 and 5 m were obtained by weighting the monthly means for each year shown in Table 4.2 according to the number of days for which calculations were made.

ments on very few selected days. However, one would not expect that an alga such as *L. hyperborea* would be much affected by the irradiance of one day; it is the mean over a period that is important. A mean for a month might be more realistic in importance. Although data are not available for any complete months, the means of sampled days for the same month in several years, based on 15 to 33 days, might give similar results. The lowest of these was in November, when 5 m received an average of 6·8% of the surface and the highest was July, receiving 19·2%, just under a 3-fold difference. The mean solar irradiance in June from 1970 to 1973 was 111·6 W/m^2 and that in December 7·6, a 15-fold difference. Thus although water type can make more difference than seasonal solar changes on a single day at 5 m below LAT it seems unlikely that it would do so over a period of a month.

Continuous measurement of underwater irradiance has also been made by Lüning (1971) at a depth of 2·5 m below mean low water springs equivalent to 2·2 m below LAT. Monthly means from his data for 1967 are plotted in Fig. 4.2. For most of the year irradiance was clearly lower than in Port Erin Bay, even at the shallower depth.

On a visit to Norway in June 1965 a number of light measurements were made near Bergen from a boat with the same instrument (equipped with three colour filters) previously used off the Isle of Man. At four sites in the open Atlantic and the outer fjords the depth receiving 1% of the surface visible irradiance was between 24 and 39 m (Fig. 3 in Kain 1971b) while June 1% figures for the Isle of Man were between 17 and 27 m. Thus on this occasion attenuation was slightly less in the Norwegian area than in the Isle of Man in June. The available evidence suggests (Kain 1971b) that attenuation is lower in the winter than in the summer near Bergen while off the Isle of Man June has almost minimal attenuation (Fig. 4.1). Thus although, because of its latitude, the Bergen area receives less solar radiation in winter, penetration may be better and deeper water consequently more irradiated.

Vertical distribution

In 1965 it was reported that the lower limit of *L. hyperborea* on rock in the Isle of Man varied from about 4 to 17 m below LAT. The puzzle as to why apparently suitable rock remained bare of this and other species at depths colonized not far away was beginning to be resolved by an experiment involving the removal of *Echinus esculentus* L. It was subsequently found (Jones & Kain 1967) that the monthly removal of *Echinus* from an area 10×12 m at 6–11 m below LAT normally bare of laminarians resulted in considerable colonization after 2 and 3 years. The part of this area adjacent to the sand bore a density of 5 *Echinus* per square metre, falling off in shallower water to between 1 and 2 per square metre, amongst the forest of *L. hyperborea*. It seems likely that the critical density of *Echinus* preventing the establishment of *L. hyperborea* will depend on the depth: grazing and irradiance working against each other.

Of the populations of *L. hyperborea* studied in the Isle of Man, the deepest is at Spanish Head where scattered plants were found at 19 m below LAT in 1968. At this depth the *Echinus* density was between 3 and 4 animals per square metre, so the scarcity of plants and their inability to grow deeper could well have been determined by grazing as well as irradiance.

The lower limit of the species has now been recorded in a number of sites. In Helgoland it is at 8 m below LAT (Lüning 1970). On the north-east coast of Britain it varies from 1–3 m below LAT in the polluted water of the Firth of Forth and the region of the Tyne and Tees rivers to 4–7 m in Yorkshire where the water is naturally turbid and 7–11 m in the relatively clear and unpolluted waters around Berwick and at Aberdeen (Bellamy *et al.* 1970). In the Hebrides also the lower limit is deeper, in 1970 it was observed at 23 m below LAT off Coll and at 26 m on the McKenzie Rock off South Uist (Kain unpublished). On the coast of Argyll and Ayrshire the species was recorded down to 24 m below the sea surface (McAllister *et al.* 1967). A similar limit, of between 24 and 27 m was found in the Isles of Scilly (Norton 1968). It appears to be deeper still in Cornwall, being recorded as 36 m below LAT in Sennen Cove by Whittick (1969). A similar limit was recorded in the Korsfjord near Bergen in Norway, this being 34 m below LAT (Kain 1971b).

Populations in natural habitats

Plants of *L. hyperborea* can be aged by counting the slow growth lines in a longitudinal section of the base of the stipe (Kain 1963). It was shown from population sampling that the growth rate (determined from stipe weight-age relationships) was slower at 5 m than at 2 m in the forest in the Isle of Man but that it was faster at a particular depth in an open than in a closed community (Kain 1966). Further sampling in the Isle of Man confirmed that individual plant production (obtained by adding the stipe increment and frond weight for each year group cumulatively) was slightly lower at 11 m below LAT than 1 m (Fig. 7 in Kain 1971b) and the stipe weights at some ages were significantly lower at 11 m than at 1 m (Fig. 9 in Kain 1967). However, considering the 10 m depth range, the difference was small, the individual production up to 6 years at 1 m being 1·5 times that at 11 m. Moreover, a similar depth range near Bergen in Norway produced no diminution in individual plant production (Fig. 7 in Kain 1971b). At Petticoe Wick Bay (55° 55′ N Lat 2° 09′ W long) the data of Whittick (1969) show that individual plant production up to 7 years at 1 m below LAT was about 1·7 times than at 9 m below LAT. All these populations consisted of closed forests.

In one site a marked reduction in apparent growth rate has been found. This was in the Korsfjord near Bergen in Norway, where at 32 m below LAT there was a sparse population of very small, but fairly old, plants. The stipe weights were very significantly smaller than those at 14 m, those of 6-year-olds being one-fiftieth of the weight (Fig. 11 in Kain 1971b). Thus where plants can become established in very deep water their growth can be severely limited by low irradiance.

The change in standing crop with depth is more spectacular. It has long been known from Walker's surveys off Scotland (e.g. Walker 1950) and Grenager's off Norway (e.g. Grenager 1953) that standing crop in terms of fresh weight per unit area decreases with depth. These surveys were carried out with a seaweed grab, involving remote sampling of the bottom, a method

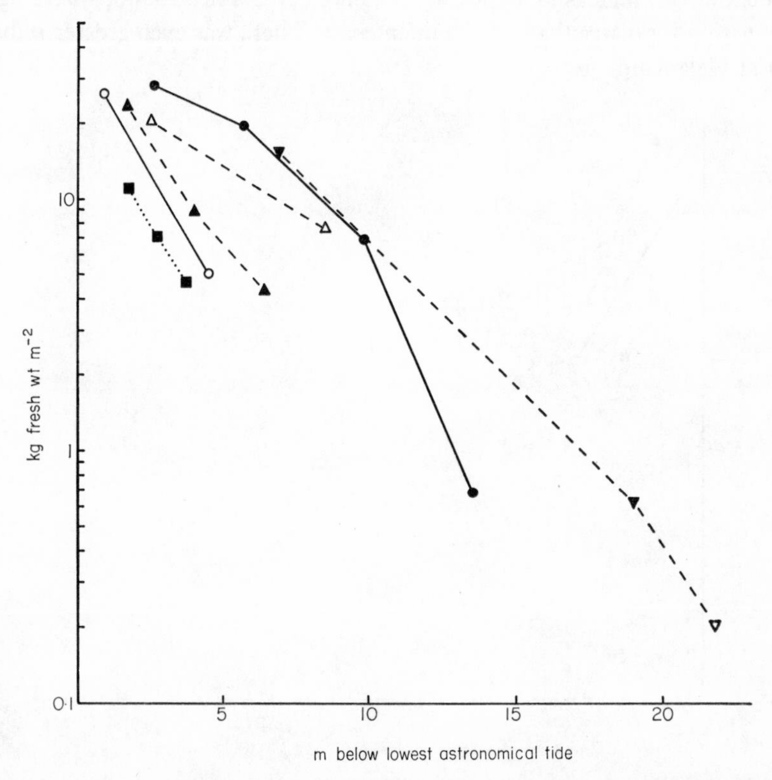

Figure 4.3. The standing crop of *Laminaria hyperborea* measured at various sites and depths by divers. ○ Port Erin; ● Spanish Head (54°03′N, 4°46′W), Isle of Man; ▲ Connel Sound (56°27′N, 5°24′W), Argyllshire (this was partly composed of *L. digitata* at the shallowest depth); ▼ Muldoanich (56°55′N, 7°27′W), Outer Hebrides; ▽ McKenzie Rock (57°08′N, 7°15′W), Outer Hebrides; △ Arisaig (56°57′N, 5°52′W), Argyllshire (Jupp 1972); ■ Helgoland (54°09′N, 7°44′E); —— Isle of Man (Kain unpublished); — — — Scotland (Kain unpublished except for Arisaig); · · · · · · Helgoland (Lüning 1969a).

known to underestimate the biomass. Divers, however, can remove every part of each plant from the rock surface. Observations on standing crop at various sites and depths obtained by diving are shown in Fig. 4.3. If standing crop is proportional to irradiance then one would expect a logarithmic decrease in standing crop with depth. Within limits, this may be the case in Fig. 4.3 but no one site has been sampled at enough depths to confirm this. At 6 m at Port Erin *L. hyperborea* is very near its lower limit which is determined by *Echinus* grazing. Here it seems highly likely that the standing crop is also affected by grazing pressure, reducing it to one quarter of that at the same depth at Spanish Head where the peak *Echinus* density is in deeper water, possibly reducing the standing crop at 13·5 m. From LAT to 10 m the crop was similar at Spanish Head, Arisaig and the Outer Hebrides. A greater reduction with increasing depth was observed at Connel Sound, where light attenuation is greater than on the open coast. There was even greater reduction at Helgoland.

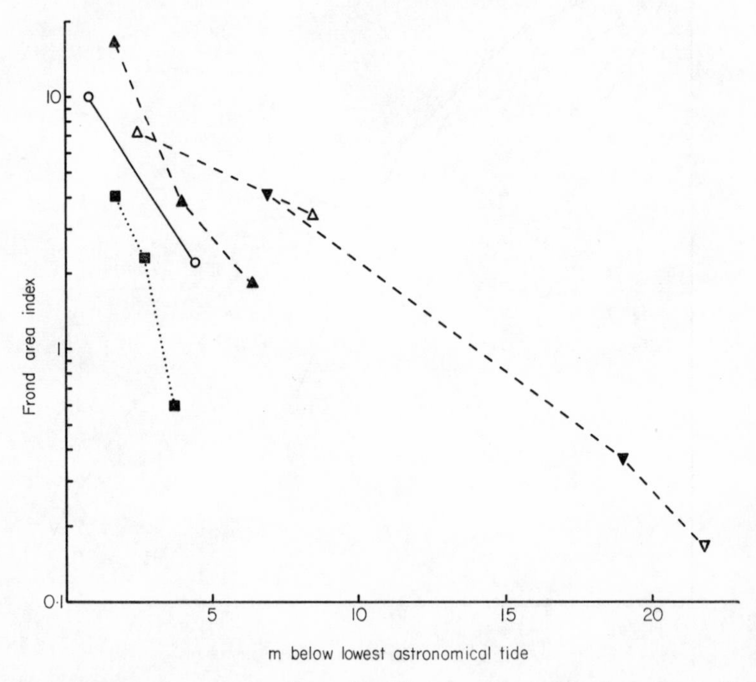

Figure 4.4. The frond area index (the ratio of frond area to area of bottom to which the plants are attached) of *Laminaria hyperborea* at various sites and depths. For symbols see Fig. 4.3.

It was stated by Walker (1954) that the decrease in standing crop (which he called density) with depth was more the result of reduction in density (individuals per unit area) than in the weight of each plant. This was confirmed by Lüning (1969a) and is supported by the above evidence that growth rate is reduced only slightly with depth in closed forests.

Another attribute, the frond area per unit area of bottom (frond area index or FAI) has been measured in some cases and some results are shown in Fig. 4.4. It would seem probable that this is reduced logarithmically with depth, but again not enough depths have been sampled at one site.

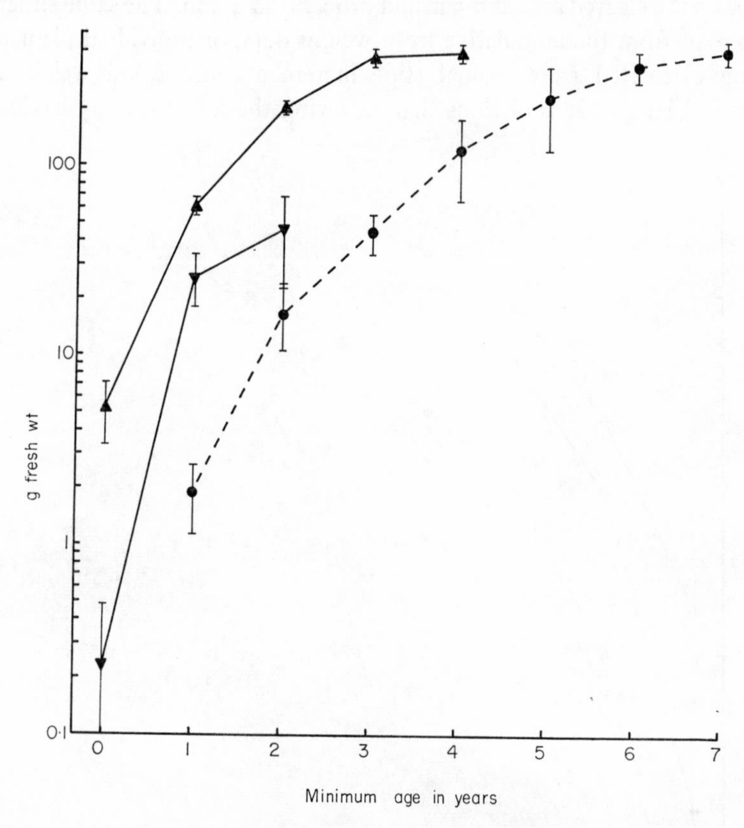

Figure 4.5. The mean stipe fresh weight of *Laminaria hyperborea*, with fiducial limits (P = 0·05) at various ages in cleared areas and virgin forest near Port Erin. ▲ cleared area at 0·5 m below LAT; ▼ cleared area at 4·5 m below LAT; ● exposed virgin forest at 1 m below LAT.

While population sampling allows an estimate of growth rate from size-age relationships, the growth of plants of known age can be studied after clearance of natural forests from rock surfaces and observation of the re-populating algae. Svendsen (1972) observed very rapid growth of *L. hyperborea* after areas had been harvested with a seaweed dredge. Clearance experiments have also been carried out in the Isle of Man (Kain unpublished). A series of the large concrete blocks making up the ruined breakwater at Port Erin were cleared of vegetation at 0·5 m and 4·5 m below LAT. Plants of *L. hyperborea* were subsequently sampled at intervals. The stipe weights are compared with those of the population previously sampled from 1 m below LAT in a similar site (population ME4 in Kain 1971b) in Fig. 4.5. Even if the age of some of the ME4 plants was overestimated by one year as was suspected (Kain 1971b), the stipe weight at each age was very significantly greater on the cleared area at 0·5 m and probably at 4·5 m. The same situation is apparent from the cumulative fresh weight data, or individual plant production calculated from annual stipe increments and frond weight and shown in Fig. 4.6. It is obvious that removing the forest canopy has had a

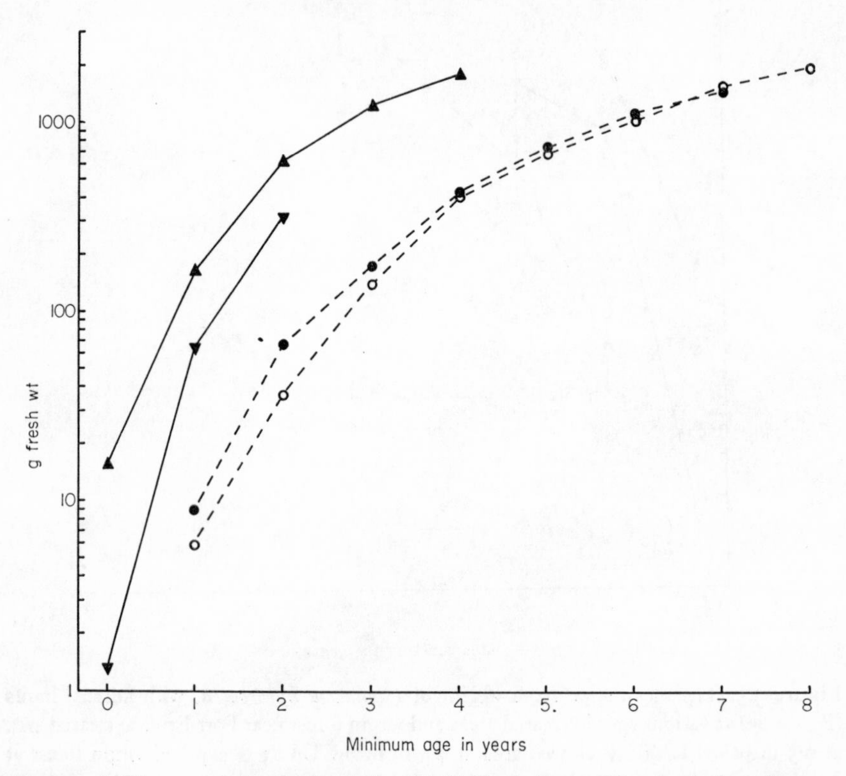

Figure 4.6. The cumulative fresh weight of individual plants of *Laminaria hyperborea* on cleared areas and in virgin forests near Port Erin plotted against minimum age. ○ sheltered virgin forest at 1 m below LAT; other symbols as in Fig. 4.5.

marked effect on the growth of *L. hyperborea* in the first few years, resulting, at 2 m, in individuals which have produced nearly 2 kg of fresh weight 2–3 years earlier than their counterparts in the virgin forest. Thus removal of the canopy resulted in a much increased growth rate.

Photosynthesis and respiration of mature fronds

The early stages of *L. hyperborea* have very low light requirements. The compensation point of gametophytes and probably of early sporophytes is between 0·04 and 0·08 W/m^2 while light saturation for growth is about 2·5 W/m^2 for both stages (Kain 1969). Early experiments indicated that the saturation level for the photosynthesis of mature fronds is much higher, in the region of 17 W/m^2 (Kain 1966). Since then a considerable amount of information has been obtained about the photosynthetic capacity of mature fronds of this species, both in laboratory experiments under controlled conditions and in *in situ* experiments in the sea.

Lüning (1971) measured photosynthetic rates in a chamber of flowing seawater, using an oxygen electrode. Measurements were made for periods of 5 minutes of irradiance up to 27 W/m^2 separated by 30 minutes of darkness. His results are shown in Fig. 4.7. A maximum rate of gross photosynthesis of about 3 ml O_2 $dm^{-2}h^{-1}$ in August was found. Assuming that mannitol is the first product of photosynthesis (Bidwell 1958) this would be the equivalent of about 15 μg C $cm^{-2}h^{-1}$. He found that to some extent the light saturated rate was dependent on temperature but there was also an ontogenetic change in potential photosynthesis resulting in a lower rate in the old than in the new frond at the same season and temperature. He concluded that light saturation was at about 17 W/m^2. From these measurements of photosynthesis and further measurements of respiration he calculated that the compensation point of the new frond was at about 2·5 and the old at about 1·3 W/m^2.

Laboratory experiments carried out by Drew (unpublished) have utilized labelled bicarbonate. Pieces of frond tissue 7·5 cm long were enclosed in 28 cm^3 glass jars containing $NaH^{14}CO_3$ in seawater and shaken at various irradiances and temperatures for one hour. Not more than 25% of the ^{14}C was used up. The tissue was killed in ethanol, the alcohol-soluble ^{14}C analysed and 10% added for the insoluble residue. The gross rates of photosynthesis of the new frond in February at 8°, 18° and 23°C are shown against irradiance in Fig. 4.8. In spite of the scatter of points it seems probable both that 18°C is more favourable than the other two temperatures and that light saturation in this material was substantially above the 17 W/m^2 found by Lüning.

Further data obtained by Drew suggest that the upper temperature and light tolerances of *L. hyperborea* may have ecological significance. A temperature maximum for photosynthesis has already been indicated and temperatures as high as 30°C appear to be lethal since only two hours exposure to this temperature reduced subsequent photosynthesis to zero.

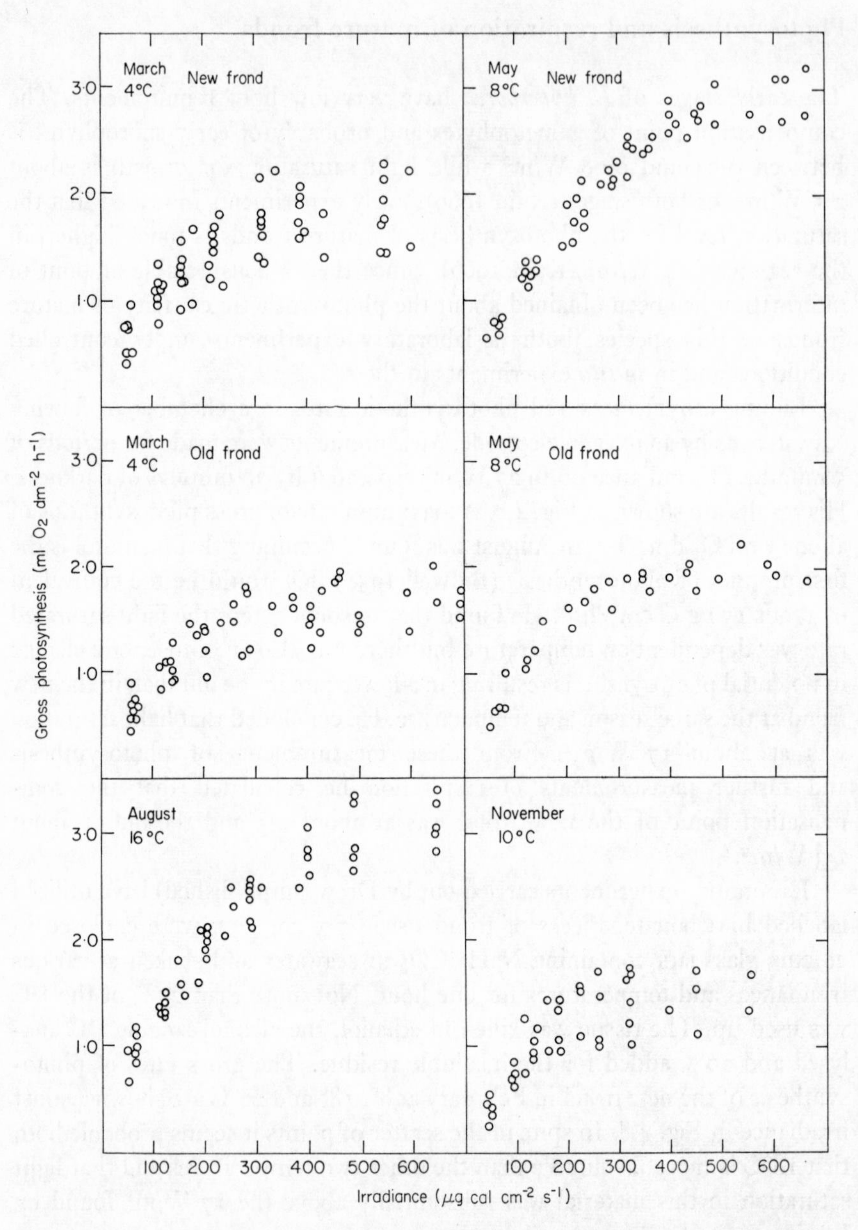

Figure 4.7. Gross photosynthesis of fronds of *Laminaria hyperborea* measured at Helgoland at various irradiances at ambient sea temperature at different seasons. Reproduced from Lüning (1971). If it is assumed that mannitol is produced: 1 ml O_2 dm^{-2} h^{-1} = 5 μg C cm^{-2} h^{-1}.

The effect of prolonged exposure to less drastic temperatures has yet to be determined.

Prolonged exposure to very high irradiance approaching that of direct solar radiation can also cause reduction of photosynthetic rate as is shown in Fig. 4.9. This may be responsible for the plant's absence from intertidal rock

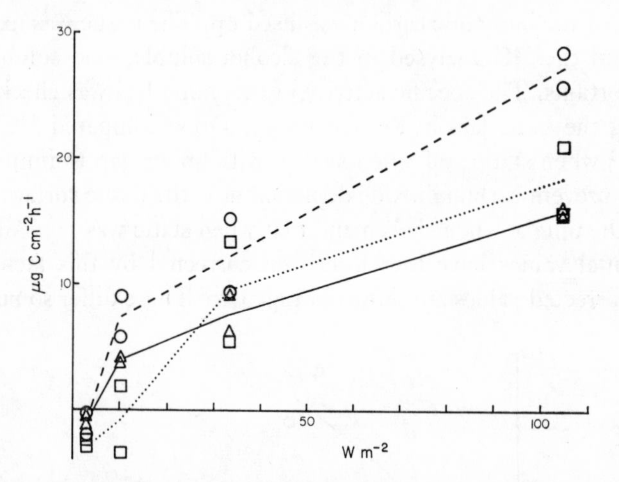

Figure 4.8. Gross photosynthesis of new frond tissue of *Laminaria hyperborea* in February at various temperatures and irradiances from a tungsten halide lamp. —△— 8°C; — —○— — 18°C; · · · · □ · · · · 23°C.

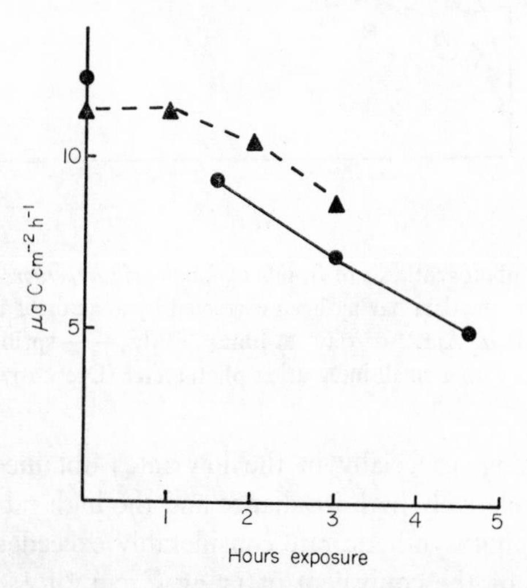

Figure 4.9. Rate of gross photosynthesis of fronds of *Laminaria hyperborea* exposed to various periods of high irradiance before measurement. —●— exposed to solar radiation of about 300 W/m² at 14°C in the field; — —▲— — exposed in the laboratory to radiation from a tungsten halide source giving about 330 W/m² at 8°C.

pools where it is replaced by *L. digitata* which has been found to be unaffected by very bright light (Drew 1974).

In situ experiments have been carried out at various times of year (Jupp 1972). Frond tissue discs of 6·5 cm diameter were enclosed each in a 460 cm^3 jar of seawater and $NaH^{14}CO_3$ for 4–6 hours. There was no mechanical shaking though some movement may have resulted from waves. Not more than 10% of the inorganic carbon was used up. The tissue was fixed in 80% ethanol and the ^{14}C analysed in the alcohol soluble, acid soluble and the residue portions. The specific activity of inorganic ^{14}C was checked in each jar. Using the same jars in the laboratory, Drew compared ^{14}C uptake in 105 W/m^2 when static and when shaken with an air gap to improve turbulence and prevent carbon dioxide depletion near the tissue surface. The ratio between the uptake when shaken and that when static was 1·5. All the *in situ* experimental values have therefore been corrected by this factor. Appropriately corrected values are shown in Fig. 4.10. These differ somewhat from

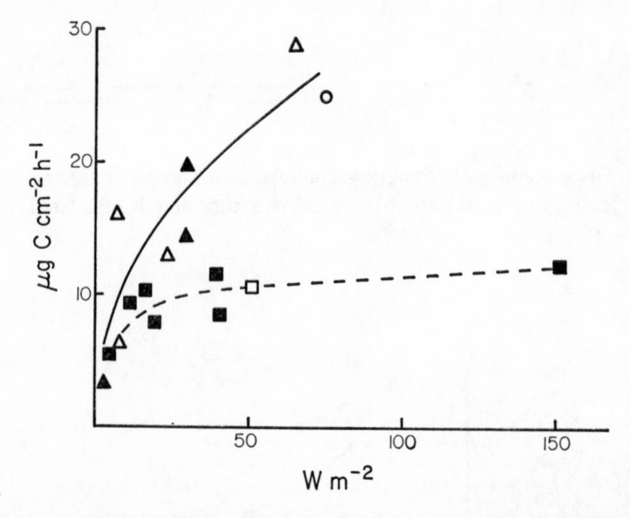

Figure 4.10. Gross photosynthesis of fronds of *Laminaria hyperborea* measured *in situ* at various times of year, the data having been corrected by a factor of 1·5 to allow for static conditions. △ March; ▲ April; ○ May; ■ June; □ July; —— spring; — — — summer. Irradiance measured with a small integrating photometer (Drew 1972).

the data of Lüning, especially in the low rates obtained in the summer months even at relatively high irradiance and the high rates in March–May when the gross photosynthetic rate considerably exceeded Lüning's which had a maximum of the equivalent of 15 $\mu g\ C\ cm^{-2}h^{-1}$. The latter would seem to support the evidence of Drew that saturation can be higher than 17 W/m^2.

It is perhaps relevant here to compare the photosynthesic rates of *L.*

hyperborea with other large brown algae such as *L. saccharina* (L.) Lamour. and *Saccorhiza polyschides* (Lightf.) Batt., both subtidal species often competing directly with *L. hyperborea*. The results of *in situ* experiments carried out in summer are shown in Fig. 4.11. *Saccorhiza* can apparently sustain a higher rate of photosynthesis in summer than its longer lived competitors. These data also indicate that the low rates measured in *L. hyperborea* at this time were not artifacts of the methods used since these were able to measure much higher rates in other plants at the same time.

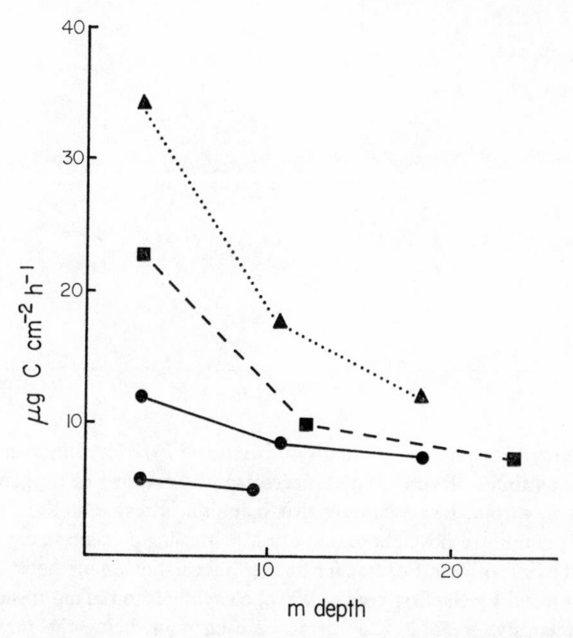

Figure 4.11. Gross photosynthesis of fronds of different species at various depths in June and July. Measured irradiances at the depths of experiments are given below, in order of increasing depth. · · · ▲ · · · *Saccorhiza polyschides* at 56, 24 and 10 W/m²; — — ■ — — *Laminaria saccharina*; —●— *L. hyperborea* at 151, 40 and 19 W/m² and at 56 and 8 W/m².

In order to calculate net photosynthetic gain, which is an indication of the growth potential of a plant, it is important to have reliable data about respiration rates at various temperatures and times of year in addition to measurements of gross photosynthesis. Determinations have been made by several workers and the respective methods used are as follows. Lüning (1971) used an oxygen electrode in a chamber of flowing seawater containing a disc of frond. Kain (unpublished) enclosed single strips of frond in static 200 cm³ bottles of seawater and measured the oxygen change by the Winkler method. Drew (unpublished) also used the Winkler method in 28 cm³ containers which were shaken. Jupp (unpublished) used a Warburg respirometer, with

15 frond discs of 7 mm diameter in each flask while Hopkin (unpublished) used a Gilson respirometer with 9 similar discs in each flask. The collected results of these workers are shown in Fig. 4.12. Again oxygen uptake has

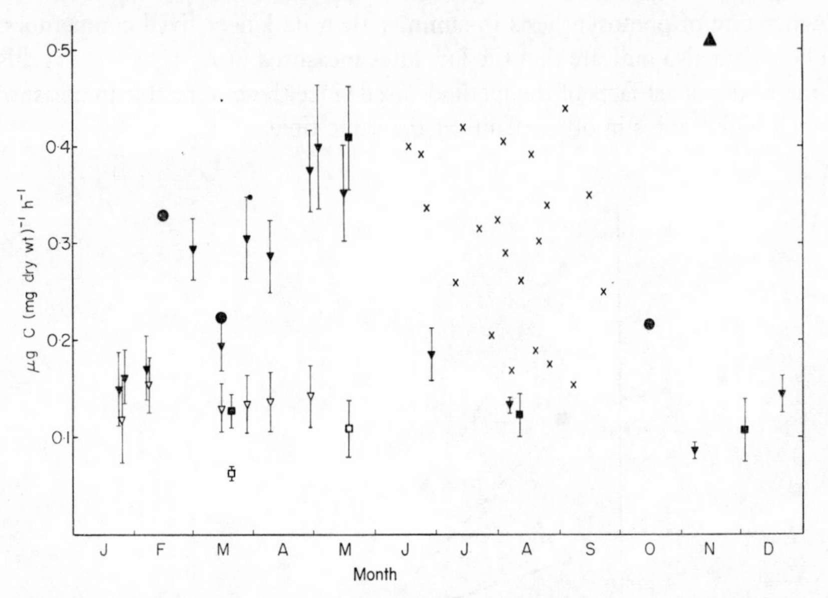

Figure 4.12. Rates of respiration of fronds of *Laminaria hyperborea* measured by various workers at various times of year. Where necessary observations of oxygen uptake have been converted to carbon loss assuming that mannitol is respired. Kain's figures were converted from per unit fresh weight to unit dry weight using percentage dry weight values appropriate to the time of year. See text for further information on methods. Solid symbols refer to the new frond for the first year of life, open symbols to the old frond when this is present simultaneously. ▼ & ▽ Kain at 10°C, each point being the mean from 5–12 separate plants, with fiducial limits (P = 0·05); ■ & □ Lüning (1971) in March at 3°C, in May at 9°C, in August at 16°C and in December at 9°C, with fiducial limits (P = 0·05); ● Drew in February at 8°C, in March at 7°C and October at 9°C; ▲ Jupp at 10°C; × Hopkin at 10°C, each point being the mean of 12 observations on one plant.

been converted to carbon loss assuming that it is mannitol that is respired. The figure serves to show the great variation in respiratory rate obtained under different conditions by different workers. However, it seems clear that the respiratory potential increases in the new frond between January and May but remains fairly low in the old frond, the difference between the fronds being statistically significant by March. Considering that in March Lüning's measurements were carried out at 3·5°C (presumably reducing the rate) and Kain's at 10°C there is good agreement between these two workers and probably Drew. Although Kain's experiments were in unshaken bottles, Drew failed to detect any difference in oxygen uptake when bottles were shaken, using the Winkler technique. It seems that Jupp's observed respiration rate in November at 10°C was much higher than other results at that time. Some of the variation apparent in Fig. 4.12 can be attributed to different methods. One pertinent factor may be the degree of wounding. A preliminary experiment by Kain indicated an increase of about 10% in respiration rate when the proportion of cut surface was doubled, while the proportion of cut surface of Hopkin's discs was 2·5 times that of Kain's strips. However, biological variation is clearly considerable and some of the variation may also be attributable to the influence of the thickness of the frond tissue. The effect of thickness on respiration rate expressed per unit dry weight was demonstrated in *Macrosystis* blades by Clendenning (1971) and pointed out by Lüning (1971). We have observed marked differences in the rates of different parts of the same frond varying in thickness.

The marked reduction in respiratory rate in summer is mainly due to an increase in dry weight of frond between April and September, a feature first observed by Black (1948) and since confirmed by Jupp (1972) and Kain (unpublished), leading to a reduced rate on a dry weight basis whereas that per unit area of fresh weight remains approximately constant.

The mid-point of the temperature range in the environment of *L. hyperborea* is probably around 10°C. From the results shown in Fig. 4.12 it can be assumed that respiration at this temperature does not exceed 0·5 μg C $cm^{-2}h^{-1}$ and will normally be considerably less. Some idea of the effect of temperature on the rate at a particular time of year can be gained from Fig. 4.13. Drew's results, giving a peak at 23°C, can be regarded as short term reactions because Jupp's longer term experiments showed a peak rate at 10°C. All these measurements refer to respiratory carbon loss in the dark and it has to be assumed in further considerations that light and dark respiration are the same, although there has been some evidence in certain experiments that there is a slight stimulation of respiration by low irradiances.

Using his own respiration data (obtained in November), Jupp (1972) calculated the compensation points of fronds from 3 m in December as 7·5 W/m^2 and from 9 m in July as 5·4 W/m^2. These levels are considerably higher than Lüning's (1971), possibly partly due to the high estimate of respiration.

Using his own physiological data, Lüning (1971) has calculated net photosynthetic gain at various times of year and his figure is shown as Fig. 4.14, with oxygen production again converted to carbon assimilation. His material, under reasonably bright light, reached a maximum assimilation rate in August but under dimmer light this maximum was in March, with a summer minimum. Because no allowance is made for night-time darkness, the effective irradiances must be considered to be means over 24 hours. However the resulting gain at saturating irradiance would not be as high as

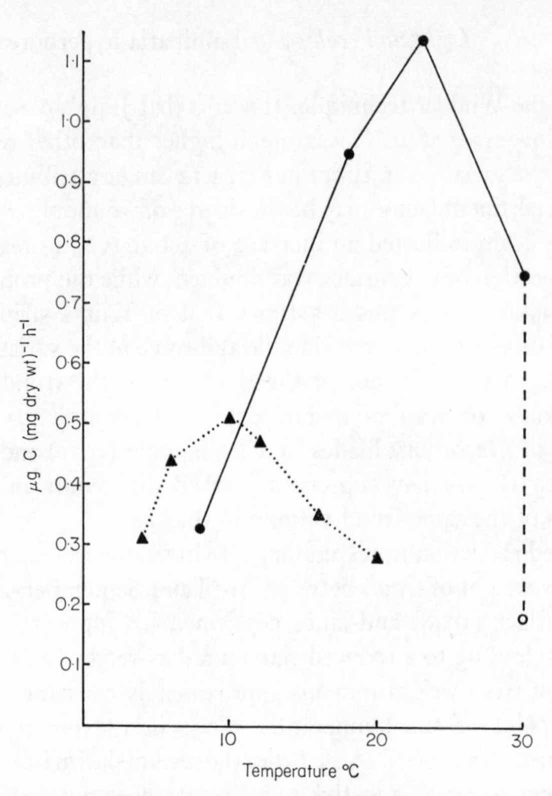

Figure 4.13. The effect of temperature on the respiration rate of fronds of *Laminaria hyperborea*. ▲ Jupp's experiment lasting 6 hours in a Warburg respirometer in November; ● ○ Drew's experiment in February, ● after 1 hour, ○ after 2 hours.

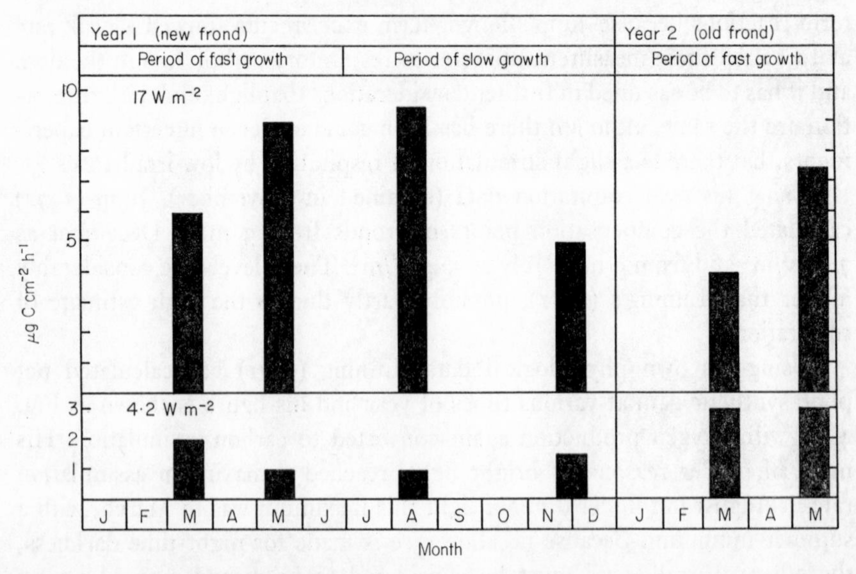

Figure 4.14. Net gain in photosynthesis in the frond of *Laminiaria hyperborea* during its 18 months of life, at two irradiances. Drawn from data supplied by K. Lüning.

shown because a higher daytime irradiance would not result in greater photosynthesis.

Translocation of photosynthate

It has long been assumed that translocation of materials between various parts of *Laminaria* plants can occur, and the trumpet hyphae of the medulla are usually implicated. A detailed study of these (Ziegler & Ruck 1967) has shown them to be living cells with normal organelles and a persistent nucleus; the end wall has 20,000 to 30,000 normal plasmadesmata with a pore diameter of about 0·06 μm. This is very much smaller than those of the sieve filaments found in one group of Laminariales, the Lessoniaceae, thus, *Macrocystis* has fewer pores in these elements but they are 2–6 μm wide. This plant has been shown to translocate photosynthate through the stipes at rates up to 78 cm h^{-1}, mainly in the peri-medullary sieve filament zone, though a minor conductive function of the medullary trumpet hyphae, also present in this plant, was not ruled out (Parker 1965).

Translocation studies in Laminaria species are few. Parker (1956) found very slow movement of phosphate (using the radioisotope ^{32}P) and fluorescein dye in *L. agardhii* Kjellm. at rates no faster than simple diffusion would provide, and he concluded that trumpet hyphae were not functional translocation elements. However, movement of some inorganic ions is difficult to demonstrate even in Lessoniaceae (Parker 1966).

Circumstantial evidence that translocation may occur in *L. hyperborea* was obtained in amputation experiments by Lüning (1969b) who showed that both the old frond and sometimes the stipe were necessary for full development of the new frond early in the year (Fig. 4.15). Jupp (1972) also found reduction of subsequent new frond development in plants with old fronds removed early in the year.

Various experiments reported by Lüning *et al.* (1972) and Schmitz *et al.* (1972) have shown that a certain amount of ^{14}C fixed photosynthetically in the old frond of *L. hyperborea* (and also in distal frond tissue of *L. saccharina*) moves into the new frond tissue and tends to accumulate in the meristematic region at the top of the stipe. Microautoradiographic evidence suggested the medulla as the route of this movement, and recent high resolution studies by Steinbiss & Schmitz (1973) have shown this to be localized in the young trumpet hyphae. Rates of movement were only about 5 cm/h with a maximum of 10 cm/h and it was principally longitudinal (Fig. 4.16), with no lateral spread (Schmitz *et al.* 1972). These rates of movement are similar to those found by Jupp (1972) who reported rates of ^{14}C movement of 1·5 to 9 cm/h from old frond to new frond tissue during the critical early months of the year. The translocation profiles were similar to those of Lüning, with a rapid drop in radioactivity immediately beyond the region of photosynthetic incorporation, whereas in *Macrocystis* this drop was less dramatic (Parker 1965). Jupp (1972) also investigated the movement of photosynthate into and out of the stipe and found some translocation each way.

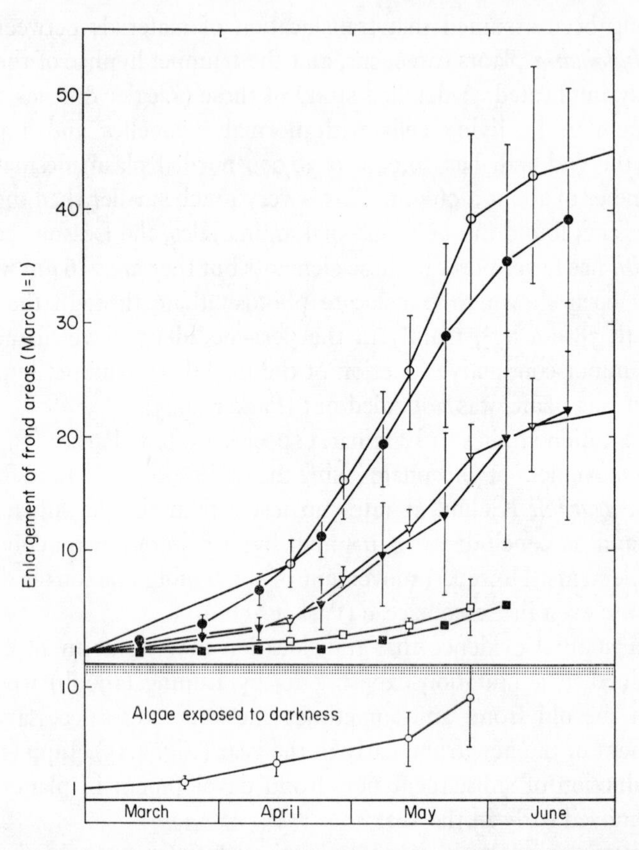

Figure 4.15. The change in frond area of *Laminaria hyperborea* with time after various treatments, with fiducial limits (P = 0·05). The mean area on March 1st was set at unity. Upper figure: Plants on frames in the sea at 2·2 m below LAT. Lower figure: Plants in dark chambers in outdoor tanks of running seawater. Solid symbols, 1967 results; open symbols, 1968 results: ● ○ normal plants; ▼ ▽ plants without old fronds; ■ □ isolated new fronds. From Lüning (1969b).

Jupp concluded that the old frond of a small 6-year-old plant of *L. hyperborea* could provide about 5 mg of carbon per day to the new frond during the early months of the year, or about 150 mg per month, whilst Lüning *et al.* (1973) suggest about 220 mg carbon per month and this could be very important to a small new frond developing under conditions of low

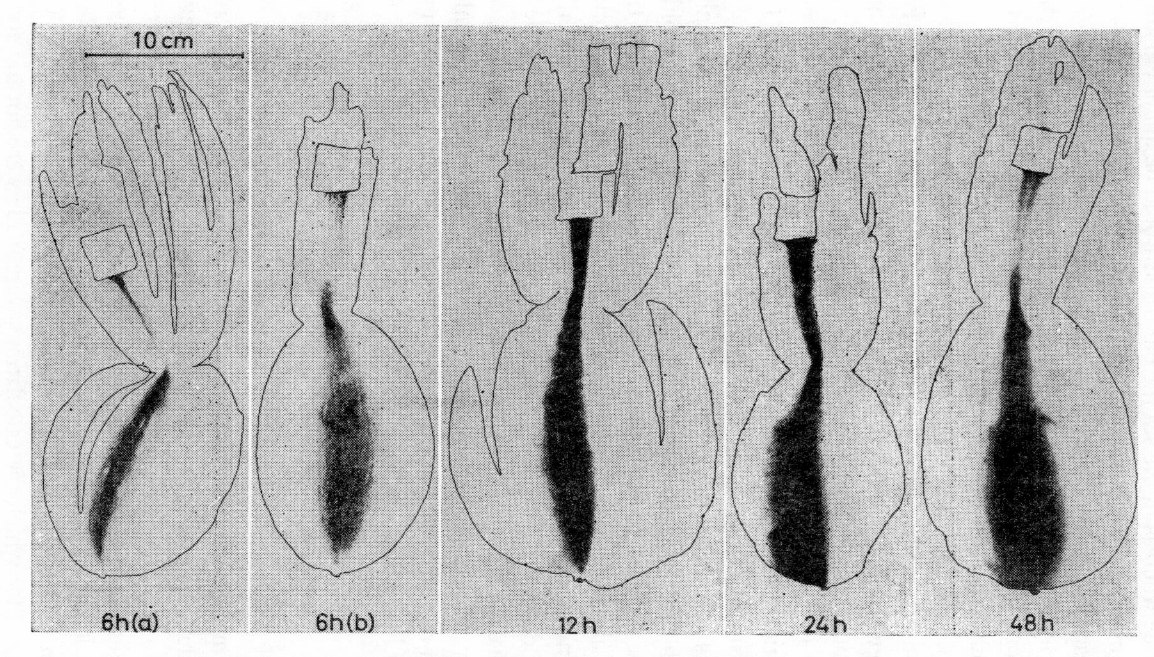

Figure 4.16. Autoradiographs of fronds of *Laminaria hyperborea*, the square parts cut out of the fronds having been previously exposed to ^{14}C for 2 hours. The times shown are from the start of the exposure. From Lüning *et al.* (1972).

irradiance due to seasonal and water turbidity factors. However, it must be remembered that both workers exposed their material to near saturating irradiance, thereby ensuring the maximal concentration gradient out of the old frond tissue whereas in the sea the old frond would be under equally low levels of illumination as the new frond. This work therefore only really demonstrates that a translocation pathway is operational in *L. hyperborea* via the trumpet hyphae. Rates of translocation determined indicate solute movement at about $3{\cdot}7 \times 10^{-6}$ cm^2 s^{-1} (Jupp 1972, assuming mannitol translocated) whilst the diffusion coefficient for mannitol in water is about $6{\cdot}8 \times 10^{-6}$ cm^2 s^{-1}, so that a simple diffusive mechanism cannot be ruled out, with the trumpet hyphae providing the least hindered route. However, by killing the new frond and stipe tissue of a plant with boiling water, leaving the old frond undamaged, Jupp was able to demonstrate the requirement for living tissue in this translocation system since no detectable movement of photosynthetically fixed ^{14}C from old to new frond was subsequently found in this experiment.

The rates of translocation reported here for *L. hyperborea* certainly do not approach the 100 cm per hour and more found in higher plant phloem for which mass transfer systems have been postulated. The amounts translocated are also relatively low, especially in view of the environmental limitations mentioned above. It must, however, be mentioned that all the experiments described here measured movement of material from the pool of recently fixed photosynthate only, and material from unlabelled pools of stored materials may also be moving undetected. This could certainly be important in the early part of the year when amounts of new photosynthate may be very low under natural conditions.

Discussion

Continuous measurement of underwater irradiance is time consuming and of limited accuracy unless highly sophisticated equipment is used. However in attempting to relate the assimilation and distribution of a plant such as *L. hyperborea* high degrees of accuracy or sensitivity are not required. The first method of calculation of underwater irradiance in the Isle of Man, allowing extrapolation to deeper water, has obvious pitfalls. It does seem, however, that the months for which the calculations were made were not too atypical, because the 5 m figures are similar to those calculated by the second method. The second method, more reliable than the first, showed that at both 3 and 5 m below LAT at Port Erin the mean irradiance can be expected to be above the highest estimate of compensation point for fronds for 6 months of the year. Taking Lüning's estimate of saturation, only at 3 m would fronds be saturated in open water for more than a few weeks, and taking Drew and Jupp's estimate they would not be saturated at all on a mean basis, though saturation would be reached during many of the daylight hours. During the six months of the year with less surface irradiance both 3 and 5 m receive more than compensating irradiance for the early stages and more than saturating irradiance for much of the time. These are open water levels, however, and there would be considerably less light on the rock surface at any time of year.

In Helgoland, in spite of a smaller tidal range, the shallower depth of 2·2 m below LAT was above compensation for only 4 months during the year that was studied by Lüning (Fig. 4.2). Underwater irradiance in Helgoland is obviously much more vulnerable to adverse weather conditions, causing marked fluctuations and very low levels in winter. Lüning (1971) correlated the low winter irradiance with seston stirred up by strong winds. It is probable that the exposed position of Helgoland and the proximity of shallow water sand banks account for the high attenuation compared with the Isle of Man.

An important effect of light in the sea on the ecology of *L. hyperborea* is on its vertical distribution. It is important, however, to distinguish between light-determined lower limits and those controlled by other factors. In the Isle of Man it has been shown conclusively that the lower limit on Port Erin breakwater is determined by the grazing pressure of *Echinus*, but it is not clear whether this is the case in other sites close by. An example is Spanish Head, where plants have been found down to 19 m below LAT. There are two pieces of evidence indicating that this also is an *Echinus*-determined limit. The first is that the *Echinus* density at this depth is between 3 and 4 animals per square metre, of the same order as that causing the limit on the breakwater. The second is that the standing crop at 13·5 m at this site showed a marked drop on a logarithmic scale, although in shallower water the crop was similar to that found in western Scotland. On the other hand, the data on the irradiance at this depth in the Isle of Man, calculated by the extrapolation method, shows that at 15 m only in June was the irradiance above even the lowest estimate of the compensation point for mature fronds, 1·3 W/m^2 (Lüning 1971, old frond). This would indicate that around 15 m that there is not sufficient light to support growth. However, as the plants from 13 to 19 m had clearly grown reasonably well there is an obvious inconsistency. This could easily be due to inaccuracy of extrapolation or it is possible that light attentuation at Spanish Head is not always the same as at Port Erin. Alternatively it may be due to applying laboratory results to the field under totally different conditions.

It would seem likely that in many sites apart from the Isle of Man irradiance and not grazing is the limiting factor in vertical distribution. On the north-east coast of Britain the vertical range seems to be determined by the sediment load of the seawater, whether the origin of this is natural or pollution (Bellamy *et al.* 1970). *Echinus* is not said to be abundant in this region. High densities of *Echinus* were not encountered in western Scotland and there, in relatively clear water, the lower limit in many suitable sites is between 20 and 30 m. Even clearer water in Korsfjord in Norway is likely to be responsible for the lower limit of more than 30 m there. In Helgoland, on the other hand, it has been shown that for most of the year, in spite of a smaller tidal range, 2·2 m receives less irradiance that 3 or even 5 m in the Isle of Man and the high lower limit of 8 m below LAT can no doubt be attributed to this.

Standing crop is clearly related to irradiance except when grazing is active. Thus, the lowest rate of reduction with depth was found in the Outer Hebrides in clear water and the highest in Helgoland, shown to have high attenuation. The small decrease in growth rate with depth when the community is closed is attributed to self-limitation by the canopy. A measure of the canopy is the frond area index which decreases with decreased irradiance. The two factors of self-shading and irradiance reduction with depth thus counteract each other and produce a similar environment within the forest, producing a relatively similar growth rate of sub-canopy plants. This idea is supported by the remarkable increase in growth rate when the canopy is removed and the then greater effect of depth on the growth rate, the latter being more dependent on open water irradiance.

Realistic measurements both of photosynthesis and of respiration are fraught with difficulties. Firstly it is virtually impossible to simulate environmental conditions exactly, especially those of appropriate water movement and hence the provision and removal of substances at the surface of the plant. Secondly the rates of both processes vary differently with temperature at different times of year. Thirdly the rates of the two processes are related in different ways to both surface and fresh or dry weight of frond tissue, making it difficult to calculate the balance. Fourthly, the measurement of effective irradiance depends on spectral activity of photosynthesis. Fifthly, because photosynthesis cannot be measured directly, the uptake of labelled carbon-dioxide or the output of oxygen have to be taken as indicators, both of them having disadvantages in their indirectness. It is possibly because of a combination of these difficulties that the levels found for compensation and saturation irradiances for mature fronds of *L. hyperborea* by Lüning and by Drew and Jupp respectively are in such disagreement. The estimation of compensation is particularly difficult because of the variation in respiration under different conditions and the necessity of estimating it for the whole plant. A particular difference in technique involves the spectral composition of irradiance. Lüning used filters to produce conditions similar to those in

Jerlov's type 6 coastal water, with a peak transmission at 550 nm, whereas Drew used (in laboratory experiments) tungsten halide lamps. From the spectral composition of the irradiance in his experiments published by Lüning (1971, Table 1), the spectral emission of the tungsten halide source used by Drew and the action spectrum for *Laminaria saccharina* found by Halldal (1969) we have calculated the relative response of photosynthesis to the two sources. The response of this species to the tungsten source is 87·7% of its response to Lüning's source. This difference, assuming that it would apply similarly to *L. hyperborea*, is not sufficient to explain the discrepancy. Alternatively, it is possible that the difference lies in the biological material. Either the Helgoland plants were considerably younger than those used by Drew and Jupp in Scotland (and both compensation and saturation must increase with age from the low levels found for early sporophytes) or the Helgoland plants may be better adapted to low light in the turbid waters there than those in Scotland. Clearly only further experimentation can clear up these points.

It was shown previously that the light requirements of the early stages of the main competitors of *L. hyperborea* were similar and therefore unlikely to be important in determining the outcome of competition. The few experiments by Jupp on the photosynthesis of mature fronds of the competing species *L. saccharina* and *Saccorhiza polyschides* seemed to show a marked difference in photosynthetic capacity, the faster growing opportunist *Saccorhiza* having the highest. If further experiments confirm this result this could explain the outcome of competition in some cases.

The seasonal picture of *L. hyperborea* assimilation and growth which emerges is as follows. After the first year, when it may be continuous, growth is mainly confined to the period between November and June, when a new frond develops above the transition zone and primary growth of the stipe occurs. During the first few months of this growth there is inadequate light in many habitats for the compensation point of the frond to be reached, so the respiration of the frond and stipe, high because of active growth, is not counteracted by assimilation and nor is the laying down of new tissue. It is clear that in many cases the energy source for these processes is the organic material stored in the old frond which must therefore be translocated. It has been shown that translocation of organic substances is possible through the medullary trumpet hyphae. By May the new frond is quite large and both fronds, if the old one is still present, are better irradiated and an assimilation surplus is likely. An interesting point here is that immediate products of photosynthesis (as distinct from storage) of the old frond have been shown to be transported to the new frond, particularly to the growing region. This fact is likely to be of more importance to the new frond in deeper, darker, water and wherever there is wave action the old frond is retained longer in deeper water than in shallow because it is less likely to be torn off there. By summer, at the time of maximum irradiance, the new frond is fully grown and from then on the products of photosynthesis are laid down as storage, being used later for the next winter's fast growth as well as for the production of sporing tissue. It is not possible yet to produce an energy budget for the whole of this cycle.

Although in the last 8 years we have learnt a lot more about *L. hyperborea*, there now seem to be at least as many unsolved problems as there seemed at the last Symposium.

Summary

Various new approaches have been made to the ecology of the species. Information on the light climate has been obtained from photocells installed in the sea near *Laminaria hyperborea* environments and the records for the Isle of Man compared with Helgoland. The lower limit may well be determined in many places by light but probably not in the Isle of Man. Within the forest growth rate is determined partly by depth and partly by the extent of the frond canopy, the two factors almost counteracting each other. Estimates of the light saturation of photosynthesis of mature fronds vary from 17 to over 40 W/m^2 and there is seasonal variation. The optimum temperature for photosynthesis is near 18°C. Respiration rates of frond tissue are variable and depend on a number of factors including season, temperature, frond tissue thickness and method of measurement. From the light data it appears that plants would be below the compensation level in most sites for at least some of the winter and certainly when new growth starts. Organic sources must therefore be utilised and these appear to be translocated from the old to the new frond. The competitors *Saccorhiza polyschides* and *L. saccharina* were found to photosynthesize faster than *L. hyperborea*.

Acknowledgements

We owe much to Dr. K. Lüning's helpful co-operation in allowing us to make so much use of his data. Some of his results are replotted in Figs. 2, 3, 4 and 12 and where necessary he provided us with the original data. We are also grateful for his permission to reproduce three of his published figures.

We are very grateful to Mr. R. Hopkin for letting us use his unpublished data in Fig. 12, and to Mr. M. Bates for underwater assistance.

The work of two of us (E.A.D. and B.P.J.) was supported by grant number GR/3/484 from the Natural Environment Research Council.

For permission to reproduce figures we are indebted to Cambridge University Press (Fig. 7), Springer-Verlag (Fig. 15) and University of Tokyo Press (Fig. 16).

References

BELLAMY D.J., JOHN D.M., JONES D.J., STARKIE A. & WHITTICK A. (1970) The place of ecological monitoring in the study of pollution of the marine environment. F.A.O. Technical Conference on Marine Pollution and its Effects on Living Resources and Fishing, Rome, 9–18 December 1970.

BIDWELL R.G.S. (1958) Photosynthesis and metabolism of marine algae II. A survey of the rates and products of photosynthesis in $^{14}CO_2$. *Can. J. Bot.* 36, 337–49.

BLACK W.A.P. (1948) Seasonal variation in chemical constitution of some of the sublittoral sea weeds common to Scotland. Part I, *Laminaria cloustoni. J. Soc. chem. Ind.* 67, 165–8.

CLENDENNING K.A. (1971) Photosynthesis and general development in *Macrocystis.* In *The biology of Giant Kelp Beds (Macrocystis) in California*, ed. W.J. North, 169–90, Lehre: Cramer.

DREW E.A. (1972) A simple integrating photometer. *New Phytol.* 71, 407–13.

DREW E.A. (1974) Light inhibition of photosynthesis in macro-algae. *Br. phycol. Bull.* 9, 217–18.

GRENAGER B. (1953) Kvantitative undersøkelser av tareforekomster på Kvitsøy og Karmøy 1952. *Rep. Norw. Inst. Seaweed Res.* 3, 53 pp.

HALLDAL P. (1969) Automatic recording of action spectra of photo-biological processes, spectrophotometric analysis, fluorescent measurements and recording of the first derivative of the absorption curve in one simple unit. *Photochem. & Photobiol.* 10, 23–34.

JONES N.S. & KAIN J.M. (1967) Subtidal algal colonization following the removal of *Echinus. Helgoländer wiss. Meeresunters* 15, 460–6.

JUPP J.P. (1972) Studies on the growth and physiology of attached marine algae. Ph.D. Thesis, St. Andrews Univ. 226 pp.

KAIN J.M. (1963) Aspects of the biology of *Laminaria hyperborea.* II. Age, weight and length. *J. mar. biol. Ass. U.K.* 43, 129–51.

KAIN J.M. (1966) The role of light in the ecology of *Laminaria hyperborea.* In *Light as an Ecological Factor*, ed. R. Bainbridge, G.C. Evans and O. Rackham, 319–34, Oxford: Blackwell.

KAIN J.M. (1967) Populations of *Laminaria hyperborea* at various latitudes. *Helgoländer wiss. Meeresunters*, 15, 489–99.

KAIN J.M. (1969) The biology of *Laminaria hyperborea.* V. Comparison with early stages of competitors. *J. mar. biol. Ass. U.K.* 49, 455–73.

KAIN J.M. (1971a) Continuous recording of underwater light in relation to *Laminaria* distribution. In *Proceedings of the IVth European Marine Biology Symposium*, ed. D.J. Crisp, pp. 335–46. Cambridge: University Press.

KAIN J.M. (1971b) The biology of *Laminaria hyperborea.* VI. Some Norwegian populations. *J. mar. biol. Ass. U.K.* 51, 387–408.

LÜNING K. (1969a) Standing crop and leaf area index of the sublittoral *Laminaria* species near Helgoland. *Mar. Biol.* 3, 282–6.

LÜNING K. (1969b) Growth of amputated and dark-exposed individuals of the brown alga *Laminaria hyperborea. Mar. Biol.* **2**, 218–23.

LÜNING K. (1970) Tauchuntersuchungen zur Vertikalverteilung der sublitoralen Helgoländer Algenvegetation. *Helgoländer wiss. Meeresunters* **21**, 271–91.

LÜNING K. (1971) Seasonal growth of *Laminaria hyperborea* under recorded underwater light conditions near Helgoland. In *Proceedings of the IVth European Marine Biology Symposium*, ed. D.J. Crisp, pp. 347–61. Cambridge: University Press.

LÜNING K., SCHMITZ K. & WILLENBRINK J. (1972) Translocation of ^{14}C-labeled assimilates in two *Laminaria* species. *Int. Seaweed Symp.* 7. Sapporo, 420–5.

LÜNING K., SCHMITZ K. & WILLENBRINK J. (1973) CO_2 fixation and translocation in benthic marine algae. III. Rates and ecological significance of translocation in *Laminaria hyperborea* and *L. saccharina. Mar. Biol.* **23**, 275–81.

MCALLISTER H.A., NORTON T.A. & CONWAY E. (1967) A preliminary list of sublittoral marine algae from the west of Scotland. *Br. phycol. Bull.* **3**, 175–84.

NORTON T.A. (1968) Underwater observations on the vertical distribution of algae at St. Mary's, Isle of Scilly. *Br. phycol. Bull.* **3**, 585–8.

PARKER B.C. (1965) Translocation in the giant kelp *Macrocystis*. I. Rates, direction, quantity of C^{14}-labeled products and fluroescein. *Jnl. Phycol.* **1**, 41–6.

PARKER B.C. (1966) Translocation in *Macrocystis*. III. Composition of sieve tube exudate and identification of the major C^{14}-labeled products. *Jnl. Phycol.* **2**, 38–41.

PARKER J. (1956) Translocation of P^{32} and dye behaviour in two species of marine algae. *Nátúrwissenschaften* **43**, 452.

SCHMITZ K., LÜNING K. & WILLENBRINK J. (1972) CO_2-Fixierung und Stofftransport in benthischen marinen Algen. II. Zum Ferntransport ^{14}C-markierter Assimilate bei *Laminaria hyperborea* und *Laminaria saccharina. Ztschr. Pflanzenphysiol.* **67**, 418–29.

STEINBISS H.H. & SCHMITZ K. (1973) CO_2-Fixierung und Stoff-transport in benthischen marinen Algen. V. Zur autoradiographischen lokalisation der Assimilattransportbahren im Thallus von *Laminaria hyperborea. Planta* **112**, 253–63.

SVENDSEN P. (1972) Noen observasjoner over taretråling og gjenvekst av stortare, *Laminaria hyperborea. Fiskets Gang* **58**, 448–60.

SVERDRUP H.U., JOHNSON M.W. & FLEMING R.H. (1942) *The Oceans*. New York: Prentice-Hall Inc. 1087 pp.

WALKER F.T. (1950) Sublittoral seaweed survey of the Orkney Islands. *J. Ecol.* **38**, 139–65.

WALKER F.T. (1954) Distribution of Laminariaceae around Scotland. *Nature, Lond.* **173**, 766–8.

WHITTICK A. (1969) The kelp forest ecosystem at Petticoe Wick Bay lat.55°55′N long. 2°09′W an ecological study. Durham University: MSc thesis. 139 pp.

ZIEGLER H. & RUCK I. (1967) Untersuchungen uber die Feinstruktur des Phloems. III. Die Trompetenzellen von *Laminaria*-Arten. *Planta* **73**, 62–73.

Question and discussion

In answer to Professor Waterman, Dr. Kain said that there was comparable data for photosynthesis by *Macrocystis* in the Pacific. Mature blades reached saturation at values of irradiance similar to those for *Laminaria*, but more light was needed in the early stages.

5 Light and plant response in fresh water

D. H. N. SPENCE, *Department of Botany, University of St. Andrews.*

All but the clearest shallow water is a shade habitat. After outlining the underwater light climate and its unique properties an attempt is made in this paper to distinguish those features of submerged attached plants or macrophytes which may be regarded as adaptations to this light climate. The significance of shade adaptation, particularly of leaves, is then assessed in relation to photosynthesis and the vertical distribution of species in standing water. The importance of shade in the life of aquatic plants has been recognized for many years and attention should be drawn to the account in Gessner (1955) of the photosynthetic response of sun and shade forms of underwater plants by Harder (1930), Ruttner (1926) and himself (Gessner 1938). Such studies complement what is discussed here.

Since, along with submerged phases in the annual growth-cycle of floating-leaved and emergent plants, submerged macrophytes to a greater or lesser extent share their photosynthetic environment with phytoplankton, the idea is then explored that a key to their performance is the extent of their competition with phytoplankton for light. 'Performance' of macrophytes as studied in this paper means the depth of the zone they colonize in standing water and the depth of occurrence and the extent of the zone of their maximal biomass; macrophytes here comprising angiosperms, vascular cryptogams, Bryophyta and Charophyta. No attempt is made to assess their productivity.

Thienemann (1926, 1950), Pearsall & Ullyott (1934), Gessner (1955) and others have claimed that there is a general relationship between amount of phytoplankton and the depth zone colonized by macrophytes. In the present study, first, it is postulated that some conditions indeed exist where the depth zone of macrophytes varies with amount of phytoplankton, or phytoplankton density, and is therefore predictable and that other conditions exist where such prediction is impossible. Both situations are defined empirically and their general validity can now be tested.

Secondly, after outlining the pattern of biomass distribution in relation to water depth, a hypothesis is advanced to account for the position of biomass maxima in any body of standing water and preliminary evidence is shown to agree with the prediction of both the relative position and size of the biomass maxima in the clearest and the most phytoplankton-dense water. Since phytoplankton can alter other environmental factors beside light, a brief analysis is made of the nature of their competition with macrophytes.

One omission in this study of light and underwater plants is any assessment of the effects on macrophytes of periphyton or epiphytic algae and bacteria. Much work has been devoted over the past 50 years to descriptions of periphyton communities in terms of species composition and distribution (references in Wetzel 1964) and recently of biomass and productivity (references in Allen 1971), the latter frequently on artificial submerged surfaces (Brown & Austin 1971, 1973). Less attention has been paid to the interaction between these epiphytes and their macrophytic hosts (Jørgensen 1957, Prowse 1959). Fitzgerald (1969) reported that cultivated populations of *Ceratophyllum* sp., *Myriophyllum* sp. and *Cladophora*, and natural lake populations of *Cladophora* remained relatively free of epiphytes if grown in conditions of apparently limiting nitrogen but bacteria-sized organisms, selectively toxic to some algae, were also present. Allen (1971) made a comprehensive study of the contribution of the periphyton of both emergent and submerged vegetation to the overall primary production of a calcareous lake (Lawrence Lake, Michigan), deliberately chosen because of the dearth there of phytoplankton, and he established the interchange of various extracellular substances between macrophytes, algae and bacteria.

Probably because of the complexity of these relationships between periphyton and macrophytes even in the presence of very little phytoplankton, no published study yet appears to exist on the quantitative, small-scale effects of epiphytic growth upon macrophyte photosynthesis either in terms of shading or of possible nutrient depletion, including depletion of inorganic carbon. As far as the consequences of high phytoplankton density and rates of photosynthesis are relevant, it can only be assumed that algal epiphytes, like macrophytes, are capable of shade adaptation and subject to competition for light and, perhaps, to adverse changes in water chemistry. Such competition between epiphytes and macrophytes may become most critical in waters of high nutrient level and phytoplankton density. This would only compound those interactions in situations which, it is argued later, already make unlikely the correct prediction of macrophyte biomass etc.

I Underwater light climate

Methods and terminology

Data on underwater light climate discussed here are mainly confined to British waters and have been derived by one of two means; firstly by use of a matched surface and underwater Evans Electroselenium photocell placed horizontally, facing upwards or downwards, a microammeter, and each of three standard optical glass broad-band-pass filters (Chance, red OR with λ_{max} of 680 nm, green OG with λ_{max} of 550 nm and blue OB with λ_{max} of 450 nm; and Schott, for which Talling (1960) by combining three factors (spectral sensitivity of cell, transmission of filters and of lake water itself) derived mean extinction mid-points of red RG1, 630 nm, green VG9, 525 nm, and blue BG12, 460 nm). Secondly, data have been derived as irradiance by use of a horizontally positioned remote probe connected to an Instrument Specialities Company Spectroradiometer by a fibre-optic cable. Both types of collector are cosine-corrected. Because reference is only made to irradiance relative to subsurface (Om) it has been unnecessary to correct values for immersion effects. Table 5.1 lists symbols used in this account

Table 5.1. Symbols used in the text, with their definitions.

Symbol	Definition
PAR	photosynthetically available radiation (400–700 nm)
Z_{eu}	depth of euphotic zone, in m; its lower limit being defined as that depth where 1% subsurface *PAR*, 1% *PAR* which has penetrated the water surface, is found (see E_{min} for empirical derivation). Primarily intended for phytoplankton: a concept based on photosynthesis—depth profiles in phytoplankton, and found broadly to coincide with the lower limit of net photosynthesis by their organisms (see e.g. Talling 1971).
Z_m	mixed depth, in m; depth to which water is mixed by turbulence: sometimes less than maximal depth (see Talling 1957).
Z_c	depth of zone colonized more or less continuously by attached macrophytes, in m. (Z_c is proposed as a new term)
B	phytoplankton population density per unit volume, in mg chlorophyll *a* m^{-3}.
E_{PAR}	vertical diffuse attenuation coefficient of downwelling *PAR* in ln units m^{-1} (proposed as a new term): written as E_{10} in $\log_{10}$ units m^{-1} (as E_e or E_{10} covering the range 390–762 nm, in Spence *et al.* 1971).
$E_R E_G E_B$	vertical diffuse attenuation measured with the appropriate Chance optical glass filters so that E_G, for example = E_G ln units m^{-1} at λ max of 590 nm; or with Schott filters (e.g. Talling 1971).
E_{min}	minimal vertical attenuation value, ln units m^{-1}, for E_e over *PAR* spectrum which is strictly correct for spectroradiometric readings but which for Chance or Schott filters is taken to be that filter (usually G) passing most light and characterized as above. Used empirically to compute a value for the euphotic zone, since $Z_{eu} \approx 3{\cdot}7/E_{min}$ (Talling 1965).

with their definitions; excepting the choice of E for attenuation coefficient, the symbols are those recommended by Winberg (1971); E_{PAR} and Zc are proposed as new terms.

Range in quality and quantity of light underwater

The first aim is to establish the range of variation in the quality and quantity of photosynthetically available radiation (*PAR*) which reaches the surface of underwater plants. Light quality measured by selenium photocells with blue, green or red filters shows two categories of standing water in Scotland and the English Lake District (Table 5.2); the locations of all sites mentioned here and subsequently are given in Table 5.3. The first category of water, exemplified by Loch Croispol (Fig. 5.1) Lochan an Eun and Loch Borralie in Scotland and (Talling 1971) by Wastwater and Ennerdale Water in England, have low attenuation values; the highest is E_R (see Table 5.1) which does not exceed 0·45, the next is E_B at not more than 0·30 whilst E_G is less than 0·19 and becomes, therefore, E_{min} (Table 5.1); E_R for pure water is 0·31 (Lythgoe 1972) which is the value given for Ennerdale. In the second category overall attenuation is higher and blue attenuation is always greater than red (Table 5.2 and Fig. 5.1, Loch na Thull). While green remains the least attenuated in relatively clear water like that of Derwentwater or Loch Uanagan, its coefficient approaches or marginally exceeds the value for red light in brown-water lakes like Loch na Thull.

Values of E_{10} per 25 nm waveband from 390–762 nm for the waters of two lochs are given in Fig. 5.2: there is a full account of their derivation in Spence *et al.* (1971). It should be noted that these spectroradiometric values, unlike those of Tyler & Smith (1967) for Crater Lake, Oregon, are based on measurements made over only 1 m depth of water and are therefore susceptible to variations in cloud cover unperceived by the human eye (Edwards & Evans 1975) and in reflection at the air-water interface during the long period needed to scan the *PAR* spectrum at different depths. Such variation was evidently present in the clear, limestone Loch Croispol (as indicated by the extrapolated subsurface values in Fig. 5.3) and could account for the sometimes sharp changes in attenuation between adjacent band-widths. Accepting the likelihood of such errors, its water reaches at some wavelengths the transparency of Crater Lake, which in turn approaches distilled water in clarity and chemical content and thus provides a graph of spectral selectivity of water itself, absorbing increasingly towards the long wave end of the *PAR* spectrum. By contrast all the curves other than that for Crater show selective absorption towards the blue end of the spectrum, a characteristic of dissolved organic matter, 'Gelbstoff', and allochthonous, peat-derived, material.

Table 5.2. Diffuse attenuation coefficients (E_e) for blue (E_B) green (E_G) and red (E_R) light measured with appropriate Chance or Schott (S)[1] optical glass filters and selenium photocell, depth (m) receiving 1% subsurface radiation with these filters; and depth of euphotic zone (Z_{eu}) (see text) in a number of English lakes and Scottish lochs (locations in Table 5.3). English lake 1% data interpolated, and E_e data calculated, from Talling (1971, Fig. 5)[1]. Lochan an Eun readings by J.J. Light (*in litt.*, and Light 1975).

Lake or Loch	Date	*E* ln units			Depth (m) with 1% subsurface radiation			Z_{eu} (m)
		E_B	E_G	E_R				
Lochan an Eun	July 1969	0·14	0·12	0·33	33·6	38·0	14·0	30·8
Wastwater (S)	June–September 1952	0·18	0·12	0·38	25·8	38·0	12·0	30·8
Ennerdale Water (S)	June–September 1952	0·25	0·13	0·31	18·5	35·8	15·0	28·7
Borralie	June 1971, 1972, 1974	0·29	0·16	0·45	15·9	28·8	10·2	23·0
Croispol	June 1971, 1974	0·29	0·18	0·43	15·9	25·6	10·7	20·8
Derwentwater (S)	June–September 1952	0·47	0·32	0·42	9·8	14·5	11·0	11·6
Windermere South (S)	June–September 1952	0·66	0·35	0·58	7·0	13·0	8·0	10·4
Uanagan	July 1967	0·62	0·37	0·43	7·4	12·4	10·7	10·0
Lanlish[2]	August 1974	0·77	0·36	0·56	6·0	12·7	8·7	10·2
Esthwaite Water (S)	June–September 1952	0·77	0·48	0·52	6·0	9·5	8·5	7·6
Clunie	June 1967	0·84	0·46	0·47	5·5	10·0	9·8	8·0
Lowes	July 1967	0·97	0·60	0·59	4·7	7·7	7·8	6·2
na Thull	June 1974	1·19	0·55	0·54	3·9	8·4	8·5	6·7
Leven	September 1974	2·01	1·07	1·04	2·2	4·3	4·4	3·5[3]

[1] Since two different sets of filters are used (λ_{max} *etc.* on p. 95), neither E_e nor 1% subsurface values are strictly comparable.

[2] Although Loch Lanlish lies (Table 5.3) near Lochs Borralie and Croispol on limestone its optically different water is perhaps a result of its limestone basin being largely overlain by sand, with a grazed fen along half its shore.

[3] Annual range 1·4 to 7·5 (Bindloss 1974).

The amount of blue absorption is by far the least in Loch Croispol, as already noted, and increases to a maximum in Loch Leven. Loch Uanagan which is illustrated in Spence *et al.* (1971) and to the eye is like Loch na Thull in Fig. 5.1, brown or peaty in colour, and Loch Leven, which is normally pale but plankton-rich, have essentially similar qualitative distribution of *PAR*, except at 675 nm. This is one of the *in vivo* absorption peaks for chlorophyll *a* where Loch Leven water shows a relative increase in attenuation, a peak that was also noted by Talling (1970) in his spectroradiometric analysis of Blelham Tarn water during a spring phytoplankton bloom (Fig. 5.2).

Table 5.3. Location, longitude and latitude of freshwater lakes of which aspects of light climate and/or vegetation are described in the text. For Scottish lochs the spelling used is that of the Ordnance Survey.

Loch Avon, Cairngorm Mountains	3° 38′ W	57° 6′ N
Loch Coire an Lochan, Cairngorm Mountains	3° 45′ W	57° 4′ N
Lochan an Eun, Lochnagar	2° 16′ W	56° 56′ N
Loch Baile na Ghobhainn ('Balnagowan'), Lismore	5° 31′ W	56° 32′ N
Loch Borralie[1], Durness	4° 47′ W	58° 33′ N
Loch Clunie, Dunkeld	3° 26′ W	56° 34′ N
Loch Croispol[1], Durness	4° 46′ W	58° 34′ N
Loch Lanlish[1], Durness	4° 47′ W	58° 34′ N
Loch Leven, Kinross	3° 23′ W	56° 12′ N
Loch of Lowes, Dunkeld	3° 32′ W	56° 34′ N
Lake of Menteith, Aberfoyle	4° 17′ W	56° 10′ N
Loch Spiggie, Shetland	1° 20′ W	59° 55′ N
Loch na Thull[1], Rhiconich	5° 0′ W	58° 24′ N
Loch Uanagan, Fort Augustus	4° 42′ W	57° 7′ N
Bassenthwaite Lake	3° 30′ W	54° 40′ N
Coniston Water	3° 5′ W	54° 20′ N
Derwentwater	3° 9′ W	54° 35′ N
Ennerdale Water	3° 24′ W	54° 31′ N
Esthwaite Water	2° 59′ W	54° 21′ N
Wastwater	3° 19′ W	54° 26′ N
Windermere	2° 57′ W	54° 21′ N
Bodensee (Lake Constance)	9° 20′ E	47° 40′ N
Lunzer Untersee, Lower Austria	15° 5′ E	47° 44′ N
Würmsee (Starnberger See) Bavaria	11° 25′ E	47° 44′ N
Crater Lake, Oregon	120° 0′ W	43° 0′ N
Crystal Lake, Wisconsin	89° 11′ W	46° 9′ N
Weber Lake, Wisconsin	89° 7′ W	46° 6′ N
Lake Bunyoni, Uganda	29° 54′ E	1° 18′ S
Lake Mutanda, Uganda	30° 40′ E	1° 10′ S
Lakes on South Georgia	36° 30′ W	54° 16′ S

[1] Alternate spelling, respectively: Borralaidh, Croisaphuill, Lon na h-Innse, na h-Ulla: in for example, Messrs. John Bartholomew's 'Half inch' series of maps.

An approximate figure for the range of *PAR* to be expected at a given depth between the clearest and darkest standing water is given by comparing a value for E_{PAR} (ln units m^{-1}, 400–700 nm) of 0·14 for Ennerdale Water, derived from optical glass filters and E_{min} (Talling 1957), with an E_{PAR}

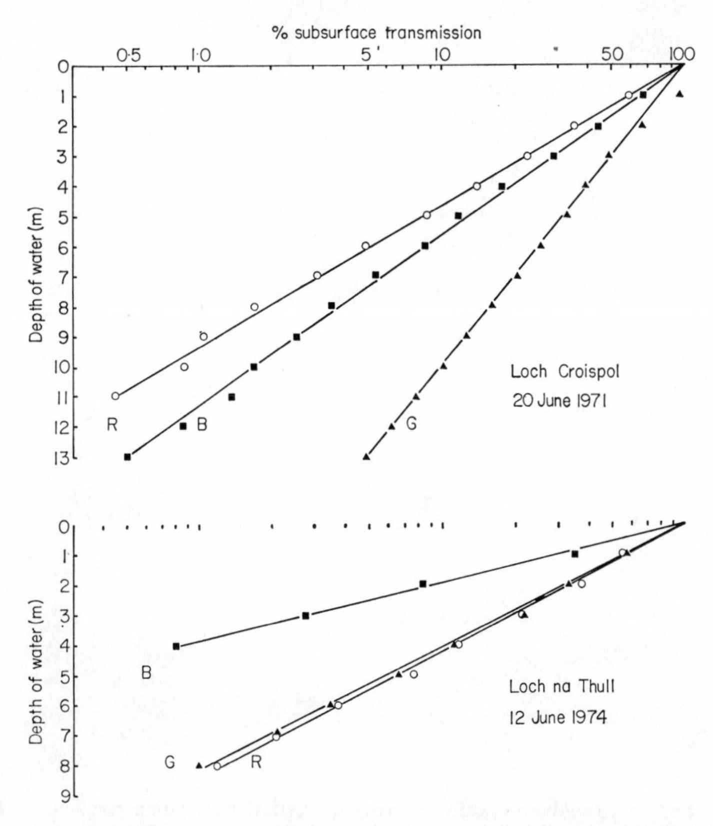

Figure 5.1. Downwelling light intensity measured using matched surface and underwater selenium photocells with Chance, optical glass, blue, green or red filters (see text). Readings have been converted to percentage subsurface transmission or percentage of transmission penetrating the water surface. Loch Croispol during bright sunshine between 16.10 and 16.45 h GMT on 22 June 1971 and Loch na Thull, with overcast sky and drizzle, between 12.15 and 12.45 h GMT on 12 June 1974. (For location of lochs and lakes in this and any subsequent figure see Table 5.3).

value of 2·9 for the over-enriched, plankton-dense Loch Leven which is based on integrated spectroradiometric data for *PAR* (Spence *et al.* 1971), like those of Fig. 5.2; this yields a 20-fold range which may be a minimal figure. To this must be added qualitative differences to which allusion has already been made in terms of colour.

In order to emphasize both these points, Figs. 5.3 and 5.4 and Table 5.4

show the qualitative and quantitative changes in light energy that occur over the depth range 0 to 1 m in the clear water of Loch Croispol which (Table 5.2) is comparable with Ennerdale, and in phytoplankton-dense Loch Leven.

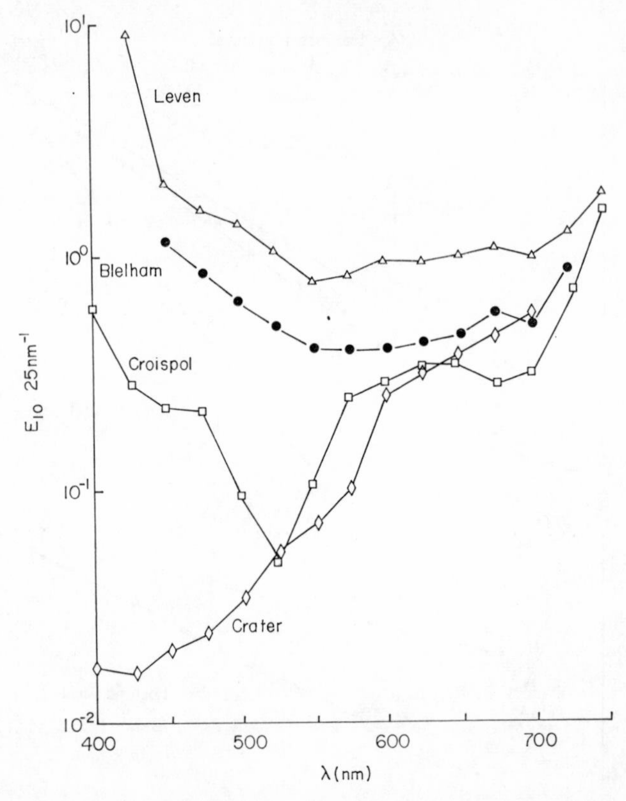

Figure 5.2. Spectroradiometrically determined vertical attenuation coefficients, E_{10}, of downwelling irradiance in Loch Croispol (4 August 1970) and Loch Leven (15 July 1970) for each 25 nm waveband from 390 or 412 nm to 762 nm; with data of Tyler and Smith (1967) for Crater Lake Oregon, where downwelling irradiance was corrected for upwelling irradiance, and converted data of Talling (1970) for Blelham Tarn, England. (Redrawn from Spence *et al.* 1971, Fig. 7).

The water values have been derived from original measurements with the spectroradiometer which gave estimates of E_{10} per 25 nm waveband of *PAR* (Table 5.1 and Fig. 5.2). Through use of the appropriate mean frequency value ($\bar{v}$) per 25 nm, derived values of irradiance in μW cm^{-2} 25 nm^{-1} (Spence *et al.* 1971) were converted to microEinsteins (μE) m^{-2} s^{-1} 25 nm^{-1}.

Table 5.4. Integrated values of *PAR*, as $\mu E\ m^{-2}\ s^{-1}$, at subsurface (Om) and 1 m in Loch Croispol on 4 August 1970, in Loch Leven on 2 July 1970, with estimates of light energy entering, and being absorbed by, a horizontally positioned, fresh leaf of *Potamogeton crispus* (for derivation of data, see text).

	$\mu E\ m^{-2}\ s^{-1}$ L. Croispol	L. Leven
subsurface Om	835	1155
1 m	485	92
entering leaf	480	90
absorbed by leaf	357	63
	percentage	
subsurface light energy absorbed by leaf	43	5

Light absorbed by underwater leaf

Estimates in Figs. 5.3 and 5.4 are also given of the distribution of light energy entering and being absorbed by a young, fresh leaf of *Potamogeton crispus* lying hypothetically in a plane parallel to, and at a depth of 1 m below, the water surface of each loch. Original transmission data are derived from those of Spooner (1971) who, with magnesium oxide as a reference standard, measured reflectivity of a few underwater aquarium plants, not including *P. crispus*. Mean reflectivity of a comparable leaf was about 1·4% with a maximal value of 3·30% at 550 nm, and a minimal red value of 0·2% at 660 nm. It is calculated that such a leaf placed at 1 m in the clear, blue-green loch would absorb about 43% of subsurface irradiance, as $\mu E\ m^{-2}\ s^{-1}$ *PAR*, compared with a mere 5% at the same depth in the darker water of Loch Leven (Table 5.4). The greater selective loss of available high energy blue light in Loch Leven is particularly notable.

Other Factors

Shade adaptations which may modify some of these effects are discussed presently; here it may be asked what other, extrinsic, modifying factors exist. One factor which could selectively reduce still further the light actually reaching the chloroplasts is the presence of periphyton but, as noted on p. 94, the photosynthetic relationship between senescence of leaves and the extent of periphyton has not yet been explored. Another factor which might be said seasonally to reduce available irradiance for wintergreen species (Sec. II) is the increased path-length of sunlight in water at low sun-angles in high northern latitudes; in Shetland for example on latitude 60 deg. N, the noon sun-angle at the winter solstice is 6·5 deg. but there is at such

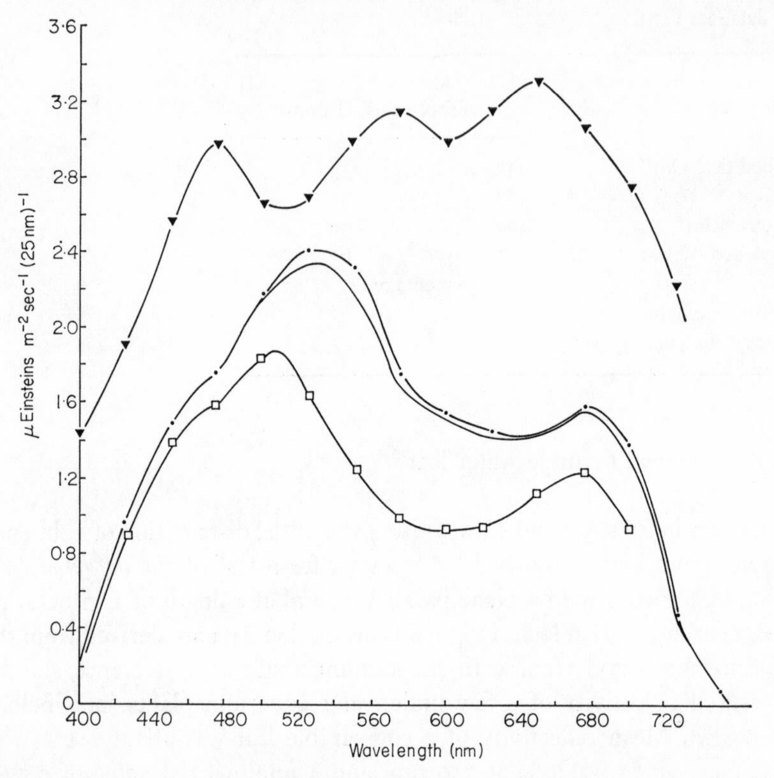

Figure 5.3. Irradiance as microEinsteins (μE) m^{-2} s^{-1} 25 nm^{-1} (400–700 nm) at sub-surface (Om) and 1 m in Loch Croispol on 4 August 1970, derived by use of appropriate mean frequency values ($\bar{\nu}$) from irradiance in μW cm^{-2} 25 nm^{-1} (Fig. 5.2). Estimates (see text) are also given of the variation per 25 nm, and totals in light energy entering and being absorbed by a young *Potamogeton crispus* leaf placed horizontally at a water depth of 1 m. Curves lie in the above order from the top to the bottom of the Figure.

angles a disproportionate amount of sky radiation. If this proportion decreases at the high solar angles of summer, so does the optical path-length of direct rays of the sun until at the sun's noon height it equals or exceeds that of the sky radiation, depending on the state of the sky (Smithsonian Meteorological Tables 1951, Hutchinson 1957). So workers like Poole & Atkins (1926, in Hutchinson 1957) ignore variation in vertical illumination due to height of the sun, and therefore regard optical light path underwater as fairly constant and the result mainly of sky radiation, and refer, as in this study, to vertical diffuse attenuation and its appropriate coefficients.

With reflectivity, however, low solar angle may be regarded as more critical. In his studies on Lake Hefner, Anderson (1952) treated reflectivity and upward scattering from sun and sky together, as R_t, in terms of solar angle, ψ_s'. For a range of meteorological conditions and solar altitudes he

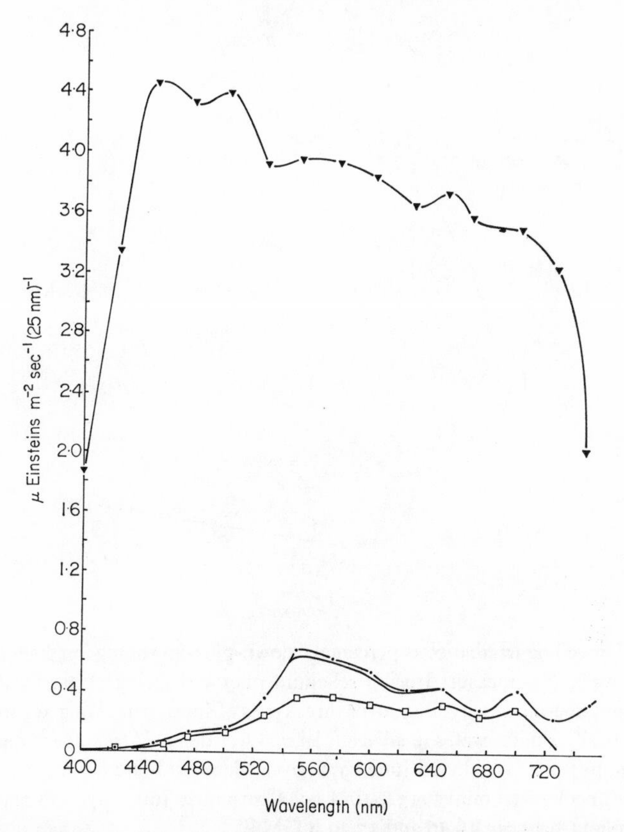

Figure 5.4. Irradiance as microEinsteins (μE) m^{-2} s^{-1} 25 nm^{-1} (400–700 nm) at sub-surface (Om) and 1 m in Loch Leven on 15 July 1970, derived as in Fig. 5.3 with, likewise, estimates of light energy entering and being absorbed by an identical *Potamogeton crispus* leaf placed horizontally at a water depth of 1 m, symbols as in Fig. 5.3. Curves lie in the above order from the top to the bottom of the Figure.

derived a series of empirical curves which related reflectivity and upward scattering as a percentage of total sun and sky radiation, of which that, say, for low cloud, overcast sky was:

$$R_t = 0{\cdot}2\ \psi_s'^{-0{\cdot}30}.$$

Under all observed meteorological conditions R_t lay around 5% at $\psi_s' > 40°$ but, at $\psi_s' < 10°$, with low cloud in an overcast sky, R_t rose to about 10% and reached as high as 30% in conditions of low scattered cloud. For considerable periods on either side of the winter solstice in northern latitudes around 57 deg., almost $\frac{1}{3}$ of total sun and sky radiation may be lost to the underwater environment. This effect of low solar angle may be countered in part by the state of the water surface as a reflector, a negligible factor where $\psi_s' > 15°$;

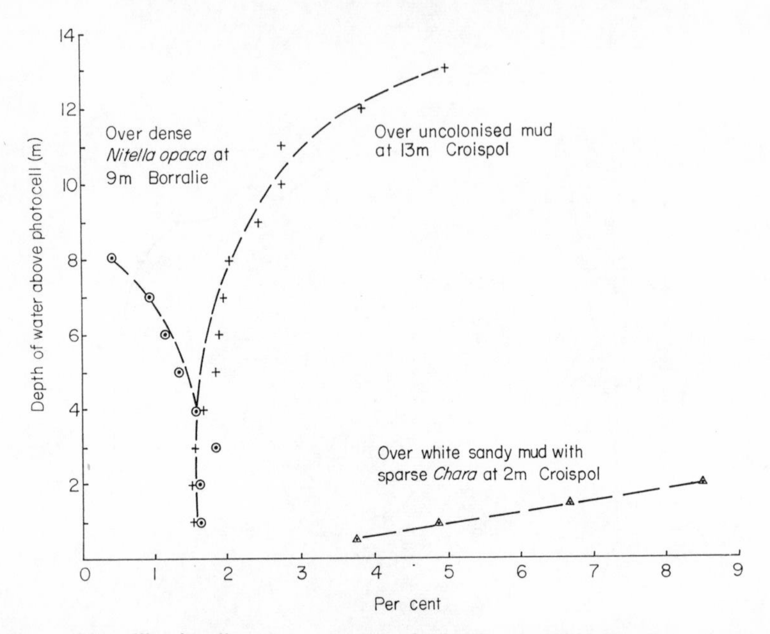

Figure 5.5. Upwelling irradiance as percentage down-plus-upwelling irradiance in relation to depth of water (m) measured using selenium photocell facing vertically downwards, with or without neutral density filter, over three types of substratum lying at three different depths in optically similar water in adjacent lakes. Over dense *Nitella opaca* bed (to 60 cm high), at 9 m, in Loch Borralie 20 June 1971; over white sandy mud with sparse *Chara* at 2 m and over uncolonized mud at 13 m, in Loch Croispol 22 June 1971. All readings taken in bright sunlight between 16.40 and 17.30 h GMT.

while Anderson (1952) could not attribute reflective differences to variation in wind velocity, Ångstrom (1925) established by experiment that strong disturbance of the water surface slightly decreased reflectivity at $\psi_s' < 15°$. But latitudinal differences of more than 10° even between European lakes (Table 5.3) do not seem to affect the depth of water reached by those macrophytes for example that penetrate furthest, species of *Fontinalis*, *Drepanocladus* and *Nitella*, and all of which are wintergreen (Sec. III and Table 5.6).

Finally there is upwelling irradiance. On a cloudless, sunny day, 22 June 1971, a selenium photocell was placed in Loch Borralie and in Loch Croispol facing vertically downward over various substrata in a range of water depths, with a matched surface cell. Upwelling attenuation was then measured. Each set of upwelling data was immediately followed by a set of downwelling readings so that upwelling irradiance could be estimated as a percentage of down- plus up-welling irradiance. Figure 5.5 shows that on light coloured sandy mud in about 2 m depth of water this percentage reaches 8, which is six times its value at the same depth in open water, and close to the eight times difference recorded by Oberdorfer (1928) between similar sites in the Bodensee. Thus scattering and upward reflection may considerably increase light availability to photosynthesizing tissue in shallow water, with open vegetation. In comparison the *Nitella opaca* and *Chara* cf *contraria* beds at 9 m in Loch Borralie are continuous and dense, with shoots about 0·6 m high. Efficiency of their light absorption may be gauged roughly by looking at the figures for uncolonized mud at 13 m in Loch Croispol which is adjacent and (Table 5.2) optically similar. Assuming the reflectivity of these uncolonized muds to be more or less alike, and a logarithmic decrease in upwelling irradiance with depth, a mud at 9 m would yield a value of around 6·5% against the figure of 0·5% for the tall, dense, *Nitella-Chara* stand at that depth.

Freshwater and woodland shade environment

Compared with woodland this shade environment has one frequent and one invariable difference. In the first case there may be wide variation within and between lakes in the season or seasons of maximal shade induced by phytoplankton and periphyton; this aspect is examined in Section III. The invariable difference, set out here, lies in the R/FR ratio underwater which may or may not influence certain photomorphogenetic responses in a fashion different from that encountered in terrestrial shade habitats. Red light (R) at 660 nm and far red (FR) at 730 nm are the absorption maxima for the interconvertible forms of phytochrome; the physiologically active form (or forms) of phytochrome (P_{fr}) being induced at 660 nm, the inactive form (P_r) at 730 nm.

Values of downwelling irradiance at these wavelengths, derived by interpolation from Fig. 5.2, are given in Table 5.5 for Loch Croispol and Loch Uanagan. In both lochs, light at 660 nm is cut out rapidly as within a tree canopy, but, unlike that situation, light at 730 nm is attenuated even more rapidly. As one consequence the R/FR ratio is always more than 1·3, the value of full sunlight, and marks a clear contrast between shade under water and a leaf canopy, where ratios range from 0·70 to 0·12 (Cummings 1963, Vèzina & Boulter 1966).

At the angiosperm depth limits of 4 m in Loch Uanagan and 6 m in Loch Croispol the extrapolated values of red (660 nm) and far red (730 nm) are

Table 5.5. Diffuse down-welling irradiance for red 660 nm and far-red 730 nm (W m^{-2} nm^{-1}) for Loch Croispol (4 August 1970) and Loch Uanagan (15 July 1970), and E_e values for 660 nm and 730 nm in these lochs.[1]

Loch	Depth (m)	W m^{-2} nm^{-1} R 660 nm	FR 730 nm	R/FR	E_e R 660 nm	FR 730 nm	R/FR (attenuation)
Croispol	0	2.32×10^{-1}	1.52×10^{-1}	1.5			
	1	1.08×10^{-1}	2.28×10^{-2}	4.7			
	4	7.20×10^{-3}	1.50×10^{-4}	48.0	0.74	1.73	0.43
	6	9.20×10^{-4}	1.60×10^{-6}	~ 575.0			
Uanagan	0	1.20×10^{-1}	8.2×10^{-2}	1.5			
	1	1.74×10^{-2}	4.8×10^{-3}	3.6	1.88	2.88	0.65
	4	7.00×10^{-5}	1.4×10^{-7}	~ 500.0			

[1] Data recalculated from Spence *et al.* 1971 p. 335.

very small because of the rapid attenuation of these wavelengths and, as calculated from Table 5.5, amount to about 0·4% subsurface red light, 0·001% far red in Loch Croispol and, in Loch Uanagan, only 0·06% subsurface red with almost no far red. In different terms the irradiance at 660 nm at 4 m depth in Loch Uanagan can at noon in summer be 10^{-5} times a surface value of 6 W m^{-2} nm^{-1}.

II Photomorphogenetic responses

The morphogenetic responses discussed here are those that may depend on light but not necessarily on day-length (or length of dark period) and the end results of which may or may not be interpreted as adaptations to this predominantly shade habitat. This should but cannot entirely eliminate bud dormancy, for example; in early autumn many aquatic plants of temperate regions produce various morphologically distinct buds, shoots with suppressed internode extension which are leafy or borne underground (Arber 1920, Sculthorpe 1967), although a number of species like *Potamogeton crispus*, *P. praelongus* (and *Nitella opaca*) can be wintergreen and form no such buds. In studies of bud dormancy in *Hydrocharis morsus-ranae*, *Utricularia vulgaris* and *U. intermedia*, Vegis (1953, 1963, 1964) has shown that growth inhibition leading to formation of these winter buds in early autumn, and the breaking of that dormancy in spring, are primarily responses to temperatures prevailing at these seasons; day-length is secondary. On the other hand, Webster (1975) has found that the predominantly vertical shoot extension of *Potamogeton obtusifolius* which signals the germination of its winter buds in spring is inhibited at natural temperatures by long day-length or high light intensity; there is both a photoperiodic and a shade response.

Anthocyanin production

There are several responses by submerged aquatic plants or by the submerged phases of floating leaved and emergent plants which may be attributed to the metabolically active (P_{fr}) form of phytochrome, the form which is induced by irradiation at 660 nm.

The first of these is the red or red brown colouring of stems and leaves produced in shallow water by, for example, *Potamogeton perfoliatus*, *P. gramineus*, *P. crispus* and *Littorella uniflora* where the plants themselves are rooted in, say, less than 25 cm water or, in the case of *P. lucens*, in leaves produced near the water surface from plants rooted in 2 or more metres depth of water. This red colour contrasts with wholly green plants or leaves produced by these species at greater depths in the same water bodies and appears to be due to anthocyanin. A number of authors have produced evidence that anthocyanin synthesis is mediated by phytochrome in a variety of plants; e.g., Vince & Grill (1966) in turnip seedlings, and, particularly, Hachtel (1972) in *Oenothera* seedlings and Mohr (1972) in subepidermal hypocotyl cells of *Sinapis vulgaris* seedlings have shown that this anthocyanin synthesis is a positive response to P_{fr} and results from short induction in red or from continuous irradiation in a standard FR source (3500 W m^{-2}) which, where photomorphogenesis is concerned, replaces high irradiation white light. The function of anthocyanin may be protective and/or the fortuituous result of photosynthesis in high light and low nitrogen conditions such as might occur in plants rooted in sandy, coarse-particled soils underlying nutrient-poor, shallow water.

Parastrophic and diastrophic position of chloroplasts

Apart from mid-ribs the leaves of submerged plants are not more than 3 cells thick, with chloroplasts in all layers. In this simple structure, parastrophic and diastrophic positions of chloroplasts can be directly observed in whole leaves under the light microscope. Thus for *P. crispus* collected from Loch Drumore, Glenshee, in March 1974, leaves in 0·25 m water had chloroplasts in the marginal or parastrophic position in surface view whilst leaves from 2·50 m depth of water exhibited the diastrophic position. In effect this means less transmission of light at depth, since spectrophotometric examination over the range 400–700 nm showed 17% transmission by a leaf with parastrophic chloroplasts, and 12% for the deep-water leaf with chloroplasts in the diastrophic condition. The parastrophic or marginal position of chloroplasts is known to be induced in certain mosses by red-promoted P_{fr} causing the appropriate movement of the cytoplasm (Mohr 1972).

Germination

A few submerged macrophytes are known to have a positive light requirement for germination and the problem was studied in some detail in *Potamogeton schweinfurthii* and *P. richardi* (Spence *et al.* 1971). The drupes or achenes of these tropical species can float for long periods, maintaining buoyancy through photosynthesis and an intact cuticle, and germination only occurs in nature once the fruits have sunk. In the laboratory neither species germinated in darkness compared with 100% germination of light-stored, sunken fruits or with similarly stored fruits from which epicarp and mesocarp had been scraped. Substantial germination of both freshly-shed and older fruit also occurred in the light after surgical exposure of the embryo. No after-ripening period was required. Whilst evidence for the promoting effect of red light on *P. schweinfurthii* germination was presumptive, it was established for *P. richardi* that germination of exposed embryos was promoted by 0·05 J cm^{-2} red light at 659 nm and reversed by a similar dose of far-red at 730 nm. Intact fruit needed much higher doses, administered for a 12 hour day over 5 days and amounting to 8 to 10 J cm^{-2}. Photoreversibility by FR then needed 13 J cm^{-2} but such a level, relative to red irradiance, could not be achieved underwater in nature; nor therefore could photoreversibility. It is possible, however, that each aquatic species has a minimal red requirement for the promotion of germination which, in the absence of a marked capacity for vegetative spread, would then influence its maximal depth in a particular light climate and, therefore, zonation.

Specific leaf area

Flowering plants have the capacity in varying degree to adjust their leaf morphology to sun and shade conditions by altering their thickness and their specific leaf area (SLA: leaf area per unit leaf dry weight). Sun leaves are thick with low SLA, shade leaves thin with high SLA. The specific leaf area of various *Potamogeton* species increases with depth of water but the incremental increase per unit depth is typical of the water body, and hence of the prevailing light climate, not of the species (Spence *et al.* 1973). One important physiological consequence of a high SLA is a lower light compensation point which is achieved largely as a result of a reduction in the rate of dark respiration of leaves per unit leaf area (Fig. 5.6) while their photochemical capacity remains relatively invariable (Spence & Chrystal 1970). *Potamogeton polygonifolius* and *P. obtusifolius* had earlier been found in nature to have mutually exclusive ranges in rooting depths: that is, ranges in depth of water above the substratum in which they were rooted. Spence & Chrystal (1970) found in addition that these species produced mutually exclusive specific leaf areas when grown in sun and shade conditions in a glasshouse (Fig. 5.6), which suggested that such inherent ranges in SLA might be of adaptive significance.

Subsequent field measurements of SLA are unavailable for two such species with mutually exclusive specific leaf areas because those studied in nature had overlapping ranges in rooting depth like *P. obtusifolius* and *P. perfoliatus* in Loch of Lowes (SLA respectively, and including standard error, of 1·72 to 1·85 and 1·20 to 1·90 cm^2 mg^{-1} in June 1970): or distinct ranges in rooting depth but overlapping ranges in vertical shoot length

(Spence *et al.* 1973). It follows however that, since magnitude of SLA is correlated with physiological function, possession by two such species of different intrinsic ranges in SLA should lead to their occupying different depth zones.

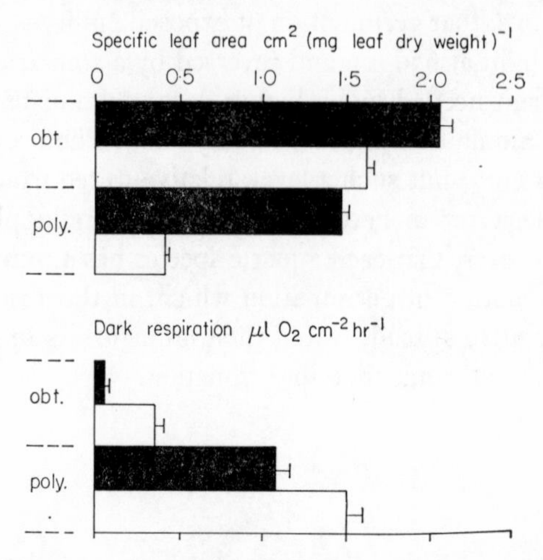

Figure 5.6. Specific leaf area (cm² mg⁻¹ leaf dry weight), with standard error, of leaves of *Potamogeton obtusifolius* (obt.) and *P. polygonifolius* (poly.) grown in uniform sun or shade (■) conditions in a glasshouse; together with their oxygen uptake (µl O_2 cm⁻² h⁻¹), with standard error, in the dark in 2–3 h experiments using sun and shade leaf discs or detached leaves: in the Warburg apparatus with 0·1 M Warburg buffer No. 11 at 20°C (data drawn from Tables 2 and 5 in Spence & Chrystal 1970).

Precisely what effect of light, or whether light alone causes variation in SLA underwater is not clear. Field evidence for the wide-ranging species *P. perfoliatus* indicates that, despite big differences in the quality of light and in particular of the R/FR ratio at the deeper limits of the species in a clear lake like Loch Croispol, and a brown-water lake like Loch Uanagan (p. 97), the maximal values of SLA are not significantly different; yet the difference is significant between the limestone population and that of another brown-water lake, Loch of Lowes (Fig. 4 in Spence *et al.* 1973). Such a difference might result from genetic or nutritional causes insufficiently expressed between plants from Loch Uanagan and Loch Croispol. However, although Loch of Lowes is the same colour as Loch Uanagan, in terms for example of its E_B/E_G ratio, it is darker (Table 5.2) and, in so far as light rather than any internal or other, external, factor determines specific leaf area, it presently seems that the amount of *PAR* exerts the predominant effect on *P. perfoliatus*. Total FR in relation to low values of R/FR ratio may be important in controlling the specific leaf area of *Impatiens parviflora* (Young 1975, this volume, 154) but it has already been noted that the amount of FR available to submerged plants in waters of the extremes of colour represented by Loch Croispol and Loch Uanagan or Loch of Lowes is negligible except near the water surface, and that the R/FR ratio in this shade habitat is greater than it is in full sunlight.

Vertical elongation of leaves, shoots, petioles

Variation of relative leaf length (length/breadth) of *Lobelia dortmanna* in light of different intensity and quality was studied under laboratory and field conditions by Åberg (1943). Relative leaf length was inversely correlated with the intensity of artificial 'white' light and, in a series of lakes with natural populations of the species, relative length increased more rapidly with depth the greater the overall light attenuation. This rise in attenuation through a series from clear to dark water also involved, predictably from Sec. I, a shift from blue-green to brown water but the results of laboratory experiments with coloured broad-band pass filters did not provide clear-cut evidence of the role of light quality.

The question of the control of leaf, shoot and petiole elongation is linked with the larger question of growth by these underwater plants. Considering only the genus *Potamogeton* and vertical length of mature shoots there are three categories of species: (1) short, less than 0·3 m, (2) tall, 1 to 3·5 m or more, and (3) variably elongate. Short-shoot species occur either in deep or in shallow water, tall-shoot species only in deep water while those in the third category have a widespread depth range. Further comment is confined here to those species with variably elongate shoots where, in consequence, mature length may be determined primarily by exogenously induced, rather than strictly endogenous, factors.

The internode elongation exhibited by terrestrial plants in the dark or in shade and the reduced elongation found in full sunlight is also shown by submerged aquatic species like *Potamogeton obtusifolius* (p. 101) and *P. crispus* which occur in both shallow and deep water. There is some evidence from plants of *P. crispus* grown experimentally at similar depths that conditions of low or high *PAR* respectively promote or inhibit internode elongation. A common feature of the laboratory shade and the natural, deep-water, environment would be low levels of red irradiation which could result in a typical phytochrome-induced response, namely, the removal or poor production of P_{fr} which in turn permits, for example, hypocotyl extension in *Sinapis* seedlings (Mohr 1972); this may be compared with P_{fr} production and the inhibition of such extension in red light or sunlight.

There is, however, an altogether different explanation for this response by variably elongate submerged aquatic plants which, like the explanation involving light, is consistent with known properties of the natural environment and of the plants themselves. Musgrave *et al.* (1972) reported that ethylene naturally promotes stem elongation in the aquatic plant *Callitriche platycarpa* which produces first submerged leaves and then, in due course, rosettes of floating leaves at the water surface. Normally short, aerial internodes elongate more, and faster, if submerged. In these reactions Musgrave *et al.* (loc. cit.) implicated ethylene and gibberellic acid, on which the response to ethylene depends, by postulating internal equilibrium concentrations of ethylene which in submerged tissue are high enough to enhance cell extension by natural levels of gibberellic acid, but which drop low enough in emergent tissue to slow down such elongation because of the increased diffusion rate of ethylene in air. Petioles of the aquatic *Ranunculus sceleratus* react to submergence like *Callitriche* stems; Musgrave & Walters (1973) found that they elongate rapidly in ethylene, and that the internal concentration in nature was high enough to account for the response to submergence for which, again, endogenous GA was essential.

Thus submergence has led aquatic plants to an environment which allows an increased equilibrium concentration of endogenous ethylene and causes a reduction in red irradiation. As with interactions of light and ethylene in the bean shoot apex (Abeles 1973) it should soon be possible to distinguish in aquatic plants between effects on stem, petiole and leaf elongation which are stimulated primarily by light and those stimulated primarily by submergence itself. Admittedly increase in atmospheric pressure with water depth must enhance the solubility of ethylene and so reduce its internal equilibrium concentration but, since angiosperms are not known to colonize any fresh waters much beyond a depth of 6 m (Section III), such a decrease in concentration can only result from an increase in external pressure of 0·6 Atmos. At least for species with mature shoots of which the length varies with water depth such variation may therefore be interpreted as a response to red light mediated through phytochrome as P_{fr}, and/or to internal ethylene levels mediated through GA. Similar possibilities exist for causes of the variable petiole and leaf length in a number of species.

Shade adaptation and zonation

Summing up the results of our discussions so far: light is known or deduced to affect a number of morphogenetic processes in submerged aquatic plants including the germination of fruits, anthocyanin production in stems and leaves, the diastrophic and parastrophic position of chloroplasts and, through its influence on such features as specific leaf area in angiosperms, on rates of photosynthesis. Vertical shoot extension, especially in variably elongate species, may be controlled by light and/or by endogenous ethylene and gibberellic acid. The forms these attributes take with increasing depth of water can be called shade adaptations. Although in terms of production it is still necessary to assess the extent to which shade adaptation in a single species improves net photosynthesis at reduced levels of *PAR* in nature over rates that would exist without such adaptation, it seems likely that, apart from limitations imposed in shallow water by excessive turbulence and in deep water by unsuitable substrata, the differing ability of a range of species to respond to shade must largely determine zonation.

Together with a given underwater light climate and pool of nutrients, including inorganic carbon, the degree of their shade adaptation may also influence the maximal depth to which attached plants penetrate in a given lake whilst the mature height and form of larger species must set for them an optimal water depth for biomass and production. How far this is achieved in a particular lake may, again, largely depend on what light is available at that optimal depth. An attempt is made in the remaining sections to embody these ideas in testable hypotheses.

III Depth of zone colonized by macrophytes Zc

Attention has already been drawn to the variable difference between the woodland and underwater shade environments, namely, in the season or seasons of maximal shade which is induced by the phytoplankton and periphyton within and between different lakes. Much quantitative phytoplankton data now exists but there is a dearth of comparable information on periphyton biomass and periodicity so that quantitative arguments are applied hereafter to relationships between macrophytes and phytoplankton only.

Hypothesis relating phytoplankton density and macrophyte performance

A hypothesis is put forward that a key to the performance of macrophytes is the extent and consequence of their competition, principally for light, with phytoplankton. It is postulated that, in a series of lakes possessing increasing quantities of phytoplankton the effects of their competition on macrophyte growth and performance, in the widest sense, will become increasingly severe and, from year to year, increasingly variable. If this postulate is correct it should be possible in some circumstances directly to relate aspects of macrophyte performance such as depth of water colonized (Z_c) or depth at which the maximal biomass occurs, to effects of the phytoplankton population, and to define the limnological conditions where Z_c or biomass maxima are predictable and those where they are not.

Light is less involved to the extent that physiological activity of the phytoplankton, such as photosynthesis, alters the water chemistry to the detriment of the higher plants; and the overall postulate is modified in as much as adaptations by higher plants to excessive or to very low turbulent flow, to shade, to variations in light quality and in form of exogenous carbon supply can be said successfully to counteract the effects of phytoplankton.

Errors in estimates of depths colonized

Bearing in mind the range in E_{PAR} from 0·14 to around 2·9 or more (p. 99), the response of attached plants is examined first in terms of the depth of the zone they colonize in standing waters in the United Kingdom. There is an initial difficulty here; to decide whether a given record refers to properly rooted or attached material which is actually growing at that depth, or whether it is a fragment which has been uprooted in shallower water and has subsequently, on being waterlogged, sunk from the surface or been carried by down-currents to greater depths where it may exist, perhaps below its light compensation point, for a considerable period during the summer. Some records obtained by grab or grapnel from a boat may well include such material (e.g. Loch of Lowes, Table 5.6).

Unless the data have been gathered on a detailed survey of small areas such as those carried out by Pearsall (1918, 1920) on parts of the Lake District lakes, the most satisfactory records for vegetation that is invisible from the water surface are probably those collected by divers who should be aided by the fact that, with uniform topography, the lower limit of more or less continuous macrophyte vegetation forms a sinuous line varying in depth by less than 0·5 m. Thus in Loch of Lowes (12 August 1969) the 1%-cover line, of *Potamogeton obtusifolius*, lay at a water depth of 3·6 m with less than 1% at 3·9 m. While at 5 m there were some rooted, at 6 m a few unrooted, fragments of *P. obtusifolius* and even fewer, sparse shoots of *Elodea* at 7·6 and 9·2 m, the macrophyte limit in this loch was set at 3·9 m.

Assessment on this basis is easiest where the slope is steep and the zone over which plant cover decreases with depth to around zero is compressed laterally into a few metres, so that vegetation 'cut-off' seems clear, but on gentle slopes the equivalent lateral distance may be 10 m or more, which introduces another subjective error, possibly compounded when the observers are not botanists and extent of rooting is questionable. Moreover ultimate depth colonized may vary widely on different shores according to the nature of their sediment, most obviously in a lake with oversteepened sides and gently sloping ends (viz. Loch Baile na Ghobhainn (p. 118) and, finally, available light and field of vision at these depths are limiting. Underwater sampling like that from a boat can therefore be inadequate but it is less likely to be incorrect.

Accuracy of the underwater estimates of Z_c can indeed be tested objectively for three Scottish lochs where biomass or % cover has been measured in relation to water depth (Sec. V and Fig. 5.10): namely, the point of intersection of the y axis (depth) by extrapolation of the straight line logarithmic plot of standing crop. The difference between the two types of estimate, in footnote to Table 5.6, lies within 10 to 30 cm, or 3 to 7%, of the estimates of Z_c by diver. None the less with the potential sources of error in data of mixed origin used in this study, any subsequently argued relationships must be treated as first approximations.

Estimates of depth zone colonized by angiosperms, and/or Z_c

In eighteen of the twenty-three lakes listed in Table 5.6, bryophytes or charophytes colonize as deeply as, and mainly beyond, higher plants and most of these records are for *Nitella opaca*. *Sphagnum subsecundum* var *auriculatum* penetrates to 15 m, the maximal depth of the east basin of Loch Avon, while *Pohlia* sp and *Nardia compressa* reach 20 m on stones in Loch Coire an Lochain (J. J. Light *in litt.*); both these are nutrient-poor lochs at altitudes over 700 m in the Cairngorms. These figures are exceeded by the depth of 30 m to which *Drepanocladus* cf *aduncus* penetrates in nutrient-poor lakes of South Georgia (Light & Heywood 1973) which are on almost the same (southern) latitude as Loch Avon (Table 5.3) and have (J. J. Light *in litt.*) quite similar ice cover; and by 35 m for *Marsupella aquatica* and *Drepanocladus exannulatus* in Latnjajaure, Sweden (Bodin & Nauwerck 1968). *Drepanocladus* sp. also forms beds with *Fontinalis* sp. at 20 m in Crystal Lake, Wisconsin (Fassett 1930, Juday 1934) some 10° south of Loch Avon, whilst *Fontinalis* reaches 14 m in the clear Lunzer Untersee (Brehm & Ruttner 1926) which too is 10° south of Loch Avon.

In Europe, neither vascular cryptograms nor flowering plants appear to exceed a depth of 6 m: like *Isoetes* in Wastwater and *Potamogeton praelongus* or *P. perfoliatus* in Loch Borralie and Loch Croispol or *Elodea* in the Lunzer Untersee. Even in high-altitude (1950 m) equatorial lakes in Uganda, these angiosperm depths are barely exceeded: low-growing *Ceratophyllum demersum* (apparently replacing mosses or *Nitella*) reaches 8 m in Lake Bunyoni and *Potamogeton schweinfurthii*, which resembles *P. praelongus*, occupies the deepest macrophyte zone at 6 m in nearby Lake Mutanda (Spence *et al.* 1971a). In the Mediterranean, by contrast, the marine angiosperm *Posidonia*

Table 5.6. Maximal depth of water (m) colonized by attached, submerged angiosperms or vascular cryptogams and Bryophyta or Charophyta in 23 freshwater lakes. The greater depth in either column for any lake represents depth zone colonzied by attached macrophytes forming more or less continuous patches of vegetation, Z_c (m), in that lake. Data gathered by diver (D), or from boat (B) by grab or (less often) by observation also. Localities of all lakes are given in Table 5.3.

Lake	*maximal depth (m)* Angiosperm or Vascular Cryptogam	*maximal depth (m)* *Drepanocladus* (D) *Fontinalis* (F) *Pohlia-Nardia* (P) *Sphagnum* (S) *Nitella* (N)	Data gathered by diver (D) or from boat (B)	Notes
Lakes on South Georgia		30·0 D	D	L. & H.[1]
Loch Coire an Lochan		20·0 P	D	J.J.L.[1,3]
Crystal Lake, Wisconsin		20·0 DF	B	F., J.[2]
Loch Avon		15·0 S	D	J.J.L.[1,3]
Lunzer Untersee	6·5	14·0 F	B	B. & R.[2]
Loch Baile na Ghobhainn		6·0 DFN	D	
	6·5	13·1 F	BD	G.W.[2,3]
Wastwater	6·0	12·2 N	D	J.W.G.L.[1]
		8·0 F	BD	W.H.P.[2], B.P.J.[1]
Würmsee (Starnberger See)		12·0 N	B	F.B.[2]
Loch Borralie	6·0	10·7 N	D	[3]
Ennerdale Water		10·0 N	B	W.H.P.[2]
Loch Croispol	6·0	(3·0)	D	[4]
	4·2	8·0 N	B	D.H.N.S.[2]
Derwentwater	5·0	6·0 N	B	W.H.P.[5]
Lake Windermere	4·6	5·7 N	D	J.W.G.L.[1]
		7·0 N	B	W.H.P.[2]
Loch Lanlish	4·5	(2·0)	D	[4]
Loch Clunie	4·0	4·3 N	D	
Loch Uanagan	3·9	4·1 N	D	[6]
Loch of Lowes	3·9	3·9 N	D	[6]
		5·0 N	B	D.H.N.S.[2]
Loch Spiggie	3·2	3·2 FN	D	[6]
Lake of Menteith	3·7	3·7 N	D	
Esthwaite Water	3·1		D	J.W.G.L.,[1]B.P.J.[1]
Loch Leven	1·0	1·0 N	D	
Lake Bunyoni, Uganda	8·0	(6·0)	B	D.H.N.S.[4,5]
Lake Mutanda, Uganda	6·0	(5·0)	B	D.H.N.S.[4,5]

oceanica forms stands at 40 m (J. N. Lythgoe and E. A. Drew, *in litt.*). From the limited evidence it seems that differences in latitude of more than 10° affect ultimate depth penetrated by mosses or *Nitella* less than light attenuation including winter ice cover (J. J. Light *in litt.*) within particular water bodies, and that the same applies, over more than 5 times the latitudinal range and with much smaller depth limits, to freshwater angiosperms.

Now Oberdorfer (1928) noted in the Bodensee that flowering plants, mosses and algae, first *Nitella* species and then micro-algae, successively replaced each other as the substratum penetrated ever deeper water. A similar sequence was observed by Brand (1896) in the Würmsee, Bavaria, and by von Brehm & Ruttner (1926) in the Lunzer Untersee, Austria, where however *Fontinalis* formed, to 14 m, the lowest macrophyte zone. Thus the British examples of Table 5.6, excepting those where the 'lower-plant zone' of *Nitella* or *Fontinalis* is apparently missing, conform to the pattern already established on the European continent; in addition attached algae (other than *Charophyta*) have been observed to form mats at 30·5 m in Wastwater, on stones (J. W. G. Lund, *in litt.*), and to 20 m in early spring, on mud, in Loch Croispol. There do not appear to be any published studies of British lakes like the sequences of attached microalgae described by Oberdorfer (1928) for the Bodensee where in 1926 they reached 30 m.

Possible limitations by substratum and/or temperature

In Loch Croispol and Loch Lanlish angiosperms form the base of Z_c in

---

[1] L. & H. (Light & Heywood 1973); J.J.L. (J.J. Light *in litt.*); J.W.G.L. (J.W.G. Lund, *in litt.*) and B.P.J. (B.P. Jupp, *in litt.*). Jupp, in surveys of the northwest and north shore of Wastwater (July 1974), puts Z_c, with *Fontinalis*, at 8 m, like Pearsall; loose, moribund fragments of *Fontinalis* and *Myriophyllum* occurred to 15 m, but no *Nitella* was noted, so area may be different from that sampled by Lund's divers.

[2] by grab, from boat: F. (Fassett 1930), J. (Juday 1934), B. & R. (Brehm & Ruttner 1926), W.H.P. (Pearsall 1918), F.B. (Brand 1896), D.H.N.S. (Spence 1964), G.W. (West 1910).

[3] data available by diving since 10 m was given as limit of Z_c in Scotland (Spence 1972, p. 491). Deeper Z_c value for Loch Balnagowan, cited by West (1910) quoted in Spence (1964 1967) and confirmed by diving in 1974; at narrow ends of loch only. Value of 15 m for L. Avon is also deepest part of east basin.

[4] figures in brackets represent maximal depth of occurrence of, respectively, *Chara* and *Nitella* in Loch Croispol and Loch Lanlish and of *Chara* in Lakes Bunyoni and Mutanda.

[5] detailed maps of Pearsall (1920) and Spence *et al* (1971a).

[6] straight line plot of log percentage maximal crop, and water depth, fitted by eye, in Fig. 5.10 gives Z_c of 4·4 m for Loch Uanagan, 3·7 m for Loch of Lowes and 3·3 m for Loch Spiggie.

areas sampled by diver. Beyond about 3 m depth of water the entire shores become very steep and this steepness may relate to the fact that the lochs are formed in limestone which means that the slopes are surfaced at least by unstable, flocculant mud, the particles of which are readily carried downslope once they are disturbed. (Little is known of rates of sedimentation under natural conditions but Pearsall (1921) noted that *Nitella opaca* in the Lake District lakes was unable to withstand too rapid sedimentation; this might even operate by inhibiting a positive light requirement for germination.)

The contrast is particularly striking between adjacent Lochs Croispol and Borralie; between 3 m and 6 m in the first loch there is a steep mud slope which supports an intermittent cover, mainly comprising tall *Potamogeton perfoliatus*, on an otherwise bare substratum while in early spring there is a fairly continuous mat of unicellular and filamentous algae on the flatter slope from about 7 to 13 m. Most of the Borralie shore slopes gently and at equivalent depths carries, all the year round, dense beds of *Chara contraria* and *Nitella opaca* among the angiosperms from 2·4m to 6 m and, beyond them, to 11 m. Now both lochs have very similar optical properties to at least 13 m depth of water (Table 5.2) so it follows that Z_c is limited in one loch by factors other than light, and the same conclusion may apply over much of Loch Baile na Ghobhainn which is optically similar to Loch Croispol (Spence *et al.* 1971).

Because West (1910) using a boat had recorded *Fontinalis* as abundant down to 13 m in limestone Loch Baile na Ghobhainn, parts of its shores excluding the narrow ends were sampled by diver in June 1969; vegetation was found to stop abruptly on the uniformly steep, very unstable, slopes at 6 m (Table 5.6, and Spence 1974). Subsequent inspection in August 1974 of one of the narrow ends of this loch showed that vegetation with 100% cover did indeed stop at around 6 m depth of water but that, on this gentler slope, *Fontinalis* continued to form diminishing numbers of discrete patches down to West's reported, though unlocalized, limits of 13 m.

On the 1969 visit and alone of a number of lochs in which summer temperature gradients were studied relative to water depth, the base of a well-defined epilimnion coincided with the apparent lower limit of vegetation, namely 6 m. On the 1974 visit, however, and in all the other cases, like Loch Spiggie, there was no thermal stratification or, as in for example Loch of Lowes, the epilimnion extended well below Z_c (Spence 1974).

Z_c, water clarity and water chemistry

Again excepting the situation where substratum may be involved, a comparison of examples common to Tables 5.2 and 5.6 suggests that Z_c relates to the clarity of the water. Thus Loch Leven which has the least Z_c in this series also achieves the highest value of E_{PAR}, 2·9. Like Talling (1971) for Esthwaite Water, Bindloss (1974) has demonstrated a positive correlation ($r = 0{\cdot}84$) for Loch Leven between E_{min} and B, where E_{min} is the minimal value for E_{PAR} (400–700 nm) in ln units per m and B is the phytoplankton density, in mg chlorophyll *a* m^{-3}. Indeed, where B exceeded 200 mg chlorophyll *a* m^{-3}, the phytoplankton absorbed more than 75% *PAR*.

From this relationship between E_{min} and B, and because the depth of the euphotic zone, Z_{eu}, approximately equals $3{\cdot}7/E_{min}$ (Table 5.1), it follows that Z_{eu} should vary with B, the phytoplankton density. As a starting point let us assume that in the broadest terms the nutrient content of the water primarily determines the phytoplankton density and that, of possible variables, the alkalinity of the water, or meq ($HCO^-{}_3$+$CO_3^=$+OH^-) l^{-1}, is likely to be a satisfactory single estimate of its nutrient status for phytoplankton (see, for example, Talling 1970 Fig. 2 p. 434) as for higher plants (Spence 1964, 1972), although it is inapplicable in marl or limestone lakes. Alkalinity of British freshwaters has a maximum approaching 3·5 meq l^{-1}.

Lakes are arranged in Fig. 5.7 in order of increasing alkalinity on the x axis with, on the y axis, values of Z_c (from Table 5.6) and of B. There are only one or two estimates of B in several lakes such as Loch Borralie, where E_{min} is consistently low, at least between February and November, and so by inference is B, while it is likely that Loch Avon and Wastwater too do not vary much in this respect. This may not, however, apply to Loch Spiggie and Loch Uanagan which are also represented here by single samples. Allowing for such uncertainties it is evident that variation within any one lake may become very large above about 20 mg chlorophyll *a* m^{-3} or, excepting limestone Loch Borralie, at alkalinities above 0·3 mM. In addition to the general increase in values and range of B with increasing alkalinity there is also, again excepting Loch Borralie, a general decrease in Z_c.

Two proposed lake categories with regard to macrophyte performance

Two basic if almost certainly intergrading categories of lake are proposed. The first comprises those lakes where phytoplankton density is always low and relatively invariable throughout the year so that interference by phytoplankton with the macrophyte environment should be minimal in terms of its altering the light reaching the shoots or (Sec. IV) the water chemistry round them. Such limitation of phytoplankton density is imposed, it is suggested, by water chemistry and/or by mixing in lakes where, because of physiography and the fact that $Z_{eu} < Z_m$ (Z_m being the mixed zone, in m), individual phytoplankton organisms are exposed to insufficient light.

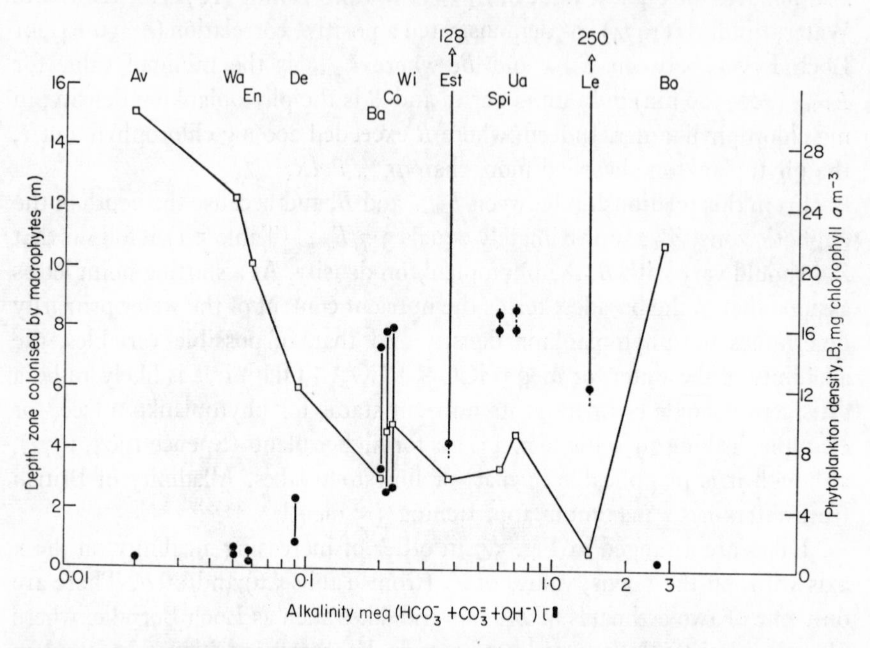

Figure 5.7. Depth zone colonized by macrophytes, Z_c, in m (□) and phytoplankton density, B, in mg chlorophyll *a* m^{-3} (●), in a number of lakes or lochs in the United Kingdom in relation to alkalinity of the water (meq $HCO_3^- + CO_3^= + OH^-$ l^{-1}). Symbols: Av (Avon), Ba (Bassenthwaite), Bo (Borralie), Co (Coniston), De (Derwentwater), En (Ennerdale), Est (Esthwaite), Le (Leven), Spi (Spiggie), Ua (Uanagan), Wa (Wastwater), Wi (Windermere). Derivation of data: for Z_c values, except those for Bassenthwaite Lake and Coniston Water, see Table 5.6; Bassenthwaite from boat, *fide* Pearsall (1918) and Coniston by diver survey (Dr. J.W.G. Lund *in litt.*). Phytoplankton density values for the English lakes from Jones (1972) and for Loch Leven from Bindloss (1974), except upper value for Esthwaite (Talling 1971) and for Coniston (Dr. J.W.G. Lund, *in litt.*); remainder, own data.

Since depth of euphotic zone, Z_{eu}, varies largely with phytoplankton density, *B*, it follows that in lakes of this first category Z_{eu} should, like *B*, have a relatively constant value. Whence Z_{eu} becomes invariable relative to the more or less static term Z_c and, at any instant of time,

$$Z_c \propto Z_{eu} \qquad \text{or } Z_c = k \, . Z_{eu}, \text{ where } k \text{ is a constant.}$$

This category is arbitrarily defined to include poor-water lakes having alkalinities <0·3 mM, and highly calcareous lakes, with maximal value of $B < 20$ mg chlorophyll *a* m^{-3} (Fig. 5.7); 'relatively invariable' phytoplankton density then means that a factor of <3 separates maximal and minimal values in a given lake in any one year. In these circumstances, as shown in Fig. 5.8, and substituting Z_c for y and Z_{eu} for x in the linear regression equation y = ax+b,

$$Z_c = a.Z_{eu}+b;$$

whence, from the values in Fig. 5.8,

$$Z_c = 0{\cdot}37\ Z_{eu}+1{\cdot}44$$

or $$Z_{eu} \approx 2{\cdot}7\ Z_c$$

and, since $Z_{eu} \approx 3{\cdot}7/E_{min}$

$$Z_c \approx 1{\cdot}4/E_{min}.$$

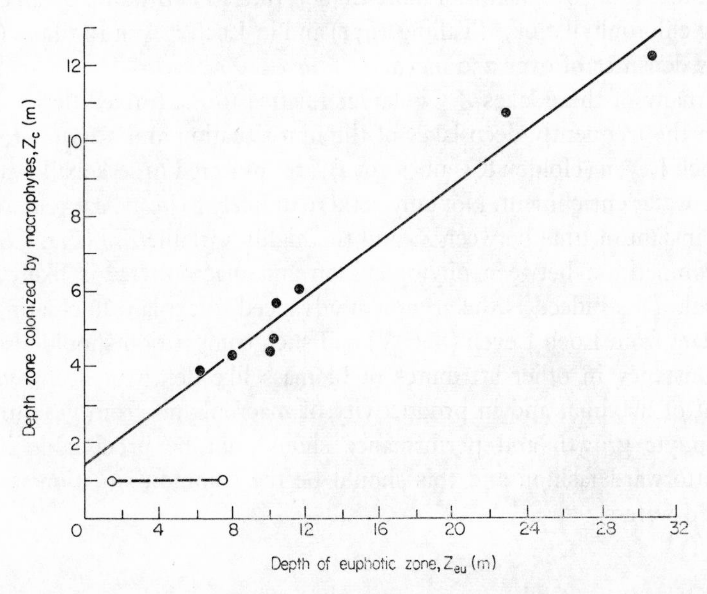

Figure 5.8. Depth zone colonized by macrophytes, Z_c (m), in relation to depth of the euphotic zone, the zone receiving an estimated 1% subsurface *PAR*, Z_{eu} (m). Z_{eu} data from Table 2 (except Leven: Bindloss 1974), Z_c data from Table 5.6 using only those of the latter established by direct observation through diving, since 1957, or in Derwentwater through detailed mapping (Pearsall 1920). Localities: Wastwater, Derwentwater and Windermere in the English Lake District and the Scottish Lochs Borralie, Clunie, Leven (○ — — — — ○), Lanlish, Lowes and Uanagan. E_{min} basis for Z_{eu} (p. 95) derived from Chance OG or Schott VG filters.

Thus the depth zone colonized by macrophytes, represented at their lower limits mainly by *Nitella* but also by *Fontinalis* and *Drepanocladus* species, is about 1/2·7 × the euphotic zone or, broadly, the depth zone occupied by phytoplankton above the light compensation point (Table 5.1). With more field data it should be possible to determine specific depth and therefore shade tolerances; and Z_c should, within the arbitrary ranges set for phytoplankton density, be predictable at any time in a given lake from measurements of E_{min}. This also indicates situations in which values of macrophyte biomass maxima (Sec. IV) and macrophyte production for a given alkalinity and concentration of total CO_2, might eventually be predicted.

In the second category of lake, it is suggested that B becomes large enough in amount and variable enough in season substantially to upset any direct relationship between Z_c and Z_{eu}; in Loch Leven for example Z_{eu} can vary over one year between 1·4 and 7·5 m (Bindloss 1974). Alkalinity in this group is more than 0·3 meq l^{-1} while B exceeds 20 mg m^{-3} and differs from its maximal value by a factor of more than 3; thus in Esthwaite, *B* can exceed 100 mg chlorophyll *a* m^{-3} (Talling 1971) and in Loch Leven Bindloss (1974) records densities of over 250 mg m^{-3}.

In many of these lakes Z_{eu} is larger relative to Z_m (mixed depth, in m) than in the frequently deep lakes of the first situation and a number, again like Loch Leven (Holden & Caines 1974), are subjected to seasonally variable surface water enrichment. Not only is there unlikely to be a clear relationship at any instant of time between Z_c and the highly variable Z_{eu} (Fig. 5.8) but, also, competition between phytoplankton and macrophytes is likely to be maximal. This indeed is the argument advanced to explain fluctuating biomass data from Loch Leven (Sec. V) and such competition should also lead to inconstancy in other attributes of biomass like depth of occurrence or amount of maxima, and in productivity of macrophytes from year to year. Macrophyte growth and performance should not be predictable in any straightforward fashion and this should be the common situation in over-enriched waters.

IV Biomass maxima and depth distribution

Hypothesis concerning depth of occurrence of maximal biomass

The biomass of submerged vegetation is shown in Fig. 5.9 in relation to depth of soil surface below the water surface in four lakes, which are all in the first category of Sec. III (p. 119). Biomass values are expressed for comparative purposes as percentages of the logarithms of the maximal oven-dry weight of whole plants in Loch Uanagan and of maximal plant cover in the remainder (from unconverted data of Spence 1974). From study of the rather similar shape of these curves in each lake, first, it is suggested that the decline in submerged biomass towards shallow water is caused by increasingly excessive turbulence operating directly on plants or, indirectly, through soil particle size and substratum chemistry; and second, it is evident that towards deeper water the decrease in biomass below its zone of maximal size is logarithmic which would follow if, for the given level of nutrients and inorganic carbon in each of these lakes, biomass were limited by light.

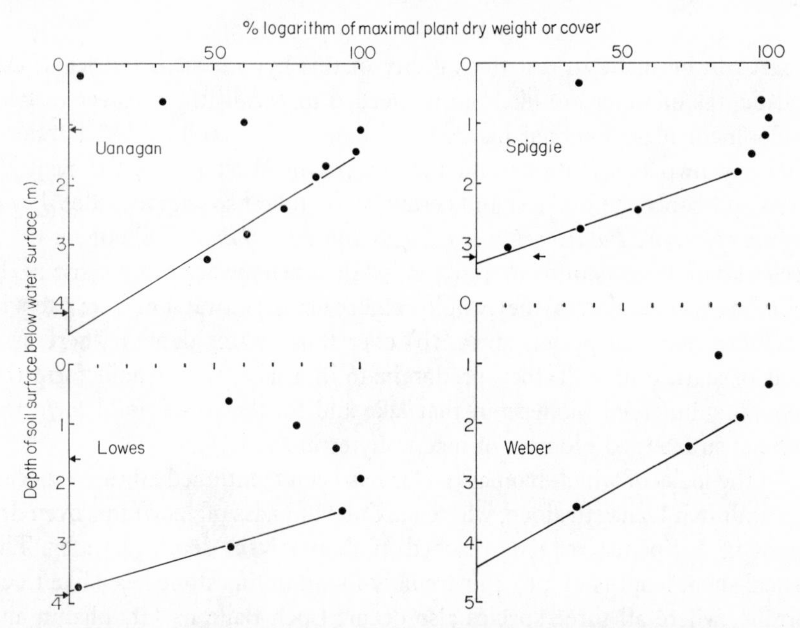

Figure 5.9. Biomass of submerged vegetation along single transects in relation to depth of soil surface below the water surface (m) in each of four lakes, expressed as percentage of the logarithms of the maximal oven-dry weight (in Loch Uanagan) or, in the remainder, of the maximal plant cover. Straight lines fitted by eye.

Data, obtained for Weber Lake (Potzger & Engel 1942) by using a grab from a boat and for the other three lakes by diving, are shown together as unconverted variables in Spence (1974, Fig. 7). Location of samples: Loch Uanagan (11 August 1967, 15 September 1968) south shore near but excluding *Schoenoplectus* beds; Loch Spiggie (8, 10, 15 July 1969) west and east shore by Littleness; Loch of Lowes (12 August 1969) south shore ¼ km from southwest end of Loch. Symbols: (←) marks water depth below which the substratum is muddy to at least 10 cm, whilst (→) indicates estimates by diver of zero colonization.

Acting through effects on summer growth or winter survival, excessive turbulence would therefore tend to force plants into deeper water where high light attenuation, augmented perhaps by enlarged boundary layers on leaves following diminished turbulent flow and by unfavourable water chemistry, would tend to force plants into shallow water. This leads to the proposition, introduced elsewhere (Spence 1974), that the resultant of the opposing factors of increasingly excessive turbulence towards shallow water and of increasing light attenuation towards deep water determines the depth at which maximal submerged biomass occurs on any shore.

Range in amount and depth of occurrence of maximal biomass

A start can be made to test the validity of this hypothesis. First let us ask what depths of water are likely to be needed to permit the greatest vertical development of submerged macrophyte biomass in standing U.K. waters. From my own observations and the records of West (1910) for Scottish waters, probably the bulkiest and certainly the tallest species include *Myriophyllum spicatum*, *Potamogeton praelongus* and *P. perfoliatus*. Shoots of these species attain heights of from 3 to 4 m, while other species can reach 3 m. It might be expected that (a) they would require depths of water of from 3 to 5 m to achieve maximal performance, (b) over those water depths where as a result of stature or bulk they predominate in a lake, they would form the maximal submerged biomass in that lake and (c) they also could form the maximal submerged biomass of macrophytes in the U.K.

Of the lochs of which biomass so far has been mentioned all three species occur in Loch Uanagan alone, where maximal biomass of 320 grams oven dry weight m^{-2} (Spence 1974) is reached in about 1·5 m depth of water. The vertical shoot lengths of 3 to 4 m are only found in limestone lakes like Loch Borralie, where all three species also occur, Loch Baile na Ghobhainn and Loch Croispol and at depths of between 3 to 5 m. As already indicated (Table 5.2 and p. 96) these are amongst the clearest fresh waters in the U.K. Preliminary evidence from an ongoing study in Loch Borralie of variation of biomass with season indicates as predicated that the maximal performance of these species is achieved between these depths and that this indeed forms the bulk of the maximal submerged biomass in that lake. The validity of the third prediction has still to be tested, namely, that the maximal biomass figure of around 450 grams oven dry weight m^{-2} approaches the largest possible figure for British fresh waters.

There are not yet enough examples to attempt an empirical model but a few general points emerge from study of the curves in Fig. 5.9 and of the data from which they were derived (Spence 1974). The gradients of the straight line plots must, like those for specific leaf area (p. 109), be a measure of relative light attenuation whilst the intersection of each extrapolation with the y axis provides an independent measures of Z_c (p. 115). Compared with Loch Borralie, maximal biomass is achieved in Loch Uanagan by the same species rooted likewise in fine-particled mud, but their substratum lies in shallower water and the maximal biomass itself is smaller. Both these features indicate limitations imposed primarily by light in this less clear, brown water (Table 5.2). With different and smaller species, maximal biomass occurs in Loch Spiggie at about the same depth as, and in Loch of Lowes at a greater depth than, that in Loch Uanagan but in both cases the actual values ($<$90 g oven dry wt m^{-2}) are far lower; in conjunction with alkalinity levels (Fig. 5.7) these facts suggest that limitations by nutrient and inorganic carbon levels may also be involved.

In Loch Leven, with Z_c around 1 m, submerged macrophyte biomass must lie mainly on exposed, coarse-particled shores, and should therefore be amongst the lowest in this series; which is indeed so. Over 1972–73, biomass did not exceed 43 g oven dry wt m^{-2} (Table 5.7) except at similar depths on a small area with fine-particled substrata north of St. Serf's Island where the maximum recorded, in 1973, was 143 g oven dry wt m^{-2} (B. P. Jupp, *in litt.*). There are large resident and migrant waterfowl populations on the loch (Allison & Newton 1974) and data from enclosures currently being studied by Dr. Jupp indicate that these biomass figures would increase without grazing by waterfowl.

Limitations on maximal biomass in nutrient-rich water

The extent of this last mentioned improvement would be limited by the depth of water column available to macrophytes, a depth that according to the hypothesis is set in this nutrient-rich loch by the more or less constant presence of dense phytoplankton; which in turn prevents larger macrophytes existing on substrata at depths far enough below the zone of excessive turbulence to permit their growth potential to be fully realized. This indeed should be a common feature of over-enriched water along with the remaining point arising from the Loch Leven data.

By contrast with the relatively invariable values over two years for Loch Uanagan (used in Fig. 5.9) and over three years for Loch Borralie (unpublished), there are large fluctuations in macrophyte biomass between one year and the next in Loch Leven (Table 5.7). These fluctuations emphasize the unpredictability, alluded to in Sec. III, of a macrophyte biomass which would follow from a variable and potentially very high phytoplankton density and they lead naturally to a concluding discussion on the nature of competition between macrophytes and phytoplankton.

V Nature of competition

Phytoplankton changes light climate

Competition from phytoplankton to the detriment of macrophytes may result from reduction of light, and from alterations in water chemistry, following algal photosynthesis. In two years for which seasonal data on biomass of phytoplankton and submerged macrophytes are available from Loch Leven, a comparison is made in Table 5.7 between mean and maximal values for the two groups during the initial and final three months' growing season of *Potamogeton filiformis*. In April, May and June 1973 there was far more phytoplankton than at the same season in 1972, implying that light absorption by algae was higher in 1973, to which the macrophytes appear to respond with a reduced biomass in both halves of the growing season. These results strongly suggest that the light climate in the first half of its growing season is a prime factor in the subsequent performance of *P. filiformis*: and it may be recalled that only 11% of the subsurface *PAR* reaching 1 m in Loch Croispol is found at 1 m in Loch Leven during a moderate phytoplankton bloom (p. 101).

Table 5.7. Mean and maximal biomass of *Potamogeton filiformis* (g dry wt. m^{-2}) and mean and maximal pH of water in Loch Leven, Kinross (shallow water by Grahamstown) during the first and last three months of the growing season in 1972, 1973; in relation to the population density of phytoplankton per unit volume of water (mg chlorophyll *a* m^{-3}) over the same periods. Macrophyte data: Jupp *et al.* (1974) and B.P. Jupp (unpublished); phytoplankton data: A. Bailey-Watts (unpublished); pH data: L.A. Caines (unpublished).

		1972 A M J	1972 J A S	1973 A M J	1973 J A S
Macrophyte biomass	mean	15·0	21·4	1·7	4·3
g dry wt. m^{-2}	max	17·8 20 June	43·5 22 Aug.	7·6 26 June	7·4 3 Aug.
pH of water	mean	8·6	8·6	9·1	8·3
	max	9·1 11 Apr.	9·3 26 Sep.	9·4 8 May	9·1 31 Aug.
Phytoplankton	mean	33·4	108·5	77·0	39·0
chlorophyll *a* mg m^{-3}	max	45·0 18 Apr.	240·0 25 Sep.	124·0 15 May	88·0 31 July

Mulligan & Baronowski (1969) reported the results of a glasshouse experiment involving interactions between phytoplankton and three rooted macrophytes including *Myriophyllum spicatum*. Uniform plant segments were placed in mud having an apparently plentiful supply of nutrients; the shoots were subsequently bathed in a solution having no nitrogen or phosphorus but an otherwise balanced nutrient complement, and in ten other solutions to which N as NH_4NO_3 and P as $CaH_4(PO_4)_2$ were added in increasing amounts. Initial pH and total alkalinity were measured. When at the end of 35 days the first plants had reached the surface of the containers all treatments were harvested. This terminal dry weight of plants, the terminal chlorophyll *a* content in mg m^{-3}, percentage transmittance and pH of water were measured.

Considering here only the *Myriophyllum spicatum* series, terminal yield decreased with an increase in level of P and N supply to the water around the shoots and with an increase in phytoplankton density as measured by a progressive fall in percentage transmittance. Figure 5.10a, is drawn from some of Mulligan and Baronowski's tabulated data; like the authors, it seems reasonable to correlate decline in terminal yield of *Myriophyllum* with a reduction in PAR at the leaf surface because of the phytoplankton.

Phytoplankton changes water chemistry

Figure 5.10b is also extracted from their tabulated data. After 35 days the initial pH of 8·0 had risen in some treatments by 1·7 pH units and this figure shows that terminal yield of *Myriophyllum* is inversely related to the magnitude of change in pH, or to the final pH, of the bathing solution. Likewise, in Loch Leven, Table 5.6 suggests higher pH values in the water during the April–June period of high phytoplankton density in 1973 than in 1972. Allen (1973) has found that, at pH 9·6, the net photosynthetic rate of *Myriophyllum exalbescens* in a total CO_2 solution of 600 mg l^{-1} is 10% of its maximal rate, which is at pH 7·0. Black (1973) had demonstrated similar effects at high pH on *Potamogeton praelongus*. In nature, Talling (in Anon 1972) has reported rises in pH in the waters of Esthwaite, with alkalinity (Fig. 5.7) around 0·4 meq l^{-1}, to more than pH 10·0 in late summer and he has established that this follows inorganic carbon depletion, a result of intense photosynthetic activity of phytoplankton (mainly *Ceratium* and *Microcystis* spp.) the density of which may exceed 100 mg chlorophyll *a* m^{-3} (Talling 1971). This raises further questions which are beyond the scope of this paper, namely the relative importance of pH and the form of exogenous inorganic carbon for photosynthesis by both macrophyte and phytoplankton; possibilities of increasing competition between these two groups when nitrogen becomes limiting (p. 94); of antagonism between, say, blue-greens (Whitton 1965) and macrophytes or, in other situations, between macrophytes and phytoplankton (Hasler & Jones 1949); and, finally, possible effects of grazing upon macrophytes (Sec. IV).

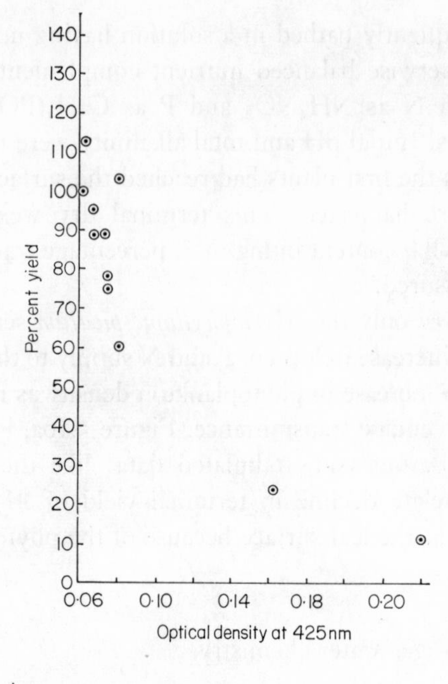

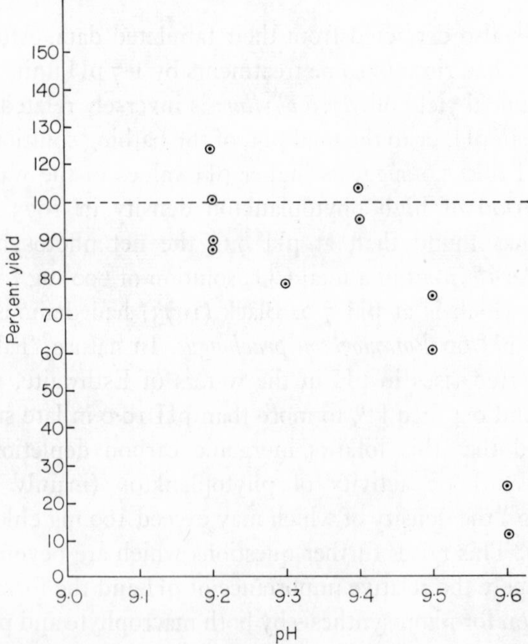

Figure 5.10. Percentage yield of *Myriophyllum spicatum* after 35 days in various treatments plotted against (a) final optical density (O.D.) of bathing solution and (b) final pH of bathing solution. Drawn from tabulated data of Mulligan & Baronowski (1969).

It has been noted that any interference by phytoplankton with the macrophyte environment is likely to be most effective during the early part of the six months' growing season of *P. filiformis*; or Z_{eu} during the period April–June determines the species' performance. As in another enriched lowland lake, Crose Mere, Shropshire (Reynolds 1973), Z_{eu} in Loch Leven even over this period is highly variable from year to year. Not all macrophytes, however, have an obligatory six months' growing season because, in Scotland for example, *Potamogeton praelongus*, *P. crispus*, *Nitella opaca* and *Chara* spp. can all be wintergreen in deep, or shallow, sheltered water: without, that is, a closed season for growth. While *P. praelongus* at least is known to photosynthesize at 4°C in laboratory conditions, there is insufficient data on photosynthesis-temperature relations of any of these species to test the possibility that winter photosynthesis may be important for, say, the slow-growing ones, and that prolonged algal interference at any time of the year may be deleterious to macrophytes. In summary, it is emphasised that, in a sequence of lake waters with an increasing and balanced ionic composition, the capacity of phytoplankton to vary in time and space will also increase, and so too will the chance of severe interference with macrophyte growth at any time of the year.

Substratum as nutrient source for macrophytes at low phytoplankton densities

It only remains to comment on an aspect of macrophyte growth in lakes with phytoplankton densities that are consistently low throughout the year, a condition attributed therefore to nutrient deficiency or imbalance in the water rather than, for example, to grazing or parasitism. Growth of macrophytes must depend here on the extent to which roots or other organs in the substratum can draw on nutrient supplies denied, because of the presence of an oxidized microzone, to the overlying water. Perhaps this conclusion is most compelling in the case of the limestone Loch Borralie which has been cited repeatedly as an exception to the correlation between phytoplankton density and alkalinity of the water and where, as I argued in the preceding section, biomass maxima of submerged macrophytes produced over certain depths approach the largest of any standing waters in the U.K.

Acknowledgements

I should like to thank the following:

for permission to quote from their unpublished data, Dr. A. Bailey-Watts,

Mr. L. A. Caines, Dr. B. P. Jupp, Dr. J. J. Light, Dr. J. W. G. Lund, and Mr. A. C. Webster;

for assistance with field work on a number of occasions, Mr. M. T. D. Carr, Mr. P. C. S. Jackson, Dr. B. P. Jupp, Mr. E. J. Torrens-Spence and, in particular, Dr. Colin Muir;

for helpful comments on various sections of the paper, Mr. E. D. Allen, Dr. J. W. G. Lund, Dr. J. F. Talling, Mr. A. C. Webster. In addition, Mr. Allen has assisted me considerably in laboratory and field. His work was funded by the Natural Environment Research Council, to whom also my thanks are due.

References

ABELES F.B. (1973) *Ethylene in Plant Biology*. New York.

ÅBERG B. (1943) Physiologische und ökologische Studien über die Photomorphose. *Symbolae Botanicae Upsalensis*, 8. Uppsala.

ALLEN E.D. (1973) *An ecophysiological study of the effects of thermal discharge on the submerged macrophytes of Lake Wabamun*. Unpublished M.Sc. Thesis, University of Alberta.

ALLEN H.L. (1971) Primary productivity, chemo-organotrophy, and nutritional interactions of epiphytic algae and bacteria on macrophytes in the littoral of a lake. *Ecol. Monogr.* 41, 97–127.

ALLINSON A. & NEWTON I. (1974) Waterfowl at Loch Leven, Kinross. *Proc. R.S.E.* (B), 74, 24, 1972/73, 365–82.

ANDERSON E.R. (1952) Energy-budget studies, Water-loss investigations: Vol. 1, Lake Hefner Studies. Tech. Rep., *U.S. Geol. Survey Circ.* 229, 71–119.

ANGSTROM A. (1925) On the albedo of various surfaces of ground. *Geogr. Ann.* 7, 323–42.

ANON (1972) *F.B.A. 40th Ann. Report*, 37.

ARBER A. (1920) *Water plants: a study of aquatic angiosperms*. Cambridge. Repr. (1963) with an introduction by Stearn, W.T., as *Historiae Naturalis Classica*, 23. Weinheim.

BINDLOSS M. (1974) Primary production of phytoplankton in Loch Leven, Kinross. *Proc. R.S.E.* (B), 74, 10, 1972/73, 157–82.

BLACK M.A. (1973) *Exogenous inorganic carbon sources in photosynthesis of submerged aquatic plants*. Unpublished M.Sc. Thesis, University of St. Andrews.

BODKIN K. and NAUWERCK A. (1968) Produktionsbiologische Studien uber die Moosvegetation eines klaren Gebirgssees. *Schweiz. und seine Grundalgen Hydrol.* 30, 318–52.

BRAND F. (1896) Über die Vegetationsverhältnisse des Würmsees. *Zeitschrift für Bot. Centralbl.* 65, 1–13.

BREHM V. and RUTTNER F. (1926) Die Biocönosen der Lunzer Gewässer. *Int. Revue ges. Hydrobiol.* 16, 281–391.

BROWN S.-D. and AUSTIN A.P. (1971) A method of collecting periphyton in lentic habitats with procedures for subsequent sample preparation and quantitative assessment. *Int. Revue ges. Hydrobiol.* 56, 557–80.

BROWN S.-D. and AUSTIN A.P. (1973) Spatial and temporal variation in periphyton and physico-chemical conditions in the littoral of a lake. *Arch. Hydrobiol.* 71, 183–232.

CUMMING B.G. (1963) The dependence of germination on photoperiod, light quality and temperature in *Chenopodium* spp. *Can. J. Bot.* 41, 1211–33.

EDWARDS D.P. & EVANS G.C. (1975) Problems involved in the design of apparatus for measuring the spectral composition of daylight in the field. *This vol.* 161–87.

FASSETT N.C. (1930) The plants of some northeastern Wisconsin lakes. *Trans. Wis. Acad. Sci. Arts Lett.* 25, 157–68.

FITZGERALD G.P. (1969) Some factors in the competition or antagonism among bacteria, algae and aquatic weeds. *J. Phycol.* 5, 351–9.

GESSNER F. (1938) Die Beziehung zwischen Lichtintensität und Assimilation bei submersen Wasserpflanzen. *Jb. wiss. Bot.* 86, 491–526.

GESSNER F. (1955) *Hydrobotanik* Vol. I. Berlin.

HACHTEL W. (1972) Der Einfluβ des Plasmotypus auf die Regulation der Aktivität der L-Phenylalanin-Ammonium Lyase (Untersuchungen an Raimannia-Oenetheren). *Planta* 102, 247–60.

HARDER R. (1930) Über die Assimilation der Kohlensäure bei konstanten Aussen Bedingungen. *Planta* 11, 263–93.

HASLER A.D. & JONES E. (1949) Demonstration of the antagonistic action of large aquatic plants on algae and rotifers. *Ecology* 30, 359–64.

HOLDEN A.V. & CAINES L.A. (1974) Nutrient chemistry of Loch Leven, Kinross. *Proc. R.S.E.* (B), 74, 7, 1972/73, 101–22.

HUTCHINSON G.E. (1957) *A Treatise on Limnology*, Vol. 1. New York.

JONES J.G. (1972) Studies on freshwater micro-organisms: phosphatase activity in lakes of differing degrees of eutrophication. *J. Ecol.* 60, 777–91.

JØRGENSEN E.G. (1957) Diatom periodicity and silicon assimilation. *Dansk. Bot. Foren.* 18, 1–54.

JUDAY C. (1934) The depth distribution of some aquatic plants. *Ecology* 15, p. 325.

JUPP B.P., SPENCE D.H.N. & BRITTON R.G. (1974) The distribution and production of submerged macrophytes in Loch Leven, Kinross. *Proc. R.S.E.* (B), 74, 12, 1972/73, 194–208.

LIGHT J.J. (1975) Clear lakes and aquatic bryophytes in the mountains of Scotland. *J. Ecol.* 63 (in the press).

LIGHT J.J. & HEYWOOD R.B. (1973) Deep-water mosses in Antarctic lakes. *Nature*, 242, 535–6.

LYTHGOE J.N. (1972) The adaptation of visual pigments to the photic environment. In: Photochemistry of Vision *ed.* H.J.A. Dartnall. *Handbook of Sensory Physiology*, Vol. 7/1. Springer-Verlag. Berlin.

MOHR H. (1972) *Lectures on photomorphogenesis*. Springer-Verlag. Berlin.

MULLIGAN H.F. & BARONOWSKI (1969) Growth of phytoplankton and vascular aquatic plants at different nutrient levels. *Verh. Internat. Verein. Limnol.* 17, 802–10.

MUSGRAVE A., JACKSON M.B. & LING E. (1972) *Callitriche* stem elongation is controlled by ethylene and gibberellin. *Nature :New Biology*, 238, 93–6.

MUSGRAVE A. & WALTERS J. (1973) Ethylene-stimulated growth and auxin transport in *Ranunculus scleratus* petioles. *New Phytol.* 72, 783–9.

OBERDORFER E. (1928) Lichtverhältnisse und Algenbesiedlung im Bodensee. *Zeitschrift für Bot.* 20, 465–568.

PEARSALL W.H. (1918) On the classification of aquatic plant communities. *J. Ecol.* 6, 75–84.

PEARSALL W.H. (1920) The aquatic vegetation of the English Lakes. *J. Ecol.* 8, 163–201.

PEARSALL W.H. (1921) The development of vegetation in the English Lakes, considered in relation to the general evolution of glacial lakes and rock basins. *Proc. Roy. Soc.* 92, 259–84.

PEARSALL W.H. & ULLYOTT P. (1934) Light penetration into fresh water III. Seasonal variations in the light conditions in Windermere in relation to the vegetation. *J. exp. Biol.* 11, 89–93.

POTZGER J.E. & ENGEL W.A. VAN (1942) Study of the rooted aquatic vegetation of Weber Lake, Vilas County, Wisconsin. *Trans. Wis. Acad. Sci. Arts Letters* 34, 149–66.

PROWSE G.A. (1959) Relationship between epiphytic algal species and their macrophytic hosts. *Nature* 183, 1204–5.

REYNOLDS C.S. (1973) The seasonal periodicity of planktonic diatoms in a shallow eutrophic lake. *Freshw. Biol.* 3, 89–110.

RUTTNER F. (1926) Über die Kohlensäureassimilation einiger Wasserpflanzen in verschiedenen Tiefen des Lunzer Untersees. *Int. Revue ges. Hydrobiol.* 15, 1–30.

SCULTHORPE C.D. (1967) *The Biology of Aquatic Vascular Plants.* London.

Smithsonian Meteorological Tables (1951) Sixth revised edition. Smithsonian Miscellaneous Collections, vol. 114. Washington, D.C.

SPENCE D.H.N. (1964) The macrophytic vegetation of freshwater lochs swamps and associated feus. In *The Vegetation of Scotland* (Ed. by J.H. Burnett), 306–425. Edinburgh.

SPENCE D.H.N. (1967) Factors controlling the distribution of freshwater macrophytes, with particular reference to Scottish lochs. *J. Ecol.* 55, 147–70.

SPENCE D.H.N. (1972) Light on freshwater macrophytes. *Trans. Bot. Soc. Edinb.* 41, 491–505.

SPENCE D.H.N. (1974) Light, zonation and biomass of submerged freshwater macrophytes. *Research Underwater* (Ed. by E.A. Drew, J.N. Lythgoe and J.D. Woods), 335–45. Academic Press, London.

SPENCE D.H.N., CAMPBELL R.M. & CHRYSTAL J. (1971) The spectral intensity of some Scottish freshwater lochs. *Freshwat. Biol.* 1, 321–37.

SPENCE D.H.N., CAMPBELL R.M. & CHRYSTAL J. (1973) Specific leaf areas and zonation of freshwater macrophytes. *J. Ecol.* 61, 317–28.

SPENCE D.H.N. & CHRYSTAL J. (1970) Photosynthesis and zonation of freshwater macrophytes II. Adaptability of species of deep and shallow water. *New Phytol.* 69, 217–27.

SPENCE D.H.N., MILBURN T.R., NDAWULA-SENYIMBA M. & ROBERTS E. (1971a) Fruit biology and germination of two tropical *Potamogeton* species. *New Phytol.* 70, 197–212.

SPOONER D.L. (1971) Spectral reflectance of aquarium plants and other natural underwater materials. *Proc. 6th Internat. Symp. on Remote Sensing of Envt.* 2, 1003–16. Ann Arbor.

TALLING J.F. (1957) Photosynthetic characteristics of some freshwater plankton diatoms in relation to underwater radiation. *New Phytol.* 56, 29–50.

TALLING J.F. (1960) Self-shading effects in natural populations of a planktonic diatom. *Wetter u. Leben* 12, 235–42.

TALLING J.F. (1965) The photosynthetic activity of phytoplankton in East African Lakes. *Int. Revue ges. Hydrobiol.* 50, 1–32.

TALLING J.F. (1970) Generalized and specialized features of phytoplankton as a form of photosynthetic cover. Proc. IBP/PP Technical Meeting, Trebon, Czechoslovakia, *Prediction and Measurement of Photosynthetic Productivity*, 431–45. PUDOC Wageningen.

TALLING J.F. (1971) The underwater light climate as a controlling factor in the production ecology of freshwater phytoplankton. *Mitt. Internat. Verein. Limnol.* 19, 214–43.

THIENEMANN A. (1926) Die Binnengewässer Mitteleuropas. *Die Binnengewässer*, Vol. 1. Stuttgart.

THIENEMANN A. (1950) Verbreitungsgeschichte der Süsswassertierwelt Europas. *Die Binnengewässer*, Vol. 18. Stuttgart.

TYLER J.E. & SMITH R.C. (1967) Spectroradiometric characteristics of natural light under water. *J. opt. Soc. Amer.* 57, 595–601.

VEGIS A. (1953) The significance of temperature and the daily light-dark period in the formation of resting buds. *Experientia* 9, 462–3.

VEGIS A. (1963) Climatic control of germination, bud break and dormancy. In *Environmental Control of Plant Growth* (Ed. by L.T. Evans), 265–87. New York.

VEGIS A. (1964) Dormancy in higher plants. *Ann. Rev. Plant Physiol.* 15, 185–224.

VÈZINA P.E. & BOULTER D.W.K. (1966) The spectral composition of near UV and visible radiation beneath forest canopies. *Can. J. Bot.* 44, 1267–84.

VINCE D. & GRILL R. (1966) The photoreceptors involved in anthocyanin synthesis. *Photochem. Photobiol.* 5, 407–11.

WEBSTER A.C. (1975) *Metabolic Basis for Control of Growth in Freshwater Plants.* Unpublished Ph.D. Thesis. University of St. Andrews.

WEST G. (1910) A further contribution to a comparative study of the dominant phanerogamic and higher cryptogamic flora of aquatic habit in Scottish lakes. *Proc. R.S.E.* 30, 65–70.

WETZEL R.G. (1964) A comparative study of the primary productivity of higher aquatic plants, periphyton and phytoplankton in a large, shallow lake. *Int. Revue ges. Hydrobiol.* 49, 1–61.

WHITTON B.A. (1965) Extracellular products of blue-green algae. *J. Gen. Microbiol.* 40, 1–11.

WINBERG G.G. (1971) *Symbols, Units and Conversion Factors in Studies of Freshwater Productivity.* International Biological Programme, London.

YOUNG J.E. (1975) Effects of the spectral composition of light sources on the growth of a higher plant. *This vol.* 135–59.

Questions and discussion

Professor Monteith expressed interest in the question of the ratio of the radiation measured above the water surface and below, clearly a function of a great complex of meteorological factors. He asked how much was known about the effect of wind on the transmission of the water surface. Professor Spence replied that very little was known: the figure of 10% reflectance at the surface was assumed in the absence of precise measurements. Other problems are connected with what may be called the 'immersion effect correction factor', the spectroradiometer having been calibrated in air, and the measurements made under water. This factor has been estimated for beam attenuation at 10 nm intervals using a collimated light source, but the relationship of these values to the corresponding ones for diffuse attenuation is unknown but obviously complex, varying amongst other things with the turbidity of the water. In the foregoing paper these problems have been avoided by relating the underwater measurements to the immediately subsurface irradiance, at 0 m. Further work on these subjects is clearly needed.

9 Effects of the spectral composition of light sources on the growth of a higher plant

J. E. YOUNG *Botany School, University of Cambridge*

Introduction

Much work has been done on the growth of the woodland annual *Impatiens parviflora* DC in artificial controlled environments (Hughes 1965b), under neutral screens in the field (Evans & Hughes 1961) and in Madingley Wood (Coombe 1966, Rackham 1966), with particular reference to the effects of the aerial environment on the pattern of growth.

Although there were large differences between the wholly artificial light climate of the growth cabinets and the semi-natural light climate under the neutral screens (for a summary see Section 18.15 in Evans 1972), the overall growth pattern in these two environments was found to be very similar at comparable values of total daily light incident on the plants. No consistent significant differences in either the pattern of growth or the absolute values of the indices of growth were observed.

When data from all three environments are compared, however, although the overall growth pattern in all environments was very similar, there were consistent significant differences in the absolute values of some indices of growth, between the Madingley Wood environment and the other two environments. For example, the specific leaf area (the ratio of leaf area to leaf dry weight for the plant as a whole) of plants grown in Madingley Wood was always greater than that of the cabinet grown ones at the same total daily light level (Fig. 6.1).

Specific leaf area is a measure of the degree of expansion of the leaf lamina, and the degree of expansion is closely coupled to the total daily light received by the plant, over a substantial range there being an inverse proportionality. Even at very low total daily light levels in the growth cabinets, very close to the compensation point for light in *I. parviflora*, the maximum observed specific leaf area was always below the maximum observed in plants growing under natural radiation in Madingley Wood, where the total daily light levels were considerably higher.

A more obvious difference between plants grown under artificial light in the growth cabinets, and plants grown under natural radiation, is the degree of hypocotyl elongation. At the same mean light intensity the hypocotyls of plants grown under fluorescent light were about twice as long as those of

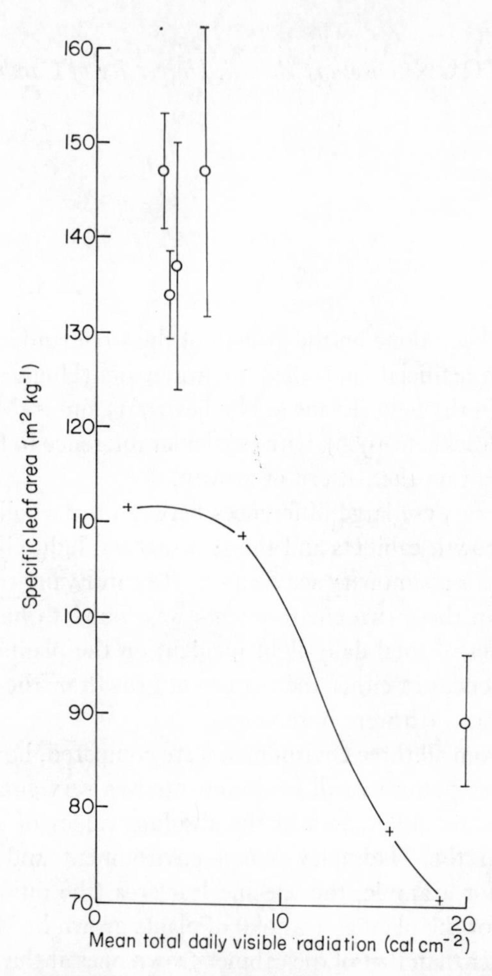

Figure 6.1. Specific leaf areas of *Impatiens parviflora* at 100 mg total plant dry weight as a function of total daily visible radiation. O, observations of Coombe (1966) from Madingley Wood with 19:1 probability limits; +, observations of Hughes (1965b) on plants under fluorescent light (after Evans 1972).

similar plants grown under natural radiation. For cabinet-grown plants Hughes (1965b) found a non-linear relationship between hypocotyl length and light intensity. He also found that changes in spectral composition of the radiation affected elongation, blue light reducing elongation whilst red light tended to increase it. There was a poor correlation between cabinet and field values, using light intensity as the basis for comparison, and Hughes suggested that hypocotyl elongation is controlled by the peak light intensity and not by either mean light intensity or total daily light (Hughes 1965b).

Hypocotyl elongation has been shown in some species to be controlled by blue light and also by far-red light (Mohr 1957, Schopfer 1969). For example, the action spectrum for the inhibition of hypocotyl elongation in *Lactuca sativa* has two broad peaks at the short wave end of the visible spectrum, one at 360 nm and the other at 420 nm, and also a sharp peak in the far-red at 710 nm (Mohr 1972). Since fluorescent tubes emit very little radiation at wavelengths less than about 420 nm, it is suggested that the poor correlation between field and cabinet values for the degree of hypocotyl elongation in *I. parviflora* may be due to the lack of short wave radiation in the cabinet illumination.

Internode elongation has also been shown to be sensitive to the spectral composition of the radiation, an increase in far-red radiation causing an increase in internode elongation (Downs *et al.* 1957, Meijer 1957, Rajan *et al.* 1971). Unfortunately no data are available for internode elongation in *I. parviflora* under either natural radiation or under artificial radiation in the growth cabinets, previous to the experiments described in this paper. Trial experiments recently done under natural radiation in Madingley Wood suggest that the degree of internode elongation, under the normal fluorescent light source used in the growth cabinets, is similar to that under natural radiation in Madingley Wood.

The suggestion has been made that the difference in specific leaf area between the cabinet-grown plants and those grown in Madingley Wood, is due to the difference in spectral composition between the radiation in the two environments (Evans 1969). The difference cannot simply be due to the use of an artificial light source in the cabinets, as a similar difference in leaf expansion is apparent between plants grown in Madingley Wood and plants grown under neutral screens in the field. Apart from the difference in form of the daily march of light intensity, and the short-period fluctuations, both of which would have been present under the neutral screens, the obvious difference between the radiation under the canopy in Madingley Wood and the radiation in the growth cabinets is the difference in spectral composition. Under a deciduous woodland canopy, due to the absorption of red light by the leaves, which are almost transparent to far-red light, the radiation reaching the woodland floor contains a much higher proportion of far-red radiation than unfiltered natural radiation. There is also very little far-red radiation in the growth cabinets, since fluorescent tubes produce very little radiation at wavelengths greater than about 720 nm in contrast to the natural radiation spectrum which extends beyond 1000 nm. This results in the ratio of red to far-red light under fluorescent tubes in the growth cabinets being higher than that under natural radiation in the open, whilst under a deciduous woodland canopy, the ratio is much lower than in the open.

This paper describes experiments designed to investigate the effects of altering the spectral composition at the red end of the spectrum on leaf expansion, hypocotyl elongation and internode elongation in *I. parviflora.*

Radiation and growth

Impatiens parviflora is capable of growing successfully under a wide range of total daily light levels, ranging from open habitats to the deep shade of a deciduous wood. Over a large part of this range the relative growth rate, that is the increase in dry weight per unit dry weight in unit time, is remarkably constant.

Instantaneously, relative growth rate (RGR) can be resolved into three components:

(1) Unit leaf rate $= \frac{1}{L_A} \times \frac{dW}{dt}$ (ULR)

(2) Leaf weight ratio $= \frac{L_W}{W}$ (LWR)

(3) Specific leaf area $= \frac{L_A}{L_W}$ (SLA)

where: L_A = total leaf area
L_W = total leaf dry weight
W = total plant dry weight
$\frac{dW}{dt}$ = rate of dry weight increase

$$\text{RGR} = \text{ULR} \times \text{LWR} \times \text{SLA}$$

Values of leaf weight ratio and specific leaf area can be determined unambiguously for individual plants at the time of a harvest. For the methods available for calculating a mean value of unit leaf rate over a period see Evans (1972, Chs 16 and 20).

The leaf weight ratio is more or less unaffected by changes in total daily light levels. It is however markedly affected by the rooting medium (Ch. 18 in Evans 1972).

In contrast, specific leaf area and unit leaf rate are markedly affected by the total daily light level. Specific leaf area is, over a substantial range, inversely proportional to the total daily light level, not light intensity (Fig. 6.1), whilst unit leaf rate is directly proportional to total daily light level. The net effect of these two relationships is that over the range for which they both hold, the relative growth rate of the plant is roughly constant.

Specific leaf area is known to have a complex ontogenetic drift and also to be affected by a variety of factors including temperature, rooting medium, daylength and total daily light (Fig. 6.2). Consequently when comparing the effects of different environmental factors on the growth of *I. parviflora* an allowance must be made for ontogenetic drift in order to be able to distinguish the effects of the environment on growth from any inherent changes which may have been going on as the plant ages.

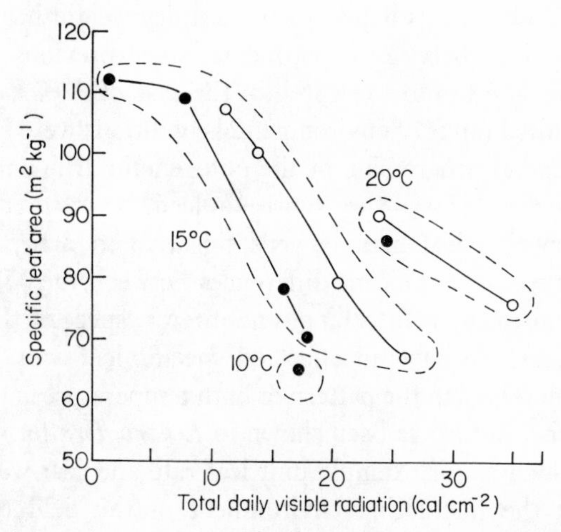

Figure 6.2. Specific leaf areas of *Impatiens parviflora* at 100 mg total plant dry weight as a function of total daily visible radiation at three temperatures and in two rooting media, sandy loam, ○, and vermiculite, sand and gravel with added culture solution, ●. (After Evans 1972.)

Some measure of the stage of the plant's development is required, so that comparisons of growth between different environments are made between plants at the same stage of development. What therefore is needed is some measure of the 'physiological age' of the plant. It seems very likely however that the concept of 'physiological age' cannot be strictly applied to the plant as a whole. Furthermore, the time taken to reach a particular developmental stage is not solely determined by the ontogeny of the plant, but is also influenced by the environment in which the plant has been developing. Thus any simple measure of ontogenetic drift will be at least partially inadequate, and the choice of measure will depend on what aspect of growth is under investigation.

The simplest measure of 'physiological age' is that of chronological age, but this suffers from the disadvantage of not being related either directly or indirectly to changes in the environment, or to the effects of environmental factors in the past on the present form and functioning of the plant. A more satisfactory basis of measurement is total plant dry weight, as this at least is related in some way to the plant's previous experience, but it is by no means wholly satisfactory. Evans and Hughes used total plant dry weight as a measure of plant age, making comparisons between experiments at a particular fixed dry weight. Evans concludes that this practice is justifiable if used with care since 'qualitative conclusions drawn from such comparisons are valid at other parts of the range of plant weights under observation' (Evans 1972, and see the discussion of possible measures in Ch. 19).

For the purposes of this paper, total plant dry weight has been used as a basis for comparison between experiments, since previous experience has shown this basis appears to work well in the case of *I. parviflora*. Furthermore, only a limited range of environmental conditions were under investigation and the general progression of the ontogenetic drift under the various experimental conditions was very similar indeed.

Although specific leaf area is very sensitive to many environmental factors, there were no significant differences between the Madingley Wood environment and the growth cabinet one in any aspect of the environment known at that time to have an effect on specific leaf area. There were of course some differences in the pattern of both temperature and light intensity change with time, but it has been shown in *I. parviflora* for some aspects of plant growth, such as for example unit leaf rate and leaf weight ratio, that changing from the fluctuating environment of the field to the constant conditions of the growth cabinets, makes negligible differences, provided that comparisons were made against some mean figure, such as average temperature over a period, or mean total daily light. For specific leaf area, Hughes & Evans were of the opinion that mean total daily light was the most important factor of the aerial environment and made allowance for this in their comparisons. However, they wondered whether the causal relationship between specific leaf area and total daily light might not operate through the availability of photosynthate, therefore explaining the connection between specific leaf area and unit leaf rate (Hughes & Evans 1962). As a consequence they often made comparisons on the basis both of total daily light and unit leaf rate (Hughes 1965b). In this way they were able to show for cabinet experiments under fluorescent tubes, that the use of blue tubes (with a negligible component in the red) and tubes with a magnesium arsenate phosphor (with a sharp peak of emission at 660 nm, but also with some breakthrough of the main mercury lines in the blue and green, see Canham 1966) produced only slight changes in specific leaf area (Hughes 1965b, Hughes & Evans 1964). All these comparisons were made at a fixed total plant dry weight, in order to minimize the disturbing effects of ontogenetic drift. Despite the great variety of temperatures and total daily light levels used in the growth cabinets, the maximum specific leaf area observed by Evans &

Hughes in all these experiments was always significantly lower than the maximum values observed in Madingley Wood.

Spectral composition

As has already been mentioned, fluorescent light sources as used in the growth cabinets are deficient in far-red radiation. The ratio of the intensity of radiation at 660 nm to that at 730 nm (hereafter referred to as the R/FR ratio) is about 3·5 for the fluorescent tubes used compared with a value of about 1·3 for sunlight (Spence *et al.* 1971).

Millener (1952, 1961) observed that seedlings of *Ulex europaeus* would not make normal growth under fluorescent lighting, but the addition of some tungsten lamps produced normal growth. Rajan *et al.* (1971) noted that abnormal leaf development in *Gossypium hirsutum* and *Phaseolus vulgaris* and abnormal pigmentation in *Zea mays* occurred under fluorescent light alone but the addition of some tungsten lamps to the light source produced normal leaves. These observations suggest that some far-red radiation is necessary for normal leaf development in at least some species.

In the case of *I. parviflora* grown under fluorescent lights, there are no signs of abnormality of leaf development or pigmentation, but the degree of expansion of the leaf lamina appears to be restricted when compared with the degree of expansion occurring under natural radiation in Madingley Wood.

Several measurements of spectral composition under vegetation have been made (Federer & Tanner 1966, Jordon 1969, Stoutjesdijk 1972a & b). All these investigations show that there is an increase in canopy transmission above about 700 nm. The size of this increase depends on the species composition of the canopy, the increase being larger under hardwood species than under conifers (Coombe 1957, Federer & Tanner 1966). The increase also depends on canopy density (Jordon 1969). Stoutjesdijk found that the increase was not so marked in the tropical rain forest that he studied (Stoutjesdijk 1972b). The data of Stoutjesdijk (1972a) for various woodland types in the Netherlands show low values of the R/FR ratio under all canopy types, in the range 0·2–0·4, compared with a value of about 1·3 for sunlight.

The close agreement between the cabinet experiments and those done under neutral screens in the field suggests that in *I. parviflora* leaf expansion is little affected by changes in the R/FR ratio in the range 1·3–3·5. The further large change of R/FR ratio from a value of 3·5 for white fluorescent tubes to around 50 for tubes having a magnesium arsenate phosphor also produces only slight changes in specific leaf area, if indeed these are significant (Hughes & Evans 1964). It therefore seems that leaf expansion is affected little, if at all, by increases in the R/FR ratio above about 1·3. It is of interest in this connection that investigations on the relationship between specific leaf area in *Potamogeton* spp. and water depth reveal no differences in specific leaf area due to differences in the R/FR ratio, although the ratio varied from 1·5 to 575 depending on the depth and type of water (Spence 1975). The Madingley results would suggest however that reducing the value below about 1·3 does cause an increase in leaf expansion in *I. parviflora*. Thus it seems likely that the effect of the R/FR ratio on leaf expansion operates over the range 1·3–0·2, values between those for sunlight and for deep shade under a deciduous woodland canopy.

The light source normally used in the Cambridge growth cabinets is a combination of blue and deluxe warm white fluorescent tubes, in the ratio of 1 blue tube to 3 deluxe warm white ones. Deluxe warm white rather than ordinary warm white tubes are used since the deluxe tubes have a significantly greater light output at wavelengths greater than about 650 nm and therefore provide a closer approximation to natural radiation (Fig. 6.3 A & B). The blue tubes are necessary in order to prevent excessive elongation of the hypocotyl which would make the plants difficult to handle and very susceptible to mechanical damage. The R/FR ratio of this combination of fluorescent tubes is 3·5.

In order to reduce the R/FR ratio below 3·5 another type of lamp must be used. Ordinary tungsten filament lamps were found to be suitable, having a stable light output throughout their life, although the intensity of radiation is much more sensitive to mains voltage fluctuations than with fluorescent tubes. For both types of lamps such fluctuations were avoided by using a voltage stabilizer. Using 40-watt lamps it was found that even illumination of the cabinet floor was possible without the use of an elaborate reflector system. Two blue fluorescent tubes were also used to prevent excessive hypocotyl elongation, as tungsten lamps produce very little radiation at the short-wave end of the visible spectrum. The R/FR ratio of this light source was 0·8 (Fig. 6.3C).

Specific leaf area, as has already been mentioned, is sensitive to total daily light levels and therefore, in order to separate the effects of total daily level from any effects of spectral composition, experiments were done in pairs, one of the pair using the standard combination of fluorescent tubes (the reference treatment) whilst the other of the pair was set up with the light climate under investigation.

As photosynthesis is primarily a photochemical reaction, it would seem likely that the rate of the reaction would be determined by the number of quanta within the effective waveband of the reaction, rather than by the total energy within the same waveband. Laboratory experiments show that the quantum yield increases across the spectrum from the blue end but from about 570 nm the quantum yield is very nearly constant until the very sharp

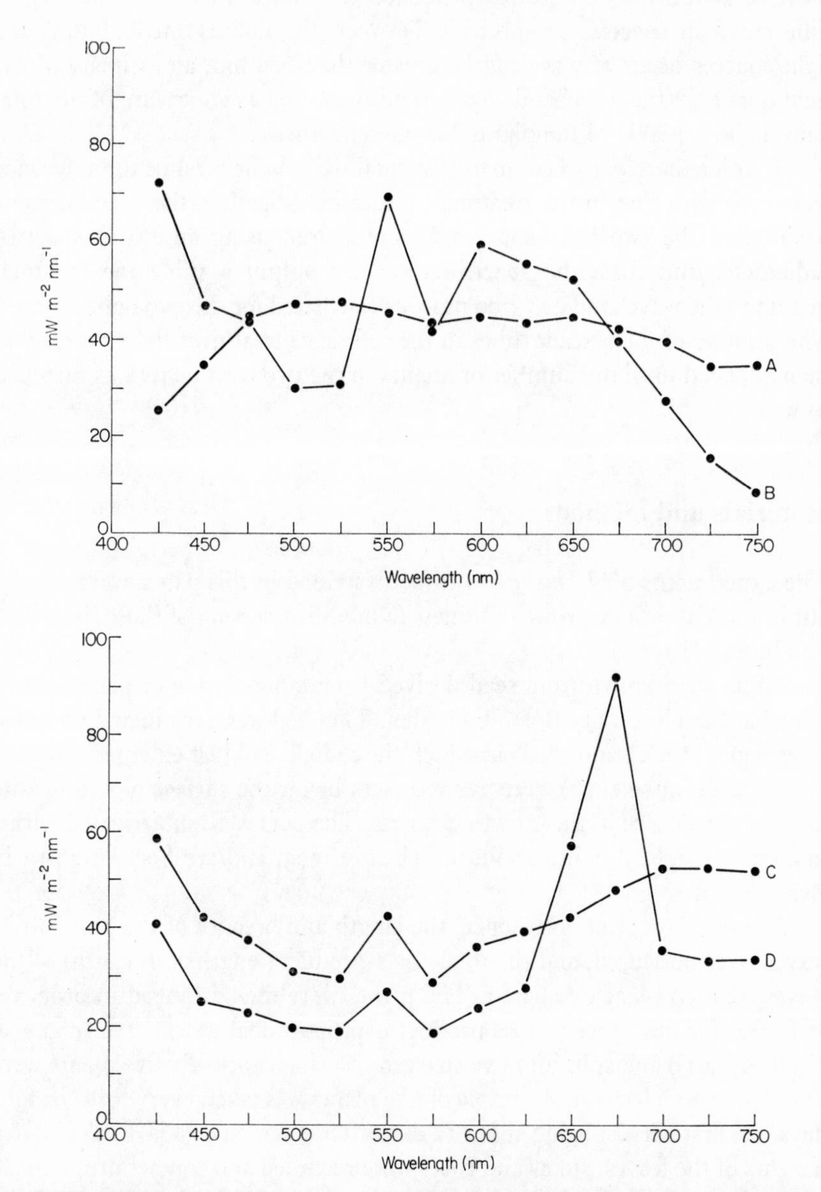

Figure 6.3. Comparison of spectral composition of daylight with various artificial light sources. A, sunny day with some cloud; B, white plus blue fluorescent tubes (Expt. 1/F); C, tungsten lamps plus blue fluorescent tubes (Expt. 1/T); D, tungsten lamps plus blue and red fluorescent tubes (Expt. 4/A).

drop at about 700 nm (Rabinowitch & Govindjee 1969). As the largest differences in spectral composition between the fluorescent and tungsten light sources occur at wavelengths greater than 600 nm, an estimate of the total quanta in the waveband 425–700 nm was used as a measure of the total daily light available to the plants for photosynthesis.

In order therefore to compare the total daily light level in the reference treatment with that in the treatment under investigation, the spectral composition of the two light sources was measured using an ISCO spectroradiometer and from the spectroradiometer output a value for the total quanta in the waveband 425–700 nm was calculated for the two light sources. The number of fluorescent tubes in the reference treatment light source was then adjusted until the number of quanta in the two treatments was matched to within $\pm 5\%$.

Materials and methods

The experiments on *I. parviflora* to be described in this paper were carried out in the Cambridge growth cabinets (a modified version of those described by Hughes (1959a)).

Plants were grown from seed derived from a population of plants maintained at the University Botanic Garden. The seed was germinated on moist filter paper at 5°C, and seeds in which the radicle had just emerged (stage C in Fig. 2, Coombe 1956) were planted 2 cm below the surface of vermiculite in 9 cm pots placed in the growth cabinets. The pots were subirrigated with a modified Hoagland culture solution (Lewis 1963, and see Section 26.13 in Evans 1972).

About 5 days after emergence, the length and breadth of the first pair of leaves were measured, and the total cross-product (length $\times$ breadth) of the leaves on each plant calculated. The plants were then arranged in order on leaf area (for this species cross product is proportional to leaf area (Evans & Hughes 1961)) and split into five size groups, since normally five plants were sampled at each harvest. A sample of five plants was taken every three or four days, the first harvest being about 12 days after emergence. The fresh and dry weights of the leaves, stems and roots were recorded and contact prints made of the leaves of each plant, thus recording leaf area, leaf number and the length and breadth of each leaf.

The cotyledons were usually removed after the second harvest when the plants were about 15 days from emergence. Hughes (1957) showed that there was no marked difference between plants in which the cotyledons had been left on, and those in which the cotyledons had been removed before the first harvest at 22 days. No account of the cotyledon weight and area was taken in the analysis of growth. The reason for leaving the cotyledons on for so long was to see whether spectral composition had any effect on cotyledon expansion, as this has been shown in some species to be a phytochrome-controlled process (Kleiber & Mohr 1964, Mohr 1959) and therefore could be affected by changes in the R/FR ratio. Unfortunately cotyledon area within a batch of plants was found to be very variable, and no significant differences in mean cotyledon area between the different light treatments were found.

All experiments were done at a constant temperature of 20°C and at a saturation deficit of 2·5 mm Hg. The photoperiod in all experiments was 16 hrs followed in most cases by an uninterrupted dark period of 8 hrs.

The area and weight data were subjected to conventional growth analysis (see Evans 1972).

Results and discussion

Low R/FR

The first comparison experiment investigated the effect of lowering the R/FR ratio on growth in *I. parviflora*. Growth under the reference light source (1/F) was compared with growth under a tungsten source (experiment 1/T) (see Fig. 6.3 B & C for the spectral composition of the two light sources).

Figure 6.4 shows the relationship between dry weight distribution as measured by weight ratios, and total plant dry weight. The dry weight distribution shows ontogenetic drift to a small degree, the drift being largest in the leaf and stem weight ratios. It can be seen from the figure that leaf weight ratio shows an initial increase to a peak at a relatively low total plant dry weight whilst shoot weight ratio shows the inverse relationship. A common feature of the growth of *I. parviflora* is this initial rapid development of leaves, at the expense of the stems. In experiments 1/F and 1/T, the root weight ratio is relatively constant although it is more often found, as in the case of experiments 4/F and 4/A (Fig. 6.4) that root weight ratio shows an initial increase before reaching a fairly constant value. As plant size increases, the proportion of dry matter in the stems increases steadily whilst the proportion in the leaves shows a parallel decline. Such relationships are commonly observed in the growth of young plants, where increasing leaf area requires a stronger and therefore thicker stem for its support.

No significant differences in the distribution of dry matter between the leaves and the rest of the plant are apparent in experiments 1/F and 1/T (Fig. 6.4). The root weight ratio under tungsten light (expt. 1/T) is however consistently lower than under fluorescent light (expt. 1/F). Also over the later part of the experiment, the shoot weight ratio is higher under tungsten light (1/T). This difference in dry weight difference is, however, barely significant.

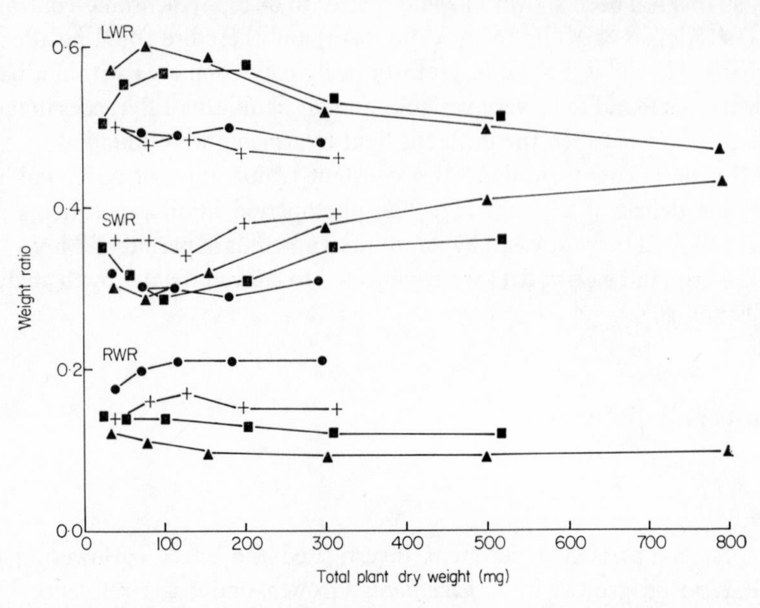

Figure 6.4. Relationship between total plant dry weight and leaf, stem and root weight ratios for *Impatiens parviflora*. ■, under white fluorescent tubes (Expt. 1/F); ▲, under tungsten lamps plus blue fluorescent tubes (Expt. 1/T); ●, under white fluorescent tubes (Expt. 4/F); +, under tungsten lamps plus blue and red fluorescent tubes (Expt. 4/A).

However, under tungsten light there is a marked increase in specific leaf area over the first part of the experiment (Fig. 6.5). Under both tungsten (1/T) and fluorescent (1/F) lights specific leaf area declines with increasing plant dry weight, as would be expected from the known form of the ontogenetic drift of specific leaf area in *I. parviflora*. As plant size increased there was a convergence of values of specific leaf area. Because of this marked ontogenetic drift shown by specific leaf area, to facilitate comparison between experiments a total plant dry weight of 100 mg has been used by previous workers as a basis for comparison of plants (see Section 28.7 in Evans 1972). At 100 mg total dry weight, the specific leaf area under tungsten light was 110 m^2/kg, whereas under fluorescent light it was only 93 m^2/kg, an increase of about 20% occurring under tungsten light.

The most obvious effect of tungsten light on the growth of *I. parviflora* was on the stem length (Fig. 6.6). By the time the plants under tungsten light (expt. 1/T) had reached 800 mg total dry weight, they were almost twice as tall as those grown under fluorescent light. This marked increase in plant height can be seen to be due to internode elongation, the hypocotyl length being unaffected. Hypocotyl elongation in *I. parviflora* appears to be controlled by peak light intensity (Hughes 1965b), although in other plants specific regions of the visible spectrum have been shown to control elongation, notably the blue and red ends of the spectrum (Mohr 1972). In the case of experiments 1/F and 1/T, no significant differences in hypocotyl length are apparent, but experiments 2/F and 2/B (Fig. 6.6), and experiments 4/F and 4/A (Fig. 6.8) show consistent differences in hypocotyl length, although the light intensities in these experiments were roughly comparable. No significant correlations between hypocotyl length and any parts of the visible spectrum have been found, and these differences within pairs of experiments are at present unexplained.

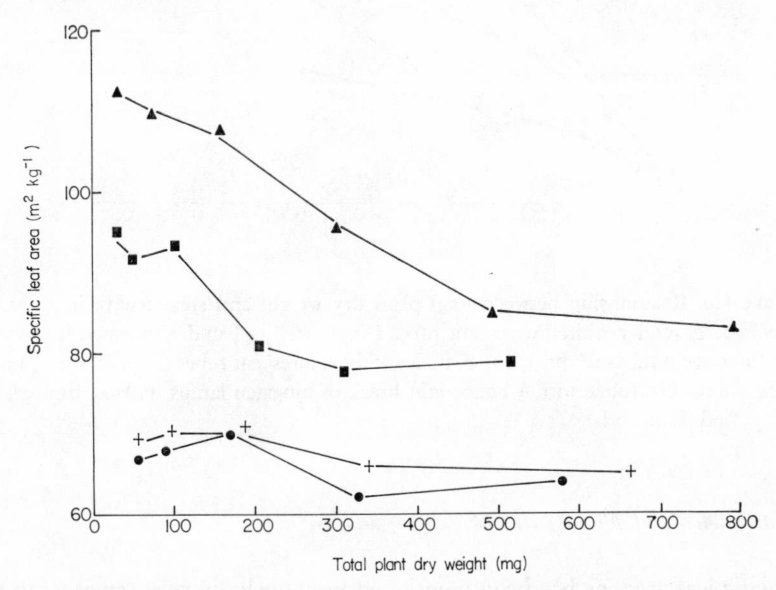

Figure 6.5. Relationship between total plant dry weight and specific leaf area for *Impatiens parviflora*. ■, under white fluorescent tubes (Expt. 1/F); ▲, under tungsten lamps plus blue fluorescent tubes (Expt. 1/T); ●, under white fluorescent tubes (Expt. 3/F); +, under white fluorescent tubes with 15 minutes' tungsten light at the end of each 16 hr photoperiod (Expt. 3/E).

The convergence of the values of specific leaf area between the two treatments, 1/F and 1/T, seen in Fig. 6.5, is almost certainly connected with the much greater internode elongation occurring under the tungsten light (expt. 1/T). This stem elongation would result in the raising of the leaves nearer to the lights thereby causing a significant increase in the total daily light received by the plants. This would cause a decrease in specific leaf area, since specific leaf area is inversely proportional to total daily light, and this decrease would be superimposed on the decrease due to ontogenetic drift of specific leaf area, resulting in a more rapid decrease in specific leaf area in plants under tungsten light (expt. 1/T) than in plants under fluorescent lights (expt. 1/F).

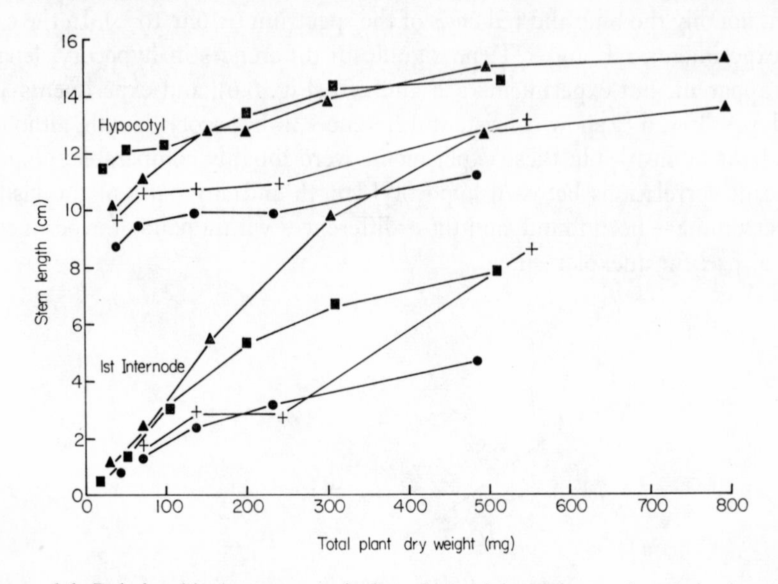

Figure 6.6. Relationship between total plant dry weight and stem length in *Impatiens parviflora*. ■, under white fluorescent tubes (Expt. 1/F); ▲, under tungsten lamps plus blue fluorescent tubes (Expt. 1/T); ●, under white fluorescent tubes (Expt. 2/F); +, under white fluorescent tubes with ½ hour night break of tungsten lamps midway through the dark period (Expt. 2/B).

Measurements of photosynthesis

Specific leaf area, as has been mentioned previously, is very sensitive to the total daily light received by the plant, and comparison experiments were therefore done at the same total daily light levels. In order to compare total daily light levels under light sources of differing spectral composition, reasons have been given above for making the assumption that it is the number of quanta in the photosynthetically active band of the spectrum that is the controlling factor. In order to check that the differences in specific leaf area observed between experiments 1/F and 1/T were due to differences in spectral composition, and not to an error in the assumption made or the method of matching the total daily light levels in the two treatments, measurements of carbon dioxide uptake of single attached leaves were made, using an infra-red gas analyser in an open current system, at a variety of light intensities. Lights of three differing spectral compositions were used, blue fluorescent tubes producing very little radiation beyond 600 nm. deluxe warm white tubes producing radiation with a R/FR ratio of about 3·5, and tungsten filament lamps producing very little radiation at wavelengths shorter than 450 nm and having a R/FR ratio of about 0·8. The number of quanta in the photosynthetically active region of the spectrum (425–700 nm) was calculated from spectroradiometric measurements of the light sources used. Both plants grown under fluorescent light (expt. 1/F) and those grown under tungsten light (expt. 1/T) were investigated. Rates of apparent carbon dioxide uptake per unit leaf area were plotted against the calculated number of quanta in the

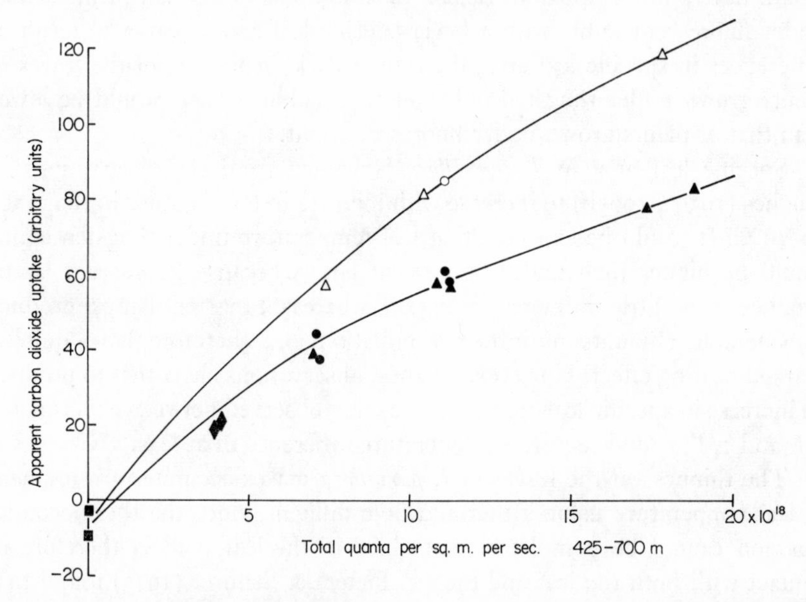

Figure 6.7. Apparent carbon dioxide uptake per unit area of leaves of *Impatiens parviflora* as a function of total quanta in the waveband 425–700 nm, under various light sources. ■, dark; ▲, tungsten lamps; ●, white fluorescent tubes; ◆, blue fluorescent tubes; open symbols, plants grown from seed under white fluorescent tubes; closed symbols, plants grown from seed under tungsten lamps.

waveband 425–700 nm, and the results are displayed in Fig. 6.7. No significant effects due to the spectral composition of the radiation under which photosynthesis was taken place could be detected in either group of plants.

There was however a highly significant difference in response between plants grown under tungsten light (closed symbols, Fig. 6.7) and those grown under fluorescent lights (open symbols). At the same total number of quanta in the waveband 425–700 nm, apparent carbon dioxide uptake per unit leaf area in plants grown under the fluorescent lights was always higher than that of plants grown under the tungsten lamps, regardless of the spectral composition of the light source under which the uptake was measured. Plants grown under tungsten lights (1/T) had a higher specific leaf area than those grown under fluorescent lights (Fig. 6.5) and therefore have a lower weight per unit area of leaf than those plants grown under fluorescent lights. This would result in a lower density of photosynthetic apparatus per unit leaf area, assuming that there were no compensatory changes in the density of photosynthetic organelles in the leaf cells as specific leaf area increases (Hughes 1959b). Hence one would expect that on an equal area basis the plants grown under tungsten light (solid symbols) with a higher specific leaf area, would have a lower apparent carbon dioxide uptake rate than plants grown under fluorescent lights with a lower specific leaf area. Also, as a result of differences in specific leaf area, the rate of dark respiration of the leaves of plants grown under tungsten light, on an equal area basis would be lower than that of plants grown under fluorescent light.

Specific leaf area in *I. parviflora* is known from the observations of Hughes (1959a, 1965b) to increase with increase in temperature in the range 10–20°C. It would be expected that leaf temperature under tungsten lamps would be higher than under fluorescent lamps because fluorescent lamps produce very little infra-red radiation whereas tungsten lamps produce considerable amounts of infra-red radiation and therefore have quite a marked heating effect. However, Hughes' observations show that to produce an increase in specific leaf area as large as that observed between experiments 1/F and 1/T, would require a temperature difference of at least 5°C.

The thinness of the leaves of *I. parviflora* makes accurate measurement of leaf temperature using a thermocouple difficult, since the thermocouple junction cannot be completely embedded in the leaf, and is therefore in contact with both the leaf and the air. Pieters & Schurer (1973) found that surface temperature measured by means of a contact thermocouple was too low, due to convective losses from the junction to the air and also because of poor heat transfer from the surface to the junction due to poor contact. The magnitude of the error depends on the temperature difference between the surface and the air. They suggest that by careful positioning of the junction, using a slight permanent force to ensure a firm contact with the surface, the error can be made less than 1°C, while for a given technique the differences between one leaf and another should be considerably less than this. This is in accord with the observations of Rackham (1966), who found the differences for *I. parviflora* to be small, and certainly much less than 1°C.

Direct measurements of leaf temperature in *I. parviflora* under tungsten and fluorescent lamps were made using a copper/constantan thermocouple in contact with the underside of the leaf, and the measurements indicate that the leaf temperature excess under tungsten lights was considerably less than 1°C, whilst leaf temperature under fluorescent lights was fractionally greater than air temperature.

In order to check the reliability of these direct measurements and to obtain a maximum possible value for the temperature rise due to tungsten lamps, an artificial leaf of metal foil, coated with matt black paint was constructed, with a copper/constantan thermocouple soldered flush on to the surface. Comparison of the temperature excess measured by means of a contact thermocouple with the temperature excess measured by the embedded thermocouple suggest that under tungsten lamps, for the artificial leaf the contact thermocouple error is of the order of 0·5°C. Taking the temperature excess indicated by the embedded thermocouple as very close to the true surface temperature, under tungsten lamps the temperature excess of the artificial leaf was 1°C. The maximum leaf temperature excess of the plants grown under tungsten lamps would therefore be considerably less than 1°C, since absorption of infra-red radiation by the leaf would be expected to be less than the absorption by the matt black artificial leaf, contact of the thermocouple would be better because of the greater flexibility of the leaf, and also cooling of the leaf would occur due to transpiration. The maximum leaf temperature difference between experiments 1/F and 1/T would have therefore been considerably less than 1°C. At most it might have been sufficient to explain roughly one tenth of the observed difference in specific leaf area observed under the two light sources.

The soil temperature over a 24-hr period was also measured to ascertain whether there were significant temperature differences of sufficient magnitude to explain the slight differences in root and shoot weight ratios observed under the two light sources. Although as would be expected, the temperature cycle under tungsten light was more marked than that under fluorescent lights, and the average soil temperature higher, the differences in root temperature appear not to be sufficient to explain the observed differences in dry weight distribution.

Night break

A night break has been found to increase both specific leaf area and internode length in the long-day plant *Callistephus chinensis* (Hughes & Cockshull 1965, Cockshull 1966). Growth of *I. parviflora* under a 16-hr photoperiod with fluorescent lights (expt. 2/F) was compared with that under a similar photoperiod and light source, but with a half hour night break of tungsten light midway through the dark period (expt. 2/B).

There was no significant effect on the distribution of dry matter between the roots and stems, and there was no effect on specific leaf area. Fig. 6.6 shows however that there was a significant increase in internode length under the night break treatment (expt. 2/B). There was also a consistent significant difference in hypocotyl length, the night break treatment producing longer hypocotyls (expt. 2/B). As has already been mentioned, a consistent difference in hypocotyl length has also been observed between experiments 4/F and 4/A (Fig. 6.8) in which there was a marked difference in spectral composition of the radiation, but no consistent difference in hypocotyl length was observed between experiments 1/F and 1/T (Fig. 6.6) although in this case also there was a large difference in spectral composition at the red end of the spectrum. These results suggest that altering the light climate does affect hypocotyl elongation, but as yet it has not been possible to explain the observed differences by correlation with any aspect of the light climate known to affect hypocotyl elongation.

I. parviflora is a day-neutral plant, unlike *C. chinensis* which is a long-day plant and one might therefore expect a night break to have no effect on the growth of *I. parviflora*. It turns out, however, that a night break does have an effect on internode elongation although not on leaf expansion.

End of photoperiod

Several workers, for example, Downs *et al.* (1957) with *Phaseolus vulgaris* and Kasperbauer (1971) with *Nicotiana tabacum*, have shown that internode elongation can be produced by giving far-red light at the end of the photoperiod. They have further shown that this effect is reversible with red light.

Changing the spectral composition at the end of the 16-hr photoperiod of fluorescent light, by giving 15 minutes of tungsten light at quite a high intensity (expt. 3/E), has no significant effect on internode elongation, on dry weight distribution or on specific leaf area in *I. parviflora* (Figs 6.5 and 6.8).

This could be because 15 minutes of tungsten light at the end of the photoperiod is not long enough to establish a new photoequilibrium, although most workers suggest that less than 10 minutes is sufficient to establish a new photoequilibrium. For example, in *Chenopodium rubrum* the process takes only 8 minutes (Kasperbauer *et al.* 1964). An alternative explanation would be that tungsten light with its low R/FR ratio does not significantly alter the photoequilibrium established under fluorescent light, which is said to be the same as that established under natural daylight in the open (Mohr 1972).

Far-red light at the end of an 8-hour photoperiod under cool white fluorescent tubes has been shown to increase both internode length and specific leaf area in *Nicotiana tabacum* (Kasperbauer 1971). Also, under field conditions, plants within the rows grow taller and have a higher specific leaf area than those at the ends of the rows and this is interpreted by Kasperbauer as possibly due to the light quality at the end of the day, although a simpler explanation is that the plants on the end of the rows received more light because they were only shaded on three sides thus causing a lowering of the specific leaf area—part, in fact, of the familiar 'edge effect'. The tobacco plants receiving far-red light at the end of the photoperiod in the growth cabinets, despite their higher specific leaf area, had a slightly lower growth rate than those receiving red light suggesting that although the specific leaf

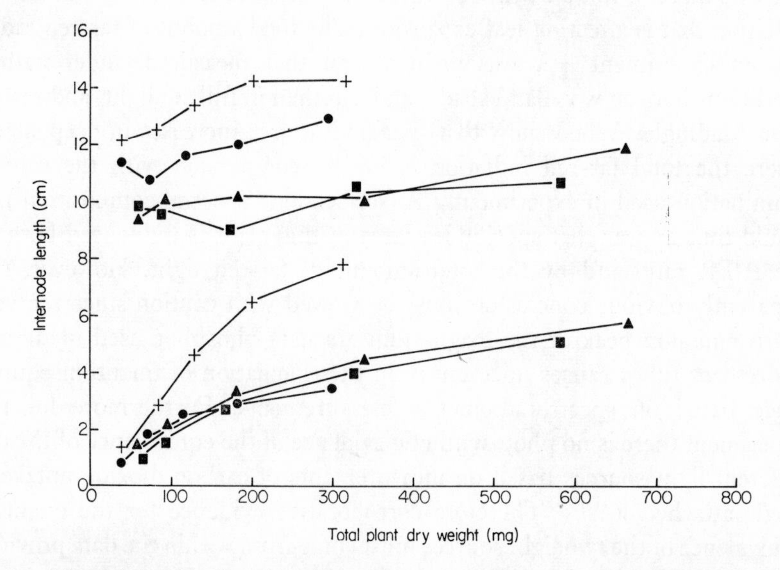

Figure 6.8. Relationship between total plant dry weight and stem length in *Impatiens parviflora*. ■, under white fluorescent tubes (Expt. 3/F); ▲, under white fluorescent tubes with 15 minutes' tungsten light at the end of each 16 hr. photoperiod (Expt. 3/E); ●, under white fluorescent tubes (Expt. 4/F); +, under tungsten lamps plus blue and red fluorescent tubes (Expt. 4/A). Upper family, hypocotyl; lower family, first internode.

area was increased by the far-red light, either the leaf area ratio had decreased or increased leaf expansion had been accompanied by a decrease in the photosynthetic capacity of the tissues.

High far-red

By use of fluorescent tubes with a magnesium arsenate phosphor having a single emission peak at about 660 nm, in conjunction with tungsten lamps, a light source with a high R/FR ratio and a high level of FR radiation can be produced. The magnesium arsenate tubes used were the same as those used by Hughes in his investigation of the efficiency of conversion of light energy by *I. parviflora* (Hughes 1965a). The spectral composition of this combination of lights is shown in Fig. 6.3D. The R/FR ratio was 3·0, about the same as that from the reference combination of fluorescent tubes, whilst the total radiation at 730 nm was nearly doubled.

Comparison of growth under tungsten lamps plus magnesium arsenate tubes (expt. 4/A) with growth under the reference combination of fluorescent tubes (expt. 4/F) showed a significant increase in specific leaf area. This result suggests that it is not only the R/FR ratio but also the total amount of far-red radiation that is affecting leaf expansion. The total amount of far-red radiation in experiment 4/A was greater than that measured under natural conditions in deep woodland shade, and less than in full sunlight, and results from Madingley Wood show that specific leaf area increases in deep shade, where the total far-red radiation is low in comparison with the cabinet illumination used in experiment 4/A (see section 'Spectral composition'). It would thus appear that specific leaf area in this plant is influenced both by the R/FR ratio and by the total amount of far-red light. However, this apparently obvious conclusion must be viewed with caution since the very sharp emission peak of the magnesium arsenate phosphor used in the red fluorescent tubes causes uncertainty in the calculation of quantum equivalence based on spectroradiometric measurements. Furthermore for this experiment there is no photosynthetic evidence of the equivalence of the two different light sources based on measurements of carbon dioxide uptake of single attached leaves. Therefore corroborative evidence for the quantum equivalence of the two light sources must come from within the data provided by the experiments themselves. It then turns out, on closer examination of the data from experiments 4/A and 4/F, that the increase in specific leaf area observed in experiment 4/A was probably due to a significant difference in the total daily light levels between experiments 4/A and 4/F, rather than to the difference in spectral composition. This conclusion follows from a comparison of the values of unit leaf rate and relative growth rate in these two experiments, and also from a comparison of the relative growth rates measured in experiments 1/T and 1/F.

In experiments 1/T and 1/F, changing the spectral composition produced an increase in specific leaf area (Fig. 6.5) and also the relative growth rate of the plants under tungsten light (1/T) was significantly greater, resulting in plants of 800 mg total dry weight after 30 days, compared with a total dry weight of only 500 mg after 30 days under fluorescent lights (1/F).

In the case of experiments 4/F and 4/A, although specific leaf areas in experiment 4/A were significantly greater than in experiment 4/F, the relative growth rates of the two groups of plants were not significantly different, the total dry weight after 26 days being 290 mg in experiment 4/F and 310 mg in experiment 4/A.

The relationship between relative growth rate (RGR), unit leaf rate

(ULR), specific leaf area (SLA) and leaf weight ratio (LWR) is (see page 138):

$$RGR = ULR \times SLA \times LWR$$

Under the conditions of growth in the experiments described here, leaf weight ratio is constant. Unit leaf rate and specific leaf area are, however, both markedly affected by the total daily light level, unit leaf rate being proportional to total daily light, whilst specific leaf area is inversely related (see section 'Radiation and Growth'). Specific leaf area is therefore inversely proportional to unit leaf rate and, as a consequence, within the range of total daily light levels used in these experiments, relative growth rate is more or less independent of total daily light level and is therefore relatively constant.

In experiments 1/T, however, not only was the specific leaf area greater than in experiment 1/F, but also the relative growth rate was significantly higher, suggesting that there had been little or no compensatory change in unit leaf rate, and that reduction of the R/FR ratio had caused an increase in specific leaf area without markedly affecting the unit leaf rate. Measurements of apparent carbon dioxide uptake of single attached leaves of plants from experiments 1/F and 1/T shows that the uptake rate per unit area in the 1/T plants is lower than in the 1/F plants (Fig. 6.5). This suggests that the rate of photosynthesis, and therefore possibly the unit leaf rate, of the 1/T plants was lower than that of the 1/F plants, although the relative growth rate data in conjunction with the specific leaf area data suggests that the difference in unit leaf rate must have been small. It must be borne in mind that unit leaf rate is a measure of the increase in plant dry weight per unit leaf area over at least a whole day, and thus measures the balance between photosynthetic gain per unit leaf area and respiratory losses in not only the leaves but also the stems and roots, whilst the apparent rate of carbon dioxide uptake of single attached leaves gives an indication of the balance between photosynthetic gain per unit leaf area and respiratory losses by the leaf per unit leaf area, during a short part of the photoperiod. Therefore no firm conclusions as to the relative magnitudes of the unit leaf rates in experiments 1/F and 1/T can be made from the measured apparent rates of carbon dioxide uptake, since this measurement gives no indication as to the rate of loss due to respiration occurring in all parts of the plant during the dark phase.

Although in experiment 4/A the specific leaf area was significantly higher than in experiment 4/F, yet the relative growth rates of the two groups of plants are more or less the same, indicating that the increase in specific leaf area was coupled with a corresponding decrease in unit leaf rate. This compensatory change is what would be expected as a result of differences in the total daily light levels between the two experiments.

Thus the similarity of relative growth rates in experiments 4/F and 4/A, and the difference in rates observed between experiments 1/F and 1/T suggests that the differences in specific leaf area observed within the two pairs of experiments were not due to the same environmental factor or combination of factors. The results strongly suggest that the difference in specific leaf area in experiments 4/F and 4/A was due to a significant difference in total daily light levels, whilst the difference observed in experiments 1/F and 1/T was due to the differences in spectral composition of the lamps used in the two experiments.

Figure 6.8 shows that there is however a significant increase in internode elongation under the combination of tungsten and magnesium arsenate fluorescent tubes with its high total of far-red light (expt. 4/A). There is also a significant change in the dry weight distribution, as can be seen in Fig. 6.4. The shoot weight ratio is increased whilst the root weight ratio is correspondingly decreased in the plants grown under the light source containing tungsten lamps (expt. 4/A). A similar but barely significant effect is observable in experiment 1/T (see Fig. 6.4). Increased internode length appears therefore to be associated with an increase in the proportion of dry matter incorporated into the stem, at the expense of the roots, suggesting that stem elongation involves both an increase in cell expansion and an increase in cell division. The longer stems also tend to be hollow indicating that there is also a change in the growth pattern of the stem tissues.

Conclusions

The results of the four comparison experiments suggest that changes in the R/FR ratio in the range 0·8–3·5 do affect the specific leaf area, and therefore provide an explanation for the observed difference in the relationship between the specific leaf area and total daily light for plants grown in Madingley Wood as compared with those grown in the laboratory in cabinets (Fig. 6.1). Internode elongation is also markedly affected by changes in the spectral composition at the red end of the spectrum. In the case of internode elongation, it appears to be the total far-red radiation and not the R/FR ratio that controls elongation.

The R/FR ratio is presumed to involve the red/far-red reversible phytochrome system and to be acting through the concentration of Pfr maintained in the tissues during the photoperiod. Other morphogenetic responses involving the intensity of far-red radiation have been proposed by some workers to be due to some excited species of Pfr (Hartmann 1966, Mohr 1972). One would expect in such a response that there would be at least a slight effect attributable to the R/FR ratio, since the concentration of the excited species (P*fr) will be a function of the concentration of the ground state form (Pfr).

$$[P^*fr] = f_1([Pfr] \cdot I_{730})$$
$$[Pfr] = f_2(R/FR)$$
$$\therefore \quad [P^*fr] = f_3((R/FR) \cdot I_{730})$$

where: I_{730} = intensity of radiation at 730 nm
Pfr = ground state of the far-red form of phytochrome
P*fr = excited species of Pfr.

The effect of a night break on internode elongation suggests that Pfr also exerts a control on internode elongation. Work on etiolated seedlings of other species suggests that both Pfr and P*fr are involved in the control of longitudinal growth, and that two distinct mechanisms are involved. The first mechanism involves the concentration of P*fr and controls the rate of growth during the photoperiod, the rate being proportional to the log. of the irradiance. The second mechanism in which only Pfr is involved, controls the rate of elongation in the dark following irradiation, by a threshold mechanism (Schopfer & Oelze-Karow 1971).

The results presented in this paper suggest that both the R/FR ratio and the total far-red radiation are involved in the control of growth in *I. parviflora*. The R/FR ratio appears to influence leaf expansion whilst total far-red radiation controls internode elongation. As a result of stem elongation, there is a change in dry weight distribution between the stems and roots.

It should be emphasized that this is an interim report on work still in progress. More experiments are needed to confirm results already obtained, and in particular investigations on the effects of a low R/FR ratio coupled with low far-red radiation have not yet been made.

Acknowledgements

I would like to thank Dr. G. C. Evans for his supervision of this work and for his helpful criticism of the manuscript. This work was carried out during the tenure of a N.E.R.C. research studentship.

References

CANHAM A.E. (1966) *Artificial light in horticulture.* Centrex Publishing Company, Eindhoven.

COCKSHULL K.E. (1966) Effects of nightbreak treatment on leaf area and leaf dry weight in *Callistephus chinensis. Ann. Bot.* 30, 791–806.

COOMBE D.E. (1956) Biological flora of the British Isles. *Impatiens parviflora* DC. *J. Ecol.* 44, 701–13.

COOMBE D.E. (1957) The spectral composition of shade light in woodlands. *J. Ecol.* 45, 823–30.

COOMBE D.E. (1966) The seasonal light climate and plant growth in a Cambridgeshire wood. In *Light as an Ecological Factor* (*Symp. Brit. Ecol. Soc.* 6, 148–66). Blackwell Scientific Publications, Oxford.

DOWNS R.J., HENDRICKS S.B. & BORTHWICK H.A. (1957) Photoreversible control of elongation of pinto beans and other plants under normal conditions of growth. *Bot. Gaz.* 118, 199–208.

EVANS G.C. (1969) The spectral composition of light in the field. I. Its measurement and ecological importance. *J. Ecol.* 57, 109–25.

EVANS G.C. (1972). *The Quantitative Analysis of Plant Growth.* Blackwell Scientific Publications, Oxford.

EVANS G.C. & HUGHES A.P. (1961) Plant growth and the aerial environment. I. Effect of artificial shading on *Impatiens parviflora. New Phytol.* 60, 150–80.

FEDERER C.A. & TANNER C.B. (1966). Spectral distribution of light in the forest. *Ecology* 47, 555–66.

HARTMANN K.M. (1966) A general hypothesis to interpret high energy phenomena of photomorphogenesis on the basis of phytochrome. *Photochem. Photobiol.* 5, 349–66.

HUGHES A.P. (1957) Plant growth and the aerial environment. Ph.D. thesis, Univ. Cambridge.

HUGHES A.P. (1959a) Plant growth in controlled environments as an adjunct to field studies. Experimental applications and results. *J. agric. Sci., Camb.* 53, 247–59.

HUGHES A.P. (1959b). Effects of the environment on leaf development in *Impatiens parviflora* DC. *J. Linn. Soc.* (*Bot.*) 56, 161–5.

HUGHES A.P. (1965a). Plant growth and the aerial environment. VI. The apparent efficiency of conversion of light energy of different spectral compositions by *Impatiens parviflora. New Phytol.* 64, 48–54.

HUGHES A.P. (1965b) Plant growth and the aerial environment. IX. A synopsis of the autocology of *Impatiens parviflora. New Phytol.* 64, 399–413.

HUGHES A.P. & COCKSHULL K.E. (1965) Interrelations of flowering and vegetative growth in *Callistephus chinensis* (var. Queen of the Market). *Ann. Bot.* 29, 131–51.

HUGHES A.P. & EVANS G.C. (1962) Plant growth and the aerial environment. II. Effects of light intensity on *Impatiens parviflora. New Phytol.* 61, 154–74.

HUGHES A.P. & EVANS G.C. (1964) Plant growth and the aerial environment. V. Effects of (a) rooting condition, (b) red light on *Impatiens parviflora. New Phytol.* 63, 194–202.

JORDON C.F. (1969) Derivation of leaf area index from quality of light on the forest floor. *Ecology* 50, 663–6.

KASPERBAUER M.J. (1971) Spectral distribution of light in a tobacco canopy and effects of end-of-day light quality on growth and development. *Plant Physiol.* 47, 775–8.

KASPERBAUER M.J., BORTHWICK H.A. & HENDRICKS S.B. (1964) Reversion of Phytochrome 730 (P_{FR}) to P_{660} (P_R) assayed by flowering in *Chenopodium rubrum. Bot. Gaz.* 125, 75–80.

KLEIBER H. & MOHR H. (1964) Der Einfluss sichtbarer Strahlung auf die Stomatabildung in der Epidermis der kotyledonen von Sinapsis alba L. *Z. Bot.* 52, 78–85.

LEWIS J.P. (1963) Plant growth and nutritional factors. Ph.D. Thesis, Univ. Cambridge.

MEIJER G. (1957) Influence of light on the flowering response of *Salvia occidentalis. Acta Bot. neerl.* 6, 395–406.

MILLENER L.H. (1952). Experimental studies on the growth forms of the British species of *Ulex* L. Ph.D. Thesis, Univ. Cambridge.

MILLENER L.H. (1961). Daylength as related to vegetative development in *Ulex europaeus* L. I. The experimental approach. *New Phytol.* **60**, 339–54.

MOHR H. (1957) Der Einfluss monochromatischer Strahlung auf das Längenwachstum des Hypokotyls und auf Anthocyanbildung bei Keimlingen von *Sinapis alba* L. *Planta* **49**, 389–405.

MOHR H. (1959) Dev Lichteinfluss auf das Wachstum der Keimblätter bei *Sinapis alba* L. *Planta* **53**, 219–45.

MOHR H. (1972) *Lectures on Photomorphogenesis.* Springer-Verlag, Berlin.

PIETERS G.A. & SCHURER K. (1973). Leaf temperature measurements. I. Thermocouples. *Acta Bot. neerl.* **22**, 569–80.

RABINOWITCH E. & GOVINDJEE (1969) *Photosynthesis.* J. Wiley & Son, Inc. New York.

RACKHAM O. (1966) Radiation, transpiration, and growth in a woodland annual. In *Light as an Ecological Factor* (*Symp. Brit. Ecol. Soc.* **6**, pp. 167–85). Blackwell Scientific Publications, Oxford.

RAJAN A.K., BETTERIDGE B. & BLACKMAN G.E. (1971) Interrelations between the nature of the light source, ambient air temperature and the vegetative growth of different species in growth cabinets. *Ann. Bot.* **35**, 323–43.

SCHOPFER P. (1969) Die Hemmung des Streckungswachstums durch Phytochrom—ein Stoffaufnahme erfordernder Prozess? *Planta* **85**, 383–8.

SCHOPFER P. & OELZE-KAROW H. (1971) Nachweis einer Schwellenwertsregulation durch Phytochrom bei der Photomodulation des Hypokotylstreckungswachstums von Senfkeimlingen (*Sinapis alba* L.). *Planta* **100**, 167–80.

SPENCE D.H.N. (1975) Light and plant response in fresh water. *This vol.* 93–133.

SPENCE D.H.N., CAMPBELL R.M. & CHRYSTAL J. (1971) Spectral intensity in some Scottish freshwater lochs. *Freshwater Biol.* **1**, 321–7.

SPENCE D.H.N., CAMPBELL R.M. & CHRYSTAL J. (1973) Specific lea areas and zonation of freshwater macrophytes. *J. Ecol.* **61**, 317–28.

STOUTJESDIJK P.H. (1972a) Spectral transmission curves of some types of leaf canopies with a note on seed germination. *Acta Bot. neerl.* **21**, 185–91.

STOUTJESDIJK P.H. (1972b) A note on the spectral transmission of light by tropical rain forest. *Acta Bot. neerl.* **21**, 346–50.

Questions and discussion

Dr. Rackham said that seed of *Impatiens parviflora* germinates in nature towards the end of March or early April, well before the deciduous canopy of the trees in Madingley Wood develops towards the end of May. Thus for the first month or so the plants will be growing under light similar in spectral composition to that in the open. Mr. Young agreed, but pointed out that the plants continued to grow, though at a slower rate, after the opening of the canopy. This later growth would be affected by the low red/far red ratio both as regards leaf expansion and stem elongation. Professor Pigott said that specific leaf area is very sensitive to changes in temperature. If, as he assumed, the temperature regime in Madingley Wood differed markedly from that in the cabinet experiments, how was this allowed for when comparing specific leaf areas of plants grown under the two conditions? Mr.

Young said that in view of uncertainty about the exact conditions in Madingley Wood, no specific allowance had been made in the comparison shown in Fig. 6.1. However, the observed difference was much too large to be accounted for by differences in temperature. In amplification, Dr. Evans said that in many series of cabinet experiments on this plant under a wide variety of temperature and other conditions, Dr. Hughes had never succeeded in raising a population of plants with a mean specific leaf area as high as 12 dm^2 g^{-1}. On the other hand, in Madingley Wood Dr Coombe's observations on plants of the same size growing in the same rooting medium, shown in Fig. 6.1, ranged around 14 dm^2 g^{-1}, and for no population there did the fiducial limits fall as low as 12 dm^2 g^{-1}. He therefore suggested that whatever caused the higher values in the wood it could hardly be temperature differences.

7 Problems involved in the design of apparatus for measuring the spectral composition of daylight in the field

D. P. EDWARDS AND G. C. EVANS
Botany School, University of Cambridge

Introduction

An analysis of data from experiments described at the previous symposium on the present subject (Coombe 1966, Hughes 1966) demonstrated the probability that changes in the spectral composition of daylight within the naturally occurring range could produce large effects on the form of a woodland plant (Evans 1969). The previous paper (Young 1975, this vol., p. 135) has established that this is indeed so, has extended the range of observed effects on plant form, and has also demonstrated how large can be the consequences for the overall rate of plant growth. So far, however, the experiments have been confined to a single species and the parallel observations on spectral composition to a single wood. In view of the potential importance of these effects it seems desirable that observations should be extended to other species and to other plant communities. The problem of the measurement of the spectral composition of natural radiation in the field is an old one, which has engaged the attention of meteorologists and ecologists intermittently for a long time (see e.g., Collingbourne 1966, Evans 1966). This problem now acquires a fresh importance, and it becomes more than ever desirable to have a measuring instrument appropriate for ecological fieldwork.

The complexity of the overall problem is sufficiently demonstrated by the fact that no official Meteorological Service in the world has published the results of systematic recordings of the spectral composition of daylight over extended periods comparable in length to the tropical year, or to the growing season in a particular part of the world.

The reasons for this state of affairs fall into two broad categories—the inherent complexities of the system being measured, which are part of the given problem: and the limitations attached to the available techniques of measurement.

Plants and animals living in a natural environment are commonly exposed to substantial short-wave radiation from sources varying in their spectral composition and extending over roughly a hemisphere or more (but see Evans 1966). When the sun is unobscured its direct radiation is approximately parallel and of high intensity compared with the other sources. During the hours of daylight there is always diffuse light from blue sky and/or from clouds, haze, mist or fog. All these sources may be modified, via reflection or transmission, by various elements of the natural environment itself, including the vegetation, the ground surface and bodies of water. What is more, unless the sky is cloudless or uniformly overcast, the spatial relations of these various sources change from minute to minute, so that it would not be profitable to attempt a complete description of the state of affairs at any instant—the various sources must be integrated to give a total comparable to the effective total at the surface of the plant or animal. Average spectral composition incident on a horizontal surface is the most convenient standard approximation, which ensures comparability with the extensive meteorological observations of total short-wave radiation incident on a horizontal surface.

But the standard methods of measuring spectral composition were all worked out using point sources or parallel light: to adapt any of them to field measurements of this kind it is necessary to interpose some device which will collect from a complete hemisphere and produce a collimated beam, without changing the average spectral composition of the collected radiation. For general purposes the most satisfactory device of this kind is some form of integrating sphere, such as that described by Collingbourne (1966). Devices involving opal glass are frequently used for cosine correction in instruments of wide acceptance angle designed to measure total short-wave radiation, or radiation in a broad band such as the visible region of the spectrum. Such devices are much less suitable for measurements involving spectral composition, as their spectral properties are liable to change with angle of incidence, so that for a particular situation it is necessary to undertake tedious checks to ensure that any consequent changes in the average spectral composition between collection and measurement fall within acceptable limits of error.

However, integrating spheres (or, indeed, any other device serving the same purpose) inevitably involve attenuation of the collected radiation by two powers of ten at least; and at the same time division of the collimated beam into spectral regions involves a further attenuation of at least one or two more powers of ten. Thus the detector is required to measure irradiance less by three or four orders of magnitude than that measured by a standard solarimeter, and the greater the spectral discrimination, the less radiation there is to measure.

Furthermore, in the case of an instrument for general ecological use, there are additional requirements over and above those demanded of a practicable laboratory instrument—requirements for robustness, stability, resistance to adverse environmental conditions, and independence of power supplies.

The present paper describes the main problems encountered in one particular attempt to design and construct a practicable field instrument within the restrictions discussed, and some of the possible methods of solving them which have been investigated. Three separate but related operations have to be accomplished: the making of field observations, the calibration of the field apparatus and the final calculation (or retrieval) of the spectral compositions. One further important consideration has to be taken into account. All three phases of the work must be within the resources of the average biological department.

Preparatory work

As a starting point in this work, an integrating-sphere radiometer was constructed (Fig. 7.1). This is fully described by Evans (1969). The radiometer

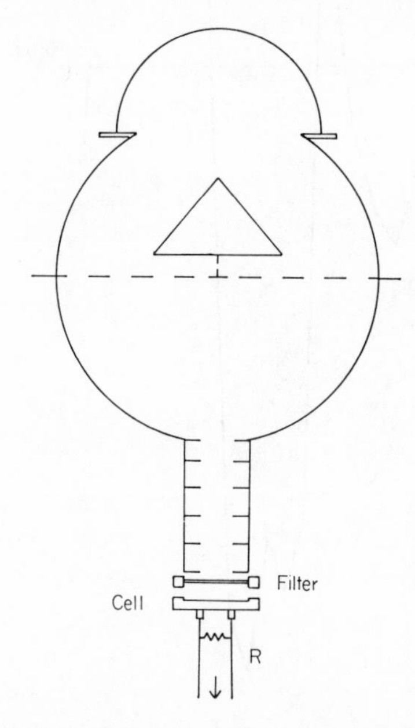

Figure 7.1. Simplified vertical section of the integrating-sphere radiometer.

consists of an integrating sphere to collect from a hemisphere, a sliding tray to interpose filters beneath the collimating tube and a selenium photovoltaic cell as the detector. The filters are a set of Schott's glass filters, which cut off on the short wavelength side. Twenty-one filters span the range 375–780 nm.

The output of the photocell is fed through a 100 ohm resistance, and the voltage drop across the resistance, which would be proportional to the photo-current, is measured by a digital voltmeter of very high input impedance. Negligible current flows in the connecting leads, which are shielded against stray voltages. If desired the readings of the voltmeter can be recorded on an electric typewriter or punched on paper tape.

Since it takes between 500 and 600 seconds to feed all 21 filters into the instrument, variations in total intensity during such measurements affect the measurements themselves. Figure 7.2a illustrates the common observation

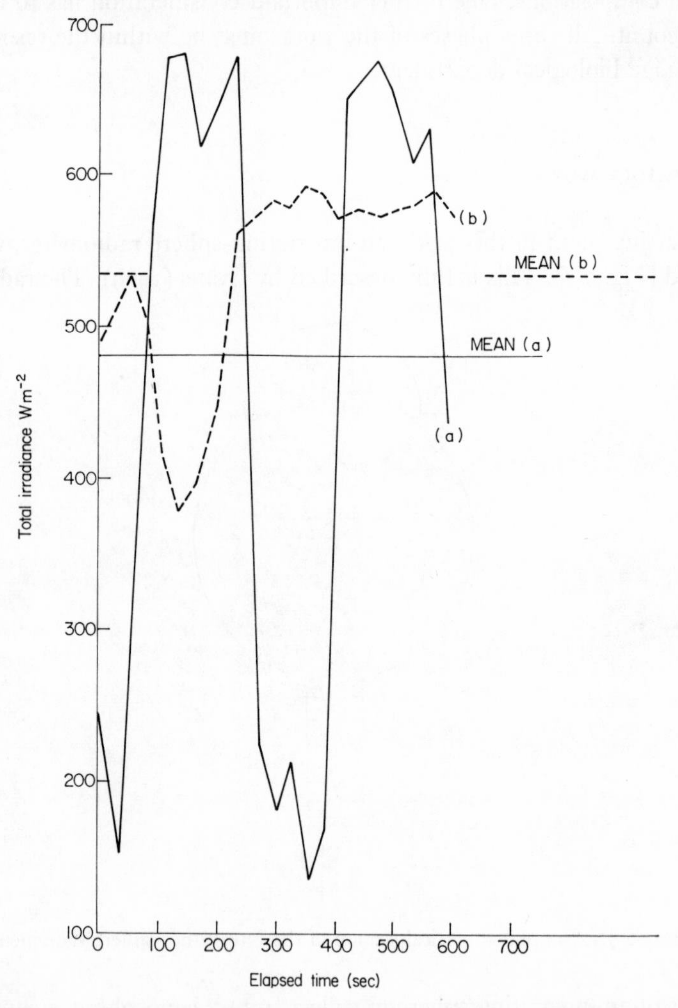

Figure 7.2. Total short-wave irradiance measured by a Kipp solarimeter at the 'large clearing' site;
a. direct sunlight, blue sky and fast-moving clouds, 2 September 1971;
b. direct sunlight and a pale blue sky without visible cloud, 7 September 1971.

that the total intensity, as measured by a Kipp solarimeter, varies greatly on a day when there is strong direct sunlight, blue sky, and fast-moving clouds. However, measurement shows that departures from the mean exceeding 20% also occurred under some conditions of clear sunlight and a cloudless pale blue sky (Fig. 7.2b). This condition is very common over East Anglia in late summer and early autumn. Equally surprising were the variations shown in Fig. 7.3. Figure 7.3a shows a variation of over ± 1% in total intensity from

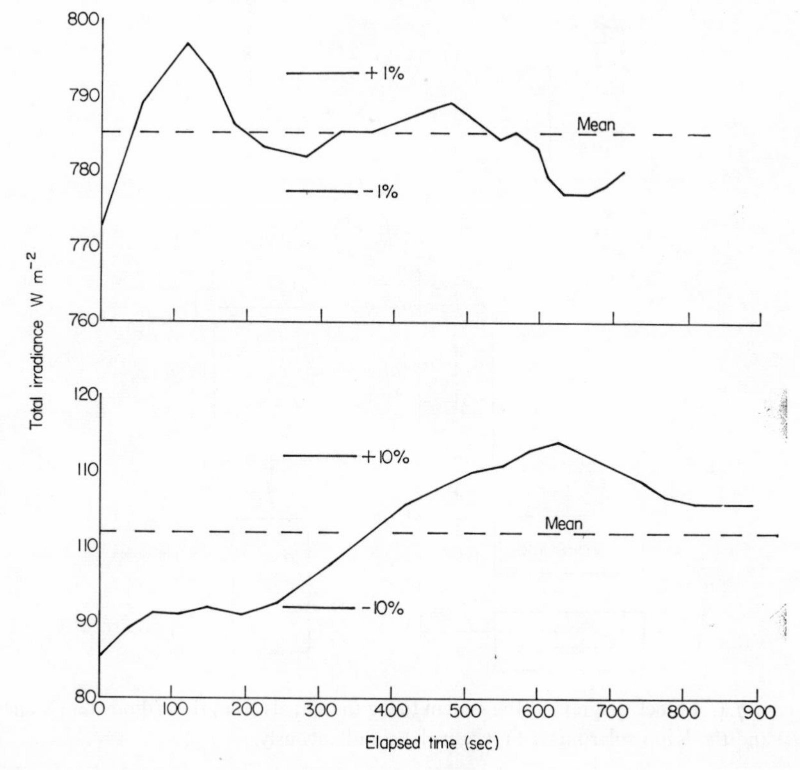

Figure 7.3. Total short-wave irradiance measured by a Kipp solarimeter at the 'large clearing';
a. direct sunlight and a deep blue sky without visible cloud, 7 July 1971;
b. complete cover of heavy uniform cloud, 11 August 1971.

direct sunlight and an apparently cloudless deep blue sky. A heavy blanket of apparently uniform cloud, during a wet August, had a variation of about ± 10%, as shown by Fig. 7.3b.

To enable an allowance to be made for these variations a Kipp solarimeter was set up 2 m south of the radiometer position in Madingley Wood near Cambridge. The sensitive area of the Kipp was on the same level as the entrance aperture of the integrating sphere. For each filter the output of the photocell and the solarimeter were logged simultaneously. They were then stored on the displays of two digital voltmeters, while being fed to a paper-tape punch. A block diagram of the layout is given in Fig. 7.4. The logging

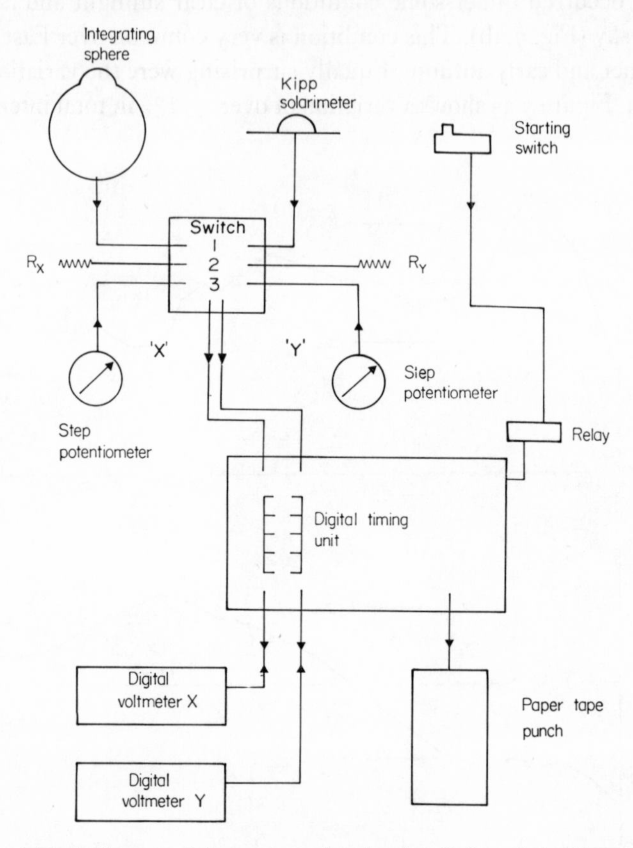

Figure 7.4. A block diagram of the system to log the signals from the radiometer (X channels) and the Kipp solarimeter (Y channels) simultaneously.

equipment was housed in a wooden hut about 80 m from the radiometer. The latter was positioned at the north end of the 'large clearing' in Madingley Wood, a position which made possible measurements both of direct sunlight and of sunlight transmitted through the woodland canopy.

By using the solarimeter measurement, each filter reading can be normalized to a constant total intensity to reduce the effect of variation. Since a large change in total intensity is very probably associated with a change in spectral composition, this is only a first approximation to a general method, but it is probably at its most satisfactory under the conditions of cloudless blue sky or uniform overcast to which observations have so far been confined.

In this field situation, damp and the very high relative humidity in the wood were the major problems. An armoured submarine cable, proof against damp and mice and terminated by military-pattern connectors, had to be used in the outdoor part to get reliable connections. Although the digital voltmeters were of a fairly modern design, they were very sensitive to damp on their circuit boards. During the season, a small dehumidifier had to be run continuously in the instrument cubicle within the hut to keep the relative humidity down and maintain serviceability.

Design requirements

From this experience, the requirements for an improved field instrument can be readily formulated.

The present radiometer is not weathertight, so that it cannot be left out between sets of observations. Thus the first requirement is a smaller weather-proof integrating sphere with a totally enclosed filter-changing mechanism. Such an instrument could then be left permanently *in situ*. Making the changing mechanism automatic to speed up the rate of change is the next improvement. With a very simple filter wheel, it should be possible to change filters at least every 20 sec., giving a time for 21 filters of 420 sec. or less.

At present the sensitivity of the instrument is fairly low, so that it cannot be used to give fine discrimination under a dense woodland canopy. This low sensitivity is partly due to the very low throughput of the integrating sphere. Only $2 - 3 \times 10^{-3}$ of the input radiation reaches the surface of the cell, so that an improvement in sphere performance would be well worth-while. Improvements in cell sensitivity are largely dependent on manufac-turing technology. Silicon photovoltaic cells and photodioides are the two obvious alternative detectors under consideration at present. The use of a photomultiplier seems to be ruled out by the complication of the high voltage supply required for the dynode chain and the great difficulty of maintaining stability under damp conditions.

The final requirement for a completely self-contained field instrument is a watertight battery-driven data-logger. One such unit has been described by Strangeways (1972). The particular unit described by that author has a scale length of 240 units, while the present digital voltmeters have one of 10,000. Thus there is some difficulty in range matching. A much greater obstacle to the use of battery-driven data-loggers is the conversion of their output (in the form of a magnetic tape cassette) to a form that is compatible with a typical computer. At present this may easily cost as much as the field instrument and calibration apparatus combined.

Calibration

The second phase of the project is the measurement of the transmission curves of the filters and the measurement of the relative spectral sensitivity of the photocell. The transmission curve of a filter can be measured to about 1 % by using a laboratory spectrophotometer (e.g. a Pye Unicam SP800). The resulting curve can be tabulated at 2 nm intervals for further calculation. This seems the best compromise between insufficient definition and picking up too much instrument noise.

The measurement of the spectral response curve for any photodetector is a far more difficult process. For the present purpose, a method of measurement must fulfil the following criteria:

a. The cell should be irradiated at a level comparable with that which it receives in the field. This avoids the assumption of linearity of response over a large power range.

b. The whole of the cell surface should be irradiated as uniformly as possible, to counteract variations in surface sensitivity.

c. This radiation should be fairly close to normal incidence, to match the conditions within the field radiometer.

d. The calibration method should be stable over a period of years, to ensure comparability between successive calibrations.

A widespread search of the literature has revealed remarkably few explicit descriptions of calibration methods. One of the best descriptions of the classical method is the one by Gillham (1961), illustrated in Fig. 7.5A. The cell under test and a black, reference, thermopile are irradiated alternately by a narrow waveband produced by a monochromator. The selenium cells used at present are about 40 mm in diameter, so that a very large monochromator would be required to irradiate one satisfactorily. As this was beyond the resources available, an alternative method described by Wyszecki (1960) was considered. This method irradiated the cell under test by a large tungsten lamp of known spectral irradiance. Instead of a monochromator, a set of interference filters were placed in the optical path between the lamp and the cell. Since the source requires recalibration against a substandard at frequent intervals, considerable auxiliary apparatus is needed.

It was decided that a hybrid method would best meet the criteria outlined above within the resources available. A block diagram of this is shown in Fig. 7.5B. The method uses a set of interference filters (Fig. 7.6) to separate the spectral intervals. The reference standard is a black detector, a compensated Moll thermopile, which is irradiated alternately with the cell under test, as in the classical method. As the Moll thermopile provides a stable long-term standard, the need for recalibration is removed and the source need not have a known spectral emittance curve. The source chosen is a xenon arc lamp. This has two advantages over a tungsten lamp. Figure 7.7 shows that the Planck curve for 3,200°K, which corresponds to a quartz halogen tungsten lamp, gives relatively little radiation below 500 nm. The xenon arc gives considerably more radiation at these shorter wavelengths. Since the bright area of the arc is very compact, about 6 × 4 mm overall, the collimator optics are much simpler than those for a filament lamp.

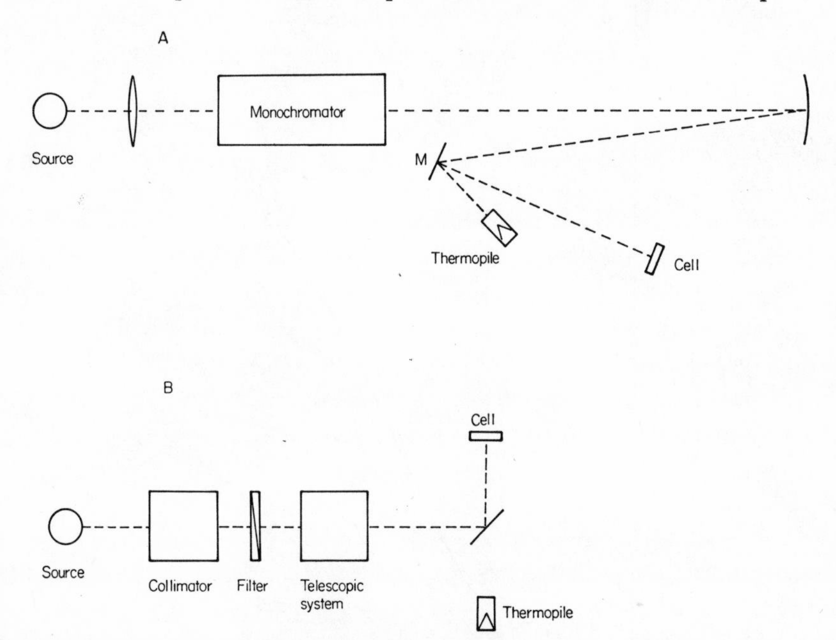

Figure 7.5. A. A block diagram of the classical method of photocell calibration. B. A block diagram of the hybrid calibration layout.

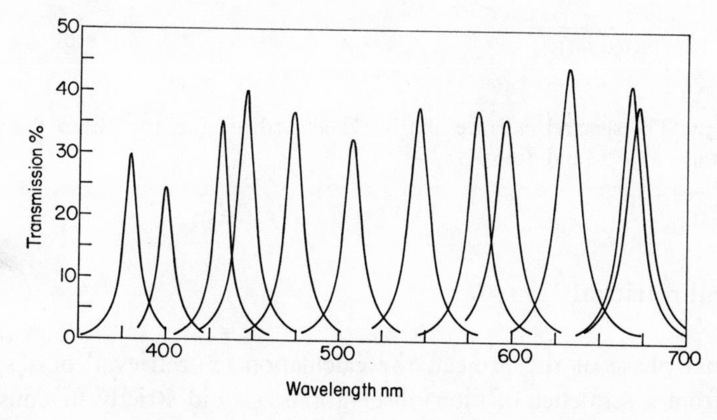

Figure 7.6. The transmission curves of the set of interference filters used in the calibration apparatus.

Using this hybrid method, a calibration apparatus has been set up and some preliminary measurements have been made. It turns out that the mechanical performance of the rotating mirror, which switches the beam between the cell under test and the reference thermopile, is rather critical. Very small wobbles can cause large fluctuations in the pile signals.

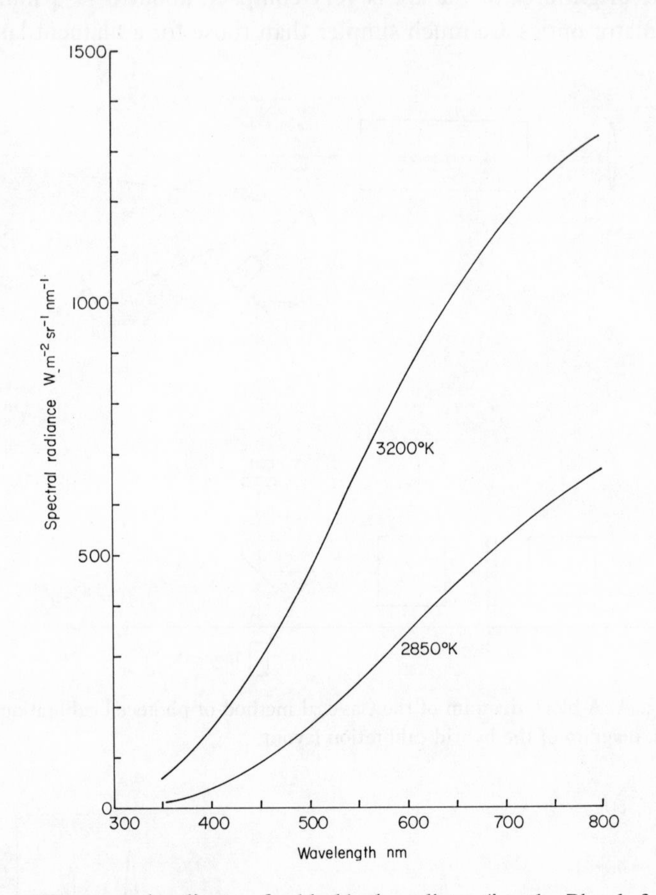

Figure 7.7. The spectral radiance of a blackbody radiator (i.e. the Planck function) at temperatures 2850°K and 3200°K.

Spectral retrieval

The final phase of the project, the calculation or 'retrieval' of a spectral curve from a sequence of filter observations, should strictly be considered first. However, as much time and heavy computation are involved it has been carried on in parallel with the other two.

One minor point has to be dealt with first. As the glass filters (Fig. 7.8) are short-wave cut-off ones, the signal for the effective filter *i* is taken as

$$\text{signal}\,(i) - \text{signal}\,(i+1)$$

working in the direction of increasing wavelength (Evans, 1969, Fig. 6). The resulting signal corresponds to the effective filter formed by

$$t_i(\lambda) = \mathrm{T}_i(\lambda) - \mathrm{T}_{i+1}(\lambda)$$

$T_i(\lambda)$ = transmission curve of the *i*th filter.
Unfortunately, it can be seen that some of these resultant filters are rather misshapen on the long-wave side.

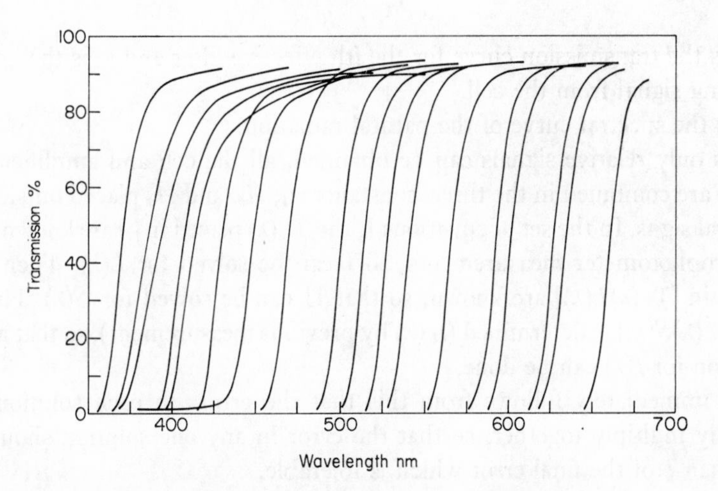

Figure 7.8. The transmission curves of the set of Schott's glass filters used in the integrating sphere radiometer.

Considering both the field and calibration apparatus, three successive sets of filter measurements are taken. For simplicity all three are considered over the same wavelength range λ_1 to λ_2.

In the calibration procedure, the thermopile is irradiated via *N* interference filters in turn, yielding *N* signals.

$$d_i = A_t \int_{\lambda_1}^{\lambda_2} [\mathrm{T}_i(\lambda)] L(\lambda) d\lambda \tag{I}$$

$\mathrm{T}_i(\lambda)$ is the transmission curve for the *i*th interference filter and d_i is the corresponding signal from the thermopile.
$L(\lambda)$ is the spectral emission curve of the source.

Similarly for the cell being calibrated

$$c_i = A_c \int_{\lambda_1}^{\lambda_2} [T_i(\lambda)L(\lambda)]S(\lambda)d\lambda \qquad \text{(II)}$$

c_i is the signal from the cell under test for the ith filter. $S(\lambda)$ is the spectral sensitivity of the cell.

Finally when the field apparatus is used to measure natural radiation, there are M signals from the M effective filters

$$g_i = A_g \int_{\lambda_1}^{\lambda_2} [t_i(\lambda)S(\lambda)]f(\lambda)d\lambda \qquad \text{(III)}$$

$t_i(\lambda)$ is the transmission curve for the ith effective filter and g_i is the corresponding signal from the cell.
$f(\lambda)$ is the spectral curve of the natural radiation.

As only relative signals can be obtained, all the cell and amplifier constants are combined in the three constants A_t, A_c, and A_g placed outside the integral signs. In the set of equations I, the T_i (λ) placed in [ ] are known from spectrophotometer measurements, so I can be solved for $L(\lambda)$. Then both terms in [T_i (λ)$L(\lambda)$] are known, so that II can be solved for $S(\lambda)$. Thus in III, [t_i (λ)$S(\lambda)$] is determined (t_i (λ) by previous measurement), so that a final solution for $f(\lambda)$ can be done.

It immediately follows from this that the errors in each solution will roughly multiply together, so that the error in any one solution should be less than $\frac{1}{3}$ of the final error which is tolerable.

If, in each of these three main sets of equations, the term in [ ] is regarded as a generalized filter function (often referred to in the literature as a kernel function) all three sets are first-order integral equations of the type

$$g_i = \int_{\lambda_1}^{\lambda_2} K_i(\lambda)f(\lambda)d\lambda$$

$K_i(\lambda)$ is the ith filter or kernel function and g_i is the corresponding signal.
$f(\lambda)$ is the spectral curve being solved for.
Equations of this type have no exact analytical solution, so that an approximate method has to be used.

Since the error analysis of possible approximate methods is either nonexistent or very involved, an experimental approach has been adopted. With some suitable trial spectral function $f(\lambda)$ and ideal filter profiles, it is possible to produce, by numerical integration, the signals corresponding to a black detector. Thus the situation represented by I can be readily modelled. Then the set of equations is solved to give a solution $f^*(\lambda)$. The difference, $f(\lambda) - f^*(\lambda)$, enables the errors of the method to be estimated.

As the solar flux incident at the top of the earth's atmosphere (Fig. 7.9)

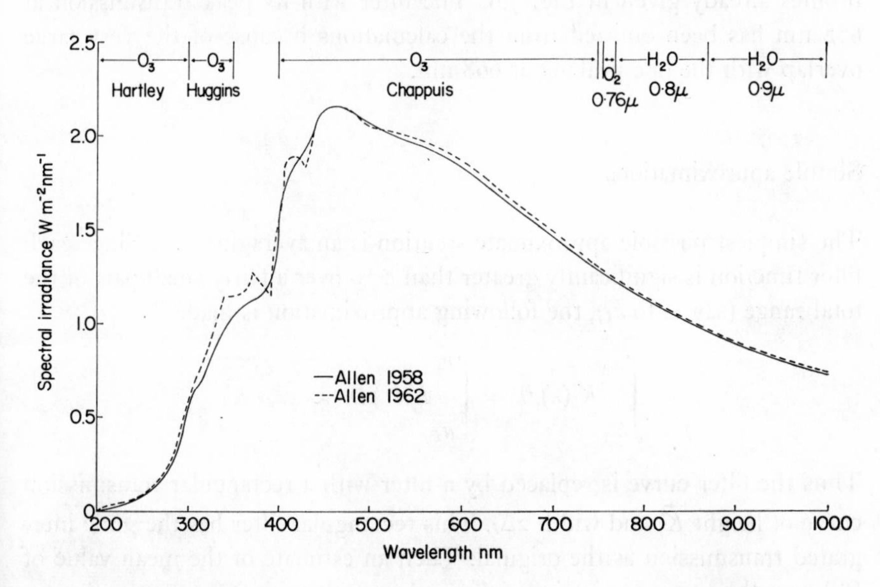

Figure 7.9. The mean total solar irradiance at the mean radius between the earth and the sun, taken from Allen (1958) and (1962). The position of the main atmospheric absorption bands is shown as well.

has a spectral composition reasonably close to that of a black body at 5,910°K, the Planck curve for this temperature has been taken as the spectral function. To model the filter functions, a set of first-order Fabry line-shapes has been taken. These are described in Appendix 1. Each Fabry shape has been

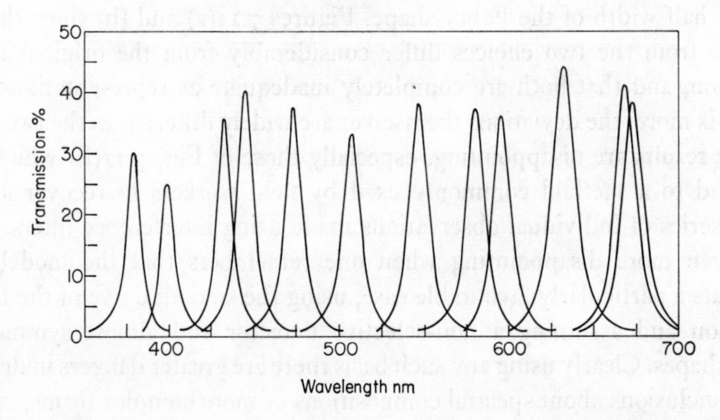

Figure 7.10. The model filter transmission curves, constructed using the Fabry formula.

assigned the same peak transmission, half-width, and overall range as the corresponding interference filter used in the calibration. The constructed shapes, shown in Fig. 7.10, are very close to the original interference filter profiles already given in Fig. 7.6. The filter with its peak transmission at 672 nm has been omitted from the calculations because of the very large overlap with the one peaking at 668 nm.

Simple approximations

The simplest possible approximate solution is an averaging one. Since each filter function is significantly greater than zero over a fairly small part of the total range (say a_i to b_i), the following approximation is made.

$$\int_{\lambda_1}^{\lambda_2} K_i(\lambda)d\lambda = \int_{a_i}^{b_i} K_i(\lambda)d\lambda \approx 2\Delta_i \overline{K_i}$$

Thus the filter curve is replaced by a filter with a rectangular transmission curve of height $\overline{K_i}$ and width $2\Delta_i$. This rectangular filter has the same integrated transmission as the original. Then an estimate of the mean value of $f(\lambda)$ over the range $2\Delta_i$ is given by

$$\overline{f_i^*} = C \cdot \frac{g_i}{2\Delta_i K_i}$$

C is a normalization constant, chosen so that $\overline{f_i^*} = f_i(\lambda_{\text{max}})$ for one particular filter. Following conventional practice (Nicodemus 1973), the filter nearest to the maximum of the spectral curve (i.e. the 508 nm one) has been chosen to fix C.

There are two obvious choices for Δ_i; it can be set either to $\frac{1}{2}(b_i - a_i)$ or to the half width of the Fabry shape. Figures 7.11(a) and (b) show that the results from the two choices differ considerably from the original Planck function, and that both are completely inadequate as representations of it. What is more, the deviations themselves are widely different in the two cases. These results are disappointing, especially those of Fig. 7.11(b) which correspond to a method commonly used by field workers to recover spectra from series of individual observations made using interference filters. They are even more disappointing when one remembers that the model used presents a particularly favourable case, using the smooth curve of the Planck function and assuming a non-selective detector and almost symmetrical filter shapes. Clearly using any such basis there are greater dangers in drawing any conclusions about spectral compositions of more complex forms, such as natural radiation. Further work, using averages, requires many assumptions as to the form of the kernel and spectral function. It is not worth pursuing since the method chosen must be able to cope with situations II and III.

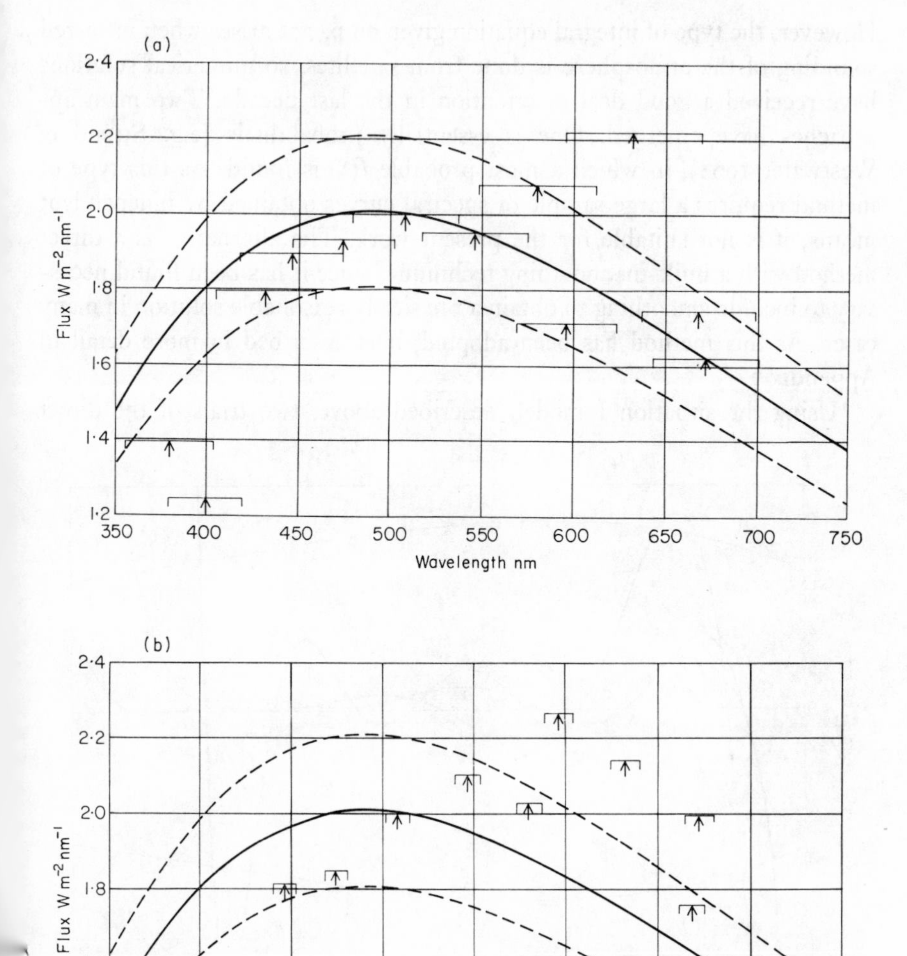

Figure 7.11. The average values of the spectral function, f^*, compared with the original spectral function. The value for the 508 nm peak filter is arranged to coincide. ———, the original spectral function; – – – –, the same, ± 10%; ↑, wavelength for maximum transmission of a particular filter; ⌐¬, $2\Delta_i$. For explanation of two choices for the value of this last, see text.

More exact methods

However, the type of integral equation given on p. 172 arises when infra-red sounding of the atmosphere is done from satellites, so numerical solutions have received a good deal of attention in the last decade. Two main approaches have emerged. One is essentially probabilistic (e.g. Strand & Westwater 1968), in which a most probable $f(\lambda)$ is found. As this type of method requires a large sample of spectral curves obtained by independent means, it is not suitable for the present work. The alternative is a direct method with a built-in smoothing technique, since it has been found necessary to include smoothing to obtain a physically reasonable solution in many cases. As this method has been adopted, it is described in more detail in Appendix 2.

Using the situation I model, described above, two trials of the direct

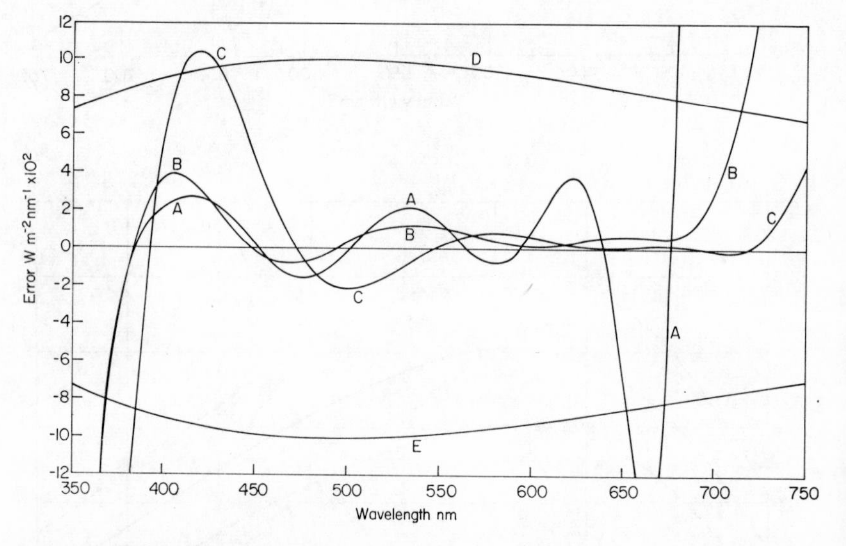

Figure 7.12. The exact spectral function minus the retrieved one when the signals are rounded to 1 in 10^3 and the weights to 1 in 10^4;
A: no smoothing;
B: small amount of smoothing, $\gamma = 1{\cdot}0 \times 10^{-10}$;
C: relatively large amount of smoothing, $\gamma = 1{\cdot}0 \times 10^{-6}$;
D: the spectral function plus 0·5% throughout the range;
E: the spectral function minus 0·5% throughout the range.

method have been done. The error curves $f(\lambda) - f^*(\lambda)$, shown in Fig. 7.12, are those obtained when the signals are rounded to 1 in 10^3 and the weights to 1 in 10^4. The former rounding is a fairly representative one for the precision of the signal measurement, while 1 in 10^4 is the integration accuracy for the weights. Thus these error curves form a lower bound on the errors for the trial. Reducing the accuracy of the weights to 1 in 10^3 gives a more representative picture (Fig. 7.13); even so this may be a little optimistic

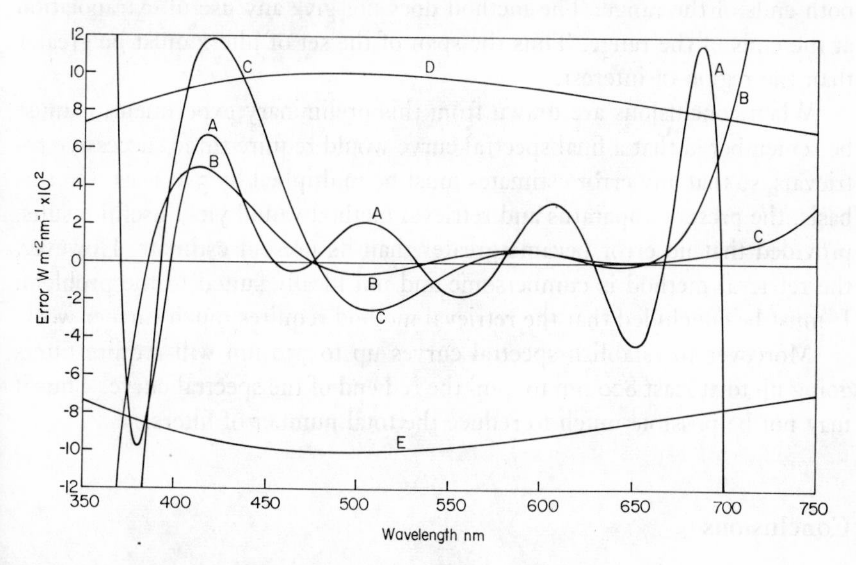

Figure 7.13. The exact spectral function minus the retrieved one when both the signals and weights are rounded to 1 in 10^3;
A: no smoothing;
B: fairly small amount of smoothing, $\gamma = 1{\cdot}0 \times 10^{-8}$;
C: relatively large amount of smoothing, $\gamma = 1{\cdot}0 \times 10^{-6}$;
D: the spectral function plus 0·5% throughout the range;
E: the spectral function minus 0·5% throughout the range.

Both sets of results show that the amount of smoothing, controlled by the parameter γ, does have some beneficial effect. Values of $\gamma = 1{\cdot}0 \times 10^{-10}$ in Fig. 7.12, and $\gamma = 1{\cdot}0 \times 10^{-8}$ in Fig. 7.13, reduce the error oscillation by a factor of 2 or 3 over the unsmoothed case. Both the oscillations in the unsmoothed case and the improvement due to smoothing are much less marked than the examples given by Twomey & Howell (1963b). The main gain is within the actual method. As described in Appendix 2, $f^*(\lambda)$ is represented by a truncated series. With smoothing the low-order coefficients are very close to their exact values calculated from $f(\lambda)$, while without it they are widely different. Thus the technique of smoothing, even at low levels, helps to produce a well-behaved solution. Beyond this, smoothing can be thought of as redistributing errors across the range, rather than further reducing them. During the earlier work, a blunder was made in a trial data set, so that filter curves were produced with an error in wavelength of 50 nm at 750 nm, decreasing to 0 at 350 nm. When smoothing was applied in this case with its much larger errors (10–15%), the redistribution rather than the reduction of error was very evident.

A second feature of all the results is their very poor behaviour at one or both ends of the range. The method does not give any useful extrapolation at the ends of the range. Thus the span of the set of filters must be greater than the region of interest.

When conclusions are drawn from this preliminary experiment, it must be remembered that a final spectral curve would require three successive retrievals, so that any error estimates must be multiplied by 3 at least. On this basis, the present apparatus and retrieval method would yield useful results, provided that no error became greater than its present estimate. However, the retrieval method is cumbersome and not ideally suited to the problem. It must be concluded that the retrieval method requires much further work.

Moreover to establish spectral curves up to 750 nm will require filters going up to at least 800 nm to 'pin' the red end of the spectral curve. Thus it may not be possible much to reduce the total number of filters.

Conclusions

This research has uncovered two formidable problems in the way of measuring the spectral composition of natural radiation in the field, in addition to all the more obvious problems discussed in the Introduction above. They are:
(i) the large changes in overall intensity of short-wave radiation which are possible within the time necessary for one spectral characterization (Figs. 7.2 and 7.3). It seems that the only conditions when such changes may be expected to be reasonably small are those of cloudless deep blue sky, when direct solar radiation makes much the largest contribution to the total. Even in Cambridge, not situated in a region noted for clarity of atmosphere, the direct beam may on occasion contribute seven times the diffuse radiation from the sky as measured on a horizontal surface. But not only are such conditions exceptional, they are also those where estimation of spectral composition presents least problems, the total radiation being dominated by solar radiation minimally altered in spectral composition by the atmosphere. The present difficulty applies to all the occasions when measurements would be most desirable.
(ii) The difficulties of recovering the original spectral composition from a series of measurements using narrow-band filters. Figure 7.11 has shown how the use of what at first sight appear to be the most obvious methods can lead to large errors, of the order of ± 20%, even in a favourable case. A method has been worked out for reducing the errors to an acceptable level,

of the order of 1%, using, say, a dozen filters spaced out over a wavelength range corresponding roughly to the visible spectrum. But the method involves much computation, and the magnitude of the possible errors increases rapidly near the ends of the wavelength range of the observations. Extrapolation is clearly impossible, and even interpolation is limited. One must plan the observations to overlap the wavelength range it is desired to study by about 50 nm at each end. This means, in effect, that the method as it stands is not useful for studying small spectral regions. (For this one would need narrower filters with peak transmissions closer together, which presents technical difficulties at present.) On both grounds it is clear that further work on spectral retrieval is highly desirable.

It might seem that the obvious way out of both these difficulties and a number of others is to use a spectrographic method depending on photographic recording. We then deal with much narrower spectral regions, so that problems of retrieval disappear, and so do the problems due to varying overall intensity during the period of observation, because we are recording the whole spectrum simultaneously. But attractive though the method appears at first sight, when translated into action it becomes impossibly complicated, especially when used in remote places far from normal laboratory facilities. The difficulties mainly stem from the fact that the density of the photographic image depends in complicated ways on so many factors (not all under the observer's control) that the only safe comparisons are those made between areas of equal density of image formed by equal times of exposure to radiation of the same wavelength on the same photographic plate of a particularly uniform, spectrographic, quality. Both areas should then have started as nearly uniform as possible, and subsequently been subjected to near-identical conditions of storage and development. This is contrived by covering the monochromator slit with an optical wedge of known gradient at all relevant wavelengths, by interspersing the experimental spectrograms with others from a standard lamp emitting radiation of known spectral composition, and by measuring the spectrograms to find areas of equal density at a given wavelength. Of course, the standard lamp is powered by a supply whose voltage is constant enough for its emission not to vary significantly during the exposure. It follows that, if the stated conditions are to be met and the spectrograms are to be interpreted with any certainty, the natural radiation to be measured should not vary during the exposure either. This rules out the method as a means of integrating over extended periods, and also introduces uncertainties under the conditions indicated in Figs. 7.2 and 7.3. Furthermore, it will be seen that to meet all the requirements for observations of this kind under the conditions of ecological field work involves many complicated arrangements, and when all is done the evaluation of the spectrograms themselves is exceedingly tedious and time-consuming. It must be concluded, regretfully, that the method is not at present suitable for ecological field work, and that some major technological advance would be needed to make it so.

It seems, then, for the reasons discussed at the beginning of this paper, that if adequate characterization of the spectral composition of natural daylight is to be achieved, we are being forced along the path discussed in this paper. None of the major problems involved is particularly straightforward, but it seems that they can all be handled. Integrating spheres combine nonselective acceptance with a satisfactory cosine response, although their throughput is low; after collimation, narrow-band filters (interference, or, for greater robustness, glass filters by difference) and a photocell as detector provide a signal which can be read on a robust field instrument (e.g. that described by Rackham, this vol., p. 447) even if sensitive digital instruments are impracticable; the photocells can be calibrated for spectral sensitivity by a method which can be set up in the laboratory without difficulty, and when calibrated, retain their characteristics for substantial periods; the problem of variation in intensity can be overcome by simultaneous measurements with a wide-band acceptor; and using the method described, spectra can be retrieved with reasonable accuracy. At the same time work is in progress on speeding up and regularizing the measuring process by an automatic filter-changer, although it seems that at present practicable field recording must wait on technological advances outside the scope of this paper. But if the major problems can all be handled, they are not as yet easy to handle, and a good deal of further development will be needed before they are. Many complications must be considered both in designing and operating a practicable method of measurement, and a strong motive is needed before it is worth embarking upon them. The preceding paper by Young provides such a motive.

Acknowledgements

We must thank Professor Brian and the staff of the Botany School, especially Mr. Freeman, Mr. Marr and Mr. Worland, whose support throughout this project has been invaluable. The earlier part of the work was supported by a grant from the Natural Environment Research Council, and the later part by a grant from the Paul Instrument Fund of the Royal Society, to whom our thanks are also due.

References

ALLEN C.W. (1958) *Astrophysical quantities.* 1st edn, Athlone Press, University of London, pp. xi + 291.

ALLEN C.W. (1962) *Astrophysical quantities.* 2nd edn, Athlone Press, University of London. pp. xi + 291.

BARR E.E. (1970) The design and construction of evaporated multilayer filters for use in solar radiation technology. *Advances in Geophysics* **14**, *Precision Radiometry* (ed. A.J. Drummond), pp. 391–412. Academic Press, New York & London.

COLLINGBOURNE R.H. (1966) General principles of radiation meteorology. In *Light as an ecological factor* (eds R. Bainbridge, G.C. Evans & O. Rackham) pp. 1–15. Blackwell Scientific Publications, Oxford.

COOMBE D.E. (1966) The seasonal light climate and plant growth in a Cambridgeshire wood. In *Light as an ecological factor* (eds R. Bainbridge, G.C. Evans & O. Rackham) pp. 148–66. Blackwell Scientific Publications, Oxford.

EVANS G.C. (1966) Model and measurement in the study of woodland light climates. In *Light as an ecological factor* (eds R. Bainbridge, G.C. Evans & O. Rackham) pp. 53–76. Blackwell Scientific Publications, Oxford.

EVANS G.C. (1969) The spectral composition of light in the field. I. Its measurement and ecological importance. *J. Ecol.* **57**, 109–25.

GILLHAM E.J. (1961) *N.P.L. Notes on Applied Science No. 23.* Radiometric standards and measurements. H.M.S.O., London, pp. iv + 23.

HUGHES A.P. (1966) The importance of light compared with other factors affecting plant growth. In *Light as an ecological factor* (eds R. Bainbridge, G.C. Evans & O. Rackham) pp. 121–46. Blackwell Scientific Publications, Oxford.

LANCZOS C. (1957) *Applied analysis.* Sir Isaac Pitman & Sons Ltd., London. pp. xx + 539.

NICODEMUS F.E. (1973) Normalisation in radiometry. *Applied Optics* **12**, 2960–73.

SNYDER M.A. (1966) *Chebyshev methods in numerical approximation.* Prentice-Hall Inc., Englewood Cliffs, N.J., pp. x + 114.

STRAND O.N. & WESTWATER E.R. (1968) Statistical estimation of the numerical solution of a fredholm integral equation of the first kind. *J. Ass. comput. Mach.* **15**, 100–14.

STRANGEWAYS I.C. (1972) Automatic weather stations for network operation. *Weather* **27**, 403–8.

TWOMEY S. (1963a) On the numerical solution of fredholm integral equations of the first kind by the inversion of the linear system produced by quadrature. *J. Ass. comput. Mach.* **10**, 97–101.

TWOMEY S. & HOWELL H.B. (1963b) A discussion of indirect sounding methods with special reference to the deduction of vertical ozone distribution from light scattering measurements. *Monthly Weather Review* **91**, 659–64.

TWOMEY S. (1965) The application of numerical filtering to the solution of integral equations encountered in indirect sensing measurements. *J. Franklin Inst.* **279**, 95–109.

TWOMEY S. (1966) Indirect measurements of atmospheric temperature profiles from satellites: II. Mathematical aspects of the inversion problem. *Monthly Weather Review* **94**, 363–66.

WYSZECKI G. (1960) Multifilter method for determining relative spectral sensitivity functions of photoelectric detectors. *J. Opt. Soc. Amer.* **50**, 992–8.

YOUNG J.E. (1975) Effects of the spectral composition of light sources on the growth of a higher plant. *This vol.*: 135–159.

Questions and discussion

Dr. J. H. S. Blaxter commented that all the papers presented so far at this session had been excessively concerned with accuracy, which in so many contexts was not an important objective in measuring radiation. For instance, many animals were unable to perceive anything less than a doubling of the brightness of a light source. In water, the bottom of the euphotic zone was usually defined as the depth at which light was reduced to one per cent of the subsurface level, and since this one per cent had not been specified in terms of colour there was little purpose in measuring it accurately. Even in the context of photosynthesis it was doubtful whether it was important to measure total radiation to an accuracy as high as 1 in 100.

Dr. G. C. Evans pointed out that, although for some purposes a 10% error of measurement might be tolerable, the cumulative effect of three separate 10% errors could be important. Professor J. L. Monteith suggested that on the whole there was a difference in the attitudes to accuracy of zoologists and of people concerned with higher plants.

Appendix 1

The Fabry line-shape

The Fabry line-shape is the transmission curve for a Fabry-Perot étalon, i.e. a single-layer interference filter (Barr 1970).

Let t_p be the maximum (i.e. peak) transmission, occurring at wavelength λ_p. The transmission at wavelength λ is then

$$t(\lambda) = \frac{t_p}{1 + F \sin^2 \delta/2}$$

F is the *coefficient of finesse*, which depends upon the reflectivity of the evaporated layers within the filter. For modern manufacturing techniques F is 10,000 or more.

For normal incidence the phase difference δ is given by

$$\delta = 2\pi \left(\frac{2nd}{\lambda}\right)$$

where nd = optical thickness of the étalon. If $nd = N\lambda_o$, where λ_o is some fixed wavelength and N is any integer,

$$\frac{\delta}{2} = \pi \left(\frac{2N\lambda_o}{\lambda}\right)$$

The simplest transmission maximum will occur when $\lambda = 2N\lambda_o$; put $\lambda_p = 2N\lambda_o$

$$\frac{\delta}{2} = \pi\frac{\lambda_p}{\lambda}$$

At the half-width, Δ, $t/t_p = \frac{1}{2}$.
As the line-shape is not absolutely symmetrical, set the half-width on the long-wave side to obtain F.

$$\frac{1}{2} = \frac{1}{1 + F\sin^2\left(\dfrac{\pi\lambda_p}{\lambda_p + \Delta}\right)}$$

Appendix 2

A direct method of solution

In order to simplify the description of the method, it is convenient to transform the wavelength λ to a variable χ, lying between -1 and $+1$, by the formula

$$\chi = \frac{2\lambda - (\lambda_1 + \lambda_2)}{(\lambda_2 - \lambda_1)}$$

where λ_1 is the lower bound and λ_2 the upper bound of the wavelength range.
Then the sets of equations I, II and III are all of the form

$$g_i = \int_{-1}^{1} K_i(\chi)f(\chi)d\chi$$

The index $i = 1, m$ where $m = M$ or N as required. $K_i(\chi)$ is the kernel or filter function with g_i the corresponding signal. $f(\chi)$ is the spectral curve that is to be found.

From this stage onwards the method closely follows that of Twomey and his co-workers (Twomey 1963a, b, 1965, 1966). The first stage in any such solution is the conversion of the integral into an approximate sum. By representing $f(\chi)$ as a table of values at the points χ_j, $j = 1$ to n, this can be done directly, using a quadrature formula. Taking the interference filter curves shown in Fig. 7.6 as typical of kernel functions for this problem, it is seen that each one is zero over at least 90% of the range. Since the number of point values of $f(\chi)$ that can be solved for must be less than or equal to m (i.e. 10 or 12), direct summation would collapse with only one or two of the $K_i(\chi_j)f(\chi_j)$ terms non-zero in each sum.

One method of overcoming this is to represent $f(\chi)$ as a truncated series of some suitable orthogonal functions in χ

$$f(\chi) = \sum_{j=1}^{n} b_j \varphi_j(\chi)$$

$$\therefore g_i = \int_{-1}^{1} K_i(\chi) \sum_{j=1}^{n} b_j \varphi_j(\chi) d\chi = \sum_{j=1}^{n} w_{ij} b_j \qquad n \leqslant m$$

$$w_{ij} = \int_{-1}^{1} K_i(\chi) \varphi_j(\chi) d\chi$$

where w_{ij} is the weight of $K_i(\chi)$ with respect to the function $\varphi_j(\chi)$.

Thus m integral equations are converted into m simultaneous linear equations, which are conveniently written in the matrix form

$$\mathbf{g} = \mathbf{W}\mathbf{b}$$

As each signal must have some measurement error associated with it, this is more exactly written

$$\mathbf{g} + \varepsilon = \mathbf{W}\mathbf{b}$$

Each signal g_i has the error ε_i associated with it.

Twomey (1963) proposed one method of solving this equation, and Twomey & Howell (1963b) showed that it gave useful results when applied to atmospheric examples. The crux of the method is that an extra physical constraint is imposed upon the solution, so that a smoothing matrix **H** is introduced into the equation together with a scalar γ, which controls the amount of smoothing. The form of **H** depends upon the constraint. Hence the matrix equation to be solved becomes

$$(\gamma\mathbf{H} + \widetilde{\mathbf{W}}\mathbf{W})\mathbf{b} = \widetilde{\mathbf{W}}\mathbf{g}$$
$$\mathbf{b} = (\gamma\mathbf{H} + \widetilde{\mathbf{W}}\mathbf{W})^{-1}\widetilde{\mathbf{W}}\mathbf{g}$$

[$\sim$ denotes the transpose of the matrix]

With the m signals $(g_1, g_2 \ldots g_m)$ and the n coefficients $(b_1, b_2 \ldots b_n)$ **W** is a $m \times n$ matrix. However, the product $\widetilde{\mathbf{W}}\mathbf{W}$ is a square $n \times n$ matrix, so that n can be selected, provided $n \leqslant m$. The case with no smoothing ($\gamma = 0$) and $n < m$ is of special interest, since it is the least squares solution for the first n coefficients.

The need for the constraint arises from two related causes. At most the equation can only be solved for $(b_1, b_2 \ldots b_m)$. Thus any two spectral functions $f(\chi)$, which only differ in their coefficients b_{m+1}, b_{m+2}, etc. when expanded as a function series, will yield the same signal vector. Hence any given signal vector of m elements can be produced by an infinite set of spectral functions; thus any solution $f^*(\chi)$ is not unique.

For many types of kernel, particularly those used in atmospheric problems, the contribution to g_i given by the b_j term in the spectral function expansion falls off as j^{-2} approximately. This rapid reduction in the contribution of the higher-order terms, plus the truncation at term m, generally results in a rapid oscillation of the solution $f^*(\chi)$ about the true value $f(\chi)$. Under these circumstances, the application of a constraint to reduce the oscillations is obviously highly beneficial.

In the actual trials described on p. 176, Chebyshev polynomials of the first kind (Appendix 3) have been chosen as the orthogonal functions for the expansions. A direct expansion of the Planck function, the trial spectral function, in terms of these polynomials, requires only 7 terms to have an error bound of less than 0·04% (Fig. 7.14), so that they are an efficient

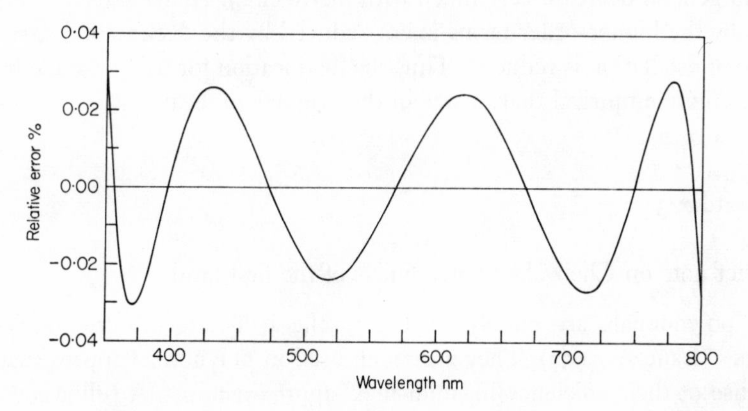

Figure 7.14. The relative error, $\frac{[\text{exact value} - \text{approx. value}]}{\text{exact value}}$, when the Planck function for 5910°K is approximated by a truncated series using $T_0(\chi)$ up to $T_6(\chi)$.

choice. The wavelength range 350–750 nm is set to correspond to the range of $-1 \leqslant \chi \leqslant 1$. This choice is a compromise between the full range of interest, 350–800 nm, and the region for which filter data are available, approximately 360–700 nm.

A straightforward Simpson's-rule integration with a strip-width corresponding to 2 nm is used to produce the trial signals g_i and the weights w_{ij}. Trials, comparing this rule with a high-accuracy adaptive Romberg procedure for calculating the w_{ij}, show that the accuracy of Simpson's rule is better than 1 in 10^4. The actual signals and weights are rounded to a specified number of significant figures. This represents a practical situation more realistically and prevents the integration errors on each side of the main equation from partially cancelling to indicate a spuriously good result.

Preliminary work has indicated that the precise form of the constraint is not critical, so the one used at present is that:

$$\text{the integral } S = \int_{-1}^{1} \frac{[f^{*''}(\chi)]^2}{\sqrt{[1-\chi^2]}}\, d\chi \text{ is minimized, subject}$$

$$\text{to } \sum_{i=1}^{m} \varepsilon_i^2 = \text{some constant.}$$

The minimization of a second differential should reduce rapid changes of gradient and hence oscillation. The term $(1 - \chi^2)^{-\frac{1}{2}}$, the Chebyshev normalization factor, gives **H** a convenient form.

When the calculated w_{ij} are examined, it is seen that for $j \leqslant 10$, w_{ij} does not in general decrease very much with increasing j. Hence the contribution to g_i by the higher-order terms is not reduced by the decline of j^{-2}, so one cause of oscillation is reduced. Thus the justification for using the method is rather more empirical than it was in the original applications.

Appendix 3

A brief note on Chebyshev polynomials of the first kind

The polynomials are closely related to the half-range Fourier series in cosines (Lanczos 1957). They are much used in polynomial approximation, because of their efficiency in 'mini-max' approximations. A full account of their properties is given by Snyder (1966).

The nth polynomial is defined over the range $-1 \leqslant \chi \leqslant 1$ by the relation

$$T_n(\chi) = \cos n\theta \qquad \text{where } \chi = \theta$$

In terms of χ:

$$T_0(\chi) = 1$$
$$T_1(\chi) = \chi$$
$$T_2(\chi) = 2\chi^2 - 1$$
$$T_3(\chi) = 4\chi^3 - 3\chi$$

This table can be extended by means of the general recurrence relation

$$T_{n+1}(\chi) - 2\chi T_n(\chi) + T_{n-1}(\chi) = 0$$

The orthogonality condition is

$$\int_{-1}^{+1} \frac{T_m(\chi)T_n(\chi)d\chi}{\sqrt{1 - \chi^2}} = \begin{cases} 0 \text{ if } m \neq n \\ \pi \text{ if } m = n = 0 \\ \pi/2 \text{ if } m = n \neq 0 \end{cases}$$

Note: There is one slight difficulty in notation. In most matrix calculations, the indices i, j etc. run from 1 to n, while the sequence of Chebyshev polynomials starts at T_0. To avoid confusion $\varphi_1(\chi) \leftarrow T_0(\chi)$, $\varphi_2(\chi) \leftarrow T_1(\chi)$, and so on.

8 The role of light in the vertical migration of fish—a review

J. H. S. BLAXTER *Dunstaffnage Marine Research Laboratory, Oban, Argyll, Scotland.*

Introduction

Vertical migration by fish and its physiological basis have been discussed by Woodhead (1966), Blaxter (1970) and Cushing (1973). Our much improved knowledge of these movements is largely due to use of echo-sounding both from research and commercial vessels since 1945. It is generally thought that no single explanation can be given for the adaptive value of vertical migration. Its most significant feature is a diel rhythm (i.e. one based on a 24-hour cycle—see Bary 1967) with movement towards the surface at dusk and towards the bottom at dawn. This rhythmical activity can sometimes be related to peaks of feeding activity at these times; it has also been suggested that the environment is exploited more fully and that vertical differences in current may result in improved horizontal distribution. It may also be involved in a form of *epideictic* display (as a form of population census—Wynne-Edwards 1964) as a means of reducing predation pressure and in prolonging the dusk and dawn periods when certain light intensities may be 'at a premium' (Ali 1959).

This review is concerned with the importance of light, both as a stimulus in initiating or releasing vertical movements and as a means of controlling their speed and amplitude once they commence. Evidence will be reviewed from less usual situations which corroborates the importance of light, and the sensory problems involved in perceiving absolute brightness and changes in brightness will be discussed. This will be done with reference first to the herring and other clupeoid fishes which are commercial species showing the clearest and best known vertical migrations in shallow seas and second, to the fish of the acoustic or deep scattering layers of the oceans. Diel vertical migration is here more protracted, giving more time for study, but identification of particular layers is rare. Indeed, the layers are themselves very much associated with the frequency of the echo-sounder in use, different species having different resonance frequencies. Thus the impression gained of the population structure and movement of the organisms beneath a research vessel will depend very much on the acoustic equipment available (see Cushing 1973). Despite this drawback over identification it is clear that fish, especially myctophids, are a major component of these layers (see, for example, Pearcey & Laurs 1966 and Taylor 1968).

The phenomenon

Diel vertical migration is found in many groups of aquatic animals (see Cushing 1951, Bainbridge 1961, Banse 1964). The general dusk ascent and dawn descent may be modulated, especially in invertebrates, by a midnight sinking and a pre-dawn rise. Amongst the clupeoids a pre-dawn rise has been described in herring (Richardson 1952, Brawn 1960) and Japanese sardine (Nomura 1960) and a midnight sinking in herring (Brawn 1960). Stickney (1972) described post-dawn and pre-dusk increases in activity by captive herring held in tanks on shore. Other workers (Dragesund 1958, Blaxter & Parrish 1965) describe the great variability of the migration in relation to age, spawning condition of the fish and season. The amplitude may also be influenced by temperature gradients (Devold 1952, Postuma 1958, Zusser 1958, Cushing 1973), by a halocline (Devold 1952) oxycline (Bary 1967) or by tides (Woodhead 1966). It is abundantly clear that though light is important, no rigid preferenda can be postulated, nor are the responses to light stimuli predictably rigid.

The deep-water scattering layers also show great variation, due no doubt in part to the abundance of species involved. The different layers are obviously existing at different levels of illumination, due partly to their preferences for light. With the variability in ocean transparency (Dickson 1972) and the existence of thermoclines which may act as barriers (see Clarke & Backus 1956) a great variety of possible migration characteristics is to be expected.

Other factors are also of importance. While clupeoids have an open swim bladder, so that pressure changes do not restrict vertical migration, many other fish are physoclistous and upward movement in particular must be carefully controlled (see Tytler & Blaxter 1973). Verwey (1966) has discussed the role of response to gravity in the control of vertical movement, especially where fish may be found off the bottom in complete darkness.

Evidence of dominant influence of light

The association of diel vertical migration with dusk and dawn is good circumstantial evidence of the role of light, especially where observations are made over a number of months (Brawn 1960, Bary 1967). Dickson (1972) found a good relationship between the depth of two scattering layers in the N. Atlantic and transparency as measured by Secchi disk, the layers being deeper in clear water. It is also unusual for veritcal migration to take place in very deep water where light cannot penetrate. Other more specific or less usual evidence is as follows:

'Following isolumes' (remaining within a preferendum)

Richardson (1952) showed how herring in the North Sea descended at dawn, keeping within a fairly narrow range of light intensity, measured in arbitary units. Postuma (1958), see Fig. 8·1, found North Sea herring ascended and

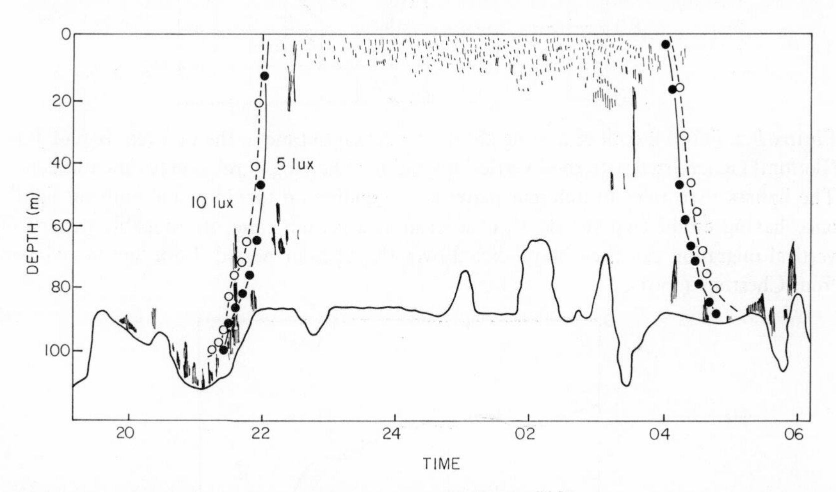

Figure 8.1. Vertical migration of herring shoals related to time and 10 and 5 mc isolumes. Redrawn from Postuma (1958).

descended 'within' the 1 and 10 mc isolumes. Chestnoy (1961) also claimed that herring followed isolumes and that it was possible to predict the pattern of vertical migration over 24 hours from knowledge of their depth at any one time. This, of course, would provide advance information for fishermen setting their nets (Fig. 8.2.).

The deep scattering layers lend themselves to an examination of this correlation over longer periods of time. Clarke & Backus (1956, 1964) and Boden & Kampa (1967), see Figs. 8.3, 8.4 and 8.5, showed how different layers moved in response to light, some keeping within a quite narrow preferendum and others migrating 'across' the isolumes. Clarke (1966), however, found that although some lantern fish in the Santa Barbara channel

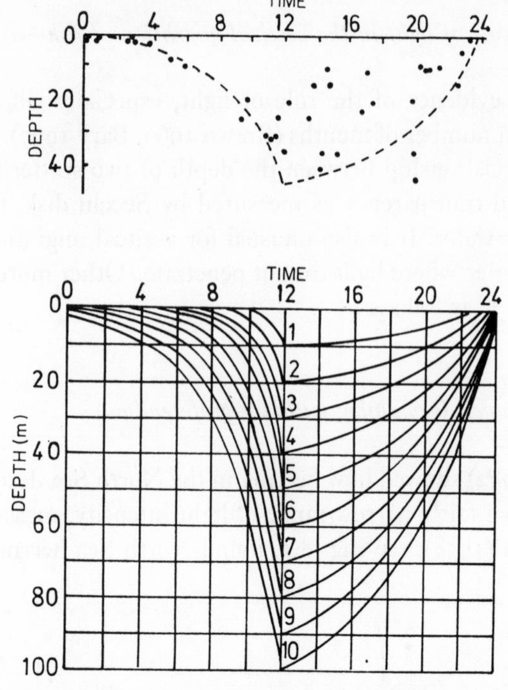

Figure 8.2. (Top) Depth of herring shoals in relation to time in the first ten days of July. (Bottom) Generalized pattern of vertical migration of herring in relation to time and depth. The figures 1–10 refer to different patterns (depending on turbidity and ambient light); once having established the depth of a shoal at a certain time, its probable pattern of vertical migration can then be predicted over the 24-hour period. Both figures redrawn from Chestnoy (1961).

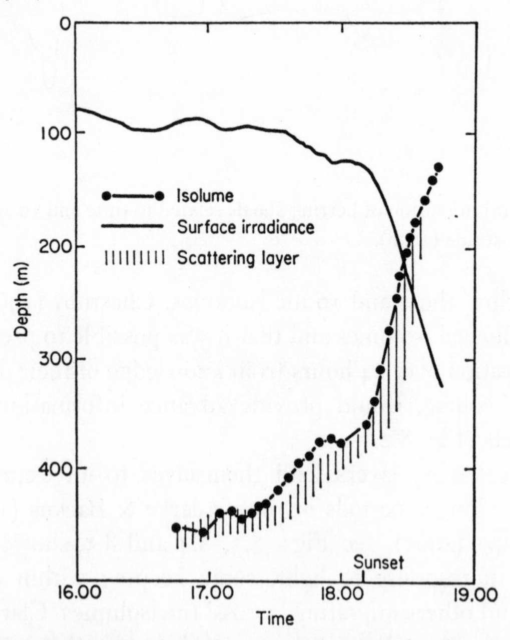

Figure 8.3. The upward migration of an acoustic scattering layer related to surface irradiance and the 5×10^{-4} μW/cm² (474 nm) isolume (N.E. Atlantic, November 1965). From Boden & Kampa (1967).

were mainly at intensities below 1×10^{-6} μW/cm² others were, nevertheless, at much higher intensities of 1×10^{-3} μW/cm². In relating the depth of scattering layers to isolumes the top of the scattering layer (leading edge) is often plotted—the actual range of depth within which a species may be found may be wide and it seems likely that individual preferenda vary greatly, even within a species at any one time.

Moonlight

The well-established role of the moon in influencing catches of herring is discussed by Blaxter & Holliday (1963). Apart from the increased night-time illumination which may affect vertical migration the moon may also operate

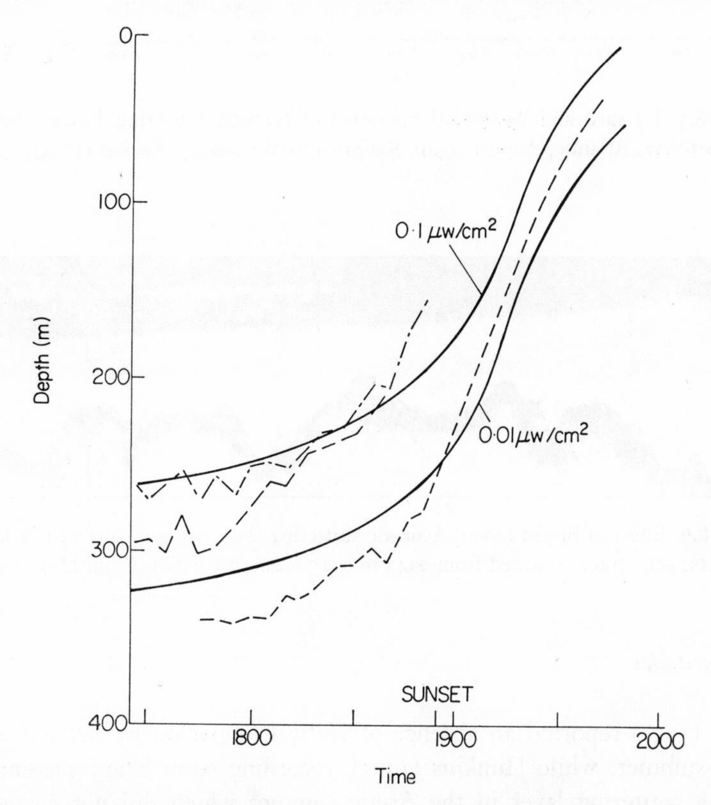

Figure. 8.4. The upward migration of three acoustic scattering layers related to isolumes (N.W. Atlantic, July 1955). Redrawn from Clarke & Backus (1956).

through a tidal influence, by increasing the visibility of nets or of the fish to each other. The oceanic scattering layers may also be 'held down' during the night by bright moonlight as shown by Dietz (1962), see Fig. 8.6.

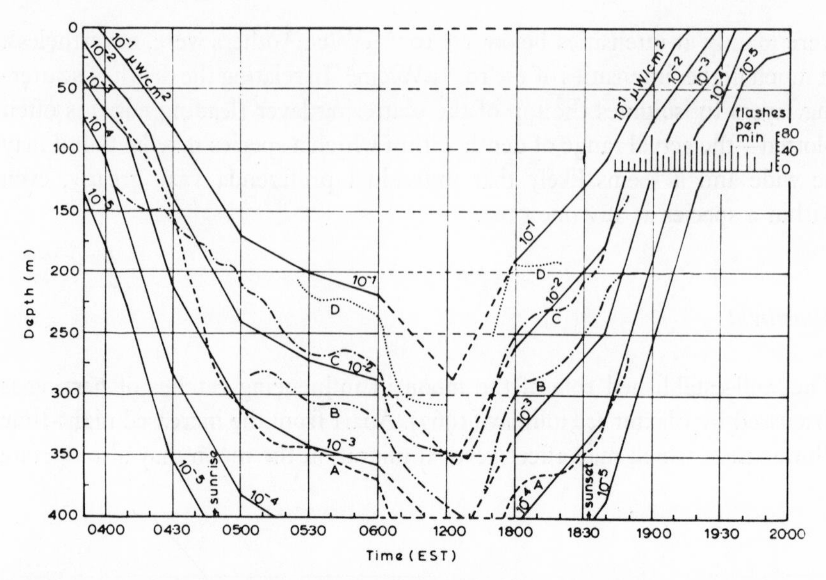

Figure 8.5. Upward and downward migration of acoustic scattering layers related to isolumes (N.W. Atlantic, August 1959). Redrawn from Clarke & Backus (1964).

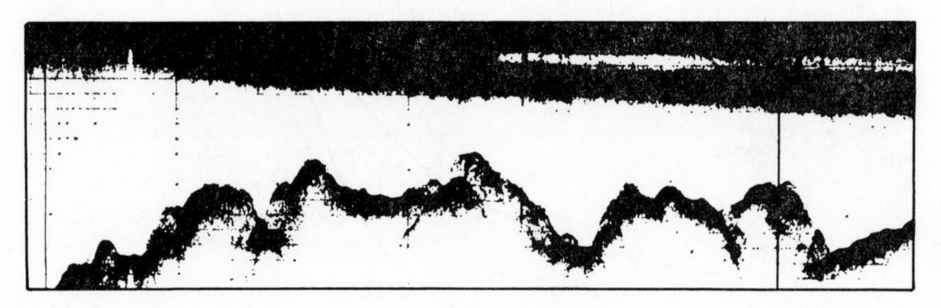

Figure 8.6. Effect of bright moon. Acoustic scattering layer rising to surface (on left) as moon sets; echo trace recorded from 2345 to 0115 (eastern Pacific). From Dietz (1962).

High latitudes

Zusser (1958) reported an absence of vertical migration by herring in the Arctic summer, while Hunkins (1965), recording from below floating ice, found a scattering layer in the Arctic summer which did not change its depth by 'day' and 'night'. In the winter there was no layer at all. Bogarov (1946) found no veitical migration (of copepods) in the polar summer although it was re-established in the autumn. Dietz (1948), in wide-ranging surveys, found that the deep scattering layer was frequently absent at high latitudes in the Pacific and Antarctic Oceans.

Thick ice

Under thick ice in Lake Baikal, Zusser (1958) reported that no vertical migration took place.

Eclipses

Schüler (1954) showed an upward movement of about 20 m of a trace, thought to be herring, during the eclipse of 1954 near the Faroe Islands (see Fig. 8.7). Skud (1968) reported changes in the behaviour of herring caged at the surface on the east coast of the U.S.A. during a partial eclipse in 1963,

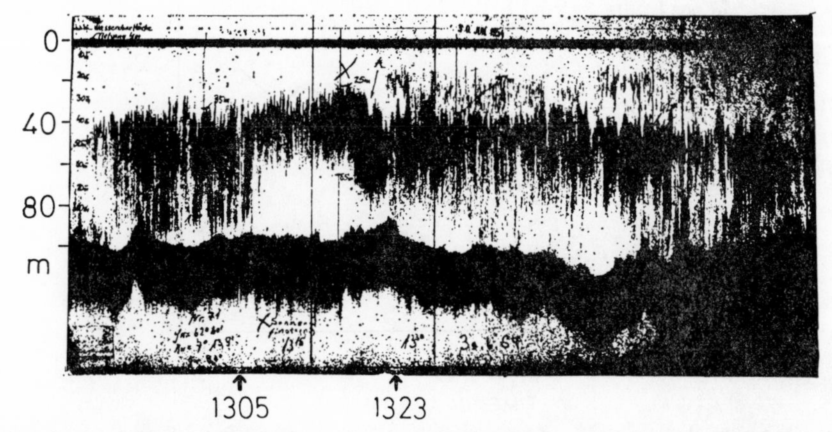

Figure 8.7. Effect of eclipse on echo traces thought to be herring near the Faroe Island, in 1954; eclipse lasted from 1305 to 1323. From Schüler (1954).

changes which were similar to ones normally made by the herring at sunset. Stickney (1972) found no effect of a partial eclipse on the activity of captive herring in 1970. Backus *et al.* (1965) showed a premature ascent by three scattering layers in the Atlantic during the 1963 eclipse (see Fig. 8.8). Both Sherman & Honey (1970) and Bright *et al.* (1972) reported changes in vertical distribution of plankton in the 1970 eclipse which suggested an ascent typical of what was normally found at dusk.

Artificial lights

These can both repel and attract clupeoids and are used in fishing practice (see Blaxter & Holliday 1963 and Woodhead 1966). There is abundant evidence that diel vertical movements in clupeoids can be interfered with by underwater and above-surface lights.

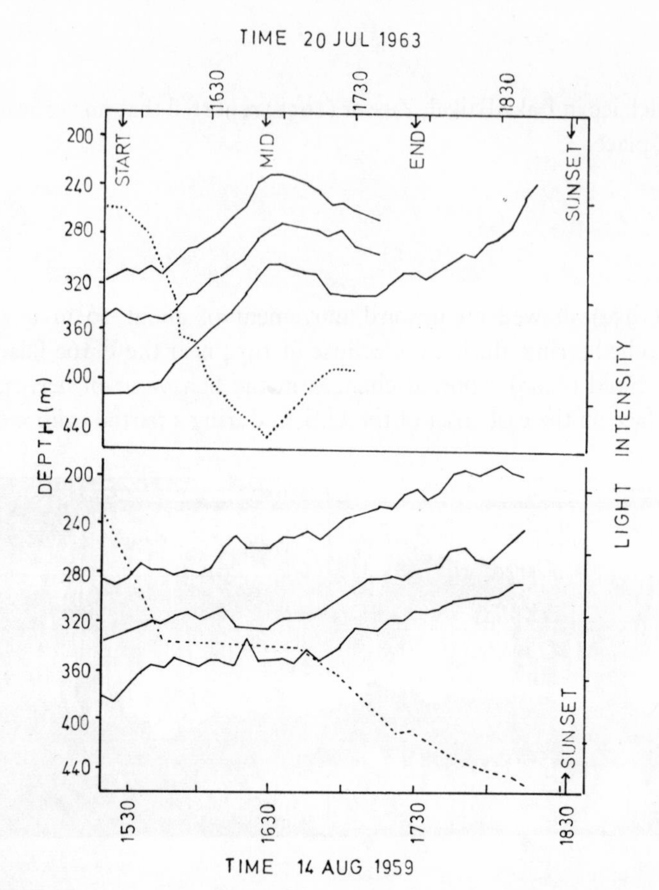

Figure 8.8. Premature ascent of acoustic scattering layers in the Atlantic in July 1963 in response to an eclipse, with trace considered to be comparable (shown below); time of eclipse given at top of figure; light intensity shown by dotted line. Redrawn from Backus *et al.* (1965).

In the oceanic scattering layers only Blaxter & Currie (1967) have attempted to modify vertical migration by artificial lights, succeeding both in delaying a dusk ascent by at least 15 minutes and driving down a layer by night from 60 to 300 m.

We may conclude from this evidence that light is the dominant stimulus in diel vertical migration, which seems to be largely an exognous rhythm. Stickney (1972) did, in fact, find evidence of a weak endogenous rhythm of activity when captive herring were held in darkness but there is no experimental (laboratory) treatment of vertical migration in fish. The effect of light can, however, be modified by physical and chemical discontinuities in the water. Organisms, even of the same species, can be found in light intensities differing by at least 10^3 units of light intensity. Preferences, where they exist, are thus widely differing both within and between species, but the net result of diel changes of light is to cause the centre of gravity of the population to ascend at dusk and descend at dawn.

The questions now arise of whether the responses to light depend on relative changes of light, the rate of relative change, the need for a sense of absolute intensity perception or the ability to discriminate changes of light intensity.

Rate of change of light intensity and the preferendum theory

Cushing (1951), Bainbridge (1961) and Ringelberg (1964, 1967) have reviewed the various theories put forward to explain how vertical migration is controlled. One of these is the 'preferendum' theory, whereby the animals are supposed to remain within a 'preferred' light intensity. As the light intensity falls at dusk the animals move up in order to keep within the preferendum. Whereas cases can be shown of this happening (see Figs. 8.1, 8.3, 8.4, 8.5) it is also quite common for animals to move 'across' the isolumes (not only into lower intensities at dusk or higher intensities at dawn, which might be explained by an inability to swim fast enough to remain within the preferendum, but also the opposite). This led Ringelberg (1964) to conclude that the dominant characteristic of light for vertically migrating organisms was not a preferendum but the rate of change of intensity. A rapid reduction in intensity would then lead to an upward movement. Ringelberg plotted speed of upward swimming against rate of change of light intensity and found a positive correlation, both in experiments on *Daphnia* and in re-working the data of Clarke & Backus (1956) on oceanic scattering layers during their rise to the surface at dusk. Clarke & Backus (1964) took exception to this analysis, saying that Ringelberg did not measure the light at the (changing) depth of the layers, but at one level. In fact if the scattering layers were remaining 'between' narrow isolumes the rate of change of light intensity would be nil. Generally speaking, vertical migration both at dusk and dawn will mean the rate of change of light intensity is being reduced or nullified.

Further unpublished data of the present author are shown in the upper part of Fig. 8.9 derived from measuring the light intensity at the depth of various scattering layers, both at dusk and dawn, off the Canary Islands and Madeira in September-October 1965. The figure shows the quite considerable change in light intensity experiencd by the organisms as the layers move up and down. There is not necessarily a close following of the isolumes (nor a consistent trend into lower or higher light intensities). If the speed of movement is plotted against rate of change of light intensity from these data

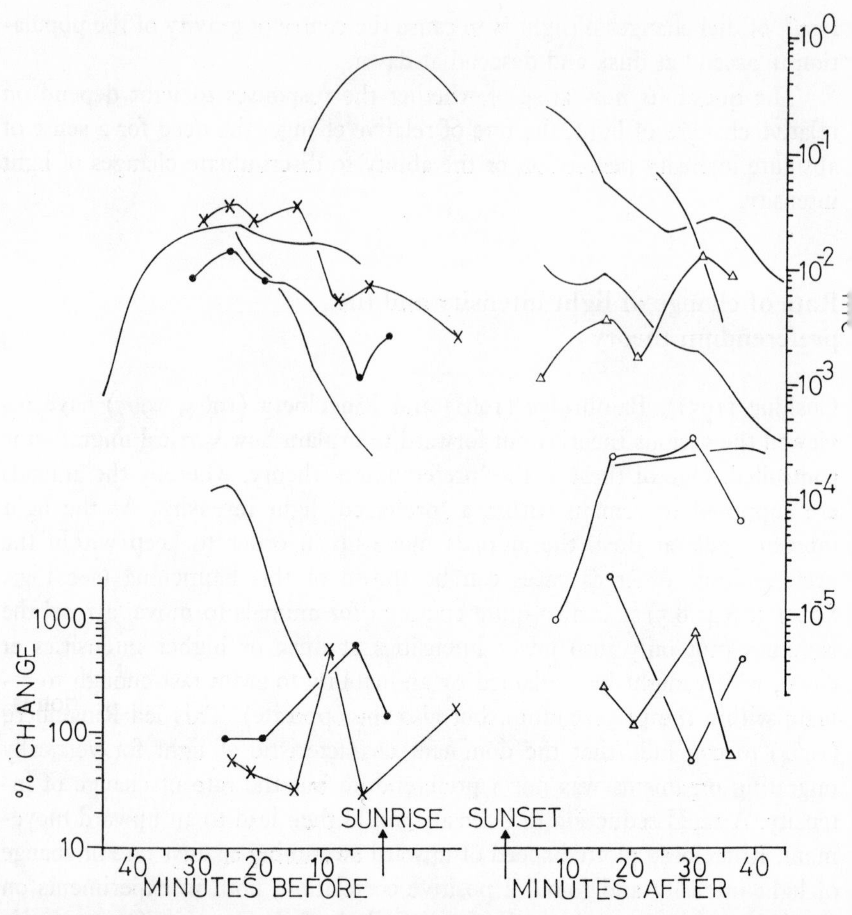

Figure 8.9. (Top) Light intensity changes experienced by six acoustic scattering layers at dawn, and six at dusk, in the N.E. Atlantic in October-November 1965. (Bottom) Two of the dawn traces (●, ×) and two of the dusk traces (△, ○) selected to show the % change in light intensity experienced during different phases of the migration. Unpublished data of the author.

(see Fig. 8.10) it is clear that there is no marked relationship between vertical displacement and rate of change of light intensity. If the position of some of the layers in relation to the isolumes in Figs. 8.3, 8.4, 8.5 and 8.9. is examined carefully there is seen to be a 'hunting 'effect, that is the layers do not remain closely within the isolumes but lag and then over-compensate; this may be a sign of inadequate discrimination of brightness changes. This means that the animals remain as near a preferendum as they are able, but with limited visual equipment they will tend to ascend (or descend) and pause, again ascend (or descend) and pause and so on. This intermittent type of movement probably indicates that the organisms start to move vertically when the light level has changed sufficiently from a reference or preferendum level.

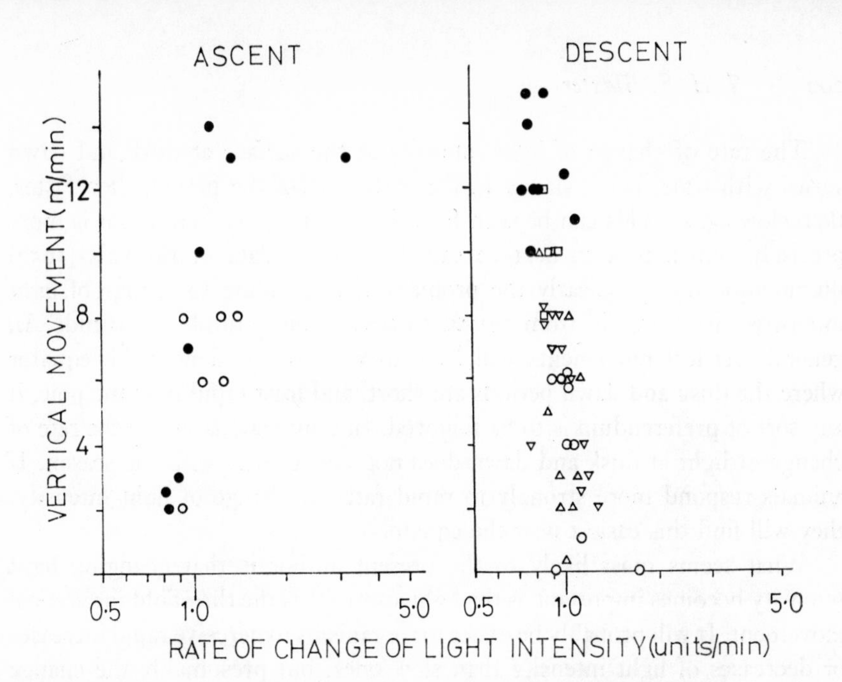

Figure 8.10. Rate of vertical displacement related to rate of change of light intensity experienced during ascent and descent of scattering layers in the N.E. Atlantic in October–November 1965; the different symbols refer to different scattering layers or different dates. Unpublished data of author.

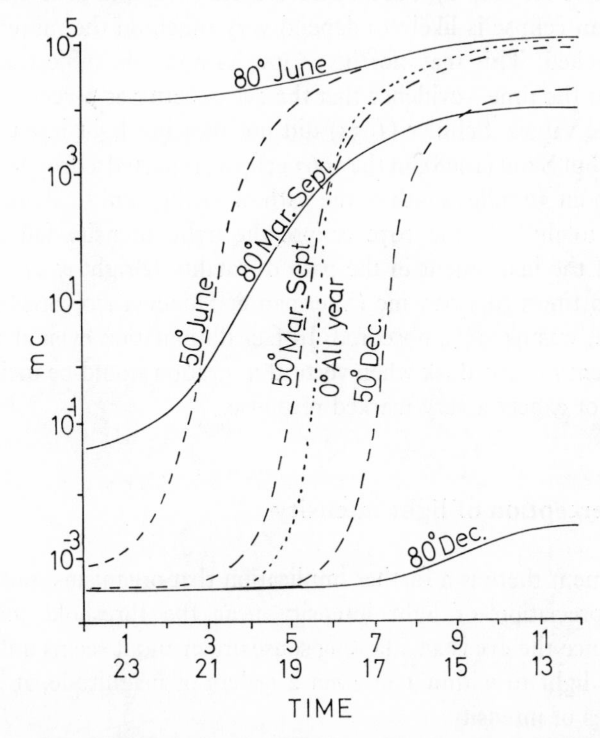

Figure 8.11. Change of light intensity at dusk and dawn at 0°, 50°N and 80°N towards the end of March–September, June and December, calculated from Brown's (1952) illumination charts.

The rate of change of light intensity at the surface at dusk and dawn varies with time, being slower in the early part of the period, then faster, then slow again. This can be seen from Bary's (1967) isolumes, but is more precisely shown in Fig. 8.11 as calculated from data in Brown's (1952) illumination charts. Clearly the problem of responding to change of light intensity varies greatly from season to season and latitude to latitude. In general, vertical movements will have to be very rapid near the equator where the dusk and dawn periods are short, and least rapid near the pole, if any sort of preferendum is to be followed. In temperate latitudes the rate of change of light at dusk and dawn does not vary greatly with the season. If animals respond more strongly to rapid rates of change of light intensity, they will find this easiest near the equator.

What seems most likely to the present author is that changing light intensity becomes important as the light approaches the threshold for vertical movement. It will probably be easier for organisms to perceive rapid increases or decreases of light intensity than slow ones, but presumably the change only becomes meaningful near the threshold. The existence of a fairly low threshold would also explain why many organisms do not respond to short-term changes in illumination caused by clouds (where rates of change of light intensity can be very rapid), but may nevertheless respond to an eclipse. The response to an eclipse is likely to depend very much on the minimum light intensity reached. This may not be as low as it seems subjectively to the human eye at the time—evidence that the eye is better at perceiving relative than absolute values. Schüler (1954) did not measure light intensity in the 1954 eclipse but Skud (1968), in the 1963 eclipse, reported a drop from 75,000 to 900 mc about 50 miles south of the path of totality and to about 10 mc in the path of totality. In the 1970 eclipse the light intensity fell below the sensitivity of the instrument in the path of totality (Bright *et al.* 1972) and by 10 to 100 times to 1,000 mc (Sherman & Honey 1970, Stickney 1972) when the sun was 94–95% obscured. In fact illumination even at totality is only equivalent to early dusk when vertical migration would be incipient and one would not expect a very marked response.

Absolute perception of light intensity

In this argument there is a further implication that organisms must have an absolute appreciation of light intensity near the threshold for vertical migration, since the eye is an adapting sense organ and it seems unlikely that it can gauge light to within 1 or even 2 orders of magnitude, at least over certain ranges of intensity.

In fish it is postulated that light is most likely to be perceived absolutely at two points in the range of vision. The first is at the absolute rod threshold and the second is at the dark to light adaptation threshold. At this latter threshold (equivalent to dusk and dawn levels of illumination at the surface) the cones take over from the rods and there is a change from monochromatic to colour vision, an increase of visual acuity and changes in such characteristics of vision as the frequency of flicker-fusion. This process does not always occur suddenly—there may be a gradual change over 2 to 3 log units of intensity but some characters such as flicker-fusion frequency may change more sharply.

There are very well marked anatomical changes taking place at the same time. These involve extension or contraction of the cone and rod myoids and movement of the retinal masking pigment (see Nicol 1963, Blaxter 1970). How the organism controls these movements is not really clear, but they seem to be intrinsic to the eye (see Blaxter 1970) and this seems again to support the idea that the eye has an absolute reference point at about this intensity.

The range of intensity over which dark and light adaptation in some relevant species take place, mainly as judged by these 'retinomotor' movements, is plotted in Fig. 8.12 which shows the most frequent range to be

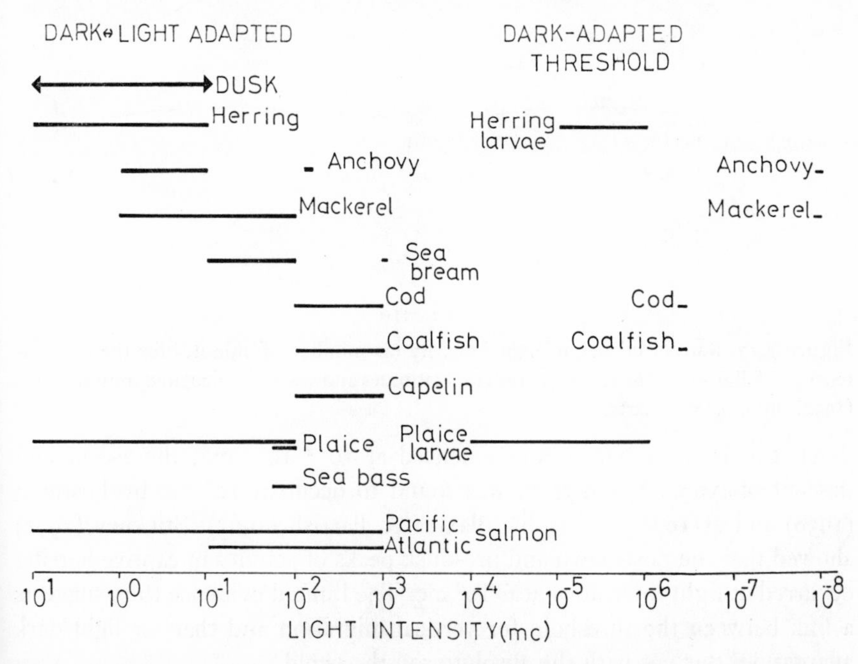

Figure 8.12. Range of light intensity over which dark and light adaptation take place as judged mainly from retinal histology (left side of figure) and dark-adapted visual thresholds (right of figure). All data are taken from review by Blaxter (1970) except dark-adapted thresholds for anchovy, coalfish, cod and mackerel by ERG technique (Protasov 1964).

10^{-3}–10° mc. The data on absolute rod thresholds available is minimal; those on fish larvae shown in the figure are for what appears to be a pure-*cone* eye and it is quite uncertain whether it may be typical of the rods of adults. Some other results by Kobayashi (1962) are so high they must almost certainly be based on light-adapted animals and are not included here. Protasov (1964) using the ERG technique shows values of the threshold from 10^{-8} to 10^{-6} mc.

How then do the light intensities at which the initial vertical movements occur, or at which organisms seem to locate a preferendum, compare with these possible absolute intensity reference points? Taking first the dark/light adaptation thresholds we may start to search for initial vertical movements as the light intensity approaches the dark/light adaptation threshold (near 10° mc at dusk and 10^{-3} mc at dawn). Nomura (1960) reported that the dawn

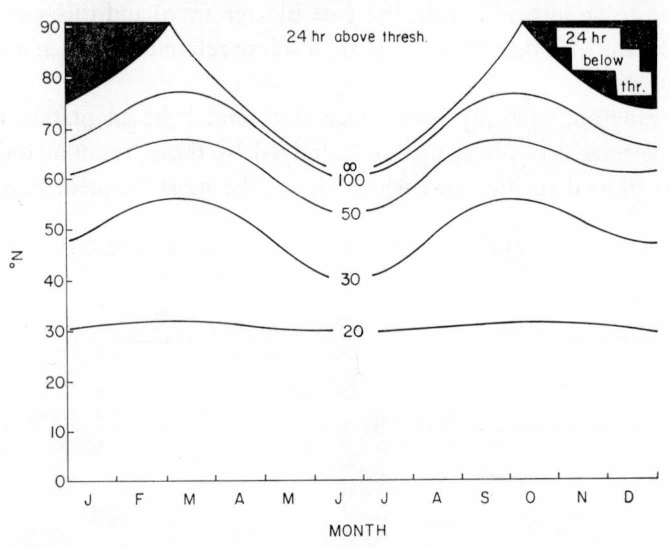

Figure 8.13. Rate of change of light intensity (as number of minutes for the light intensity to fall from 10^{2} to 10^{0} mc) at different latitudes and seasons, calculated from Brown's (1952) illumination charts.

descent of Japanese sardines commenced at 10^{-2}–10^{-1} mc; the ascent and descent of North Sea herring was found to occur at 10° mc by Postuma (1958) and at 10^{-1}–10^{1} mc by Blaxter & Parrish (1965). Stickney (1972) showed that the post-dawn and pre-dusk peaks of activity in captive herring occurred at light intensities near 10^{2} mc. The limited evidence thus suggests a link between the threshold for vertical migration and that for light/dark adaptation, but not with the absolute rod threshold.

In shallow seas fish may well be in light intensities above the preferendum by day. Thus Nomura (1960) found Japanese sardines at 10^{1}–10^{3} mc. Cushing (1960) reported traces, thought to be pilchards, in the English

Channel at light intensities estimated at $2{\cdot}5 \times 10^1$–1×10^3 mc. Postuma (1958) and Blaxter & Parrish (1965) found that North Sea herring were in light intensities of $> 10^1$ and 10^0–10^3 respectively. Chestnoy (1961) does not quote the values for the isolumes but it is clear from his Figs. 1 and 3 that the variation in depth at any one time is quite large and would suggest a variation in ambient light of 2 or even 3 log units of intensity.

In the deep-water scattering layers all that can be said at present, without fuller knowledge of the organisms and their physiology, is that at least some of the organisms may be moving 'within' isolumes of 10^{-2}–10^{-1} $\mu W/cm^2$ (equivalent at the surface to about 2×10^{-2}–2×10^{-1} mc) which might be near the dark/light adaptation range, and others near 10^{-5} $\mu W/cm^2$ may be almost at their lowest threshold.

In Fig. 8.13 rate of change of light intensity is plotted for different latitudes and seasons in the northern hemisphere as the number of minutes of time required for the light to change from 10^2 to 10^0 mc, i.e. on the approach to the light/dark adaptation threshold. This extension of Fig. 8.11 shows that it should be relatively easier for organisms to combine the stimuli associated with rate of change of intensity with the dark/light adaptation threshold near the equator throughout the year and at higher latitudes at the equinoxes.

Brightness discrimination

An organism can only stay closely within its preferendum if the ability to perceive a change of light intensity caused by a change of depth or by a change of downwelling light is sufficiently good. This brightness discrimination is normally quantified by the Weber Fraction (if the reference intensity is I, and $I + \Delta I$ the minimum increase in brightness perceptible, then $\Delta I/I$, is the Weber Fraction, usually expressed as a percentage). The Weber Fraction is not necessarily constant over the whole visible range—there may be points where it is minimal. Blaxter (1970) reviewed the literature on brightness discrimination in fish and the information for relevant species is shown in Fig. 8.14. It should be clear that I has always been taken arbitrarily as the *lower* intensity in calculating the Weber Fractions in this paper. For example if a fish can just perceive a rise in intensity of 0·5 units compared with reference level of 1 unit (i.e. from 1·0 to 1·5 units), then the Weber Fraction would be expressed as $0{\cdot}5/1{\cdot}0 \times 100 = 50\%$. Alternatively if the organism perceives a fall in intensity of 0·5 units compared with a reference level of 1·5 units the Weber Fraction is again $0{\cdot}5/1{\cdot}0 \times 100 = 50\%$. If the reference level is taken as the *higher* intensity the Weber Fraction is never higher than 100% and in the second example would be 33%.

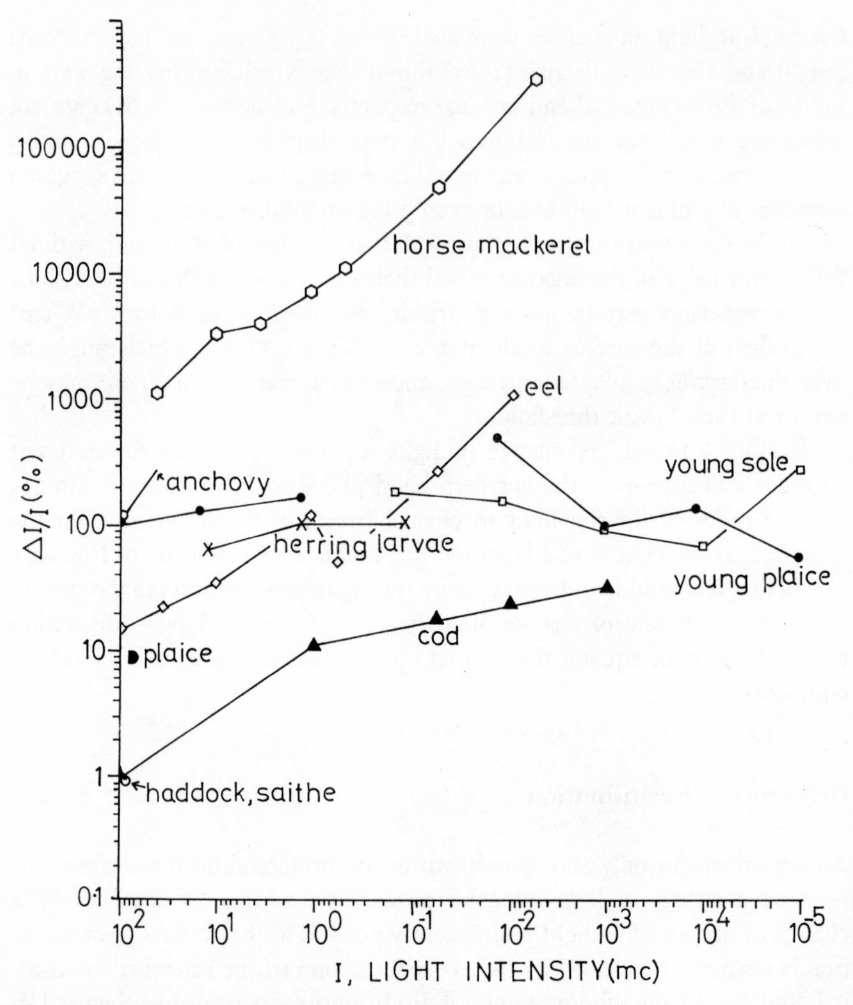

Figure 8.14. Brightness discrimination as the Weber Fraction ($\Delta I/I$) expressed as a percentage at different reference levels (I). All data taken from review by Blaxter (1970), Kobayashi (1972) and Blaxter (1972).

The brightness discrimination data in Fig. 8.14 show a great variability between species, which may, in part, be due to the experimental methods used by the authors. In general the Weber Fractions lie between 10 and 100%, i.e. the organisms can perceive brightness increases 10 to 100% above the reference level (e.g. with a reference level of 1 unit the next brightness intensity perceived would vary between 1·1 and 2 units).

The herring in Fig. 8.1. show a rather precise following of isolumes which suggests that brightness discrimination is good, although the movement is rather rapid and analysis thus more difficult. It is more instructive to examine Figs. 8.3, 8.4, 8.5 and 8.9 where we find that there is a tendency for some of the deep-water scattering layers to 'hunt', as mentioned earlier. It is therefore of interest to calculate the percentage change of light intensity experienced by the organisms as they over- and under-compensate on their way up or down. This is shown in Figs. 8.15 and 8.16 for some data of Clarke & Backus (1956,

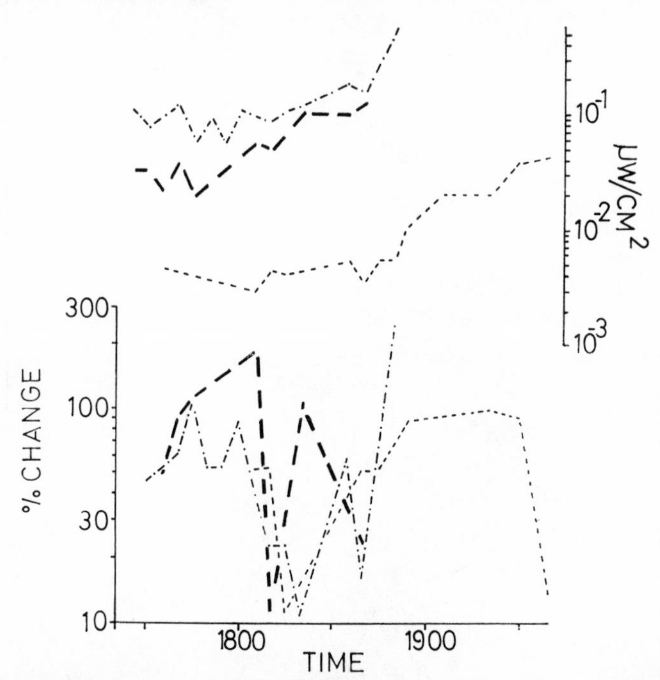

Figure 8.15. Actual and percentage change in light intensity experienced by the organisms in the three acoustic scattering layers during different phases of the vertical migration shown in Fig. 8.4.

1964) where the percentages lie mainly between 10 and 100% and Fig. 8.9 for Blaxter's data where the values are more centred round 100%. There is thus some circumstantial evidence that the changes of light intensity experienced during over- and under-compensation agree roughly with the intensities which are likely to be discriminated.

Two points need clarification. The first is that sometimes physiological experiments on brightness discrimination have involved the *simultaneous* comparison of intensities, whereas vertically migrating organisms must make a *sequential* or *temporal* comparison, i.e. a memory factor for the reference level or preferendum is required. This may be easier if the comparison is made with an intensity where there is absolute perception. The second point is the apparent tendency for the scattering layers (as typified by the leading edge or top of the trace) to move together. This is rather difficult to explain, except as resulting from an interaction between the organisms.

Almost nothing is known about behaviour of vertically migrating fish,

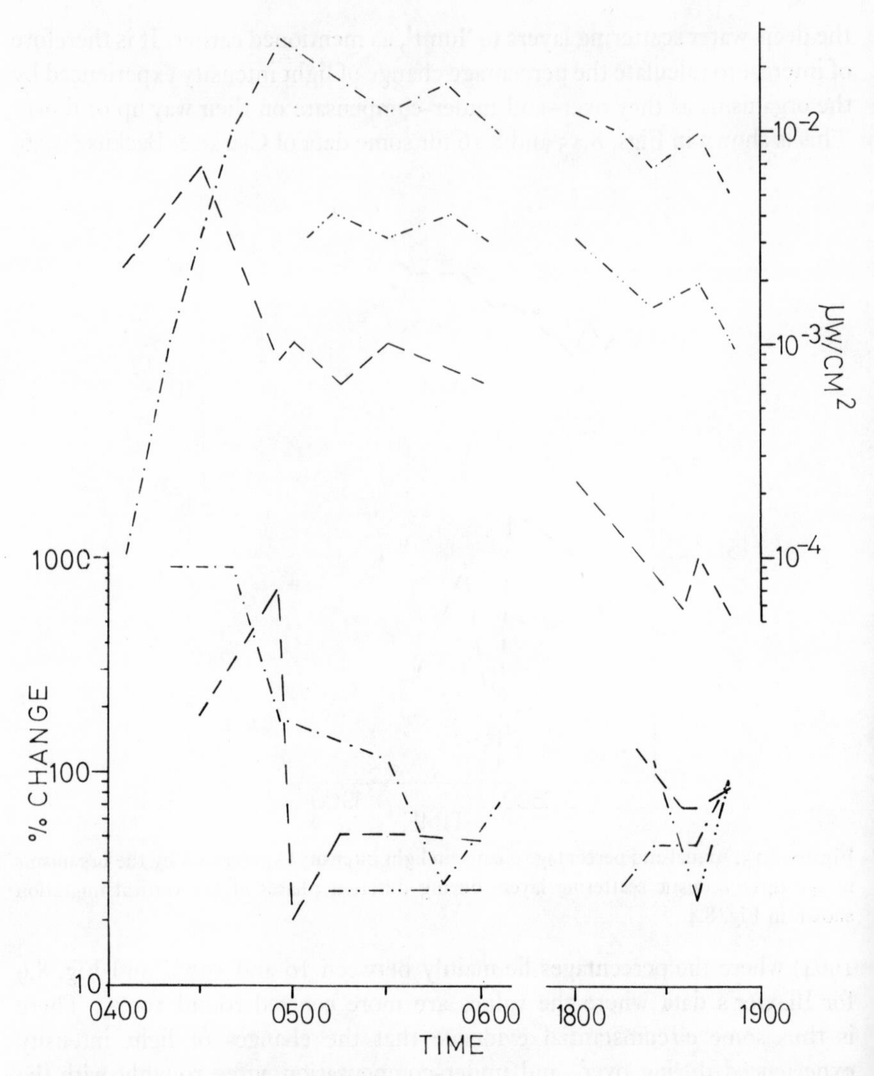

Figure 8.16. Actual and percentage change of light intensity experienced by the organisms in three acoustic scattering layers during different phases of vertical migration in Fig. 8.5. – – – layer A, – · · · – layer B, – · · – layer C.

although Bainbridge (1961) summarized some of the observations made on invertebrate plankton underwater. The echo-sounding records suggest cluepoids move very quickly once the threshold is reached. Only Barham (1966, 1970) has reported observations on fish, especially myctophids, in deep-water scattering layers. In his earlier report he found that the fish were concentrated in the upper components of the scattering layer and were sometimes actively swimming at 45° up at dusk and 45° down at dawn. Later, he showed that a large percentage of myctophids were orientated vertically, head up, especially in deep water by day and night. The myctophids near the surface by night were much more often in a horizontal posture. Barham suggests that the fish are remaining in the vertical position by use of respiratory currents from the opercula. No evidence is available as to whether the fish in a scattering layer might move in concert, although they have the ability to signal to each other with photophores. None of the methods of observation to date has lent itself to determining whether vertical migration is intermittent.

Discussion

The ideas put forward in this review must remain highly speculative until further information is obtained on the visual threshold and brightness discrimination of fish, especially those of the deep sea. The general concept of an absolute sense of intensity perception has not been clearly stated, even in organisms higher than fish. Many of the sense organs are adapting and the reference points, if they exist, other than the rod threshold, need to be defined. To some extent the need to use light as a depth-regulating stimulus is to use it in a proprioceptive role, and of all the sense organs the proprioceptors are least likely to be adapting. The mechanisms of the duplex retina do, indeed, seem to provide some intermediate reference point in the visual range, because there is here a quite complex change in both anatomical and physiological relationships.

Water has a special characteristic, in that the range of light intensity available to aquatic organisms is very great. Without doubt light does control the depth of organisms and they have the possibility, if the water is deep enough, to remain within a given light intensity throughout the day and night. That possibility is only available to terrestrial animals in high latitudes and then only for a limited season. The inshore fish, like the herring, might well seek lower light intensities by day if the depth of water were greater, but they are normally subjected to a much wider range of light intensity than the deep-sea organisms. It is possible then that the deep sea organisms have a more precise response to light. There can be no doubt that fish must have a preferred light intensity of some sort. This may vary by day and night and in some cases the preferred intensity may not be 'available' in shallow water. The evidence points to the preferred level being rather variable from species to species and from individual to individual. The weakness of the normal echo-sounder lies in its inability to follow individual organisms during a vertical migration. It is ideal at looking at a population as a whole.

Summary

1. The dominant role of light is discussed as a stimulus influencing the vertical migration of clupeoid fishes and organisms of the oceanic acoustic scattering layers.

2. Circumstantial evidence from the time of vertical migration in relation to dusk and dawn, the lack of vertical migration at high latitudes as well as experiments showing the following of light preferenda and the interference of depth-holding by the use of artificial lights or during eclipses, show the importance of light.

3. Information on the light intensity at which organisms are found by day, as well as the thresholds at which they commence to migrate, suggests that vertical migration must be linked with an absolute appreciation of light intensity. The eye is an adapting sense organ and absolute appreciation of intensity may only be possible at the rod threshold and at the transition from light to dark adaptation.

4. Depth-holding and vertical migration seem to be linked, also, with a need to discriminate brightness. In this way movements away from a preferendum can be monitored.

Acknowledgments

The author is grateful to Dr. Boden and to the Zoological Society of London for permission to reproduce Fig. 8.3 and to Dr. R.S. Dietz for permission to reproduce Fig. 8.6.

References

ALI M.A. (1959) The ocular structure, retinomotor and photobehavioural responses of the juvenile Pacific salmon. *Can. J. Zool.* 37, 965–96.

BACKUS R.H., CLARK R.C. & WING A.S. (1965) Behaviour of certain marine organisms during the solar eclipse of July 20th 1963. *Nature, Lond.* 205, 989–91.

BAINBRIDGE R. (1961) Migrations. In *Physiology of the Crustacea* Vol. 2, pp. 431–63; edited by T. Waterman. Academic Press, London & New York, 681 pp.

BANSE K. (1964) On the vertical distribution of zooplankton in the sea. *Prog. Oceanogr.* 2, 53–125.

BARHAM E.G. (1966) Deep scattering layer migration and composition: observations from a diving saucer. *Sci.N.Y.* 151, 1399–403.

BARHAM E.G. (1970) Deep sea fishes—lethargy and vertical orientation. In *Proceedings of an International Symposium on Biological Sound Scattering in the Ocean.* Pp. 100–18; edited by G.B. Farquhar, Maury Center for Oceanic Science, Department of the Navy, Washington D.C.

BARY B.McK. (1967) Diel vertical migrations of underwater scattering, mostly in Saanich Inlet, British Columbia. *Deep Sea Res.* 14, 35–50.

BLAXTER J.H.S. (1970) 2. Light 2.3. Animals 2.32. Fishes. In *Marine Ecology*, Vol. 1, Pt 1, pp. 213–320; edited by O. Kinne, Wiley Interscience, London.

BLAXTER J.H.S. (1972) Brightness discrimination in larvae of plaice and sole. *J. exp. Biol.* 57, 693–700.

BLAXTER J.H.S. & CURRIE R.I. (1967) The effect of artificial lights on acoustic scattering layers in the ocean. *Symp. zool. Soc. Lond.* 19, 1–14.

BLAXTER J.H.S. & HOLLIDAY F.G.T. (1963) The behaviour and physiology of herring and other clupeids. *Adv. mar. Biol.* 1, 261–393.

BLAXTER J.H.S. & PARRISH B.B. (1965). The importance of light in shoaling, avoidance of nets and vertical migration by herring. *J. Cons. perm. int. Explor. Mer*, 30, 40–57.

BODEN B.P. & KAMPA E.M. (1967) The influence of natural light on the vertical migration of an animal community in the sea. *Symp. zool. Soc. Lond.* 19, 15–26.

BOGAROV B.G. (1946) Peculiarities of diurnal vertical migrations of zooplankton in polar seas. *J. mar. Res.* 6, 25–32.

BRAWN V.M. (1960) Seasonal and diurnal vertical distribution of herring (*Clupea harengus* L) in Passamaquoddy Bay, N.B. *J. Fish. Res. Bd Can.* 17, 699–711.

BRIGHT T., FERRARI F., MARTIN D. & FRANCESCHINI G.A. (1972) Effects of total solar eclipse on the vertical distribution of certain oceanic zooplankters. *Limnol. Oceanogr.* 17, 296–301.

BROWN D.R.E. (1952) Natural illumination charts. U.S. Dept. Navy, Bureau of Ships, Project NS 714–100, Rept No 374–1.

CHESTNOY V.N. (1961) Vertical migrations of herring and the methods of their utilization in fishing practice. (Russ.) *Ryb. Khoz.* 37 (6), 53–7.

CLARKE G.L. & BACKUS R.H. (1956) Measurements of light penetration in relation to vertical migration and records of luminescence of deep sea animals. *Deep Sea Res.* 4, 1–14.

CLARKE G.L. & BACKUS R.H. (1964) Interrelations between vertical migration and deep scattering layers, bioluminescence and changes in daylight in the sea. *Bull. Inst. Océanogr. Monaco* 64 (1318), 36 pp.

CLARKE W.D. (1966) Bathyphotometric studies of the light regime of organisms of the deep scattering layer. General Motors Defense Research Lab., Santa Barbara. Mimeo Rept UC48, Tid-4500, 47 pp.

CUSHING D.H. (1951) The vertical migration of planktonic Crustacea. *Biol. Rev.* 26, 158–192.

CUSHING D.H. (1960) Fishing gear and fish behaviour. *Proc. World Sci Meeting on Biology of Sardines*, 3, 1307–26.

CUSHING D.H. (1973) 'The Detection of Fish'. Pergamon Press, Oxford, 200 pp.

DEVOLD F. (1952) A contribution to the study of migrations of the Atlanto-Scandian herring. *Rapp. P.-v. Réun. Cons. perm. int. Explor. Mer* 131, 103–7.

DICKSON R.R. (1972) On the relationship between ocean transparency and depth of some scattering layers in the North Atlantic. *J. Cons. perm. int. Explor. Mer* 34, 416–422.

DIETZ R.S. (1948) Deep scattering layers in the Pacific and Antarctic Oceans. *J. mar. Res.* 7, 430–42.

DIETZ R.S. (1962) The sea's deep scattering layers. *Sci. Amer.* August, 8 pp.

DRAGESUND O. (1958). Reactions of fish to artificial light with special reference to large herring and spring herring in Norway. *J. Cons. perm. int. Explor. Mer* 23, 213–27.

HUNKINS K. (1965) The seasonal variation in the sound scattering layer observed at Fletcher's Ice Island (T-3) with a 12 Kc/s echo sounder. *Deep Sea Res.* 12, 879–81.

KOBAYASHI H. (1962) A comparative study on electronetinogram in fish with special reference to ecological aspects. *J. Shimonoseki Coll. Fish.* 11, 407–538.

KOBAYASHI H. (1972) Some aspects on the relationship between light intensity. *Bull. Jap. Soc. sci Fish* 38, 913–20.

Nicol J.A.C. (1963) Some aspects of photoreception and vision in fishes. *Adv. mar. Biol.* 1, 171–208.

NOMURA M. (1960) Some knowledge on behaviour of fish schools. *Proc. Indo-Pacific Fish. Coun.* 8th Session (111), 95–6.

PEARCEY W.G. & Laurs R.M. (1966) Vertical migration and distribution of mesopelagic fishes off Oregon. *Deep Sea Res.* 13, 153–65.

POSTUMA K.H. (1958) The vertical migration of feeding herring in relation to light and the vertical temperature gradient. Int. Coun. Explor. Mer Herring Cttee., mimeo rept. 8 pp.

PROTASOV V.R. (1964) Some features of the vision of fishes. (Russ.) *Inst. Morfologii Zhivotrykh Akad. Nauk. S.S.S.R.* pamphlet. Pp. 29–48.

RICHARDSON I.D. (1952) Some reactions of pelagic fish to light as recorded by echo-sounding. *Fishery Invest. Lond* (*Ser* 2). 18 (1), 20 pp.

RINGELBERG J. (1964) The positively phototactic reaction of *Daphnia magna* Straus: A contribution to the understanding of diurnal vertical migration. *Neth. J. Sea. Res.* 2, 319–406.

RINGELBERG J., Kasteel J.V. & SERVAAS, H. (1967) The sensitivity of *Daphina magna* Straus to changes in light intensity at various adaptation levels and its implication in diurnal vertical migration. *Z. vergl. Physiol*, 56, 397–407.

SCHÜLER F. (1954) Über die echographische Aufzeichnung des Verhaltens von Meeres-fischen während der Sonnenfinsternis vom 30 Juni 1954. *Dt. hydrogr. Z.* 7, 141–3.

SHERMAN K. & HONEY K.A. (1970) Vertical movements of zooplankton during a solar eclipse. *Nature, Lond.* 227, 1156–8.

SKUD B.E. (1968) Responses to marine organisms during the solar eclipse of July 1963. *Fishery Bull. Fish Wildl. Serv. U.S.* 66, 259–71.

STICKNEY A.P. (1972) The locomotor activity of juvenile herring in response to changes of illumination. *Ecology* 53, 438–45.

TAYLOR F.H.C. (1968) The relationship of midwater trawl catches to sound scattering layers off the coast of northern British Columbia. *J. Fish. Res. Bd Can.* 25, 457–72.

TYTLER P. & Blaxter J.H.S. (1973) Adaptation by cod and saithe to pressure changes. *Neth. J. Sea Res.* 7, 31–45.

VERWEY J. (1966) The role of some external factors in the vertical migration of animals. *Neth. J. Sea Res.* 3, 245–66.

WOODHEAD P.M.J. (1966) The behaviour of fish in relation to light in the sea. *Oceanogr. mar. Biol. Ann. Rev.* 4, 337–403.

WYNNE-EDWARDS V.C. (1962) *Animal Dispersion in Relation to Social Behaviour*. Oliver & Boyd, Edinburgh, 652 pp.

ZUSSER S.G. (1958) A study of the causes of diurnal vertical migrations in fishes. (Russ.) *Trudy Sovesch. ikhtiol. Kom.* 8, 115–20.

9 The ecology function and phylogeny of iridescent multilayers in fish corneas

J. N. LYTHGOE *MRC Vision Unit, Centre for Research on Perception & Cognition, University of Sussex, Falmer, Brighton*

Introduction

The study of corneal iridescence in fishes does not have the same venerable history that is usually associated with the functional morphology of the eye. Indeed, systematic observations of the phenomenon do not seem to have been made until those made quite recently by Lythgoe (1971) on the cornea of coral reef fishes. The fine structure of one of these, *Nemanthias carberryi*, was investigated by Locket (1972) who showed that the iridescent layer in this particular case was the regularly arranged endoplasmic reticulum situated in the corneal endothelium.

The cornea is a rather unrewarding organ to study with the light microscope since its major component, the substantia propria, or stroma, is made up of collagen fibrils that are much less than the wavelength of light in diameter and, therefore, cannot be resolved by it. A layer that it is now believed almost certainly causes interference reflections in the cornea has been described by Walls (1963) who named it the antochthonous layer, but since the thickness of structures causing interference are also less than a wavelength of light, little information was gained from them.

The second reason why corneal iridescence has been neglected in the past is that it is generally only visible when the eye is illuminated from above and viewed from the side (Lythgoe 1971). These are the conditions that naturally occur under water but are perhaps less common in aquaria.

The only function previously suggested for the iridescence in fishes is that of a sophisticated sunshade (Lythgoe 1971) and some evidence is discussed here that supports this view. Diptera possess conspicuously coloured iridescence on the eye facets. Bernard (1971) suggests that these may act as coloured interference filters that enhance the animal's perceptions of colour contrasts.

Description of the light climate

The characteristics of natural light in water have been reviewed by Tyler &

Preisendorfer (1962) and by Jerlov (1968). The characteristics that may be relevant to the function of corneal multilayers are described in this section.

Light from the sun and sky is modified in three distinct ways as it passes from the air into the water. Firstly, its direction is changed by refraction at the air-water interface. Once it has entered the water the light is scattered by particles suspended in the water and by the water molecules themselves. The light is also absorbed so that it is reduced in intensity as it travels through the water. Some wavelengths are absorbed more than others with the result that the ambient light under water is coloured.

Since we are concerned with the visual ecology of fishes, it seems reasonable to consider the underwater scene as it is presented to a fish swimming in its natural habitat. On looking upwards it will be able to see details of the land and sky through the water surface, but instead of subtending a solid angle of 180° as it would at the surface, it subtends a solid angle of only 98°. (Fig. 9.1). This circular window is known as 'Snell's' window since its presence and dimensions can be predicted from Snell's Law.

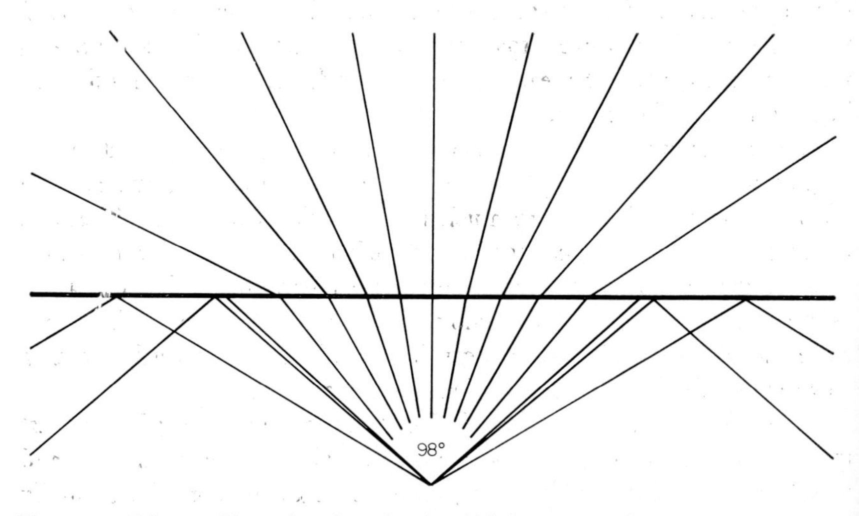

Figure 9.1. Diagram illustrating the refraction of light at the surface of calm water. The diagram has been calculated from Snell's Law assuming the refractive index is 1 for air and 1·33 for water. A fish looking upwards will see the entire hemisphere above the water condensed into a solid angle of 98°.

$$\text{Sin } V_a n_a = \text{Sin } V_b n_b \qquad — 1$$

where n_a and n_b are 1 and 1·33 and are the approximate refractive indices of air and water respectively and V_a and V_b are the corresponding angles that the light ray makes to the normal. A feature of Snell's window is that it always subtends the same angle no matter what the depth, although the actual area of surface that it includes does increase with depth.

If water only absorbed light and did not scatter it Snell's window would be visible to all depths provided, of course, that there remained enough light to see by, although at very great depths the direct light from the sun alone would be visible and all the underwater light would come from that direction. At any depth Snell's window would be the only light visible since the rest of the water volume would apear black in the same way that the sky above the atmosphere appears black.

In reality the water molecules themselves scatter light as do the minute particles of plankton, sand and decay products that natural water contains. If there was no absorption of light but only scatter a fish in deep water would receive light equally from all directions.

Since there is both absorption and scatter, light reaches the fish from all

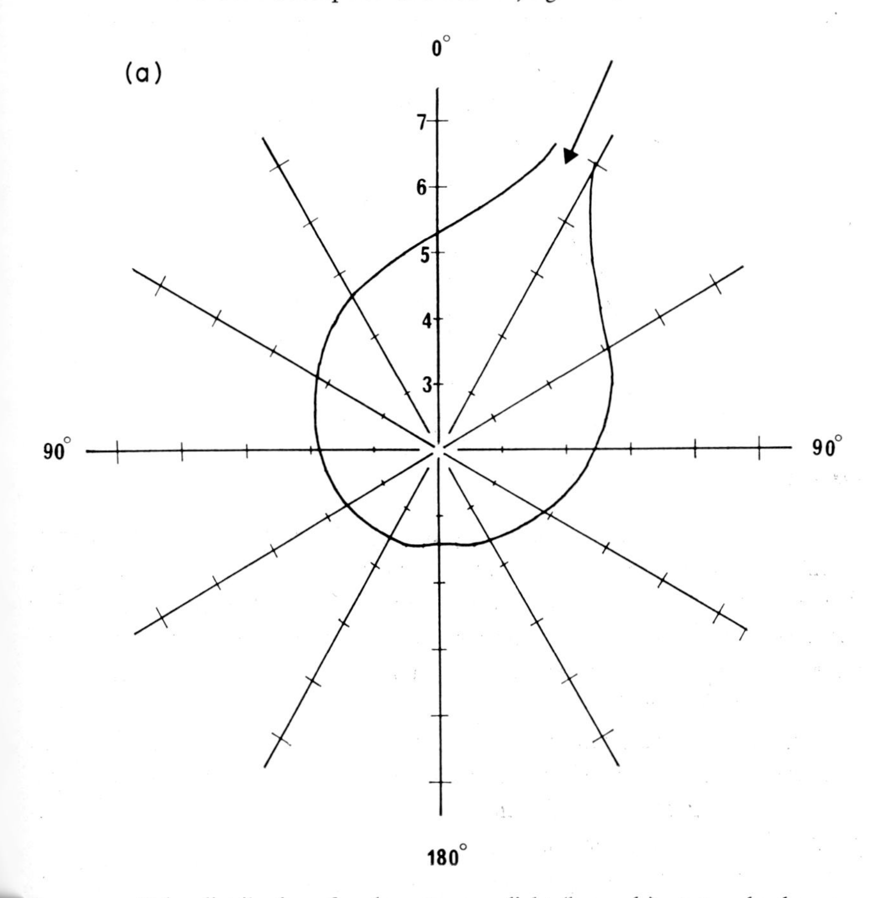

Figure 9.2. Polar distribution of underwater spacelight (log scale) at two depths on a sunny day in a freshwater lake. The measurements are in the plane of the sun; the arrows indicate the sun's direction. Note how the dominating effect of the sun is reduced at the deeper station. (a) 4·24 m depth (b) 41·3 m depth. Data plotted from Tyler & Preisendorfer, 1962.

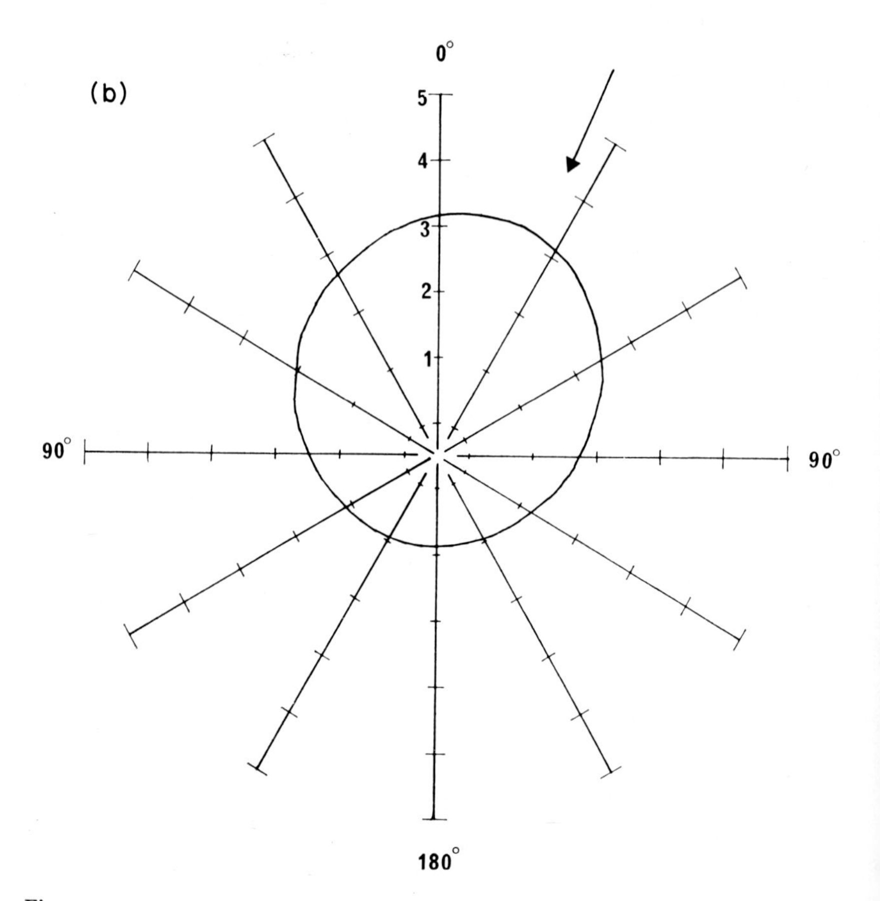

Figure 9.2

directions but chiefly from above. Near the surface the light distribution is dominated by Snell's window and particularly by the sun if it is out. In deeper water the brightness of Snell's window relative to light from other directions is reduced by the scattering action of the water (Fig. 9.2). As the water gets deeper the angular distribution of light settles down to a characteristic asymptotic distribution.

Light that is scattered from particles less than a wavelength of light in diameter is plane polarized and the plane of maximum polarization is at right angles to the direction of the light. The underwater spacelight is thus plane-polarized down to all depths.

Light from sun and sky is reflected from the surface of the sea and the more oblique the light ray the more is reflected. Light that is reflected back from the surface does not enter the water. At high solar elevations the loss due to reflection is only of the order of 2% or 3%, but at more oblique angles the reflection losses are considerable. The actual amount can be calculated from Fresnell's equations. In these equations the two planes of polarized light have to be calculated separately.

$$r_{11} = \frac{-\tan(V_a - V_b)}{\tan(V_a + V_b)} \quad — 2$$

and

$$r_1 = \frac{+\sin(V_a - V_b)}{\sin(V_a + V_b)} \quad — 3$$

Snell's and Fresnell's equations, which only apply to flat surfaces, rarely describe the actual situation at the surface of the sea. The edges of Snell's window are broken up by the dark and light pattern of waves and ripples, whilst the waves trap oblique sunlight more efficiently than would be predicted from Fresnell's equations.

Ecology of fish with iridescent corneas

Fishes that posses iridescent corneas are typically shallow water benthic or demersal marine teleosts. Where their habits are known, they are diurnal or crepuscular in activity. There is no obvious correlation between feeding habits and iridescent corneas.

Depth of habitat

The normal depth range of all species that are known to possess iridescent corneas extends to the surface or to only a few metres below it (see Table 9.1). Some species, such as the Clingfish, *Lepadogaster lepadogaster*, are confined to very shallow water, but others, such as the Plaice, *Pleuronectes platessa*, have a downward range that extends to some 400 m in the Mediterranean (Bini 1967).

Iridescence is not equally bright in all individuals of the same species, but, as yet, the reasons for the variation are not known. It might be that individuals from deep water show less iridescence, but it should be pointed out that all the fishes in this study were caught in relatively shallow water. There are three likely functions for iridescence, namely, concealment, display and reduction of intra-ocular flare. Concealment and display are functions that remain equally important at all depths, but the reduction of flare would become somewhat less important at depths where the light from Snell's window no longer dominates the visual scene.

Type of habitat

With a single exception no pelagic or epipelagic fish is known to possess an iridescent cornea. The exception is the Sargassum fish, *Histrio histrio*. However, this fish lives a somewhat peculiar pelagic life since it is only found living amongst floating clumps of Sargassum weed.

All the other species are either benthic or demersal. Some, such as the Plaice, *Pleuronectes platessa*, and the Weever fish, *Trachinus vipera*, bury themselves in the sand with only the eyes showing. Others, such as the Sappharine gurnard, *Trigla lucerna*, and the Lizard fish, *Synodus intermedius*, rest on the sand, or like the Sea Scorpion, *Myxocephalus scorpius*, spend much of their time resting motionless on a rock.

The demersal species are those that spend most of their active time swimming above the bottom but rarely venturing into mid-water. These demersal species with iridescent corneas typically live near to rock or coral reef. Their food is either browsed or grazed from the substrate or consists of small fish and invertebrates that are themselves closely associated with the substrate.

The majority of these demersal fish swim well down between the rocks and corals and will take refuge in caves and crannies when they are disturbed.

Activity period

Many species of fish are active either by day or night. The phenomenon has been well studied on the coral reef where the clear water and abundance of species makes direct study possible. Here there is an almost complete changeover from a daytime population to a night-time one (Hobson 1968, 1972, Starck & Davis 1966). Iridescence has not been observed in any nocturnal species but is the rule rather than the exception in the large diurnal families of the Wrasse (Labridae), Trigger fish (Balistidae) and Parrot fishes (Scaridae). (Hobson 1965). It is interesting to note in this context that many species that possess yellow intra-ocular filters and which are diurnal (Muntz this volume) also possess an iridescent layer in their cornea (Table 9.1).

Feeding

There is no very clear correlation between feeding habits and corneal iridescence. Browsing and grazing species, such as the Trigger fish (Balistidae), Puffers (Tetraodontidae), Rabbit fishes (Siganidae) and Parrot fishes

(Scaridae) (Hiatt & Strasburg 1960) show it, possibly because they are often species of shallow water.

Slow predators on other fish, such as the Trumpet fish, *Aulostomus maculatus*, the Dragon fish, *Pterois volitans*, and the John Dory, *Zeus faber*, also have iridescent corneas. None of the very swift predators, such as the Barracudas (Sphyraenidae), Mackerel (Scombridae) or Horse Mackerel (Carangidae) show it. Some of the less active predators, particularly the Sea Perches and Groupers (Serranidae) show iridescence, but others with somewhat similar habits, such as the Snappers (Lutjanidae) do not.

Salinity

Iridescence is rare amongst freshwater species. It has been observed in the green Puffer fish, *Tetraodon samphongsi*, which lives in brackish water in S.E. Asia, and in the European fresh water Millers Thumb, *Cottus gobio*. It might be that the reason for this rarity is physiological. Water is continually being pumped out of the cornea by the corneal endothelium (Maurice 1972). It is possible that iridescent layers present a barrier to the movement of water which can only be overcome when there is a relatively shallow concentration gradient between the stroma and the surrounding saline water.

Iridescence and water colour

The colour of iridescence depends upon the angle of the illuminating light and upon the angle of view. However, when directly observed under water, corneal iridescence can be pinkish bronze, green or blue, according to the species.

The colour of the water in which these fish swim may be yellow-green (maximum transmission greater than 550 nm) in some European inshore waters to blue (maximum transmission less than 480 nm) in tropical waters (see Jerlov 1968). There is, however, no correlation between water colour and cornea colour of the fish that swim in it.

Internal structure

Six distinct types of structure that are sufficiently regular and of about the right thickness to result in interference reflections are presently known in teleost cornea. It is expected that further work will reveal more. These six

Table 9.1.

	Colour	Type	Period of activity	Depth	Habit	Substrate	Yellow filter
Super order PROTACANTHOPTERYGII							
Order SALMONIFORMES							
Sub-order MYCTOPHOIDEA							
Family Synodidae							
Synodus indicus	Green				Benthic	Sand	
Synodus intermedius	Green	1b			Benthic	Sand	Lens
Super order PARACANTHOPTERYGII							
Order BATRACHOIDIFORMES							
Family Batrachoididae							
Histrio histrio	Blue green	1b		0–1	Pelagic	Sargassum weed	None
Order GOBIESOCIFORMES							
Family Clinidae							
Lepadogaster lepadogaster	Green			0–1	Benthic	Rocks Stones	
L. candollei	Green			0–30	Benthic	Weed and rocks	
Super order ACANTHOPTERYGII							
Order ZEIFORMES							
Family Zeidae							
Zeus faber	Blue green	1a		2–200	Dermersal	Sand rock	None

Order GASTEROSTEIFORMES							
Family Aulostomidae							
Aulostomus maculatus	Green	(3)?			Demersal	Coral reef	None
Order SCORPAENIFORMES							
Family Triglidae							
Aspitrigla cuculus	Blue	1c		5–250	Benthic	Sand and gravel	None
Trigla lucerna	Blue	1c		5–200	Benthic	Sand and gravel	None
Family Cottidae							
Taurulus bubalis	Blue green	1c		0–30	Benthic	Rock	Cornea
Family Agonidae							
Agonos cataphractus				1–500	Benthic	Sand and mud	
Family Cyclopteridae							
Cyclopterus lumpus				0–300	Benthic	Rock	
Order PERCIFORMES							
Family Serranidae							
Epinephelus merra	Pink-bronze		Crepuscular	1–30	Demersal	Coral reef	
E. guttatus	Pink-bronze	3				Coral reef	None
Cephalopholis fulva	Green	3			Demersal	Coral reef	None
Hypoplectrus puella	Faint	2b			Demersal	Coral reef	None

	Colour	Type	Perod of activity	Depth	Habit	Substrate	Yellow Filter
Nemanthias carberryi	Blue green	2b			Demersal		None
Anthias sp.	Blue green	2b					None
Family Labridae							
Crenilabrus melops	Green	3	Diurnal	0–20	Demersal	Rock	Cornea
Halichoeres scapularis	Blue green		Diurnal		Demersal	Rock	Cornea
Crenilabrus rupestris	Faint	3	Diurnal		Demersal	Rock	
Thalassoma pavo	Green		Diurnal		Demersal	Rock	Cornea
Cheilinus diagrammus	Blue green		Diurnal		Demersal	Rock	Cornea Lens
Family Scaridae							
Scarus taeniopterus	Green	3	Diurnal		Demersal	Coral reef	Lens
Callyodon viridifucatus	Blue green		Diurnal		Demersal	Coral reef	Lens
C. ghoban	Blue green		Diurnal		Demersal	Coral reef	Lens
Xanothon bipallidus	Blue green		Diurnal		Demersal	Coral reef	Lens
Family Trachinidae							
Trachinus vipera	Green	3		0–50	Benthic	Sand	None
Family Blenniidae							
Blennius sphinx	Blue green			0–	Benthic	Rock	
B. pholis			Diurnal	0–	Benthic	Weed rock	None

B. rouxi			0–7	Benthic	Rock Sand	
Family Tripterygiidae						
Tripterygion tripteronotus			0–10	Benthic	Rocks	
Family Pholididae						
Pholis gunnellus	Green		0–40	Benthic	Stone and sand	None
Family Gobiidae						
Gobius paganellus	Blue green	3	0–9	Benthic	Rocks	None
G. niger	Blue green	3	0–75	Benthic	Sand	None
Pomatoschistus pictus	Blue green		0–20	Benthic	Gravel and sand	None
P. minutus	Blue green	3	0–20	Benthic	Sand	None
P. microps	Blue green	3	0–	Benthic	Sand and mud	None
Family Acanthuridae						
Siganus rostratus	Blue green			Demersal	Rock and weed	
S. stellatus	Pink-bronze			Demersal	Coral reef	
S. oramin	Blue green			Demersal	Weeds, coral, rock	Vitreous
S. vulpinus	Green	2a				None
Order PLEURONECTIFORMES						
Family Pleuronectidae						
Pleuronectes platessa	Bronze	3	0–120	Benthic	Sand gravel	None

	Colour	Type	Period of activity	Depth	Habit	Substrate	Yellow Filter
Scopthalmus maximus				0–80	Benthic	Sand Mud	
Family Soleidae							
Solea solea				1–183	Benthic	Sand Mud	
Order TETRAODONTIFORMES							
Family Balistidae							
Balisatapus undulatus	Green yellow	2a	Demersal	1–3	Demersal	Coral reef	Cornea
Family Tetraodontidae							
Tetraodon samphongsi	Yellow green	2a				Brackish water	Cornea
Arothron citrinellus	Yellow green				Demersal	Coral reef	Cornea
Family Canthigasteridae							
Canthigaster valentini	Green yellow				Demersal	Coral reef	Cornea

Notes on **Table 9.1.** Data on the activity period were taken from Hobson 1965, 1972 and for Epinephelus merra from Hiatt & Strasburg 1960. Data on depth range were taken from BINI 1967 and WHEELER 1969. In some cases our own observation suggest a slightly extended upward range. The biological information on the Gobiidae is from Miller 1971. More details on yellow filters is to be found in the chapter by Muntz in this volume.

types can be classified according to the type of tissue that is modified and to the position of the iridescent layer in the cornea.

All vertebrate corneas have a basically similar structure (Fig. 9.3). It is bounded on the inside by an endothelium a single cell thick and on the out-

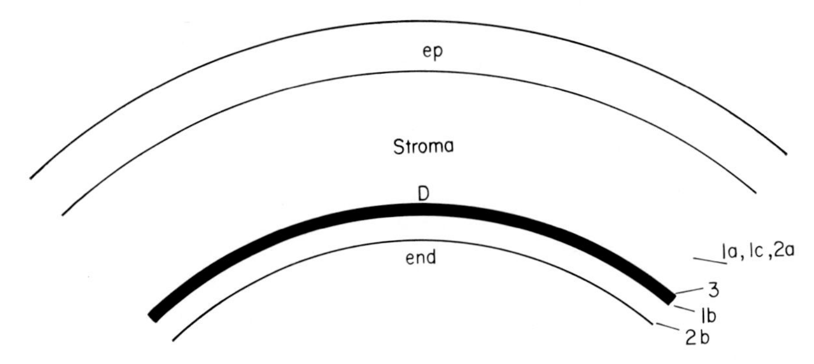

Figure 9.3. The positions that the various types of reflecting multilayers occupy in the typical teleost cornea. end, endothelium; D, Descemet's layer; ep, epithelium.

side by an epithelium which is several cells thick. In mammals, birds and some fish the endothelium has a basement layer known as Descemet's membrane. The body of the cornea is called the stroma or *substantia propria* and is made up of collagen fibrils embedded in a ground substance of mucopolysaccharide. In bottom-living fishes there is an outer protective covering to the cornea that is called the spectacle. The spectacle is usually several times thicker than the cornea but is morphologically analagous to the epithelium and the stroma (Walls 1963). The orientation of the reflecting plates have been described for the Indo-Pacific reef fish, *Nemanthias carberryi*, by Locket (1972) and for the Sand goby, *Pomatoschistus microps*, by Lythgoe (in press). In both cases, the lamellae are arranged in regular oblique layers (Fig. 9.4).

The six known types of iridescent structure are classified below. A more detailed classification is given by Lythgoe (in press).

1. Connective tissue

Type 1a (Plate 9.1)
Plates of electron-dense staining connective tissue, perhaps related to collagen, embedded in the stroma matrix.

Type 1b (Plate 9.2)
Electron-dense amorphous material condensed to form lamellae within a matrix of electron-rare material which corresponds in position to Descemet's layer. This matrix also contains randomly oriented fibrils. In the matrix also

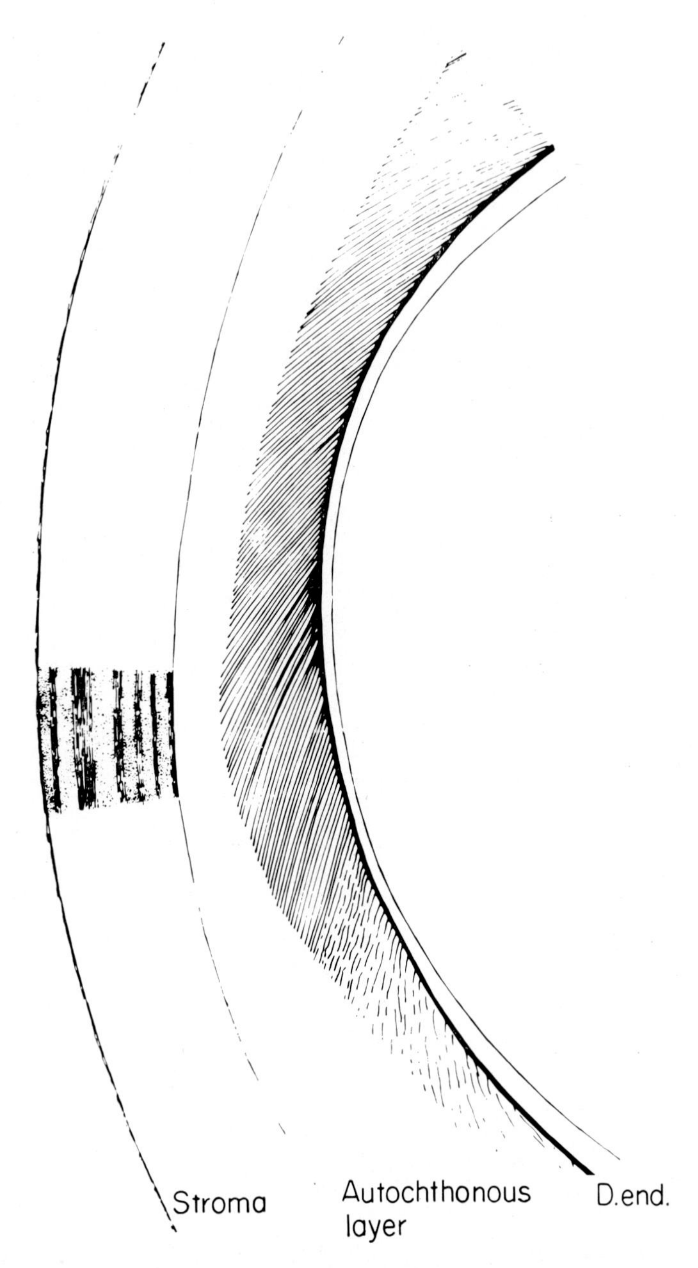

Figure 9.4. Orientation of reflecting plates in the cornea of the Sand goby, *Pomatoschistus minutus*. The spectacle is not shown in the diagram. The horizontal scale is exaggerated. The true appearance of the reflecting plates is shown in Plate 9.8.

contains randomly oriented fibrils. In the Sargassum fish, *Histrio histrio*, the endothelium also contains ordered lamellae within the protoplasm.

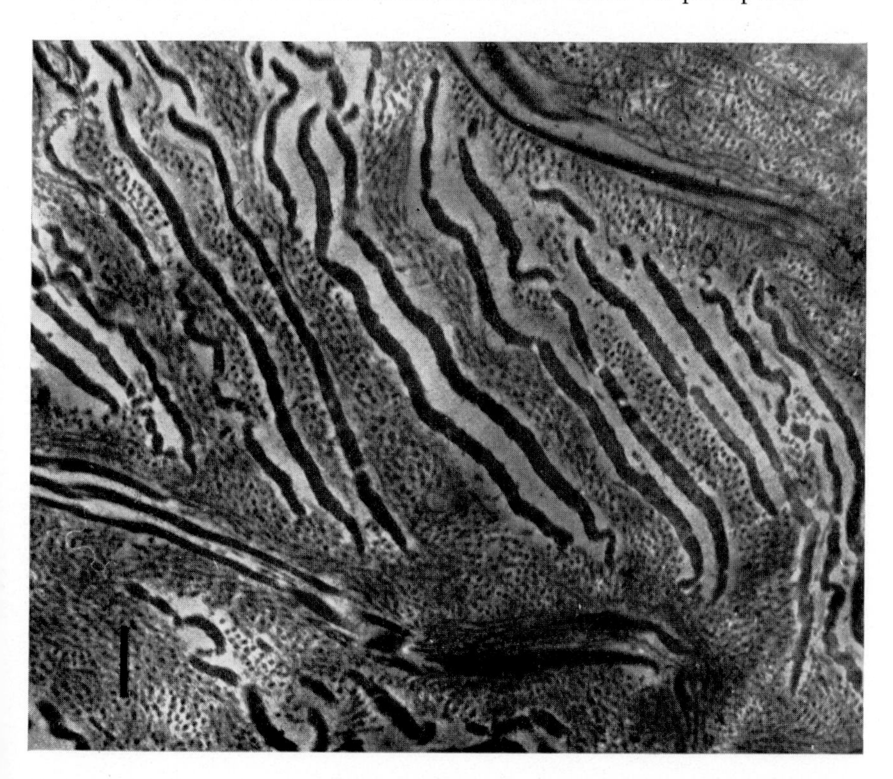

Plate 9.1. John Dory, *Zeus faber*. Iridescent layer in the corneal stroma. Plates of electron-dense staining materials are embedded in clear matrix. In some places the plates appear to be continuous with the collagen fibrils. Type 1a. Bar equals 500 nm.

Plate 9.2. Lizard fish, *Synodus intermedius*. Iridescent layer in the cornea is situated between the endothelium and the stroma. See Plate 9.9. Amorphous electron-dense staining material is condensed into lamellae. These are separated by electronrare material apparently containing fine fibrils. Type 1b. Bar equals 500 nm. ir, iridescent layer; end, endothelium.

Type 1c (Plates 9.3, 9.4, 9.5)
Enlarged collagen fibrils in ordered ranks usually associated with a matrix that contains amorphous electron-dense material.

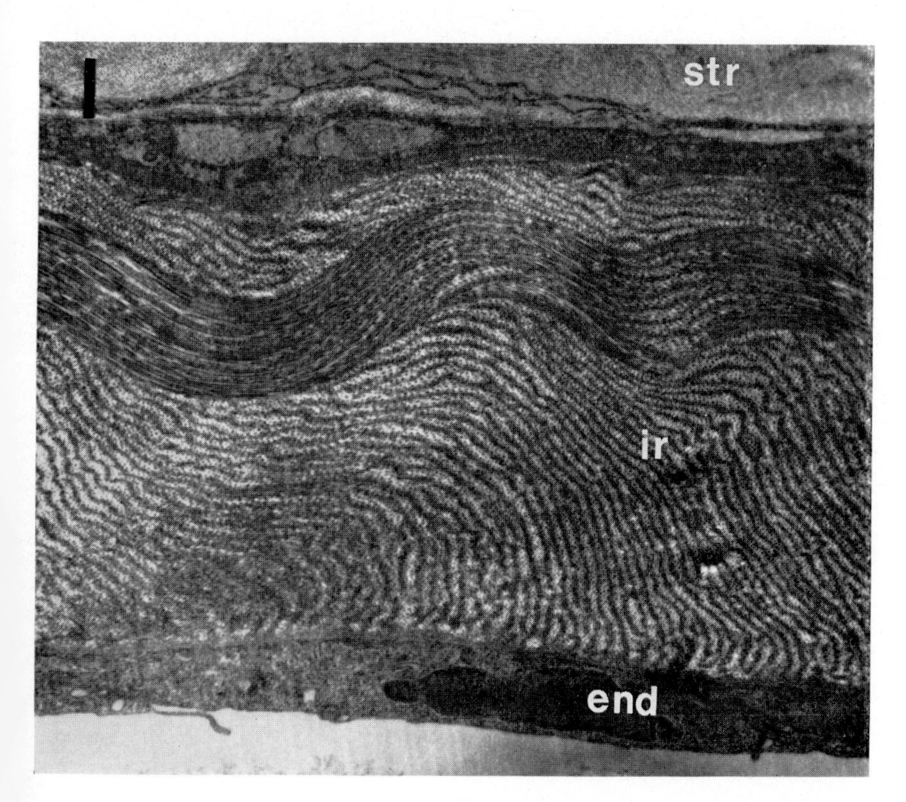

Plate 9.3. Red Gurnard, *Trigla cuculus*. Iridescent layer in the cornea between the epithelium and the stroma. Descemet's layer is apparently lacking in this specimen. A detail of the iridescent layer is shown in Plate 9.4. Type 1c. Bar equals 1,000 nm. end, endothelium; ir, iridescent layer; str, stroma.

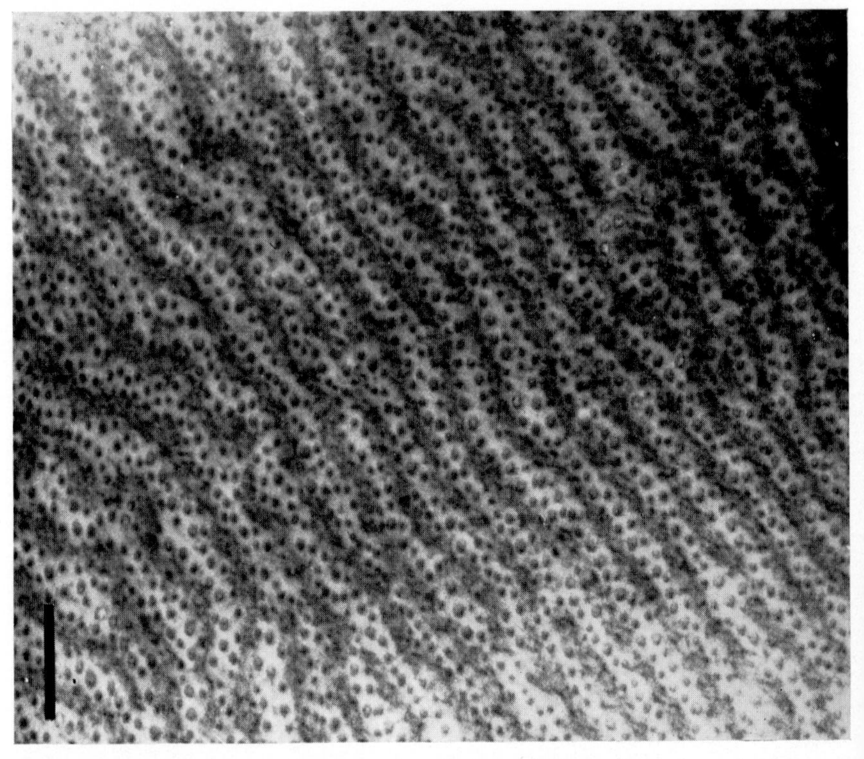

Plate 9.4. Red Gurnard, *Trigla cuculus*. Detail of the iridescent layer shown in Plate 9.3. The collagen fibrils are enlarged. Electron-dense staining material is condensed into lamellae within the collagen matrix. Type 1c. Bar equals 500 nm.

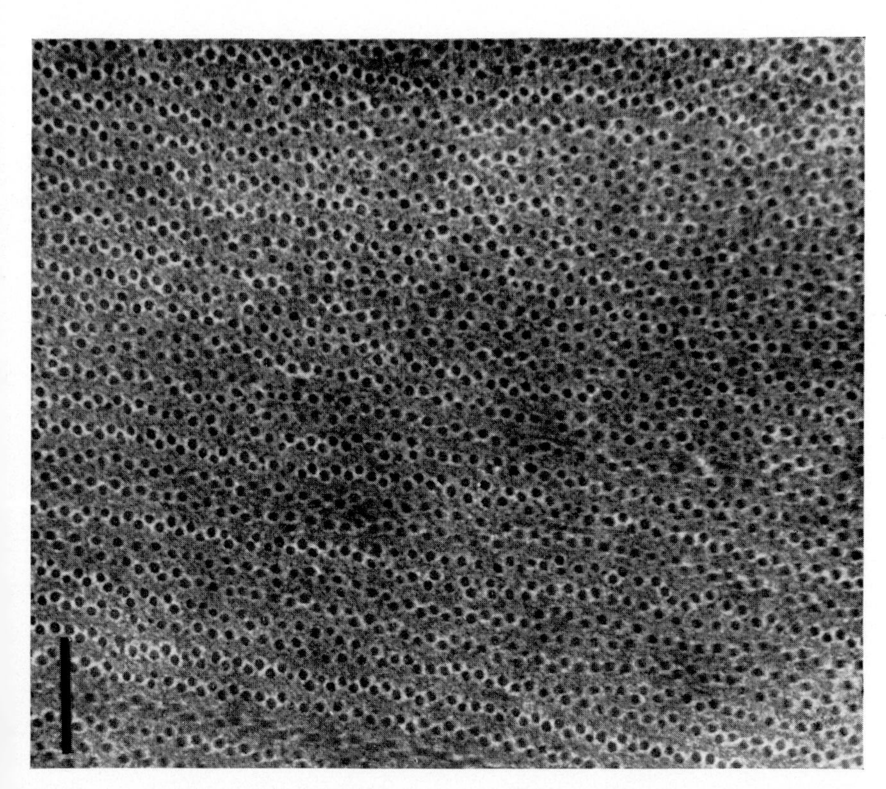

Plate 9.5. Saphirine gurnard, *Trigla lucerna*. Detail of the iridescent layer which is situated within the stroma. The enlarged collagen fibrils are arranged in ranks. Amorphous electron-dense staining material is condensed in the collagen matrix. Type 1c. Bar equals 500 nm.

2. Endoplasmic reticulum

Type 2a (Plate 9.6)
A layer of cells within the stroma nearly filled with regular layers of rough endoplasmic reticulum.

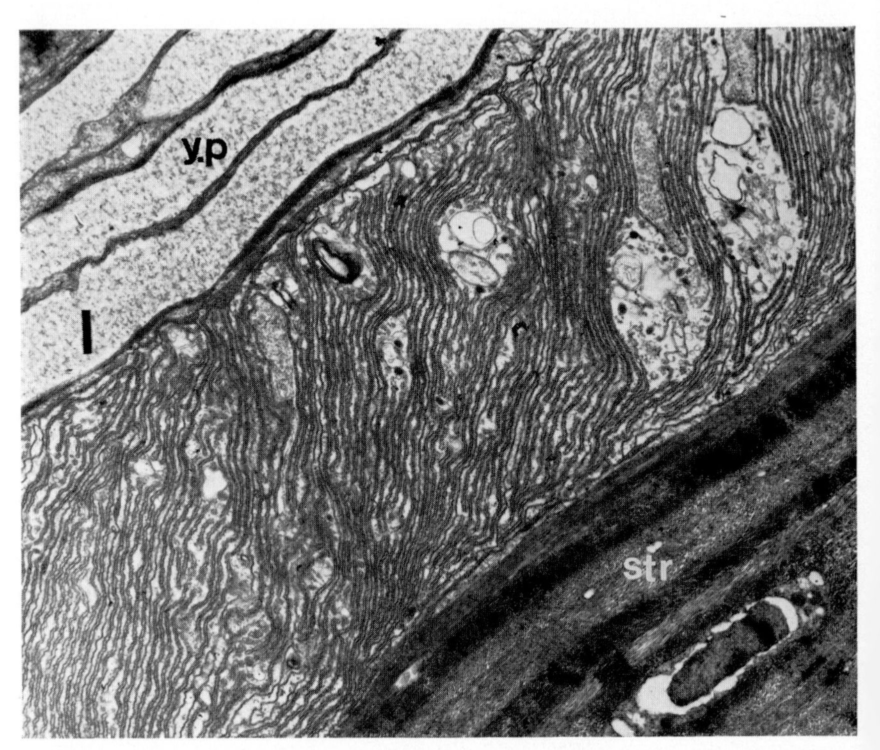

Plate 9.6. Green spotted Puffer, *Tetraodon samphongsi.* Iridescent layer situated in the corneal stroma. Adjacent to the iridescent layer lies the yellow pigment layer. The parallel lamellae are rough endoplasmic reticulum. There are several cells in this area of iridescent layer. Type 2a. Bar equals 500 nm. str, stroma; y.p., yellow pigment layer.

In the Rabbit fish, *Siganus vulpinus*, the cells are extremely slender and are separated from each other by stroma collagen fibrils.

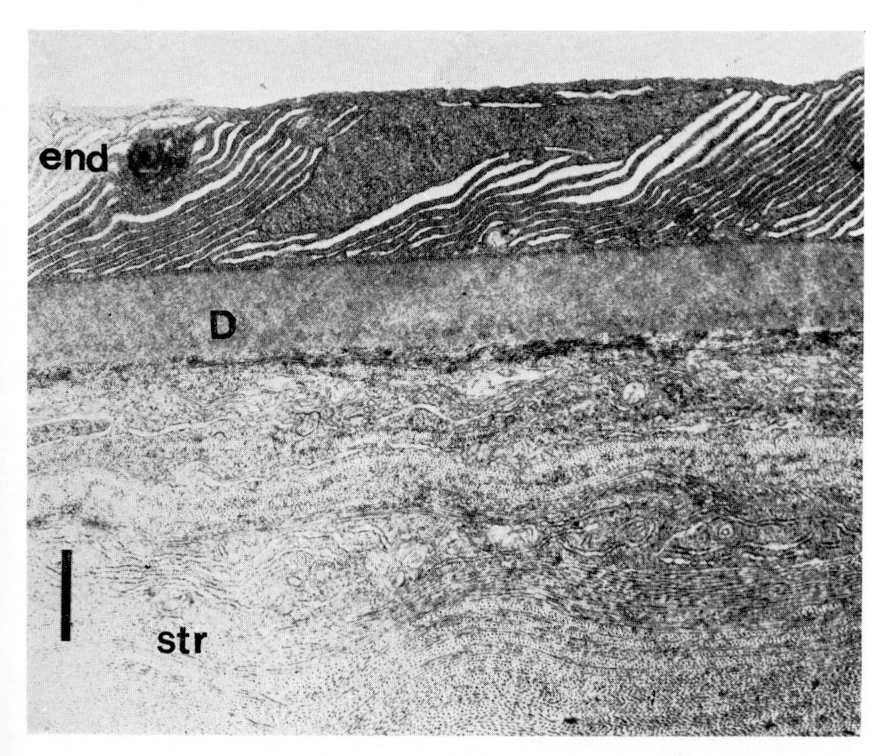

Plate 9.7. *Anthias* sp. The iridescent layer situated in the corneal endothelium. The lamellae consist of modified rough endoplasmic reticulum. More detailed pictures of a similar structure in the related *Nemanthias carberryi* have been published by Locket 1972. Type 2b. Bar equals 1,000 nm. str, stroma; D, Descemet's layer; end, endothelium.

Type 2b (Plate 9.7)
Modified rough endoplasmic reticulum nearly filling the endothelial cells. This structure was first described by Locket (1972).

3. Whole Cell

Type 3 (Plate 9.8)

Plates of protoplasm, probably whole cells, separated by an extra-cellular electron-rare ground substance that is continuous with the ground substance surrounding the collagen fibrils in the stroma.

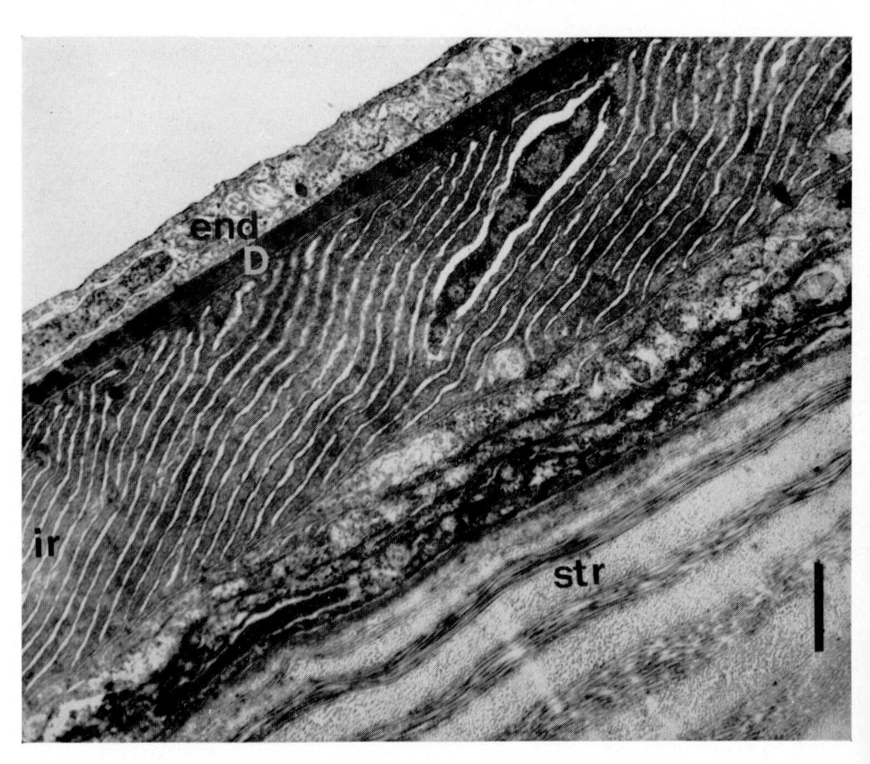

Plate 9.8. Sand goby, *Pomatoschistus minutus.* The iridescent layer situated between Descemet's layer and the stroma. The layer consists of plates of protoplasm embedded in an electron-rare staining matrix that is continuous with the collagen matrix. Type 3. Bar equals 1,000 nm. end, endothelium; D, Descemet's layer; ir, iridescent layer; str, stroma.

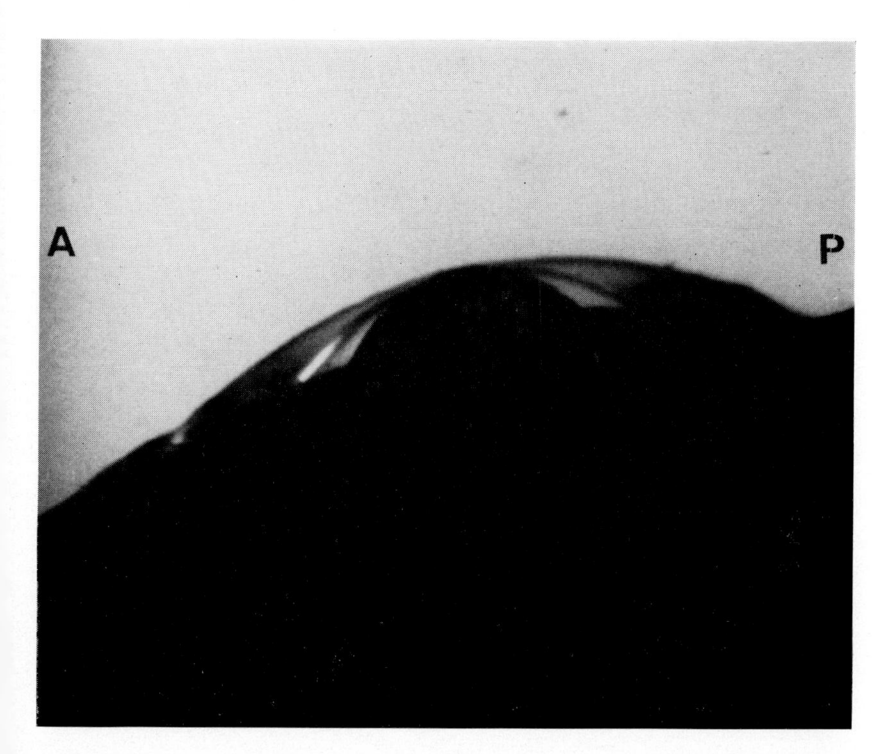

Plate 9.9. The Lizard fish, *Synodus intermedius*. Eye photographed from below. The fish is illuminated from a dorsal-anterior direction by a collimated light source just outside the frame of the photograph. The reflection from the iridescent layer (see Plate 9.2.) is clearly seen on the inner portion of the cornea.

Phylogeny

In Fig. 9.5 is reproduced Greenwood *et al.*'s (1966) provisional scheme to show the evolutionary relationships of recent teleost fishes. Superimposed on this scheme are the types of iridescent cornea as classified above that are found in the various taxa.

It is too early to say whether the cornea can be used as a taxonomic indicator although present indications are that it can. It is clear, however, that in evolutionary terms iridescence is a recent innovation, and that several different methods of achieving the same results (whatever these might be!) have been independently evolved. The situation amongst the Perciformes and their derivatives is of particular interest. Three types of iridescence are known from the Perciformes and their derivatives; two based on endoplasmic reticulum or the whole cell embedded in stroma (Type 2a and 2b) and one

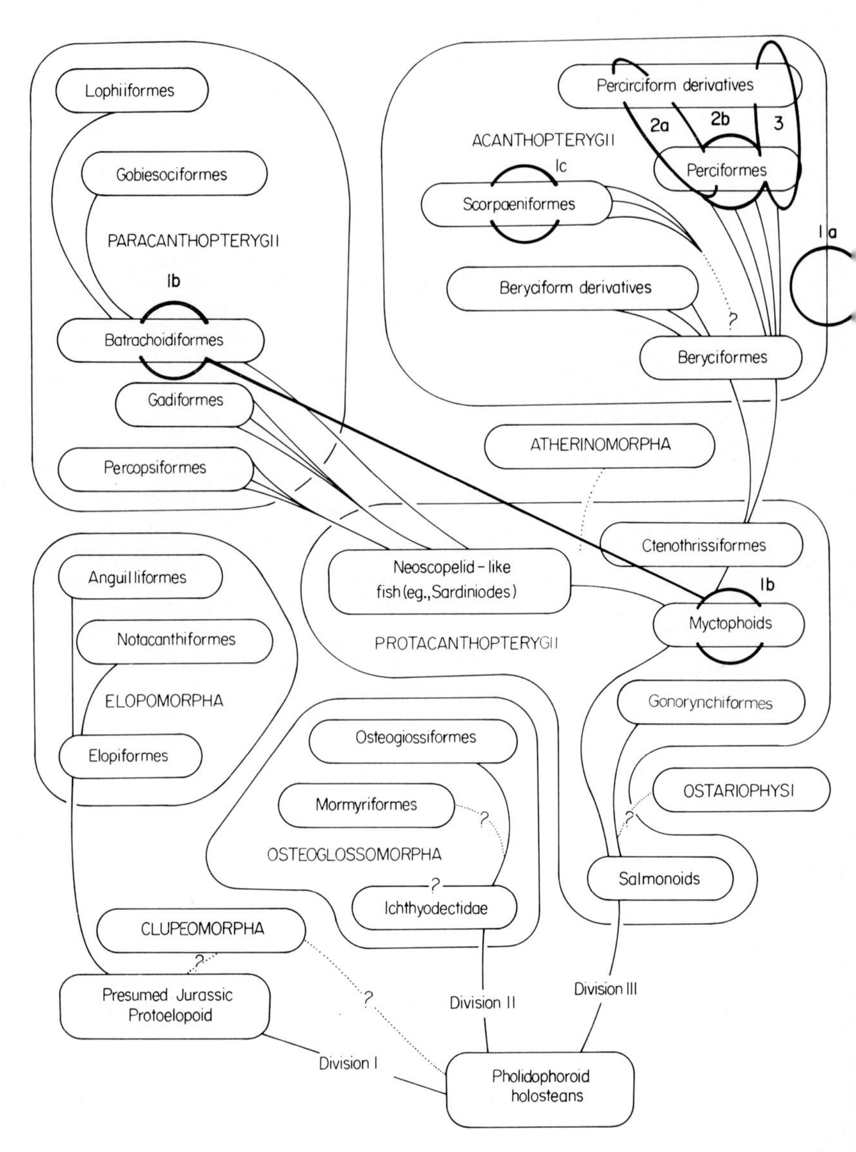

Figure 9.5. Schematic representation of the evolutionary relationships of teleost fishe after Greenwood *et al.* (1968). The dark outlines indicate the taxa that possess iridescen corneas of known type.

based on the whole cell (Type 3). The distribution of these three types is summarized by families in Table 9.2 using the data in Table 9.1.

Table. 9.2.

	Type 2a	Type 2b	Type 3
PERCIFORM DERIVATIVES	Tetraodontidae Canthigasteridae Balistidae Acanthuridae		Pleuronectidae
PERCIFORMES		Serranidae (*Nemanthias Hypoplectrus*)	Serranidae (*Cephalopholis Epinephelus*) Labridae Scaridae Trachinidae Gobiidae

The Tetraodontiformes are thought to be derived from the Acanthuroidei (Romer 1945). The Pleuronectiformes are also derived from some Perciform stock but their exact origins are less clear. Amongst the Serranidae there are two distinct cornea types. *Epinephelus* and *Cephalopholis* have type 3 and *Hypoplectrus*, *Anthias* and *Nemanthias* have type 2b. It is interesting in this context that in a personal communication Lovett Smith has pointed out that *Anthias* and *Hypoplectrus* are synchronous hermaphrodites whilst *Epinephelus* (including *Cephalopholis*) is protogynous.

The Theory of Iridescent Reflection

The theory of iridescent reflections for biological material has been worked out by Huxley (1968); and Land (1972) has reviewed the optics and anatomy of the many iridescent structures found in animals. It is thus only necessary to treat the matter in a simplified manner here.

Whenever there is a boundary between two transparent materials of different refractive index there will be some reflection at the interface. The equations that describe this reflection have already been given in relation to the reflection of light at the water's surface (Equations 1–3). Provided that the refractive index is independent of wavelength, the reflected light will be of the same colour as the incident light.

If there is a thin transparent layer of the order of a wavelength of light in thickness, the light rays reflected from the top and the bottom of the stack may mutually interfere. If the two waveforms are in phase they will interfere constructively, i.e. the layer will reflect light very strongly and transmit very little. If the two waveforms are exactly out of phase they will interfere destructively and there will be no reflection. For any given wavelength of light the presence of destructive or constructive interference depends upon the length of the optical path through the layer and back again. The length of the optical path is

$$2nd \cos V$$

where n is the refractive index of the medium, d is its thickness and V is the angle of incidence on the stack.

The phase delay, ø, is given by

$$ø = 2\left(\frac{2\pi}{\lambda}\right) nd \cos V \qquad — 4$$

where λ is the wavelength of light.

Where there are two thin layers of different refractive index in conjunction, there will be two values for ø. Cos V in each case can be found from Snell's equation (Equation 1).

When the difference in ø in the two media is equal to π, 3π, 5π, etc., there will be destructive interference. When the difference is equal to 0, 2π, 4π, etc., there will be constructive interference.

Several generalizations can be made about the nature of reflected light from biological multilayers.

1. The reflection is usually coloured.
2. At a particular angle of incidence, which depends upon the refractive index of the two media, the reflected light will be completely plane-polarized.
3. A multilayer can act as an anti-reflection coating.
4. Oblique light will be highly reflected if it is not damped out by destructive interference.

From these generalizations it follows that a regular multilayer in the cornea could act in the following ways:

1. Act as a coloured reflector and/or as a coloured transmission filter.
2. Act as a polarizing reflector and/or as a polarizing filter.
3. Act as an anti-reflection coating.
4. Act as an unselective reflector and hence as a neutral transmission filter.

Possible functions of corneal iridescence

Coloured reflector and transmission filter

It is notoriously difficult to judge the reflecting efficiency of an iridescent surface by eye. This is because they are nearly always set against a non-reflecting background. The pupil of the eye usually appears black and the iridescence is subjectively brighter than objective measurement would confirm. It is extremely difficult to measure the proportion of the incident light that is reflected from a curved surface since the reflected light must be collected and integrated over a solid angle of at least 180°.

The measurement of the transmissivity of the cornea is considerably easier. The main problems here are that the isolated cornea swells and loses its transparency and imperfections on both surfaces of the cornea caused when it is removed also tend to reduce transparency.

When an iridescent cornea is examined by transmitted light no pronounced colour is visible. Furthermore, measurements using a small-spot spectro-photometer show that the iridescent Sand Goby (*Pomatoschistus minutus*), cornea has a transmission exceeding 90% throughout the visible spectrum except for the usual increased absorbance at shorter wavelengths. It is, therefore, unlikely that iridescent corneas act significantly as colour filters in the way that the yellow pigment layer present in many fish corneas almost certainly do (Muntz, this volume).

A more likely possibility is that iridescent reflections from the eye reduce the visibility of the pupil by reducing its blackness to a brightness, and perhaps colour, near to that of the rest of the fish. Examples are the Scorpion fish *Scorpaena scrofa* and the Crocodile fish *Platycephalus indicus.*

A problem in discussing questions of display and camouflage is that we must rely upon the evidence of our own eyes rather than the response of the fishes themselves. Problems concerned with conspicuous display colorations are difficult because they are linked to the behaviour of the fishes concerned. Nevertheless, the vision of fishes is not greatly different from our own, especially in the light conditions under water (see Lythgoe & Northmore 1973, for a review), and it is perhaps relevant to discuss the observations of underwater swimmers.

One such observation made off the Cornish coast (McDonald 1972) describes very well the appearance that iridescent corneas can have to the diver:

'There, hovering like a desert-camouflaged helicopter only a foot away . . . was a John Dory. His fins almost hummed as he manoeuvred a foot this way then a foot that. And all the time his glowing sapphire eyes never left us.'

The John Dory (*Zeus faber*) has a large dark spot on its flanks and opinions differ whether it is in general to be considered a cryptic or a conspicuous fish. However, some fishes, such as the Gobies, are generally very difficult to detect on the bottom yet their eyes do shine brightly. Members of the Siganidae (Rabbit fish) often have conspicuously iridescent blue-green eyes underwater and yet their general body colour is a drab grey.

If an object is to be perfectly camouflaged its spectral radiance must match that of its background. When that is the case no eye will be able to detect it. Silvery pelagic fish come near achieving this radiance match using interference reflections from guanine crystals in the scales (Denton & Land 1971, Denton & Nicol 1965). Yet the pupils of the fish in question do not show iridescence and, indeed, appear black to the observer.

The pupils of some species, such as the Plaice, *Pleuronectes platessa*, and the Red Hind, *Epinephelus guttatus*, have bronze or pinkish-bronze iridescence. Near the surface this iridescence is not cryptic to the human eye. However, at depths of several metres the longwave component of daylight is much reduced by absorption of the water and in this case the iridescence may well serve to reduce the conspicuousness of the pupil.

Polarizing filters

The reflection from the interface between two transparent materials is plane-polarized except at normal incidence and when reflection is total. This follows from Fresnell's equations (2 and 3) which have been plotted in Fig. 9.6. At a particular angle of incidence which depends upon the refractive index of the two materials, light polarized in one plane only is reflected. The iridescent layer of the cornea has thus the potential to act as a linear polarizing filter. However, the actual transmission of the cornea is so high that the transmitted light will be only partially plane-polarized.

The underwater spacelight is plane-polarized with the plane of maximum polarization at right angles to the direction of maximum illumination. This polarization is a result of scattering from particles in the water which are of the order of the wavelength of light in diameter or less and the water molecules themselves. Polarizing phenomena resulting from surface reflection are negligible (see Jerlov 1968 for a review).

Selective sensitivity to the plane of polarized light can lead to improved visual range underwater. The underwater spacelight is linearly polarized, but large diffuse reflecting objects depolarize the incident light. Hence a linearly polarizing filter can change the apparent luminance of the background light relative to the light from the object. And it is in part their relative difference in brightness (i.e. contrast, discussed in a later section) that governs the

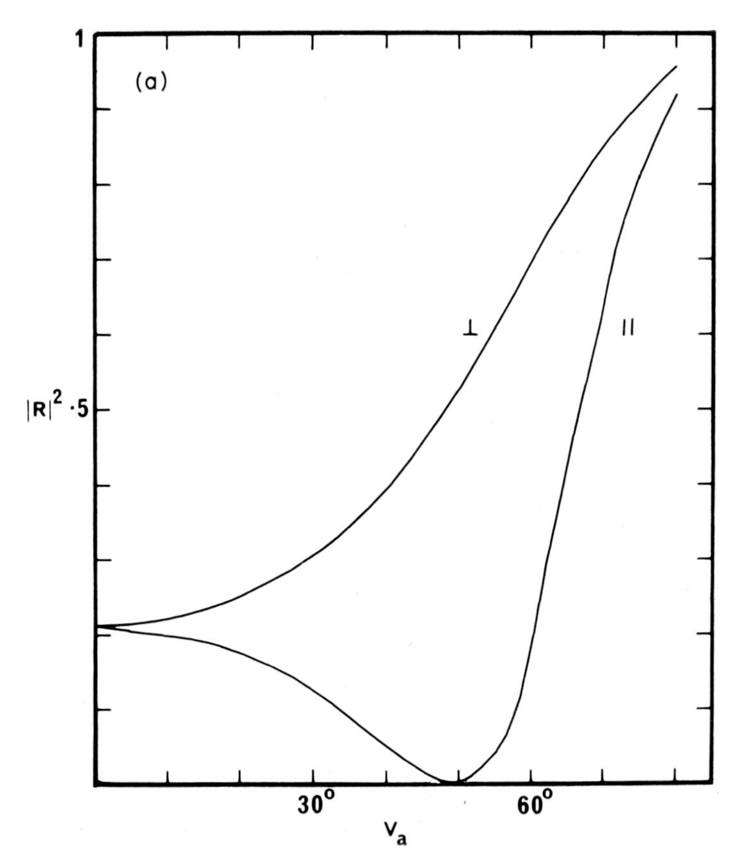

Figure 9.6. The proportion of light reflected, R^2, from a stack of 20 thick plates (no iridescence) for two planes of polarization. V_a is the angle of incidence of the light. The refractive index of the two materials are 1·62 and 1·374. (a) Optically dense plates within a rare matrix. (b) Optically rare plates within a dense matrix.

distance from the eye that an object can be seen (Lythgoe & Hemmings 1967).

However the polarizing nature of the cornea does not seem sufficient to have a visual function similar to this.

An anti-reflection coating

Most optical lenses now have a thin coating on them which is of a different refractive index to the glass and of an optical thickness that leads to destructive interference of the reflected light. These anti-reflection properties only work for light passing through the lens at nearly normal incidence. Oblique rays show constructive interference and reflection is enhanced giving the characteristic 'bloom' to the lens.

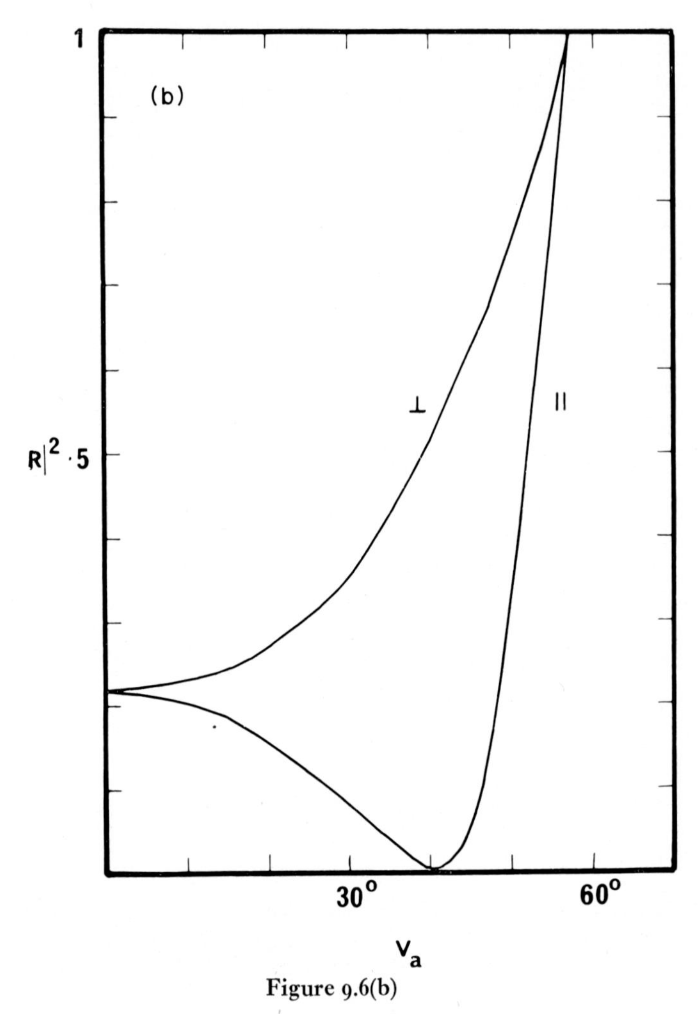

Figure 9.6(b)

The measured transmission of the Goby cornea (Fig. 9.7) together with a theoretical study of the probable reflecting properties of the multilayer (Lythgoe, in press), leads to the belief that for image-forming light near to the optical axis the multilayer exhibits destructive interference of the reflected light. This will not lead to greater transparency of the cornea as a whole since the multilayer is embedded within the cornea but it will reduce reflection losses at the multilayer. The Serranid teleosts, *Anthias*, *Nemanthias* and *Hypoplectrus*, are exceptions since the multilayer is situated in the endothelium. In this case, the multilayer could act as an anti-reflection coating.

Unselective reflection

It has already been mentioned that a ray of light shining obliquely on an interface between two materials of different refractive index will be strongly reflected. If the ray is passing from the high refractive index to the low refractive index material, there may be total reflection at the interface (Fig. 9.6b).

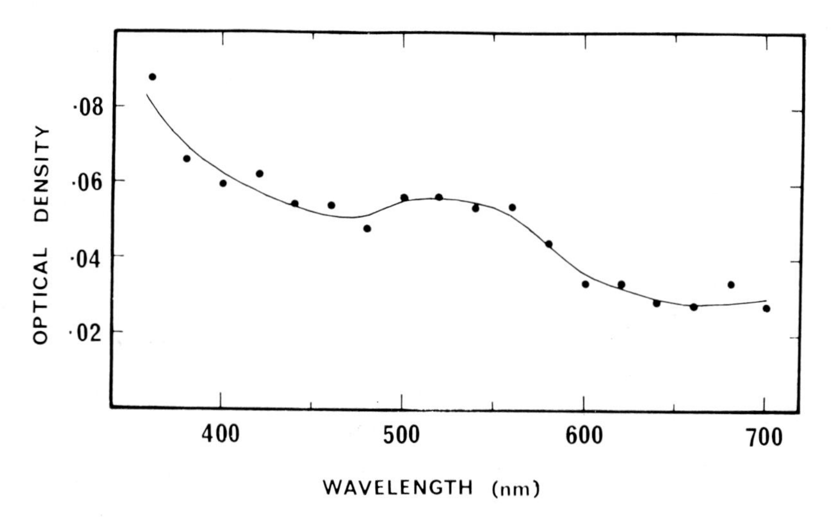

Figure 9.7. The spectral transmission of the Sand goby (*Pomatoschistus minutus*) cornea at normal incidence.

If both layers are thick, the reflection will depend upon the number of layers, the refractive index of each and the angle of incidence of the light. See, for instance, Land 1972.

When we come to calculate the theoretical reflection near to the critical angle for layers less than about two or three times the wavelength of light in thickness, we find the existing mathematical treatment such as is proposed by Huxley less satisfactory. This is because reflection is not total when the layer below the interface is sufficiently thin. (Born & Wolf 1970.)

Nevertheless, it can be demonstrated that at grazing incidence there is a layer in the cornea, corresponding in position to the iridescent layer that acts as an efficient reflector of light. In the Sand goby, this reflection is an unsaturated blue. In the Red Hind and the Lizard fish, *Synodus intermedius* (Plate 9.10), it is white. In the Wrasse, *Crenilabrus melops*, the reflection is yellow, presumably because the iridescent layer lies inside the yellow pigment layer. It is perhaps reasonable to assume that it is only in the Goby that at oblique angles of incidence, the layers are spaced in such a way as to produce iridescence. In the other fishes the reflection is simple and does not involve interference phenomena.

The Cornea as a Sunshade

It has been observed that in most fishes it is the downwelling light that is reflected back off the cornea (Lythgoe 1971). Since this downwelling light

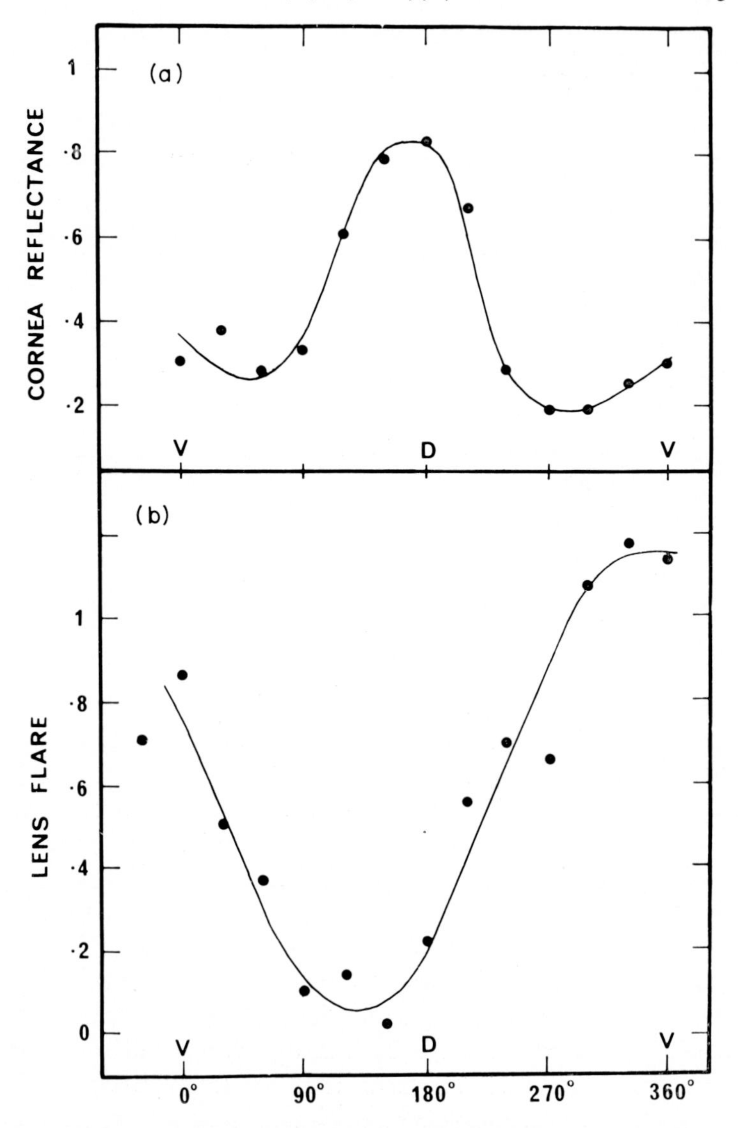

Figure 9.8. Comparison between the angular distribution of corneal reflectance and lens flare in the Sand goby. (a) Relative reflection of green light (λ_{max} 525 nm) directed onto the cornea from different directions but always at 75° to the optic axis. (b) Flare measured at the back of the lens resulting from green light (λ_{max} 555 nm) directed onto the eye at 90° to the optic axis. Expressed as a proportion of the incident intensity $\times 10^{-4}$. D light from the dorsal direction; V, light from the ventral direction.

can be several log units brighter than light from other directions (Fig. 9.2.) (see Jerlov 1968 for a review) it is reasonable to speculate that the iridescent layer acts as some kind of sophisticated sunshade.

This belief was confirmed by an experiment that involved cutting away the back of an excised eye and measuring the 'flare' caused by non-image forming light in the cornea and lens (Lythgoe, in press). It was found that in the Rudd (*Scardinius erythropthalmus*) eye, which has no multilayer structure, the flare produced by a beam of light that just grazed the cornea was independent of direction. In fishes, such as the Sand goby (Fig. 9.8a), Weever and *Anthias*, which possess an iridescent layer, grazing light from the dorsal direction results in much less flare than grazing light from the ventral direction.

The effect of lens flare on visual range

Most questions of visual performance are concerned with visibility of actual objects of known optical characteristics. These considerations are much simplified in the underwater environment because the optical characteristics of the water itself are of paramount importance whereas on land the characteristics of the atmosphere are relatively less important.

The central question is always whether the perceived colour of brightness of an object or pattern is sufficiently different from its background to be visible. In man we can make this calculation for both colour and brightness, but in all other animals, where precise details of colour vision are less well known, we must be content with considerations of brightness alone. Underwater this is less of a limitation than may at first sight appear since natural water acts as an efficient colour filter (Tyler 1958), and the most important visual differences are those of brightness.

Brightness contrast is defined for these purposes as

$$C = \frac{T-B}{T} \qquad \text{— 5}$$

where T is the brightness of the object and B is the brightness of the background. Man can detect contrasts of about 0·004 (Blackwell 1946) whilst a fish (the Goldfish) has been measured as having a contrast discrimination nearer 0·05 (Hester 1968).

If there is light flare (f) in the lens the apparent contrast becomes

$$C_f = \frac{(T+f)-(B+f)}{(B+f)} \qquad \text{— 6}$$

Of course, T, f and B should really be expressed in terms of the number of quanta absorbed by the visual pigment in the retina and corrected for the ambient light level, angular subtense and shape of the target, etc. (Lythgoe 1968). For our present purposes, these precise calculations, which can only be based on estimates, are unnecessary.

The important fact that enables calculations of visual range to be made is that the contrast presented by an object decreases exponentially with distance if the path of sight is horizontal (Duntley 1962, 1963). This relationship has both a mathematical and an observational basis.

$$C_r = C \exp(-\alpha r) \qquad — 7$$

where r is the visual range and α is the narrow beam attenuation coefficient of the water.

The quantity of flare at the lens was found by numerically integrating at 10° intervals. Tyler and Preisendorfer's (1962) data for the radiance distribution at depth of 4·2 m on a sunny day with the measured lens flare in the Sand goby (Fig. 9.8b) resulting from tangential illumination at 90° to the optical axis.

The reduction in apparent contrast presented by an object when it is 'veiled' by flare can then be calculated from equation 6 and the resulting reduction in visible range calculated using equation 7. The results of one such calculation are shown graphically in Fig. 9.9.

The effect of the measured flare when the cornea is *in situ* causes little reduction in visual range. But if, for purposes of comparison, it is assumed that the percentage of the light incident on the eye that results in flare is equal from all directions and numerically equal to the mean of the actual percentages observed, it is found that flare causes a severe reduction in visible range.

Conclusions

The widespread occurrence of iridescence in shallow water marine fishes and the quite different morphological structures that produce it lead to the belief that corneal iridescence must confer some advantage to the fishes that possess it.

Iridescence is only the most obvious manifestation of interference. The multilayers in fish corneas could also act as wavelength selective filters, birefringent filters and anti-reflection coatings. In fact, the percentage reflection of the iridescent layer at near normal angels of incidence is low and it is unlikely that the layer acts as a colour filter or as a polarizing filter unless it is in some way used to detect the plane of polarized light.

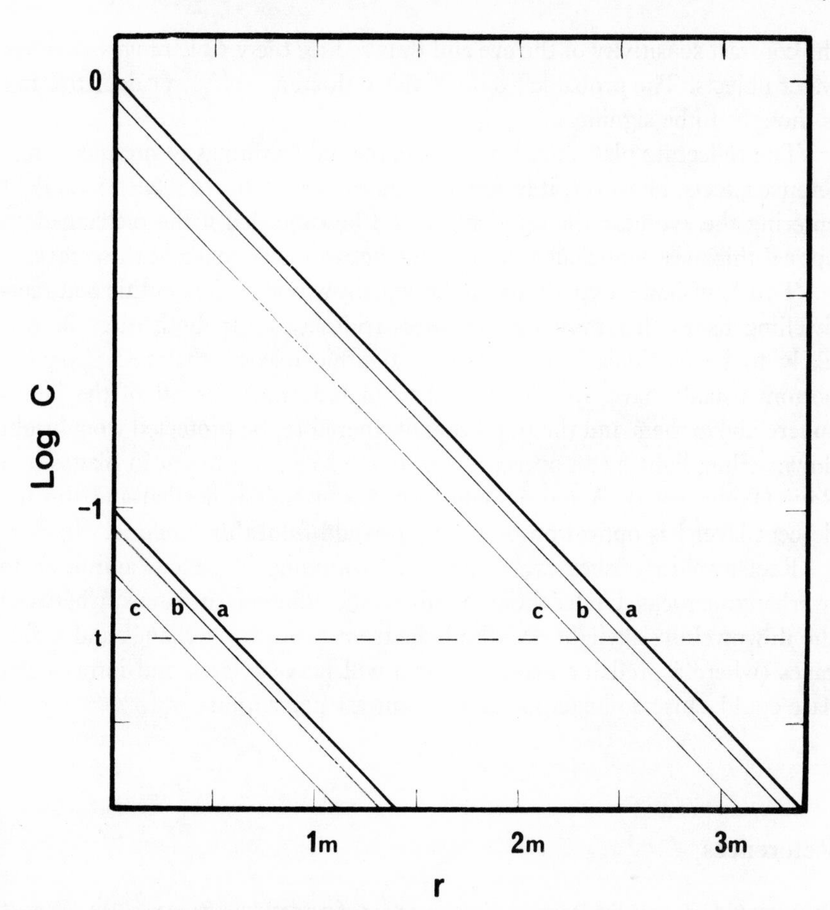

Figure 9.9. The rate that the apparent contrast of a large black object (contrast C = –1 at zero range, log –1 = 0) and a grey object (log C = –1) is reduced as the horizontal range (r) increases. When the contrast falls below some threshold value, here taken as 0·05 (log ·05 = –1·30) the object becomes invisible. a, No intra-ocular flare. b, Measured intra-ocular flare for the Sand goby. c, Non-direction intra-ocular flare. A diver has a contrast perception threshold of about 0·2 = –1·70) and in these waters could see a large black object at a horizontal range of 3·4 m.

It is not thought that the iridescence often has the effect of camouflaging the pupil. Whether or not it is used in behavioural display remains an open question.

The most likely and widespread function of the iridescent layer is strictly not an interference effect at all. The reflecting plates in the eye are arranged in such a way that the very bright downwelling light present under water is reflected out of the eye before it is reflected from the inner surface of the cornea and the outer surface of the lens and scattered by the ocular media. This would help to reduce intra-ocular flare, which in its turn, would reduce the contrast sensitivity of the eye and thus reduce the visible range of underwater objects. The probable extent of this reduction has been calculated and is thought to be significant.

The reflecting plates need not, of course, be so thin as to produce interference effects. However, it is possible that the losses due to reflection of light entering the eye near the optic axis could be cancelled if the plates had an optical thickness sufficient to cause destructive interference of these rays.

This 'sunshade' hypothesis might explain why it is the benthic and reef-dwelling fishes that most often possess iridescence. In both cases flare is likely to be particularly troublesome. Benthic teleosts that rest upon the bottom usually have eyes positioned to include most or all of the hemisphere above them and the pupil cannot, therefore, be protected from bright downwelling light by an operculum of tissue like that present in Skates and Rays (Walls 1963). A more sophisticated sunshade is needed and the iridescent layer has optical properties that would fulfil this function.

Reef-dwelling fishes are constantly swimming in gulleys and near to overhanging rocks. Under these conditions the difference in intensity between the downwelling skylight and the light from beneath overhangs and inside caves (where a predator may well lurk) will be very large and intra-ocular flare could cause a dangerous loss in contrast perception

References

BERNARD G.D. (1971) Evidence for visual function of corneal interference filters. *J. Insect. Physiol.* 17, 2287–300.

BINI G. (1967) Atlante dei Pesce delle Coste Italiane. Vol. 2–8 Mondo Sommerso, Rome.

BLACKWELL H.R. (1946) Contrast thresholds of the human eye. *J. opt. Soc. Amer.* 36, 624–43.

BORN M. & WOLF E. (1970) *Principles of optics.* 4th Ed. Pergamon, Oxford.

DENTON E.J. & LAND M.F. (1971) Mechanism of reflexion in silvery layers of fish and cephalopods. *Proc. roy. Soc. B.* 178, 43–61.

DENTON E.J. & NICOL J.A.C. (1965) Reflexion of light by external surfaces of the herring *Clupea harengus. J. mar. biol. Assoc. U.K.* 45, 711–38.

DUNTLEY S.Q. (1962) Underwater visibility in *The Sea*, pp. 452–5. Ed. by M.N. Hill, Interscience publishers, New York & London.

DUNTLEY S.Q. (1963) Light in the sea. *J. opt. Soc. Amer.* 53, 214–33.

GREENWOOD P.H., ROSEN D.E., WEITZMAN S.H. & MEYERS G.S. (1966) Phyletic studies of teleostean fishes, with a provisional classification of living forms. *Bull. Am. Mus. Nat. Hist.* 131, 341–455.

HESTER F.J. (1968) Visual contrast thresholds of the goldfish (*Carassius auratus*). *Vis. Res,* 8, 1315–36.

HIATT R.W. & STRASBURG D.W. (1960) Ecological relationships of the fish fauna on coral reefs of the Marshall Islands. *Ecol. Monogr.* 30, 65–127.

Hobson E.S. (1965) Diurnal-nocturnal activity of some shore fishes in the Gulf of California. *Copeia*, 3, 291–302.
Hobson E.S. (1968) Predatory behaviour of some shore fishes in the Gulf of California. *U.S. Fish Wildl. Serv., Res. Rep.* 73, 92 pp.
Hobson E.S. (1972) Activity of Hawaiian reef fishes during the evening and morning transitions between daylight and darkness. *Fish. Bull. Nat. Mar. Fish. Ser.* 70, 715–740.
Huxley A.F. (1968) A theoretical treatment of the reflexions of light by multilayer structures. *J. exp. Biol.* 48, 227–245.
Jerlov N.G. (1968) Optical oceanography. Elsevier, London, Amsterdam, New York.
Land M.F. (1972) The physics and biology of animal reflectors. *Prog. in Biophys. Mol. Biol.* 24, 75–106.
Locket N.A. (1972) The reflecting structure in the iridescent cornea of the serranid teleost *Nemanthias carberryi. Proc. roy. Soc. B.* 182, 249–54.
Lythgoe J.N. (1968) Visual pigments and visual range underwater. *Vis. Res.* 8, 997–1012.
Lythgoe J.N. (1971) Iridescent corneas in fishes. *Nature, Lond.*, 233, 205–7.
Lythgoe J.N. & Hemmings C.C. (1967) Polarized light and underwater vision. *Nature, Lond.*, 213, 893–4.
Lythgoe J.N. & Northmore D.P.M. (1973) Colours underwater in *Colour* '73. Adam Hilger, London.
Maurice D.M. (1972) The location of the fluid pump in the cornea. *J. Physiol.* 221, 43–54.
McDonald K. (1972) *Fish Watching & Photography*. John Murray, London.
Miller P.J. (1971) Gobies. In *Fishes of the Sea*. J. and G. Lythgoe pp. 259–78. Blandford, London.
Muntz W.R.A. (1973) Yellow filters and the absorption of light by the visual pigments of some Amazonian fishes. *Vis. Res.* 13, 2235–54.
Muntz W.R.A. (1975) The visual consequences of yellow filtering pigments in the eyes of fishes occupying different habitats. In *Light as an ecological factor*. II.
Romer A.S. (1955) *Vertebrate Palaeontology*. University of Chicago Press, Chicago, U.S.A.
Stark W.A. II & Davis W.P. (1966) Night habits of fishes of Alligator Reef, Florida. Ichthyologia. *The Aquarium Journal*, 38, 313–56.
Stewart K.W. (1962) Observations on the morphology and optical properties of the adipose eyelids of fishes. *J. fish. Res. Bd. Canada*, 19, 1161–2.
Tyler J.E. (1958) Natural water as a monochromator. *Limnol. Oceanog.* 4, 102–5.
Tyler J.E. & Preisendorfer R.W. (1962) 'Light' in *The Sea, I.* edited by M.N. Hill. John Wiley & Sons, New York.
Walls G.L. (1963) *The vertebrate eye*. Hafner Publishing Co., N.Y. and London.
Wheeler A. (1969) *The fishes of the British Isles and North West Europe*. Macmillan, London, Toronto, Melbourne.

Questions and discussion

In answer to a question, Dr Lythgoe explained the absence of iridescence in the eyes of freshwater fish on physiological grounds; as the complex layers generating the iridescence would interfere with the important activity of clearing water from the cornea and therefore keeping it transparent.

10 The visible spectrum during twilight and its implications to vision

W. N. McFARLAND *Section of Ecology and Systematics, Division of Biological Sciences, Cornell University, Ithaca, New York 14850, U.S.A.*

and

F. W. MUNZ *Department of Biology, University of Oregon, Eugene, Oregon 97403, U.S.A.*

Introduction

The fleeting dimness prior to sunrise and following sunset—twilight time—separates human and most animal activities into two distinctive patterns. Few organisms are active day and night, but tend to specialize in one or the other. Active diurnal or nocturnal behaviour is the rule, arrhythmic and crepuscular visual activities for the most part are severely restricted. One of the more precise studies that relates behaviour to the time of day is that of Hobson (1965, 1972, 1973) on the diel activities of reef fishes. An ordered temporal exchange without intermixing of the diurnal and nocturnal fishes occurs before sunrise and after sunset. In his vivid description, Hobson (1972) emphasized that a period of transition separates the movement of these faunal elements between the substrate and the water column at both dawn and dusk. During this transitional 'Quiet Period', neither diurnal nor nocturnal fishes remain in the water column where they normally forage. Hobson suggested that the changing levels of illumination during twilight triggered these daily movements. Central to Hobson's observations is the *thesis* that predation peaks during twilight. Therefore, crepuscular predation against reef fishes succeeds best, diurnal and nocturnal predation least.

In a coincident investigation of vision in coral reef fishes (Munz & McFarland 1973), we developed a *hypothesis* that the militant response of reef fishes to twilight stemmed from the consequences of light and dark adaptations in the vertebrate eye. In this view crepuscular predators succeed only because diurnal and nocturnal reef fishes are at a visual disadvantage during twilight. A primary goal of our study was to relate the absorption spectra of the visual pigments of reef fishes to the spectrum of underwater light. We hoped to determine what was most important in scotopic vision, maximum photosensitivity (Munz 1965) or contrast enhancement (Lythgoe 1966). We found, as day passed into twilight, that the spectrum of underwater light above the reef shifted from a broad flat spectrum (450–600 nm) to a narrow spectrum (450–500 nm). Thus, during twilight the illumination available to the fishes was decidedly bluer and more monochromatic than during the day. The maximum absorption (λ_{max}) of the scotopic visual pigments obtained from a variety of reef fishes centered between 485 and 500 nm. It is our contention that these particular wavelengths were selected for by the interplay between the effects of the twilight spectrum on vision and the predation pressures associated with this time of the day. The scotopic visual pigments of reef fishes, therefore, have been moulded through evolution to produce maximum photosensitivity to available light.

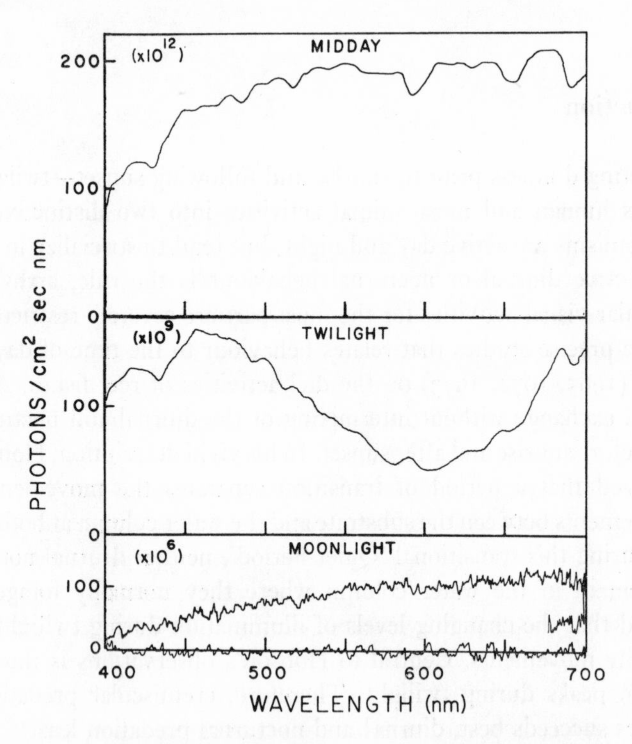

Figure 10.1. Spectral irradiance during midday, evening twilight and full moonlight a the earth's surface, Eniwetok Atoll, Marshall Islands, summer 1970. A high, thin cirru cover and low cumulus clouds (ca 30% cover) occurred during midday. Sun unobscure when recording; low cumulus also were present at twilight and moonlight. Total irradianc between 400 and 700 nm is 53,000 × 10^{12} (midday), 32 × 10^{12} (twilight) and 30 × 10 photons/cm² sec (moonlight).

The blue shift in underwater light during twilight results from the higher relative attenuation of yellow-green to orange light in the atmosphere as compared to light for other visible wavelengths (Fig. 10.1). Several factors contribute to this loss: the increased horizontal 'reddness' and the 'blueness' of diffuse skylight jointly yield less yellow and orange light and, the longer atmospheric path length through which sunlight passes causes it to penetrate more ozone which, in itself, absorbs broadly around 600 nm. This spectral loss is unique since it is limited primarily to twilight. Both the daytime and night skies contain relatively more yellow and orange light (Fig. 10.1).

The effect of these spectral changes on twilight vision in reef fishes seems clear (Munz & McFarland 1973). But what effect(s) do they have on vertebrates living in other habitats? We propose to evaluate this question here. Specifically, we present twilight spectra for several different habitats and relate the photic changes, where possible, to visual properties of the vertebrates that dwell therein.

The visible spectrum during twilight

Methods

The spectral distribution and intensity of visible light were measured with a Gamma Scientific 3,000 R recording spectroradiometer. The instrument and its use have been described in detail elsewhere (Munz & McFarland 1973). The instrument may be calibrated in either energy units ($\mu W/cm^2$ nm) or in quanta (photons/cm^2 nm sec). Units reported here are in photons except for data recorded at Douglas Lake, Michigan where calibration was in $\mu W/cm^2$. Two types of radiometric light measurements were used: (1) spectral irradiance and, (2) spectral radiance. Spectral irradiance recordings resulted from use of a detector head which yielded a cosine response. Spectral radiance was measured by restricting the field of view of the detector to a cone with a plane angle of 15° (ca 0·05 steradians) and recording the spectrum in particular directions. Multiplication of the recorded value at a given wavelength by 20 yielded radiance. To obtain intensity of visible light (total quanta) individual spectra were digitized at 5 nm intervals between 400 and 700 nm and then summed for total photons/cm^2 sec. A FORTRAN program (TRAPEZOID) which performed the summation also calculated the wavelengths at which 25, 50 and 75 % of the total photons occur. These wavelengths are designated as λP_{25}, λP_{50} and λP_{75}.

Because of the methodological value, we emphasize that the Gamma spectroradiometer provides a record of light intensity against wavelength that does not require further data reduction to correct for differential photomultiplier sensitivity. As a result, when in the field, recordings provide an immediate graphic display of the spectral distribution of light (Fig. 10.1).

Spectral Changes at the Earth's Surface

When recording atmospheric spectra at twilight, the reduction of yellow and orange light was evident in spectral irradiance and in spectral radiance measurements of different quadrants of the sky (Munz & McFarland 1973).

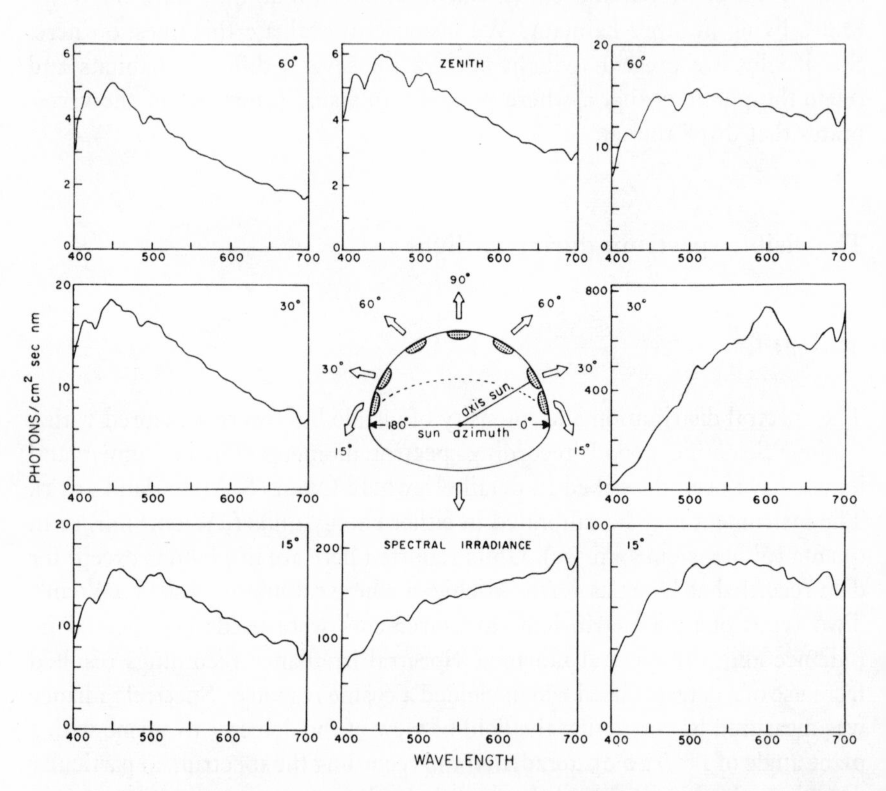

Figure 10.2. Spectral irradiance and spectral radiance during midday for different portions of a clear sky. Recordings were made on 30 January 1974 at Ithaca, New York. Sun altitude 30°, sky cloudless. All values are photons/cm^2 sec nm $\times$ 10^{12}. Spectral radiance values result from multiplication of spectral data recorded for a cone of 15° plane angle (0·05 steradians) by a factor of 20. Total irradiance and λP_{50} values in the visible spectrum for the spectral irradiance curve are: 42,710 $\times$ 10^{12} photons/cm^2 sec (400–700 nm) and 576 nm. Total radiance and λP_{50} values for the various radiance spectra are: azimuth 0°–15° = 21,500 $\times$ 10^{12} photons/cm^2 sec, 560 nm; 30° (directly at sun) = 141,900 $\times$ 10^{12} 590 nm; 60° = 4,060 $\times$ 10^{12}, 550 nm; zenith = 1,320 $\times$ 10^{12}, 523 nm; azimuth 180°–60° = 970 $\times$ 10^{12}, 509 nm; 30° = 1,890 $\times$ 10^{12}, 517 nm; 15° = 3,550 $\times$ 10^{12}, 528 nm.

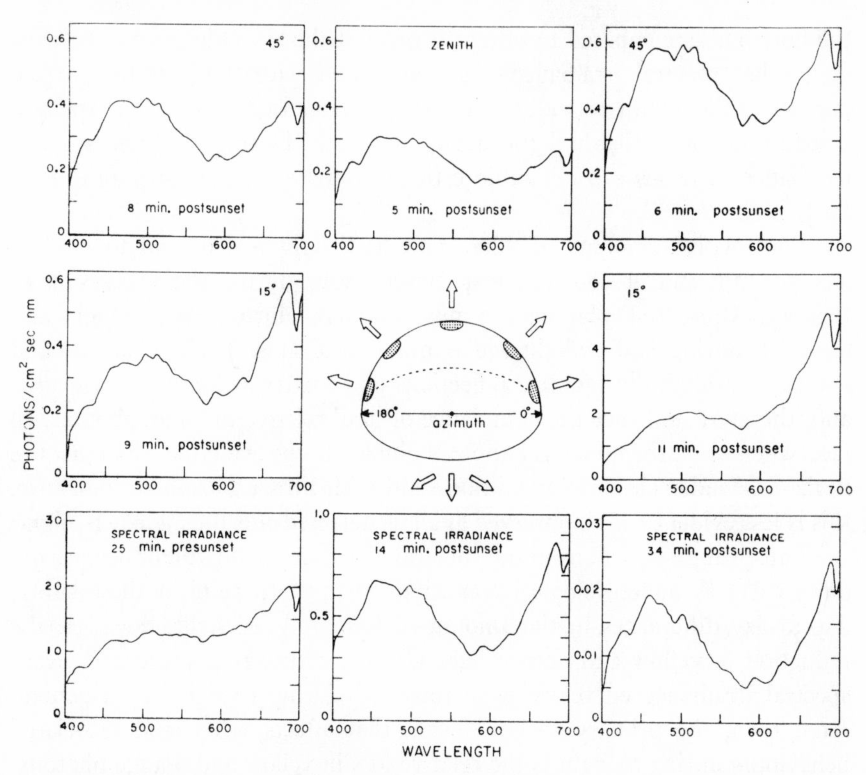

Figure 10.3. Spectral irradiance and spectral radiance for different portions of a clear, winter sky during evening twilight. Recordings were made on 30 January 1974 at Ithaca, New York. All values are photons/cm² sec nm × 10^{12}. Spectral radiance values as in Figure 10.2. Note the change in the spectrum of irradiance in the pre- and post-sunset spectra. Total irradiance and λP_{50} values in the visible spectrum for the three spectral irradiance curves are: 25 min pre-sunset = 3,820 × 10^{12} photons/cm² sec (400–700 nm), 574 nm; 14 min post-sunset = 670 × 10^{12}, 542 nm; 34 min post-sunset = 3·7 × 10^{12}, 527 nm. Total radiance and λP_{50} values for the various radiance spectra are: azimuth 0°–15° = 330 × 10^{12}, 606 nm; 45° = 143 × 10^{12}, 530 nm; zenith = 71 × 10^{12}, 530 nm; azimuth 180°–45° = 97 × 10^{12}, 560 nm; 15° = 97 × 10^{12}, 566 nm. Note—λP_{50} for the bimodal spectra at twilight are not a reasonable index of the wavelength region containing the most abundant light.

Furthermore, and as expected, the loss occurred under clear or overcast conditions and, at different latitudes and seasons. Because this spectral phenomenon is unique, worldwide and sparingly documented in the literature a recent series of atmospheric radiometric spectra are included (Figs. 10.2 and 10.3). The spectral differences between day and twilight and for different portions of the sky are best appreciated through comparison of these spectra. As expected, spectra of small portions of the sky vary considerably. Daytime radiance levels and spectral distribution vary most when recorded toward and away from the sun (Fig. 10.2). At twilight, however, differences in radiance levels are less and spectral differences are associated mainly with the horizons as compared to other portions of the sky (Fig. 10.3). We conclude that spectral irradiance when measured with the detector surface normal to the zenith provides a reasonable overall estimate of the spectral distribution of available light during twilight. During daylight spectral irradiance provides a comparable estimate only when the sun's altitude is fairly high.

The relative amounts of 'blue' and 'red' light which tend to peak at 455–500 nm and at 680 nm, respectively, vary during the procession of twilight. Thus, 'red' photons are most abundant during early twilight and diminish during late twilight (30 minutes post sunset). Haze, clouds and overcast dramatically increase reflections and scatter at longer wavelengths and, therefore, enhance the abundance of 'red' relative to 'blue' photons. In fact, when hazy, the upper sky appears pinkish to the eye rather than just the horizons. Under clear sky conditions and with little atmospheric moisture, this is less evident and, if observed at all, is detected only during late twilight.

Thus, subtle variations in the spectral distribution of twilight occur from day to day as meterorological conditions change. In spite of these subtle day-to-day differences in the amount of blue and red light, however, the reduction in yellow and orange light always occurs. It is present in every spectral irradiance curve we have recorded during twilight. In a general sense then, the primary spectral event that affects vision and associated behaviours during twilight is the relative loss in yellow and orange photons.

Spectral Changes in Aquatic Habitats

It is widely accepted that the spectral distribution of light in fresh waters varies more than in other habitats (Munz 1965, Hutchinson 1957). In appearance open seas are characterized as blue, coastal waters as blue-green to green, and brackish and fresh waters most often as green, but with tendencies to yellow, orange and even red. These differences in general hue are associated with a decrease in transparency that results from dissolved and/or particulate materials. High spectral variation occurs even in closely confluent bodies of water (Fig. 10.4). In these examples, the Carrabelle River, which is rusty red to the human eye, transports large quantities of dissolved and particulate humic inclusions from the extensive coastal pine-swamps of North Florida into St. George Sound. Much of this material precipitates upon the surface of the estuarine sediments. Saint George Sound, therefore, is characteristically green as shown by the spectrum of downwelling light (Fig. 10.4). The spectrum of downwelling light, however, is much broader in the open ocean. Light in the horizontal and nadiral directions for St. George Sound and the Carrabelle River, although less intense is spectrally similar to downwelling light (not figured). In the open ocean the spectra of horizontal and upwelling light are not as broad as for downwelling light, but are very monochromatic and blue (peak 460–490 nm). As a result, fishes inhabiting these three areas encounter different photic conditions characterized either by 'blueness', 'greenness' or 'reddness'.

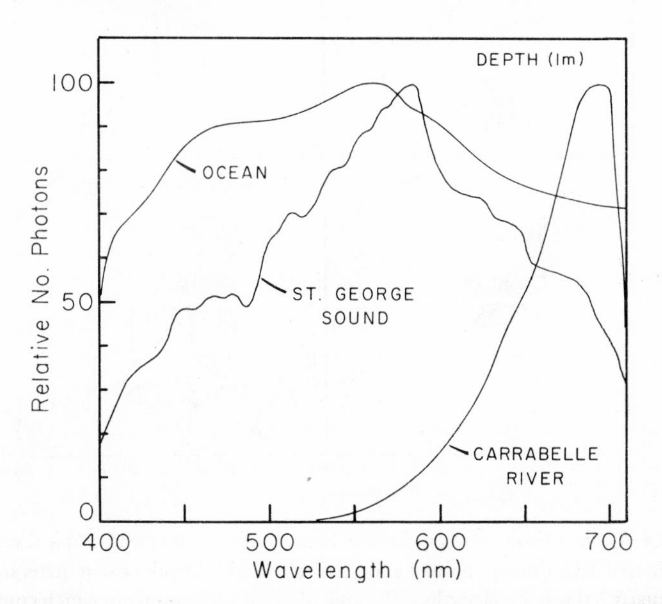

Figure 10.4. Relative number of photons in downwelling irradiance in a clear sea, an estuary and tidal river. Depth of detection 1 meter. All spectra recorded for a clear sky. Ocean, summer 1971, Eniwetok Atoll; Saint George Sound and Carrabelle River, Florida, July 1973.

The examples of underwater spectra for different Florida waters (Fig. 10.4) are for daytime, i.e., photopic conditions. What happens in non-oceanic habitats during twilight? Since the reduction in yellow and orange light during twilight is worldwide, beneath the water surface a shift from the daytime spectrum toward shorter wavelengths can be anticipated. But this will depend on the presence of light at wavelengths below 600 nm. In the Carrabelle River, for example, there is little light below 600 nm. As a result, the major reduction in light about 600 nm would virtually eliminate green, yellow and orange light and narrow even further the sharp daytime red peak (Fig. 10.4). As yet we have not measured this red shift and, indeed, empirical data on fresh waters and brackish estuaries are limited. But a series of day-time and twilight spectra for Silver Lake near Ann Arbor, Michigan which we obtained in collaboration with Dr. Donald Allen, allow an initial comparison with spectra for a clear tropical sea (Fig. 10.5).

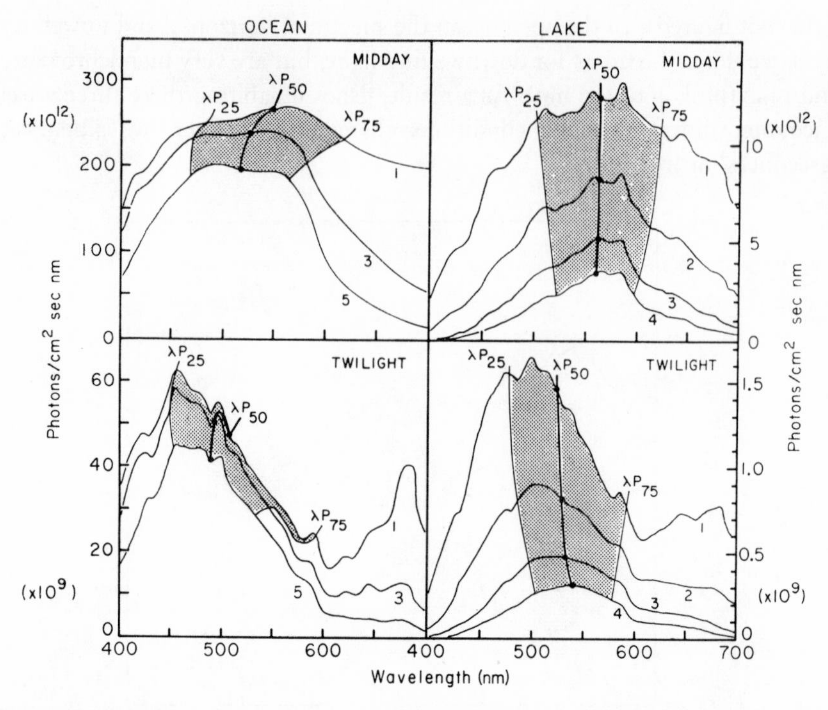

Figure 10.5. A comparison between downwelling irradiance in a clear, tropical sea and in a 'green' coloured lake during midday and during twilight. Depths are in meters. Stippled area encompasses those wavelengths (λP_{25} and λP_{75}) for each spectrum which contain 50% of the total number of photons in the visible (400–700 nm). The heavy line connects the wavelengths which evenly divide the visible light in each spectrum (λP_{50}). Note the blue shift in λP_{50} in the sea during midday and during twilight with increased depth as compared to the lake (see text).

Daylight

With increasing depth, the downwelling spectral irradiance in a clear sea shifts toward the blue. This blue shift, over just a few meters, changes the λP_{50} index by 18 nm (Fig. 10.5). In Silver Lake the spectral irradiance at 1 m is more monochromatic than for clear seas. It is likely that this results from the presence of more phytoplankton and dissolved substances which cause an increase in absorption of blue light. Unlike a clear sea, however, a significant blue shift does not occur with slight increases in depth. The λP_{50} index, for example, only shifts from 563 to 560 nm as depth increases from 1 to 4 meters. But, in both the ocean and Silver Lake, the spectrum narrows with depth (compare changes in λP_{25} and λP_{75}, Fig. 10.5)

Twilight

At twilight the reduction in yellow and orange light shifts the underwater spectrum toward shorter wavelengths. At 1 m λP_{50} for the ocean is 508 nm; for Silver Lake it is 524 nm. This represents a downward shift from daytime to twilight of 44 and 39 nm, respectively. With increased depth, however, the spectral shifts are different from daytime. In the sea at 5 m λP_{50} shifts downward to 490 nm; in Silver Lake λP_{50} shifts upward to 540 nm.

In summary, during twilight as compared to daylight, in both the ocean and in a 'greenish' lake downwelling light just beneath the surface 'blue-shifts'. With increased depth, however, light in the sea becomes bluer and in the lake, greener.

Spectral Changes in Forests

What spectral changes occur within forest communities? Attempts to characterize the twilight spectrum within a forest are fraught with difficulties. Yet the forest, as compared to open country, provides a distinctively different photic environment for vertebrates. The most obvious difference is the severe attenuation of light from the top of the canopy to the forest floor. For most spectroradiometers attenuation during twilight will severely test their sensitivity limits. Also, potential heterogeneity from direct skylight through openings in the canopy may cause inconsistencies in recordings.

We measured spectral irradiance outside of and within a heavy maple forest at Douglas Lake, Michigan (Fig. 10.6). With a cosine detector at ground level and oriented toward the zenith, the difference in visible energy was approximately 100-fold (Fig. 10.6, day curves). Although both spectral energy curves were reasonably flat from 450 to 700 nm, the spectrum within the forest peaked at 560 nm and the open sky spectrum peaked below 500 nm. The spectral irradiance obtained for horizontal orientation of the detector was slightly less than half the intensity of downwelling light within the canopy (Fig. 10.6, day curves). The energy peak at 560 nm, however, is obvious and there is relatively less energy at shorter and at longer wavelengths. Of course, these curves merely quantify the general feeling of dimness and 'greenness' that a person has when standing within a forest. However, the human eye also detects discrete objects, e.g., tree trunks, branches, etc., that have different spectral qualities and degrees of 'brightness'. Does then, spectral irradiance provide a measure of the background spacelight of the forest against which objects are viewed? Of course no single measurement can; but, at least for rather large objects, we feel it is a reasonable approximation. For example,

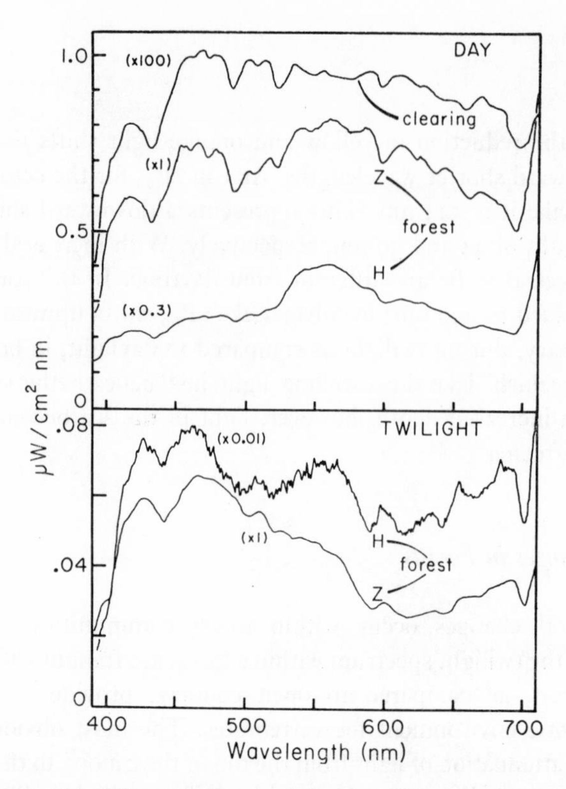

Figure 10.6. Spectral irradiance during the day and twilight in a dense maple forest. All values are in energy units, not photons: Note scaling factors for each curve. Z is zenith and H horizontal orientation of the detector surface. In the clearing, orientation is zenith. Spectra recorded adjacent to Douglas Lake, Michigan. 6 July 1971.

the light transmitted through a maple leaf and that detected horizontally within the forest community, for different fields of view have similar qualities (Fig. 10.7). Spectral irradiance contains relatively more 'blue' and 'red' energy than the 'yellow-green' light transmitted through the maple leaf. However, restriction of detector view also reduces the relative amount of 'blue' and 'red' energy. Narrower detector fields, therefore, would reduce the included direct light that passes through the canopy openings and the brighter reflections from branches, etc., that would be more prevalent in the upper portion of the detector's field of view when measuring irradiance. As a result, the spectrum becomes 'greener' (Fig. 10.7). Obviously, the spectrum reflected from a tree trunk would differ dramatically from light transmitted through or reflected from leaves. To obtain, therefore, a general measure of background spacelight the detector field should not be restricted too greatly. Since we wished to measure twilight effects, we chose to merely measure spectral irradiance for downwelling and horizontal light. Because of the 100-fold attenuation in light intensity within the canopy (Fig. 10.6), this choice provided for maximum impingement of light on the detector surface and, therefore, increased the probability that we could record the spectrum during twilight.

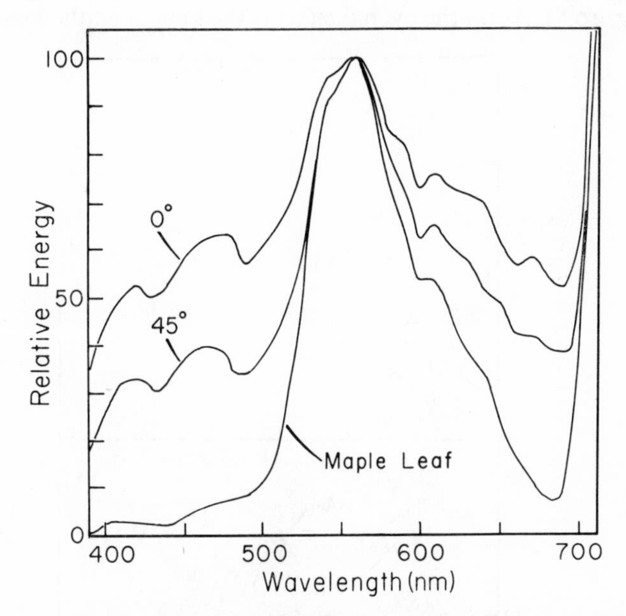

Figure 10.7. Comparison of the spectral distribution of visible light transmitted through a maple leaf and horizontal light in a maple forest. Spectra were recorded in $\mu W/cm^2$ nm, during midday at Douglas Lake, Michigan. Zero (0°) is the spectrum for an unrestricted cosine detector oriented normal to the horizontal line of sight; 45° is the spectrum with the detector restricted to a cone with a plane angle of 45°; the maple leaf was placed over the unrestricted detector.

Spectra of downwelling and horizontal irradiance during twilight are shown in Fig. 10.6. The curves were obtained about 10 minutes before sunset, but the reduction of light at 600 nm is evident in the downwelling spectrum (Fig. 10.6). The effect of this loss on the horizontal spectrum indicates that the 560 nm peak typical of daytime, although still obvious, has been attenuated relative to energy in the 'blue' and 'red' wavelength regions. In fact, peak energy is now at 460–470 nm. At twilight, greens no longer represent the dominant wavelengths of background spacelight in a forest.

To be sure, variance in canopy density, the type and the thickness of the understorey and other physiognomic features of forests should affect the spectral changes that occur during twilight. What happens in a lush tropical-type situation is illustrated in Fig. 10.8. The presunset downwelling and horizontal curves (30 min. presunset, Fig. 10.8), however, do not differ greatly from those measured at Douglas Lake for a maple forest (Fig. 10.6) when allowance for the difference in radiometric units is made. Horizontally light peaks at 550 nm and is essentially flat to 700 nm. During twilight the spectrum of downwelling light followed that expected in an open situation, but the horizontal spectrum flattened throughout almost the entire visible range (Fig. 10.8). Again the overall effect is the same, i.e., the loss of light at

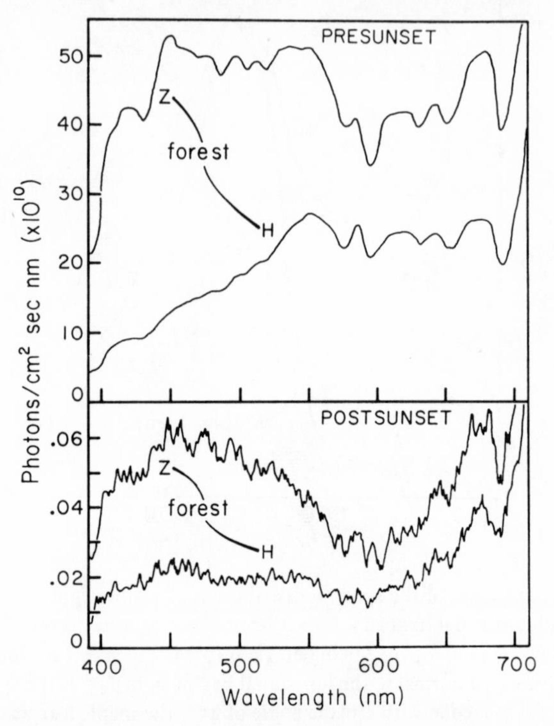

Figure 10.8. Spectral irradiance during the day and twilight in a tropical forest. Z is zenith and H is horizontal orientation of the cosine detector. Records were made in the dense 'rain-forest' surrounding a reservoir in the Nuuanu Pali, on Oahu, Hawaii, on 30 July 1971.

600 nm causes a blue shift which is reflected in the flat spectrum of horizontal light. Downwelling light, in contrast, is blue and red shifted. Although we made no measurements, the tropical forest tended to have many more large openings in the overstorey than the maple forest. Most likely this caused the marked similarity of the downwelling curve to a more open situation.

Significance of twilight to vertebrate vision

To assess the effects of spectral changes during twilight on vertebrate vision one needs information on (1) the actual spectral distribution of light in the animal's habitat, (2) the absorption spectrum of the scotopic visual pigment(s) and, (3) familiarity with critical visual behaviours. Only in a few instances are all of these attributes adequately known. An attempt to relate visual pigment absorption to photic changes during twilight for fishes and terrestrial vertebrates follows:

Fishes

Tropical marine fishes

The concept that the scotopic visual pigments of fishes and other vertebrates match the spectrum of underwater light where they live has been based on the broad correlation found between λ_{max} and the general 'blueness', 'greenness' or 'reddness' of the habitat (Munz 1957, 1965, Denton and Warren 1957, Wald *et al.*, 1958, McFarland 1970, 1971).

As satisfying, and even as generally correct as the concept might be, the correlation is misleading. Correlative statements of the predominant hue, i.e., 'blueness', 'greenness', etc. refer mostly to data that characterize the wavelength of maximum transmittance for a particular body of water (McFarland & Munz 1974, ms). Such data are almost invariably descriptive of daytime conditions and bear little relationship to the photic context in which a fish actually detects a target. For most of the visual pigments reported, the λ_{max} values were ascertained from retinal extracts (Lythgoe 1972). They represent, therefore, rod and not cone pigments since in general, the latter have defied extraction (Munz & McFarland 1973). As a result, the correlations relate maximum absorbance of scotopic visual pigments to the maximally transmitted light during the photopic period of the day. Obviously this is an inappropriate, if not inaccurate, comparison. Only for deep sea fishes does the correlation between λ_{max} and maximum transmittance seemingly make sense. Here solar light is blue and, importantly, at depth very dim. During the daytime then, scotopic vision can prevail. But, the correlation can as well be made with the bluish light available from bioluminescent organisms (Munz 1958, McFarland 1971).

Only Lythgoe (1966, 1968) and Munz & McFarland (1973) have attempted to relate maximal visual pigment absorbance to spectral distribution for those periods of the day when scotopic vision prevails, i.e., at twilight and at night. For marine tropical fishes, as we indicated earlier, the blue shift in underwater light during twilight in conjunction with the higher predation at twilight has selected for rhodopsins with λ_{max} values between 485–500 nm. This waveband coincides with those wavelengths (λP_{50}) on each side of which visible light is equally distributed (see Fig. 10.5) and, usually, also most abundant. There is, therefore, a close spectral match between the visual pigments and photic conditions which *must*, in turn, maximize photosensitivity at twilight.

Freshwater fishes

The spectral differences at twilight between oceanic and freshwaters are of consequence to the differences in the spectral location of scotopic visual pigments of not only tropical marine but also, we think, freshwater fishes. The range for the λ_{max} values of visual pigments of fishes, including both rhodopsins and porphyropsins, extends from 467 to 544 nm (Lythgoe 1972). It is apparent that the λ_{max} values for these scotopic visual pigments approximate the span of λP_{50} values encountered in clear seas and greenish lakes during twilight (Fig. 10.5). Recall that twilight is a unique period of the day when, presumably, visual behaviours are critical to survival. If maximum photosensitivity is important to scotopic vision then the λ_{max} values should approximate closely the wavelengths at which light is most abundant during twilight. For tropical marine fishes this is true (Munz & McFarland 1973). But is it also true for fishes in other aquatic habitats? In Silver Lake the fish fauna is composed largely of centrarchid and ostariophysan fishes, especially cyprinids. These fishes have either pure porphyropsins or mixtures of rhodopsin and porphyropsin in their rods. As a pigment class porphyropsins range from λ_{max} values of 517 to 544 nm (excluding the strictly diurnal wrasse family) with the greatest number occurring between 520 and 535 nm. As a rule, porphyropsins are limited to freshwater or to euryhaline fishes. Only a few are present in marine fishes. The various freshwater centrarchids, for example, possess only porphyropsins with λ_{max} values between 525–527 nm. Cyprinids of the types encountered in Silver Lake usually contain pigment mixtures that maximally absorb light close to 520 nm—but pigment balance probably changes seasonally.

Although the visual pigments for most fishes in Silver Lake are as yet unknown, we feel confident that most, if not all, will approximate the λ_{max} values obtained for some of their close relatives (Lythgoe 1972). We can suggest, therefore, that the rhodopsins and porphyropsins of fishes in Silver Lake provide a reasonable match to the spectral distribution of available light and, especially, during twilight (Fig. 10.5). In fact, the porphyropsins of most freshwater and euryhaline fishes (517–544 nm) coincide with the twilight spectrum that would typify most bodies of water whose 'greenness' results largely from phytoplankton and dissolved materials derived from plankton, benthic plants and run-off from terrestrial areas. Even coastal areas where the waters are usually green, such as St. George Sound (Fig. 10.4) probably can be included in this category.

In nature, however, conditions are seldom stable. During our attempts to characterize twilight in St. George Sound, strong winds developed for several days subsequent to the intial measurements of daytime photic conditions. The winds hampered field measurements and, importantly, whipped the sound into a frenzy. As a consequence, the humic materials from the pine-swamps that settle to the bottom of the sound were mixed into the water column. Saint George Sound changed from its normal 'greenness' to an 'orangy-brownness' (Fig. 10.9). The green peak between

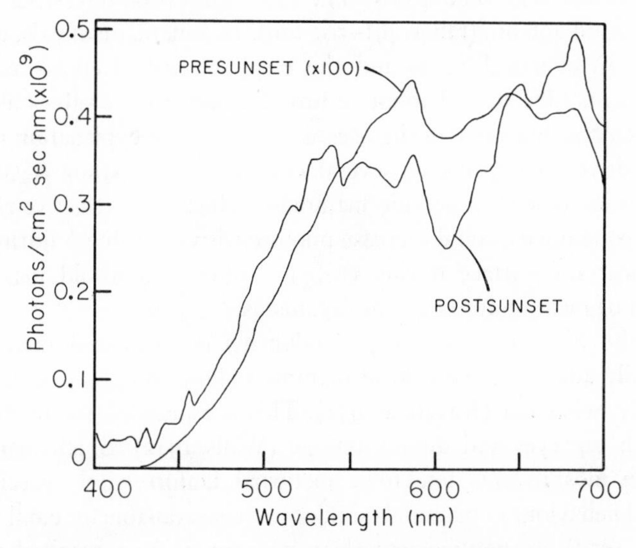

Figure 10.9. Comparison of spectral irradiance before and following sunset in St. George Sound, Florida. Depth 2 metres. Note the 'redness' of the water during the day and only slight blue shift following sunset. The normal 'greenness' of the water in the sound was affected by several days of very strong winds (see text).

550 and 600 nm typical of calmer periods (Fig. 10.4) was present, but the spectrum was also rich in red light. At sunset a blue shift occurred, but it was less dramatic than that recorded in Silver Lake. One might suggest that visual pigments that characterize fishes typical of 'greenish' waters would be of lesser value in St. George Sound under these photic conditions. But this does not follow. Examination of the twilight spectra does show a slight reduction of light at 600 nm (Fig. 10.9). Light below 600 nm did blue shift, although less than expected in the absence of the humic materials. Presumably, at increased depth there would be a slight green shift as there was in Silver Lake. In either case, visual pigments with λ_{max} values above 500 nm would tend to provide for maximum photosensitivity. Importantly, estuarine

fishes also tend to possess visual pigments above 500 nm, although more data are needed (Lythgoe 1972).

Mammals

Aquatic and terrestrial mammals

In distinction to fishes, the extractable visual pigments of mammals show little diversity in λ_{max}. Over 35 species have been analysed (Lythgoe 1972) and all possess only rhodopsins. The λ_{max} values for terrestrial mammals cluster around 498 nm (range 491–502 nm). In aquatic mammals, as in some cetaceans (McFarland 1971) and the elephant seal (Lythgoe & Dartnall 1971), λ_{max} is blue shifted about 12 nm. The lower λ_{max} values likely relate to the essential 'blueness' of the oceanic habitat. The explanation for similar blue sensitive rhodopsins in several oceanic fishes (Munz & McFarland 1973) applies as well to oceanic mammals. Thus, blue sensitive rhodopsins in oceanic mammals would increase photosensitivity at depth during the day and/or near the surface during twilight. They also would maximize the detection of most bioluminescent organisms.

Are the blue-green sensitive rhodopsins of terrestrial mammals also ecologically adaptive? Terrestrial mammals show considerable plasticity in their daily behaviour (Kavanau 1971). This is especially true of large mammals with big eyes and duplex retinae (Walls 1942). Small mammals, as typified by most rodents, tend to be nocturnal. Unfortunately, precise studies of natural behaviour in mammals similar to those available for coral reef fishes (Hobson 1972) are non-existent (Kavanau 1971). As a result, behavioural interpretations must rely largely on a multitude of field notes or experimental data obtained for individual species outside of their natural habitat. Our general conclusions about daily behaviour, which follow, are largely intuitive.

Diurnal mammals in contrast to nocturnal and crepuscular forms tend to be less affected by changes in illumination, especially when presented out of temporal context (Kavanau 1971). Nocturnal mammals seem more rigidly bound by illumination, and activity becomes maximal at low light levels. In Kavanau's study predators, like red fox and cats, were maximally active during twilight periods. Indeed, Kavanau states that:

'The fact that twilights are prime times for hunting probably accounts for the tendency of many animals to be active during these periods. Twilights generally would be the best times for predation on small nocturnal animals because visibility would be better than at night.'

To us a parallel exists between mammalian and fish behaviour. Obviously

rictly diurnal mammals separate their activities from strictly nocturnal
ıammals, as do fishes. But, apparently predatory pressures on mammals are
ighest during twilight transitions as they are for reef fishes. We suspect that
ıe underlying reasons are similar, i.e., both diurnal and nocturnal animals are
t a visual disadvantage during twilight. We are not sure whether an equi-
alent 'Quiet Period' separates the transition between diurnal and nocturnal
ıammals as effectively as it does in reef fishes. Quite likely physiological
ifferences do occur since the retinal processes involved in dark and light
daptation differ somewhat between the two groups. Certainly, exacting
tudies of behavioural transitions in temperate and tropical terrestrial
ıammalian communities would be rewarding. Taken overall, however,
wilight does seem to be as critical for mammals as for fishes. For, regardless
f any physiological differences, twilight defines that period when vision in
ertebrates usually changes from photopic to scotopic processes or *vice versa*.

The effects of changes in the levels of illumination are well documented.
'he more subtle effects of changes in spectral balance, however, have been
irtually ignored, especially for terrestrial species (Munz & McFarland
973). For mammals that dwell in open habitats the spectral changes that occur
uring twilight provide a photic environment rich in blue and blue-green
ght and in deep red light (Figs. 10.1, 10.2 and 10.3). Only toward the
orizons and, particularly, the western horizon is red light more abundant
han blue light (Fig. 10.3). But as twilight deepens, red light wanes. To
rovide maximum photosensitivity the absorption maxima of scotopic visual
igments could be centered either at 460–500 nm or near 690 nm. Although
/e can offer no compelling reason as to why, amongst vertebrates, scotopic
isual pigments match only the shorter wavelength region. Perhaps red-
ensitive rhodopsins are chemically forbidden or highly unstable. As an
lternative ecological reason, diurnally blue-sensitive scotopic pigments, if
sed, would better match skylight (Fig. 10.2). Whatever the reason, selec-
ion of pigments with λ_{max} values near 500 nm would maximize photo-
ensitivity at twilight.

Twilight photic conditions in more closed terrestrial habitats, such as
orests, also show a reduction in yellow-orange light (Figs. 10.6 and 10.8). A
cotopic visual pigment would trap light most effectively if its λ_{max} was
ocated near or below 500 nm. For horizontal lines of sight, the background
pacelight is spectrally broad (Fig. 10.8). It would matter little where λ_{max}
/as centered between 450 and 650 nm. But downwelling light which filters
hrough the canopy and openings in the forest crown is 10 times as intense
nd contains relatively more blue and blue-green light (Figs. 10.6 and 10.8).
'hotosensitivity for all lines of sight would be maximal for visual pigments
/ith λ_{max} values near 500 nm.

To be sure, as in surface fishes, visual pigments that absorb maximally

near 600 nm might augment absorption during moonlight nights (Fig. 10.1
or starry nights which, presumably, are 'redder'. But the resultant loss in
photosensitivity during twilight would more than offset the gain by matching
pigment to moonlight.

Twilight vision in man

We are most aware of the visual difficulties encountered at twilight, perhaps
when driving, for in this nether land of uncertainty objects are discriminatec
poorly and, as a result, sometimes with 'instant' finality. Although inherently
a diurnal primate, man does possess a duplex retina and a scotopic rod pigment. Compared to other vertebrates, such as fishes, however, the *in situ*
pigment absorbance is low (O.D. 0·1 in man; 0·5 in fish; Denton and Nico
1964). Relative to fishes and to nocturnal mammals, man's nocturnal capabilities are somewhat limited. Nevertheless, man does accomplish visual task
during twilight and at night. How do the spectral changes affect man's visua
capabilities? Although the question is impossible to answer quantitatively, a
qualitative answer is possible and, interestingly, provides an ecologica
explanation for some of the heretofore inexplicable attributes of dark adaptation in humans.

The λ_{max} of man's rhodopsin has been difficult to specify. Values var
from 493 to 505 nm (Lythgoe 1972). Liebman's (1972) estimate of 498 nm
which was obtained by microspectrophotometry, agrees quite well with the
λ_{max} of 497 nm first reported by Crescitelli and Dartnall (1953). But scotopi
spectral luminosity (V_λ) for a standard observer peaks at 507 nm (Le Grand
1972) and this agrees more closely with the λ_{max} of 505 nm reported by Brown
& Wald (1963). Whatever the exact λ_{max} value, obviously, it falls within th
blue-green region of the spectrum. As in other terrestrial mammals, man'
rhodopsin matches the blue to blue-green peak of twilight (Fig. 10.3). I
must, therefore, provide for a maximum absorption of available light durin
twilight.

The photopic spectral luminosity (V_λ) of a standard observer peaks a
about 558 nm (Le Grand 1972) a value, interestingly, not far displaced from
λP_{50} for a clear, tropical summer sky ($\lambda P_{50} = 570$ nm, Fig. 10.1). Throughou
the day there is little change in spectral irradiance, other than a reduction in
intensity, until a few minutes before sunset. Since the daytime luminanc
exceeds cone thresholds photopic vision prevails. The luminance threshold
used to define photopic, mesopic and scotopic vision in man (ca 5 and 10^{-}
cd m^{-2}, Le Grand 1972) approximate to the illuminance levels at sunset an
about 40 minutes following sunset (clear summer sky). The difficulties o
'seeing' during evening twilight could result, therefore, not only from th

diminution in light, but also from the spectral changes that occur at this time. As twilight encroaches, the reduction in yellow-orange light about 600 nm must decrease photopic vision beyond the total diminution in light intensity. At 560 nm, for example, the decline in yellow-green photons is equivalent to a reduction of 75% relative to the decline in blue photons at 460 nm, and of about 50% relative to the decline in blue-green photons at 500 nm (Fig. 10.1).

In terms of behaviour this relative reduction in yellow light is best translated as a temporal effect. For example, the photopic threshold, which is minimal at 558 nm (V_λ), occurs earlier (ca 8 to 10 minutes) than it would if spectral change did not take place at twilight. At dawn, of course, the onset of photopic vision would be delayed. This effect is an important one. Consider its influence on scotopic processes. Man's rhodopsin is light saturated until sunset or even slightly after sunset and, presumably, is of little visual consequence. Furthermore, it is well established that the higher the intensity of a preadaptive light the longer it takes to achieve a given level of dark adaptation (Barlow 1972). Dark adaptation during twilight must be delayed since light is not suddenly extinguished at sunset. Also the absorption band of man's rhodopsin does not coincide with the relative reduction of light in the yellow wavebands at twilight. The reduction, therefore, could hardly accelerate the normal decline in the rate of bleaching that occurs during twilight. Indeed, a reduction in the rate of bleaching during twilight must be controlled by the rate of reduction in blue and blue-green light during twilight which is, of course, more abundant (Fig. 10.1). As a result, the spectral changes at twilight effect a significant reduction in man's photopic vision, but have little effect in accelerating the onset of his scotopic vision.

Human difficulties in easily distinguishing objects during twilight then are partly the result of the loss in yellow light and a consequent reduction in excitation of the green and, especially, the red-sensitive photoreceptors. Temporally, this must magnify the separation between effective photopic and scotopic vision. Consequently, the period of mesopic vision is prolonged and dependent mostly on the blue and green sensitive cones.

Two unusual aspects of human vision, which have been experimentally demonstrated, fit the sequence of events at twilight and may be of ecological and behavioural significance: (1) the blue shift in V_λ $(_M)$ for mesopic vision (Hough 1968) and, (2) the sensitizing effect of red light over darkness in accelerating the rate of dark adaptation of rods (Brown 1968). At levels of illumination just above the accepted threshold for human scotopic vision (10^{-3} cd m^{-2}), Hough (1968) and Hough & Ruddock (1969) have demonstrated a reverse Purkinje shift, i.e., a shift in sensitivity to shorter rather than longer wavelengths. Indeed, V_λ for these mesopic light levels is at about 450 nm and is believed to be affected by 'blue' cones. They postulate that the shift results from cone-rod interactions, since for tritanopes V_λ at mesopic levels of illumination is near 500 nm, i.e., at the scotopic V_λ function. Whatever the mechanism, the shift makes ecological sense because under natural mesopic levels of illumination, which occur during twilight, blue and not yellow-green light is most available (Figs. 10.1 and 10.3). It would be of interest to know if this effect occurs in other vertebrates with duplex retinae.

As indicated, the rate of dark adaptation under natural conditions is slowed, at least during early twilight, because of the bleaching of rhodopsin. The fact that red light is abundant, during twilight in all quadrants of the sky (Fig. 10.3) might counteract this effect by sensitizing the rods (Brown 1968). Whether red light is operative in the presence of blue light, and if so, how effective it might be remains conjecture. But, if operative, the sensitization would reduce the temporal separation between photopic and scotopic vision, i.e., the period of mesopic vision when neither cone nor rod receptors function very well.

Conclusions

We conclude by re-emphasizing that the spectral changes which occur during twilight provide a natural explanation for the Purkinje shift associated with most heterochromatic vertebrates (Munz & McFarland 1973). The suggestions made here, that the spectral effects in different habitats during twilight explain the spectral location of rhodopsins and prophyropsins in the rods of a variety of vertebrates, extends the concept that evolutionary forces have acted largely to enhance photosensitivity at critical periods of the day. A more detailed examination of twilight effects for different vertebrate species might, as postulated here for man, provide natural explanations for many of the paradoxical responses described for the vertebrate system. Thus, in addition to experimental studies, whether during daylight, twilight or at night the vision of vertebrates must be studied in their natural, physical and biological context if we are to establish a unified theory of visual function.

Acknowledgements

This research was supported by U.S. Public Health Service Grants from the National Eye Institute, EY-00323 and EY-00324. We should like to thank Dr. Donald Allen of the University of Michigan for writing the Fortran programme used in the study and for his collaboration in collection of the light spectra from Silver Lake in Michigan that, in part, are presented here.

We are most grateful for the assistance of Ms. Phyllis Toyryla, who under a limited and pressured schedule, somehow, with a smile, produced the finished manuscript.

References

BARLOW H.B. (1972) Dark and light adaptation: psychophysics. In *Visual Psychophysics*, Vol. VII/4, *Handbook of Sensory Physiology* (edited by D. Jameson and L.M. Hurvich). Springer-Verlag, New York.

BROWN J.L. (1968) Problems in the specification of luminous efficiency. *Techn. Rep.* 5, Department of Psychology, Kansas State University.

BROWN P.K. & WALD G. (1963) Visual pigments in human and monkey retinas. *Nature* (*Lond.*) **200**, 37–43.

CRESCITELLI F. & DARTNALL H.J.A. (1953) Human visual purple. *Nature*, *Lond.* **172**, 195–6.

DENTON E.J. & NICOL J.A.C. (1964) The chorioidal tapeta of some cartilaginous fishes (Chondrichthyes). *J. mar. biol. Ass. U.K.* **44**, 219–58.

DENTON E.J. & WARREN F.J. (1957) The photosensitive pigments in the retinae of deep sea fish. *J. mar. biol. Ass. U.K.* **36**, 651–62.

HOBSON E.S. (1965) Diurnal-nocturnal activity of some inshore fishes in the Gulf of California. *Copeia* No. 3, 291–302.

HOBSON E.S. (1972) Activity of Hawaiian reef fishes during the evening and morning transitions between daylight and darkness. *Fish Bull. Nat. Mar. Fish. Ser.* **70**, 715–40.

HOBSON E.S. (1973) Diel feeding migrations in tropical reef fishes. *Helgolander wiss. Meeresunters* **24**, 361–70.

HOUGH E.A. (1968) The spectral sensitivity functions for parafoveal vision. *Vision Res.* 8, 1423–30.

HOUGH E.A. & RUDDOCK K.H. (1969) The Purkinje shift. *Vision Res.* **9**, 313–15.

HUTCHINSON G.E. (1957). *A Treatise on Limnology*, *Vol. I Geography*, *Physics*, *and Chemistry*. John Wiley & Sons Inc., New York.

KAVANAU J.L. (1971) Locomotion and activity phasing of some medium-sized animals. *J. Mamm.* **52**, 386–403.

LEGRAND Y. (1972) Spectral luminosity. In *Visual Phychophysics*, Vol. VII/4, *Handbook of Sensory Physiology* (edited by D. Jameson and L.M. Hurvich). Springer-Verlag, New York.

LIEBMAN P.A. (1972) Microspectrophotometry of photoreceptors. In *Photochemistry ol Vision*, Vol. VII/1, *Handbook of Sensory Physiology* (edited by H.J.A. Dartnall). Springer-Verlag, New York.

LYTHGOE J.N. (1966) Visual pigments and underwater vision. In *Light as an Ecological Factor* (edited by R. Bainbridge, G.C. Evans, and O. Rackham). Blackwell, Oxford and Edinburgh.

LYTHGOE J.N. (1968) Visual pigments and visual range underwater. *Vision Res.* 8, 997–1012.

LYTHGOE J.N. (1972) List of vertebrate visual pigments. In *Photochemistry of Vision*, Vol. VII/1, *Handbook of Sensory Physiology* (edited by H.J.A. Dartnall). Springer-Verlag, New York.

LYTHGOE J.N. & DARTNALL H.J.A. (1970) A 'deep-sea rhodopsin' in a mammal. *Nature*, *Lond.* **227**, 955–6.

McFarland W.N. (1970) Visual pigment of *Callorhinchus callorynchus*, a southern hemisphere chimaeroid fish. *Vision Res.* **10**, 939–42.

McFarland W.N. (1971) Cetacean visual pigments. *Vision Res.* **11**, 1065–76.

Munz F.W. (1957) Photosensitive pigments from retinas of deep-sea fishes. *Science* **125**, 1142–3.

Munz F.W. (1958) Photosensitive pigments from the retinae of certain deep-sea fishes. *J. Physiol., Lond.* **140**, 220–35.

Munz F.W. (1965) Adaptation of visual pigments to the photic environment. In *Colour Vision: Physiology and Experimental Psychology*. CIBA Foundation Symposium (edited by A.V.S. DeReuck and J. Knight). *J. & A. Churchill*, London.

Munz F.W. & McFarland W.N. (1973) The significance of spectral position in the rhodopsins of tropical marine fishes. *Vision Res.* **13**, 1829–74.

Wald G., Brown P.K. & Brown P.S. (1957) Visual pigments and depths of habitat of marine fishes. *Nature, Lond.* **180**, 969–71.

Walls G.L. (1942) *The vertebrate eye and its adaptive radiation.* Reprinted 1963, Hafner Publ., New York.

11 The visual consequences of yellow filtering pigments in the eyes of fishes occupying different habitats

W. R. A. MUNTZ *Laboratory of Experimental Psychology, University of Sussex, Falmer, Brighton*

Introduction

It has long been known that the corneas and lenses of some vertebrates are yellow in colour, and therefore affect the spectral composition of the light reaching the retina. Other intraocular structures that must similarly affect vision by preferentially absorbing or reflecting certain wavelengths are also common, such as the oil droplets of reptiles and birds, the macular pigment of man, the retinal capillary network of many mammals and of eels, and the reflecting tapeta found in most vertebrate phyla (see Walls & Judd 1933, Muntz 1972, Nicol *et al.* 1973, for reviews). Such structures are also usually either yellow or red, and so will decrease the relative amount of short wavelength light reaching the receptors. This led Walls & Judd (1933) to propose that they all have similar functions and they suggested that these may include the reduction of chromatic aberration, scattered light, and 'glare', as well as an increase in the contrast of objects viewed against certain backgrounds.

The problem may be approached by attempting to correlate the presence of coloured filters and tapeta with the characteristics of the animals' light environments. Yellow corneas and lenses are particularly common among the fishes, and these animals appear to be ideal subjects for such a study since they are also exposed to a very variable light environment. These variations include factors such as the total quantity and spectral quality of the light, as well as the degree of scattering and absorption due to the water, all of which will strongly affect the nature of the visual tasks facing the animals.

The fishes have also attracted attention because they possess a much wider variety of visual pigments than the other vertebrate groups, and several attempts have been made to find a relationship between the spectral absorbance of these pigments and the spectral characteristics of the water (e.g. Lythgoe 1972, Munz & McFarland 1973). In some cases, such as the deep sea fishes, a clear correlation has been demonstrated (Denton & Warren 1956, 1957, Munz 1957, Wald *et al.* 1957), but in other cases the relationship, if any, remains obscure. Clearly, any such attempt to correlate the absorptive properties of visual pigments with the environment must depend on a knowledge of the distribution and properties of any intraocular structures that affect the spectral quality of the light reaching the receptors, and in fishes, the characteristics of the lens and cornea are of obvious importance.

Characteristics of the yellow filtering pigments of fish eyes

Background

The early work of Walls & Judd (1933) was qualitative, and does not give spectrophotometric data on the absorbances at different wavelengths, so that it cannot be used to assess the effect of the lens and cornea on the animal's spectral sensitivity. In 1968, however, Moreland & Lythgoe presented spectral absorbance curves for the corneas of three species of wrasse and the perch, obtained by a photographic densitometry technique. The results showed that the maximal absorption occurred between about 400 and 500 nm, with three clear submaxima at around 425, 450, and 482 nm, which suggests that the pigment is a carotenoid. Moreland & Lythgoe also confirmed in detail Walls & Judd's observation that the density of the pigment is greater in the dorsal part of the cornea. The corneal pigment has also been extracted and studied in solution, with similar results (Bridges 1969).

Detailed spectrophotometric measurements of the lenses of various animals, including some fishes, have long been available (Kennedy & Milkman 1956); and Denton (1956) and Motais (1957) have obtained further data on a wide variety of fishes by photographing a mercury spectrum through the lenses. The lenses in all cases act as cut-off filters, passing long wavelengths and absorbing short wavelengths, and the spectral position at which they start to absorb varies greatly between species. Only one species (*Chlorophthalmus agassizi*) was reported to have a visibly coloured lens, cutting off at about 440 nm: this species is however particularly interesting because it lives at great depths (over 100 m) where the light levels are very low. Recently it has become clear that yellow lenses are common in fishes (see below).

The spectral data summarized in the above two paragraphs were either obtained by indirect means (such as densitometry), or by using commerical spectrophotometers. The latter method is not well adapted to measuring the absorbances of small areas, and is therefore difficult to use on structures such as the cornea, in which the depth of pigmentation varies greatly with position. Recently we therefore built a spectrophotometer in which the measuring spot was 0·8 × 0·1 mm, and which could be used for measuring the spectral absorbance of the intact retina with its photosensitive pigments, the lens, and small areas of cornea (Muntz 1973). This allows the effect of the lens and cornea on the absorbance of the pigment to be calculated. The general procedure was to dark adapt the fish for at least 3 hours, after which it was killed by decapitation or immersion in Tricane Methanesulfonate (MS222). The eyes were then removed, the lens and cornea cut away, and the retina carefully freed from the underlying pigment epithelium. The retina was placed, receptor side up, in glycerol, in a shallow well on a microscope slide, which was then mounted on the spectrophotometer and positioned so that the measuring spot fell on the area of the specimen whose spectral absorbance was to be measured. The absorption of the retina measured in this way is almost entirely due to the visual pigment it contains (Denton 1959). The absorption of the lens and cornea were similarly measured after mounting in glycerol.

The spectrophotometer was also designed to be portable, and to require a minimum of supporting facilities, so that it could be used in areas where suitable laboratory facilities are lacking. The results that are summarized in the next section were obtained using this apparatus at Manaus, near where the Rio Negro and the Solimoẽs join to form the Amazon proper. This area is particularly interesting because there are wide variations between the colours of adjacent rivers (see Myers 1947, and Sioli 1967, for general description of the rivers in this area).

Yellow filters in Amazonian cichlids

Photostable yellow pigments were found in the corneas, lenses, and retinas of many cichlids. The corneas absorbed maximally between about 400 nm and 500 nm, and measurements made at 5 nm intervals showed the three characteristic submaxima previously reported by Moreland & Lythgoe (1968) and Bridges (1969) (Fig. 11.1). The optical density at the wavelengths of maximal absorption was often very high, reaching a value of 1·4 in a specimen of *Astronotus ocellatus* (Fig. 11.2, Table 11.1). The pigmentation was always deeper in the dorsal part of the cornea, though the manner in which it varied along the vertical axis varied considerably (Fig. 11.3).

The lenses, in contrast, acted as cut-off filters, passing long wavelengths and absorbing short wavelengths. The spectral position at which the cut-off occurred varied between species (Fig. 11.4, Table 11.1), but they invariably absorbed at shorter wavelengths than the cornea. Since in general yellow lenses and corneas occurred together, this means that in life light of wavelength shorter than about 500 nm will be almost totally prevented from

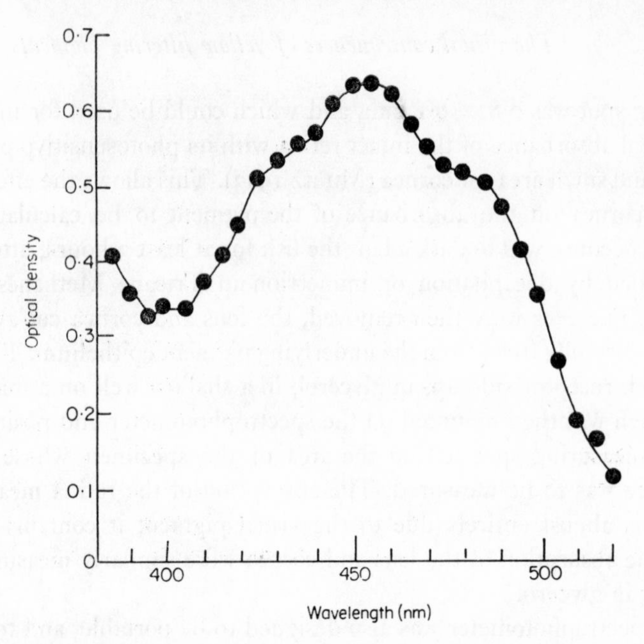

Figure 11.1 Spectral absorbance of the cornea of a specimen of *Cichlasoma festivum*.

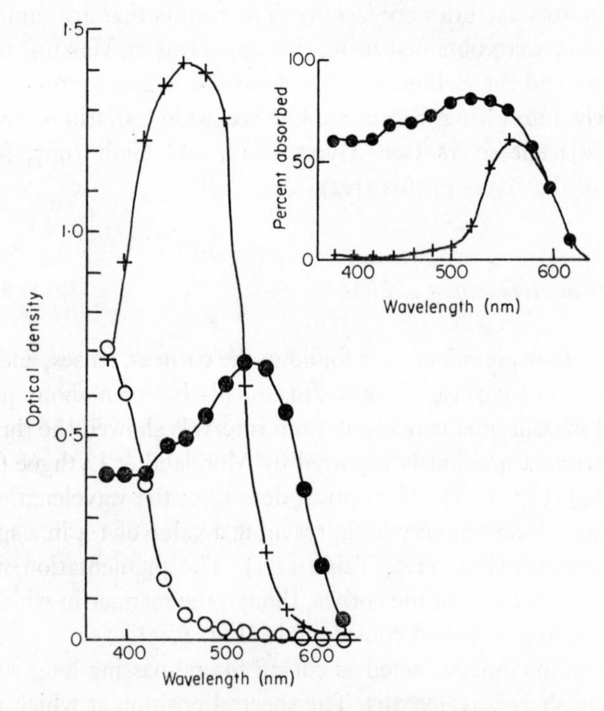

Figure 11.2. Spectral absorbance of cornea, lens, and retina of *Astronotus ocellatus*. In the main part of the figure the crosses show the absorption of the cornea, the empty circles of the lens, and the filled circles of the retina. The inset shows the percentage of light absorbed by the retina alone (filled circles), and by the retina in life when the lens and cornea have the characteristics shown in the main part of the figure.

reaching the retina, with the result that the effective absorption of the visual pigment will be markedly shifted towards long wavelengths. This is illustrated in Fig. 11.2 which shows spectral absorption curves for the lens, cornea, and retina of a specimen of *Astronotus ocellatus*. It can be seen that the retinal absorption was maximal at about 520 nm. The inset to Fig. 11.2 shows the percentage of the incident light that would be absorbed by the retina alone, and the percentage that would be absorbed when the lens and cornea have the characteristics shown in the main part of the figure: the effective absorption is much reduced, and the position of maximal sensitivity shifted to about 560 nm.

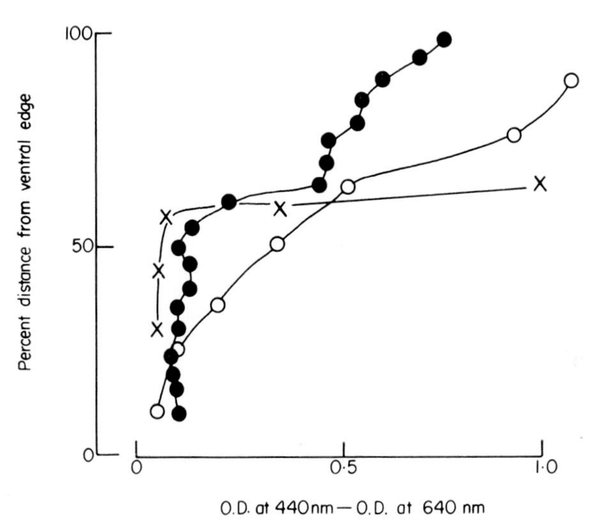

Figure 11.3. Depth of pigmentation, expressed as the difference in optical density at 440 and 640 nm, at different distances along the vertical axes of corneas from specimens of *Astronotus ocellatus* (empty circles), *Cichla ocellaris* (filled circles), and *Aequidens tetramerus* (crosses).

Table 11.1 summarizes the data obtained on the lenses, corneas, and retinas of 12 species of cichlid fish. The visual pigments are characterized by the point on the long wavelength side of the absorption curve at which the optical density has fallen to 50% of its maximum (the λ_{50}), since this is easier to estimate accurately that the wavelength of maximal absorption (λ_{max}). It seems likely from the work of Schwanzara (1967) that the retinas of most South American cichlid fishes contain two extractable visual pigments, absorbing maximally at 500 and 522 nm and based on vitamin A_1 and vitamin A_2 respectively. The Table shows the percentage of the A_1-based pigment in each case assuming that this is so, and finding the best fit using the computer program described in Muntz & Northmore (1971). The same computer program was used to estimate the λ_{max} by the local fitting of a fourth-order

Table 11.1.

Species	Lens cut-off (nm)	O.D. cornea	Visual pigment λ_{50} (nm)	Visual pigment λ_{max} (nm)	% A_1 pigment	Comments
Astronotus ocellatus (Cuvier)	425	1·4	582	522 ·3	0	Colour extends over whole cornea.
Aequidens tetramerus (Heckel)	440	0·74, 0·96	585,560 573	522·3,505·0 516·4	0,83·7, 23·1	Colour in dorsal part of cornea, extending over part of pupil.
Uaru amphiacanthoides Heckel	440	0·87	563	511·5	46·9	
Petenia spectabilis (Steindachner)	430	0·53	550	502·7	96·5	Colour extends over top third of pupil.
Cichlasoma festivum (Heckel)	455	0·63	576	522·2	0·5	Colour only in dorsal part of cornea, not reaching pupil.
Cichla ocellaris Schneider	NVC	0·50	567	512·2	43·2	Another small specimen had no visible colour in the cornea.
Crenicichla lenticulata Heckel	NVC	0·40	unknown	unknown	unknown	Corneal colour restricted to a patch over dorsal edge of pupil: retina deeply pigmented yellow.
Acaronia nassa (Heckel)	NVC	NVC	557	507·9	68·2	
Acarichthys heckelii (Mueller & Trotschel)	NVC	NVC	565	509·3	60·2	
Geopagus jurupari Heckel	NVC	NVC	565,570	510.9,513·4	50.2,35·4	
Geophagus surinamensis (Bloch)	NVC	NVC	550	502·1	100	
Geophagus sp	NVC	NVC	570	514·1	33·5	

The lens cut-off wavelengths are those at which the optical density of the lens was 0·3 greater than it was at 550 nm. Corneal optical densities were measured at 460 nm at the dorsal margin of the pupil, except in *Astronotus ocellatus* where it was measured at the centre of the pupil, and in *Cichlasoma festivum*, where it was measured near the dorsal edge of the cornea. NVC means that there was no visible colour. For the visual pigment λ_{50} refers to the point on the long wavelength side of the retinal spectral absorption curve at which the optical density has fallen to 50% of the maximum absorption, and λ_{max} is the calculated wavelength of maximal absorption. The Table also shows the percentage of A_1 -based pigments in each case, assuming that ... figure is given in the Table the results refer to

polynomial to the retinal absorption curve, followed by a numerical differentiation procedure.

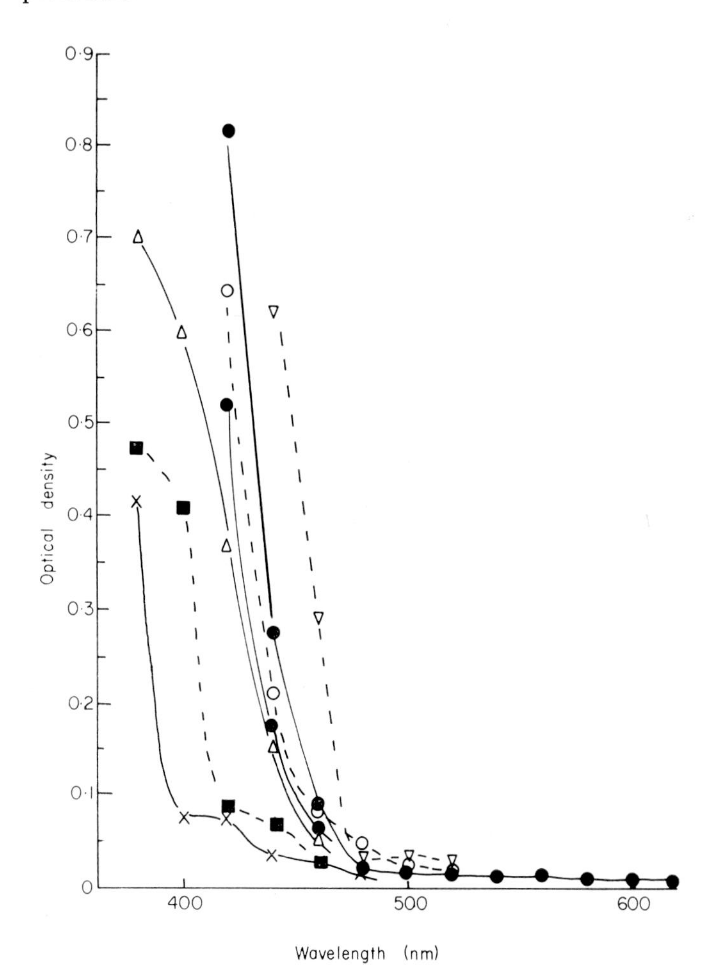

Figure 11.4. Spectral absorbances of lenses. From right to left the symbols on the curves are as follows: △, *Cichlasoma festivum*; ●, *Uaru amphiacanthoides*; ○, *Aequidens tetramerus*; ●, *Petenia spectabilis*; △, *Astronotus ocellatus*; ■, *Crenicichla lenticulata*; ×, *Acarichtys heckelii.*

Finally, it is clear that many of the cichlids also had yellow pigment in the retina itself. This manifested itself as a hump on the retinal absorption curve, located between about 440 and 480 nm, which was photostable and did not disappear on bleaching. In one species (*Crenicichla lenticulata*) much larger quantities of retinal yellow pigment were present; the retina when dissected out was a vivid yellow all over and this colour was photostable. The presence of the yellow pigment made it impossible to detect any visual pigment in this species. Measurements at 5 nm intervals showed that, although the yellow retinal pigment absorbed over the same spectral range as the corneal pigment of other species, the three submaxima were not apparent (Fig. 11.5), and subsequent ethanol extracts confirmed that a different

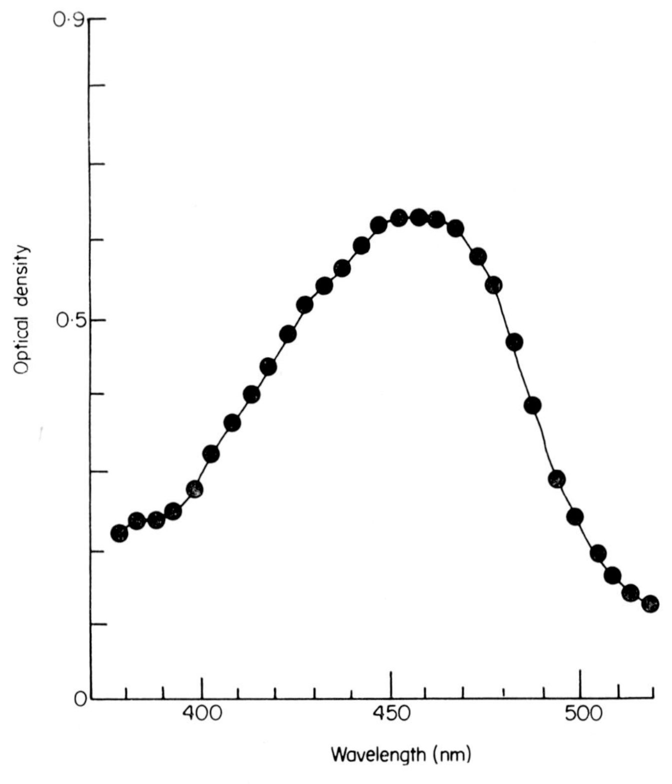

Figure 11.5. Spectral absorbance of the retina of *Crenicichla lenticulata.*

pigment is involved (Muntz 1973). Amazonian cichlids may thus possess three quite distinct yellow filtering pigments. The effects of the lens and corneal pigments on the absorbance of light by the receptors may be calculated (e.g. Figs. 11.2, 11.6), and is very large. The effect of the retinal pigment is, however, less easy to assess, since it is not known exactly where it is located in relation to the receptors.

Distribution of yellow filtering pigments of fishes

As we have seen, yellow filtering pigments are common among the Amazonian cichlids, occurring in seven out of twelve species. They did not occur in any

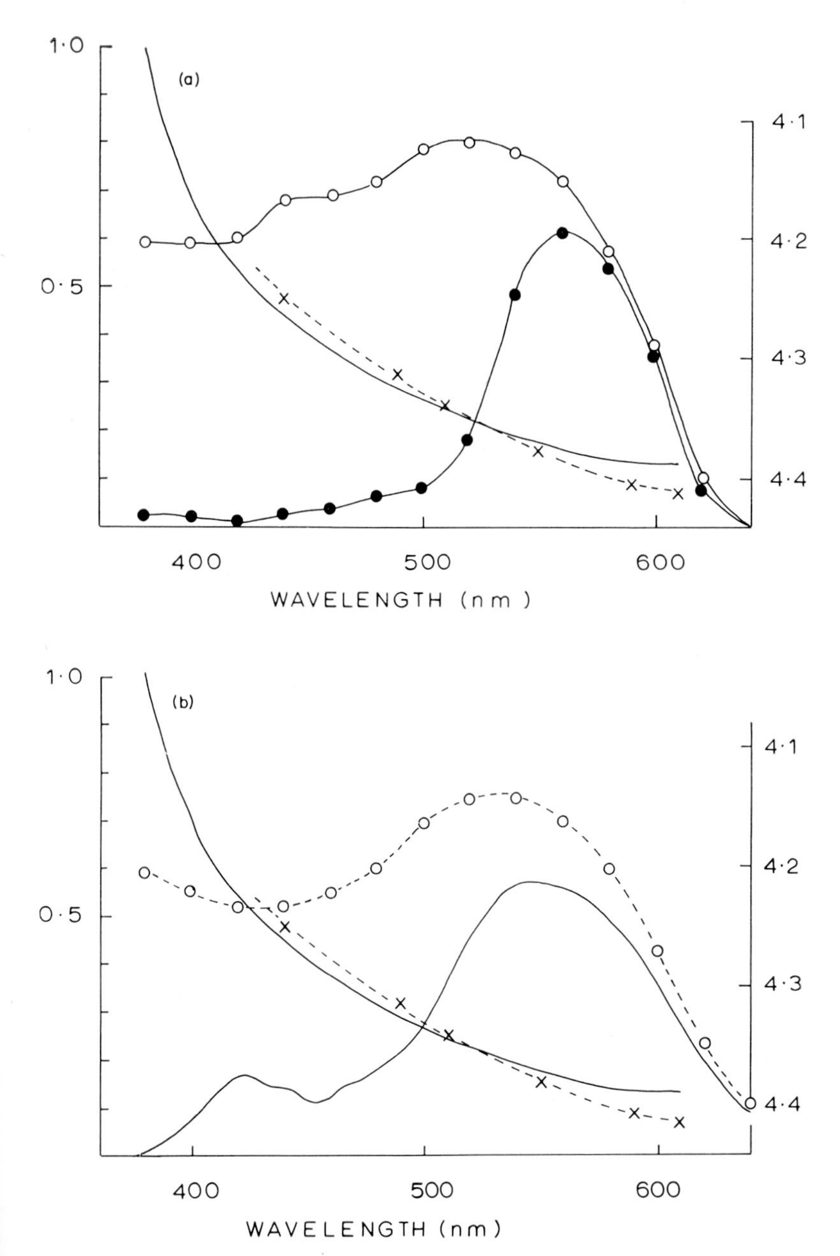

Figure 11.6. Effects of yellow lenses and corneas on Rayleigh scattering and chromatic aberration in (a) *Astronotus ocellatus*, and (b) *Perca fluviatilis*. In both parts of the figure the left-hand ordinate gives the proportion of light absorbed by the visual pigment alone (unfilled circles and dashed line) and by the visual pigment when the cornea and lens are present (filled circles and continuous line in (a), continuous line in (b)), as well as the amount of Rayleigh scattering as a proportion of the amount at 380 nm. The right-hand ordinate gives the focal length in nms of a 3·5 mm fish lens (crosses and dashed line).

of the characins (seven species) or catfish (four species) studies. Since yellow corneas have also been reported for four African species of *Tilapia* (Moreland & Lythgoe 1968), we may conclude that they are a very frequent characteristic of the cichlids. Yellow intraocular filters are also especially common in two other perciform groups, the wrasses (Labridae) and parrot fish (Scaridae). Thus Moreland & Lythgoe (1968) reported yellow corneas from six species of wrasse, and yellow corneas or lenses have also been found in eleven further species of wrasse and four species of parrot fish collected at Aldabra Island in the Indian Ocean (Lythgoe & Muntz unpublished observations). In this collection from Aldabra, which totalled about two hundred species in all, yellow filters were largely restricted to these two families, being otherwise only found in three trigger fish (Balistidae), two puffer fish (Tetradonitidae), and one sharp-nosed puffer (Canthigasteridae) (Table 11.2), three groups that are probably all derived from the Perciformes (Greenwood *et al.* 1966).

Yellow filters thus appear to be especially common in these groups of perciform and perciform derived fishes, although they do of course sometimes occur in other groups as well (see Walls & Judd, 1933, and Bridges 1969, for examples). The wrasse, parrot fish, trigger fish and cichlids at least have the common characteristic that most of their members are highly diurnal. Thus Hobson (1972) has described how, on the coral reefs of Hawaii, the diurnal fishes desert the water column at sunset and are replaced by the noctural fishes, with the reverse process occurring at sunrise. Prominent among the diurnal fishes were the wrasse, parrot fish, and trigger fish. Two of the species with yellow filters collected at Aldabra are also present in Hobson's diurnal group (*Scarus soridus*, *Rhinecanthus rectangularis*), while another genus is represented by a different species (*Thalassoma*). The wrasse and parrot fish also often show specialized methods of passing the night inactive, the former lying on their sides and burying themselves in the sand, and the latter remaining immobile after secreting a mucuous envelope around themselves (Winn 1955).

A similar situation occurs among the freshwater fishes of South America (Lowe-McConnell 1964, 1969). Here again there is a striking change over in the fish fauna between night and day, the nocturnally active fishes emerging from crevices at night to occupy and feed in the waters used by the diurnal fishes in the daytime. The cichlids are prominent among the diurnal fishes, and at night remain motionless, usually pressed up against the bank or some other submerged object.

There thus appears to be a relationship between diurnal habits and the presence of yellow filters. A similar relationship has also been suggested for terrestrial animals: among the squirrels, for example, there is a rough correlation between the depth of the pigmentation of the lens and the light levels normally encountered in the animal's environment (Walls & Judd 1933). Some of the possible beneficial consequences of yellow filters will be considered in the next section, but one inevitable detrimental consequence of *any* intraocular filtering structure will be a loss of sensitivity. It may be

Table 11.2. Teleost species from Aldabra Island (Indian Ocean) that have yellow corneas or lenses (J. N. Lythgoe & W. R. A. Muntz, unpublished observations). A plus sign indicates the presence of clearly visible yellow coloration. For the cornea a single plus sign shows that the colour was restricted to the top of the cornea, and a double plus sign shows that it extended further down, past the dorsal margin of the pupil.

	Cornea	Lens
Wrasse, Labridae		
Gomphosus caeruleus Lacépède	++	+
Lepidaplois axillaris Benn	++	+
Lepidaplois hirsutus Lacépède	+(pale)	o
Coris formosa Benn	++	o
Coris frerei Gunther	++	+
Coris angulata Lacépède	++	+
Halichoeres centriquadrus (Lacépède)	++	+
Halichoeres scapularis (Benn)	++	+
Anampses caeruleopunctatus Rüppell	++	+
Cheilinus diagrammus (Lacépède)	++	+
Thalassoma hebraicum (Lacépède)	++	+
Parrot fish, Scaridae		
Scarus sordidus Shultz	o	+
Callyodon urbanus Smith	+	+
Callyodon viridifucatus Smith	o	+
Xanothon bipallidus Smith	+	+
Trigger Fish, Balistidae		
Rhinecanthus aculeatus (Linn.)	++	o
Rhinecanthus rectangulus (Schneider)	+	o
Balistapus undulatus (Mungo Park)	++	o
Puffer fish, Tetraodontidae		
Arothron citrinellus (Gunther)	++	o
Arothron meleagris (Shaw)	++	o
Sharp-nosed puffers, Canthigasteridae		
Canthigaster valentini (Bleeker)	++	o

that the wrasses, parrot fish, and cichlids remain inactive at night in order to avoid the consequences of this sensitivity loss.

The fact that the depth of pigmentation is greater in the dorsal part of the cornea may similarly be related to the greater intensity of downwelling as opposed to upwelling light, in which case it could have the same significance as the predominance of yellow filters in diurnal fishes. Finally, it is possible that it is related to the depth at which the fishes live: *Geophagus* at least, three species of which have been shown to lack filters, is a bottom living genus. This again may be related to the lower light levels that bottom living fishes are exposed to.

The presence of yellow intraocular filters does not appear to correlate with any other obvious aspect of the fishes' ecology. In particular, as noted by Moreland & Lythgoe (1968), there is no correlation with the colour of the water; for yellow filters have been found in fishes from both very highly coloured (165 platinum units) and relatively clear (20 platinum units) waters in the Rio Negro region (Muntz 1973), and they are also common in fishes from the very clear Jerlov Type 1 waters of coral reefs (e.g. Table 11.2). Similarly, there is no clear correlation with diet, for they occur in both predatory cichlids, such as *Cichla ocellaris* and *Crenicichla lenticulata*, and in leaf chopping and deposit eating species of *Tilapia* (Moreland & Lythgoe 1968, Fryer & Iles 1973).

Visual consequences of yellow filters

The most obvious effect of yellow intraocular filters will be on the fishes' spectral sensitivity, for very little light of wavelengths shorter than about 500 nm will reach the retina at all. Thus the open circles of Fig. 11.6a show the percentage of incident light that was absorbed by the retina of a specimen of *Astronotus ocellatus*, for which the maximum optical density of the visual pigment was 0·68, and the solid circles show the percentage of the incident light that would be absorbed in life when the lens and cornea have the characteristics shown in Fig. 11.2. Since the absorption of light by the visual pigment usually correlates well with the animal's dark adapted spectral sensitivity curve (e.g. Muntz & Northmore 1973), this latter curve is probably a good approximation to the scotopic visual sensitivity at different wavelengths. Fig. 11.6b shows the results of a similar calculation for the perch, *Perca fluviatilis*. The visual pigments of the perch absorb maximally at 535·5 nm (Bridges 1972), and it has been assumed that the maximal optical density in the retina is 0·6, which is a fairly typical value for freshwater fishes (Denton 1959, Denton *et al.* 1971, Muntz 1973). The data for the cornea have been taken from the curve published in Moreland & Lythgoe (1968), and for the lens from Kennedy & Milkman (1956). Both the lens and cornea are less heavily pigmented in the perch than they are in *Astronotus ocellatus*, but there is still a very marked reduction in the absorption at short wavelengths.

The presence of yellow filters of this type obviously makes any attempt to correlate the visual pigments of such fishes with the spectral characteristics of the environment extremely difficult, especially since the depth of pigmentation varies over the cornea so that spectral sensitivity will be different for different parts of the visual field. There is also no correlation between the presence of such filters and the spectral absorbance of the visual pigments, which we should expect if the pigments were adapted to the spectral quality of the incoming light; for as far as the retina is concerned the effects of living in coloured water or having a yellow cornea should be much the same. Thus (Table 11.1) cichlids with yellow filters do not have pigments absorbing at appreciably longer wavelengths than those that lack them, and the data presented in Muntz (1973) show that if anything their pigments absorb at shorter wavelengths than those of characins occupying the same habitats. Similarly, Bridges (1969) has reported that two holosteans, *Amia calva* and *Lepisosteus platyrhincus*, have yellow corneas, and it is also known that their visual pigments are based on vitamin A_2 and absorb maximally at 525 and 523 nm respectively (Bridges 1964), which are comparatively short wavelengths for freshwater fish pigments. Finally, the wrasse, parrot fish, trigger fish, and puffer fish that were investigated by Muntz & McFarland (1973) did not have pigments absorbing at longer wavelengths than those of fishes of other groups: the parrot fish indeed tended to have pigments absorbing at noticeably shorter wavelengths than expected. We do not know which of the fishes in this study had yellow corneas or lenses, but since a number of the genuses used by Munz & McFarland also occurred in the Aldabra sample, and were found to have yellow filters, it is likely that they were common.

The relationship between light intensity and yellow filters suggests the possibility that these filters are not functional, but are merely a consequence of high light intensities causing some change in the cornea and lens. For example, the yellow pigmentation of the human lens, which increases with age, is apparently due to degradation products of proteins (McEwen 1959), and it is difficult to believe that this has any function. It might be suggested that intense light causes similar changes in the eyes of fishes.

The presence of three distinct yellow pigments in cichlid eyes makes this idea unlikely, however, and it is also striking that the yellow lenses often start to absorb at exactly those wavelengths where the corneas start to transmit again (e.g. Fig. 11.2), so that together they filter out all the short wavelength light. It is also difficult to see on this view why a fish such as *Crenicichla* should have a high degree of pigmentation in the retina, but little colour in the lens or cornea. Finally, light levels in the Amazon river system are seldom excessively high, especially since the fish often live in the flooded forest under the tree canopy; and for *Chlorophthalmus agassizi* at least, which lives at depths of over 400 ft and has a strongly yellow lens (Denton 1956), the light levels must be extremely low.

There are two aspects of vision which are very probably improved by yellow filters, namely the reduction of contrast through scattering, and the effects of chromatic aberration. The contrast between an object and its background is smaller underwater than it is in air. This occurs because the image forming light from the object is attenuated in its passage through the water by being both absorbed and scattered out of the light beam, while at the same time the water between the object and the observer scatters diffuse veiling light into the eye (Duntley 1962). In pure water the scattering component follows Rayleigh's law, and is proprotional to λ^{-4}; this relationship is shown in Figs. 11.6a & b. Under these circumstances yellow filters should increase the contrast of objects against their backgrounds, since the scattering is occurring predominantly at short wavelengths. When, however, the water becomes turbid and the scattering particles are large, scattering becomes independent of wavelength, and yellow filters would no longer be beneficial in this way.

The situation is analogous to that obtaining in the atmosphere, where again scattering results in loss of contrast. On clear days the scattering follows Rayleigh's law and yellow filters improve the contrast of distant objects, but on hazy or foggy days scattering becomes spectrally neutral. Middleton (1952) has presented a summary of data showing that, as we should expect, the visual range of objects is greater for long wavelength light under conditions of good visibility, but that this no longer holds once the visual range drops below about 1km. For terrestrial animals therefore yellow filters will only improve the visibility of very distant objects on clear days, and while this might be important for some birds it is difficult to believe that it is important for most of the other terrestrial animals that have yellow filters.

Underwater, however, the situation is different, for here the amount of absorption and scattering is so much greater that visibility seldom exceeds about 30 m even on the clearest days, and under these conditions scattering will be heavily dependent on wavelength. Objects 30 m away could well be relevant to the survival of a fish. Furthermore, a visual range of 30 m refers to the detection by divers of objects of optimal size that have optimal intrinsic contrast with the background. Objects that are smaller (or larger) than the optimal size, or have low intrinsic contrast (such as grey objects), will only be visible at smaller ranges; but since the point at which they disappear still depends largely on scattered light, yellow filters should be beneficial in such cases also.

When the visibility is less good the scattering will be mainly caused by larger particles, and will be less dependent on wavelength. The exact point at which scattering becomes spectrally neutral is unclear, but it is likely that a component of Rayleigh scattering will be present even when the water is relatively coloured. The polarization of light underwater depends on Rayleigh scattering, and many animals use this phenomenon as an aid to navigation (Waterman 1966 for a review): whenever this is possible it is also possible that yellow filters will be beneficial.

A second way in which yellow filters may be beneficial is through the reduction of chromatic aberration. Although an early report (Pumphrey 1961) suggested that chromatic aberration is not a problem in the fish lens, more recent studies have shown that substantial amounts of aberration may be present in many species, including the perch (Sivak 1974). In Figs. 6a & b the chromatic aberration of a 3·5 mm diameter lens from a rudd (*Scardinius erythrophthalmus*) is shown, taken from unpublished data of Dr. J. H. Scholes. It is clear that the presence of yellow filters will substantially reduce the effects of aberration, both by shifting the sensitivity to longer wavelengths, where the chromatic aberration is less marked, and by narrowing the spectral range over which the fish are sensitive. Campbell & Gubisch (1976) have shown that in man it is not the perception of very fine detail (high spacial frequencies) that is impaired by chromatic aberration but the perception of medium detail at low levels of contrast. Since this effect is a consequence of the optics of the eye, and not of the properties of the retina or brain, we should expect it to apply to fish as well; and since fish live in a low contrast world, it may well be especially important in their case.

Summary and conclusion

Yellow filtering pigments are common in the corneas and lenses of many teleost fishes, especially among the perciformes and perciform derivatives. It is probable that such filters improve vision by reducing the amount of scattered light reaching the retina (which is particularly important underwater) and by reducing chromatic aberration. Both these effects should improve visual resolution. Yellow filters will, however, also inevitably result in a loss of sensitivity.

The presence of yellow filters correlates with conditions of high light intensity, when the loss of sensitivity may be relatively unimportant. Thus they are particularly common in the wrasses, parrot fish, trigger fish and cichlids, which are four groups of predominantly diurnal fishes that often show specialized methods of passing the night inactive, and there is also a tendency for them to be absent in bottom living species. Finally the pigmentation is denser in the dorsal part of the cornea and will therefore affect upward vision more than downward vision, which agrees with the greater amount of light available for vision in an upward direction. It is likely therefore that yellow pigments occur under conditions where the loss of sensitivity is less important than the gain in visual resolving power.

There is no correlation between the presence of yellow pigments and the colour of the water, nor does their presence correlate with the spectral absorptive properties of the visual pigments. This is not surprising if the function of such filters is to gain resolution at the expense of sensitivity.

References

BRIDGES C.D.B. (1964) Periodicity of absorption properties in pigments based on vitamin A_2 from fish retinae. *Nature* (*Lond.*) **203**, 303–4.

BRIDGES C.D.B. (1969) Yellow corneas in fishes. *Vision Res.* **9**, 435–6.

BRIDGES C.D.B. (1972) The rhodopsin-porphyropsin visual system. In *Handbook of Sensory Physiology* Vol. VII/I, *Photochemistry of Vision*, ed. H.J.A. Dartnall, Springer-Verlag; Berlin, Heidelberg, New York.

CAMPBELL F.W. & GUBISCH R.W. (1967) The effect of chromatic aberration on visual acuity. *J. Physiol.* (*Lond.*) **192**, 345–58.

DENTON E.J. (1956) Recherches sur l'absorption de la lumière par le cristallin des poissons. *Bull. Inst. oceanogr. Monaco* **1071**, 1–10.

DENTON E.J. (1959) The contributions of the orientated photosensitive and other molecules to the absorption of whole retina. *Proc. Roy. Soc. B* **150**, 78–94.

DENTON E.J., MUNTZ W.R.A. & NORTHMORE D.P.M. (1971) The distribution of visual pigment within the retina in two teleosts. *J. mar. biol. Ass. U.K.* **51**, 905–15.

DENTON E.J. & WARREN F.J. (1956) Visual pigments of deep sea fish. *Nature, Lond.* **178**, 1059.

DENTON E.J. & WARREN F.J. (1957) The photosensitive pigments in the retinae of deep sea fish. *J. mar. biol. Ass. U.K.* **36**, 651–62.

DUNTLEY S.Q. (1962) Underwater visibility. In *The Sea*, Vol. 1, ed. M.N. Hill, Wiley & Sons, Inc.

FRYER G.H. & ILES T.D. (1972) *The Cichlid Fishes of the Great Lakes of Africa Their Biology and Evolution.* Oliver & Boyd, Edinburgh.

HOBSON E.S. (1972) Activity of Hawaiian reef fishes during the evening and morning transitions between daylight and darkness. *Fish. Bull. Nat. Mar. Fish. Ser.* **70**, 715–740.

KENNEDY D. & MILKMAN R.D. (1956) Selective light absorption by the lenses of lower vertebrates, and its influence on spectral sensitivity. *Biol. Bull.* **111**, 375–86.

LOWE-MCCONNELL R.H. (1964) The fishes of the Rupununi savanna district of British Guiana, South America. Pt. I. *J. Linn. Soc.* (*Zool*) **45**, 103–44.

LOWE-MCCONNELL R.H. (1969) The cichlid fishes of Guyana, South America, with notes on their ecology and breeding behaviour. *Zool. J. Linn. Soc.* **48**, 255–302.

LYTHGOE J.N. (1972) The adaptation of visual pigments to the photic environment. In *Handbook of Sensory Physiology*, Vol. VII/I, *Photochemistry of Vision*, ed. H.J.A. Dartnall, Springer-Verlag; Berlin, Heidelberg, New York.

MCEWEN W.K. (1959) The yellow pigment of human lenses. *Amer. J. Ophthal.* **47**, No. 5 pt II, 144–6.

MIDDLETON W.E.K. (1952) *Vision through the atmosphere.* Toronto University Press, Toronto

MORELAND J.D. & LYTHGOE J.N. (1968) Yellow corneas in fishes. *Vision Res.* 8, 1377–80.
MOTAIS R. (1957) Sur l'absorption de la lumière par le cristallin de quelques poissons de grand profondeur. *Bull. Inst. Oceanog. Monaco*, **1094**, 1–4.
MUNTZ W.R.A. (1972) Inert absorbing and reflecting pigments. In *Handbook of Sensory Physiology*, Vol. VII/I, *Photochemistry of Vision*, ed. H.J.A. Dartnall, Springer-Verlag; Berlin, Heidelberg, New York.
MUNTZ W.R.A. (1973) Yellow filters and the absorption of light by the visual pigments of some Amazonian fishes. *Vision Res.* **13**, 2235–54.
MUNTZ W.R.A. & NORTHMORE D.P.M. (1971) Visual pigments from different parts of the retina in rudd and trout. *Vision Res.* **11**, 551–61.
MUNTZ W.R.A. & NORTHMORE D.P.M. (1973) Scotopic spectral sensitivity in a teleost fish (*Scardinius erythrophthalmus*) adapted to different daylengths. *Vision Res.* **13**, 245–52.
MUNZ F.W. (1957) Photosensitive pigments from retinas of deep sea fishes. *Science, N.Y.* **125**, 1142–3.
MUNZ F.W. & McFARLAND, W.N. (1973) The significance of spectral position in the rhodopsins of tropical marine fishes. *Vision Res.* **13**, 1829–74.
MYERS G.S. (1947) The Amazon and its fishes. Part I. The River. *Aquar. J.* **18**, 4–9.
NICOL J.A.C., ARNOTT H.J. & BEST A.C.G. (1973) Tapeta lucida in bony fishes (Actinopterygii): a survey. *Can. J. Zool.* **51**, 69–81.
PUMPHREY R. J. (1961) Concerning vision. In *The cell and the organism*, ed. J.A. Ramsey & V.B. Wigglesworth, Cambridge University Press.
SCHWANZARA S.A. (1967) The visual pigments of freshwater fishes. *Vision Res.* 7, 121–48.
SIOLI H. (1967) Studies in Amazonian waters. *Atlas do Simpósio sôbre a Biotica Amazônica* 3, 9–50.
SIVAK J.G. (1974) The refractive error of the fish eye. *Vision Res.* **14**, 209–13.
WALD G., BROWN P.K. & BROWN P.S. (1957) Visual pigments and depth of habitat of marine fishes. *Nature, Lond.* **180**, 969–71.
WALLS G.L. & JUDD H.D. (1933) The intra-ocular colour-filters of vertebrates. *Brit. J. Ophthal.* **17**, 641–75; 705–25.
WATERMAN T.H. (1966) Systems analysis and the visual orientation of animals. *Amer. Sci.* **54**, 15–45.
WINN H.E. (1955) Formation of a mucous envelope at night by parrot fishes. *Zoologica (N.Y.)* **40**, 145–8.

12 Polarized light sensitivity in arthropods

R. MENZEL *Technische Hochschule, Darmstadt, West Germany*

Polarized light as a visual cue

The human eye cannot perceive all aspects of light in the visual world—we cannot detect polarized light. Natural light is not always equally polarized in all directions. For example reflected light differs from incident light in that certain directions of polarization are suppressed after reflection. If we could see the polarization pattern, as well as being able to discriminate intensity and wavelength, the visual world would be full of additional impressions.

A specially impressive pattern of linearly polarized light that occurs regularly in our environment is that in blue sky. This polarization pattern results from the scattering of light by small particles (Tyndall's effect) as well as by refraction in the sky. Plate 12.1 (facing p. 292) is an attempt to show with colours how the polarization pattern looks to an eye which can perceive polarized light. Here we have assumed that polarization detectors are similar to colour receptors. We have presumed that there are polarization detectors for 3 angles of polarized light, with preferred sensitivity axes to the E-vector 60° apart, similar to the three colour receptors blue, green and red. Directions of polarized light between these three main classes are illustrated by mixtures of the corresponding colours. The different amount of light polarized in any direction is shown by altering the intensity of the colour in the diagram.

There are four points in the sky, apart from the sun, where there is no polarized light: Babinet's point (20° above the sun), Brewster's (20° below the sun), the point directly opposite the sun (opposite point) and Arago's point (20° above the opposite point). Only three of these points are visible at any time. Polarization is maximal at 90° to the sun. The direction of polarization is vertical to the plane containing the observer, the observed point in the sky and the sun. As this pattern has a fixed relationship to the sun, animals with polarization detectors can orient towards it even when the sun is not actually visible. If only a small part of the sky is visible the animal cannot interpret the pattern unambiguously, for there are two or three points with the same polarization pattern at each altitude (see Plate 12.1). However, when a large part of the sky can be seen the animal is able to use the asymmetry of the overall pattern to orient correctly. There are only two occasions when the pattern is not asymmetric: (1) shortly before sunrise (sun at 10°) when Babinet's and Arago's points are equally distant from the horizon and (2) when the sun lies at the zenith.

Behavioural experiments with arthropods have shown without doubt that arthropods orient towards the polarization pattern of the sky. This orientation was discovered in bees (v. Frisch 1967), and has also been studied in ants (*Formica rufa*, *F. fusca*, *Lasius niger*—Carthy 1951, Jander 1957,

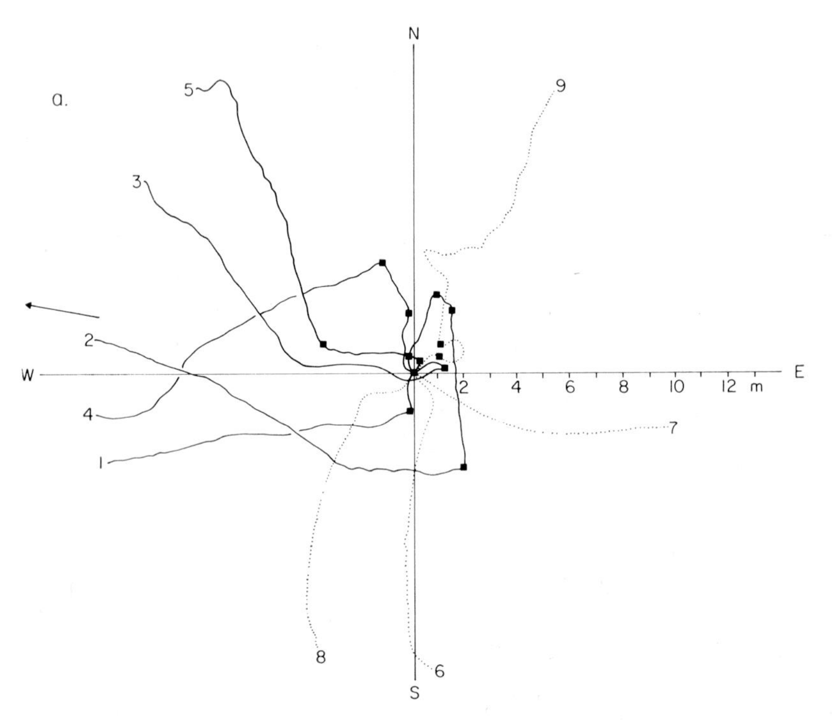

Figure 12.1. Polarization orientation of the Isopod *Ligia italica*.
a. Individual runs by **Ligia** in the natural environment (near Umag, Yugoslavia). The animals were transferred from the shore to a flat field unknown to them. With sun and blue sky, their paths (run 1 + 2) are oriented exactly towards the shore (the arrow points in the direction of the shore). In cloudy conditions, when less than one-third of the sky is visible, the runs are still well oriented (runs 3, 4, 5). When the sky is completely covered with clouds, the runs are disoriented (runs 6, 7, 8, 9). ■ indicates that the animal paused.
b. and c. Runs of *Ligia italica* in an arena where most of the sky was visible but landmarks were hidden (see text). The animals were released in the centre of the arena facing different directions in subsequent runs. The directions of the runs are quantified by measuring the position at which the animals crossed the 4 concentric circles (marked by a dot).

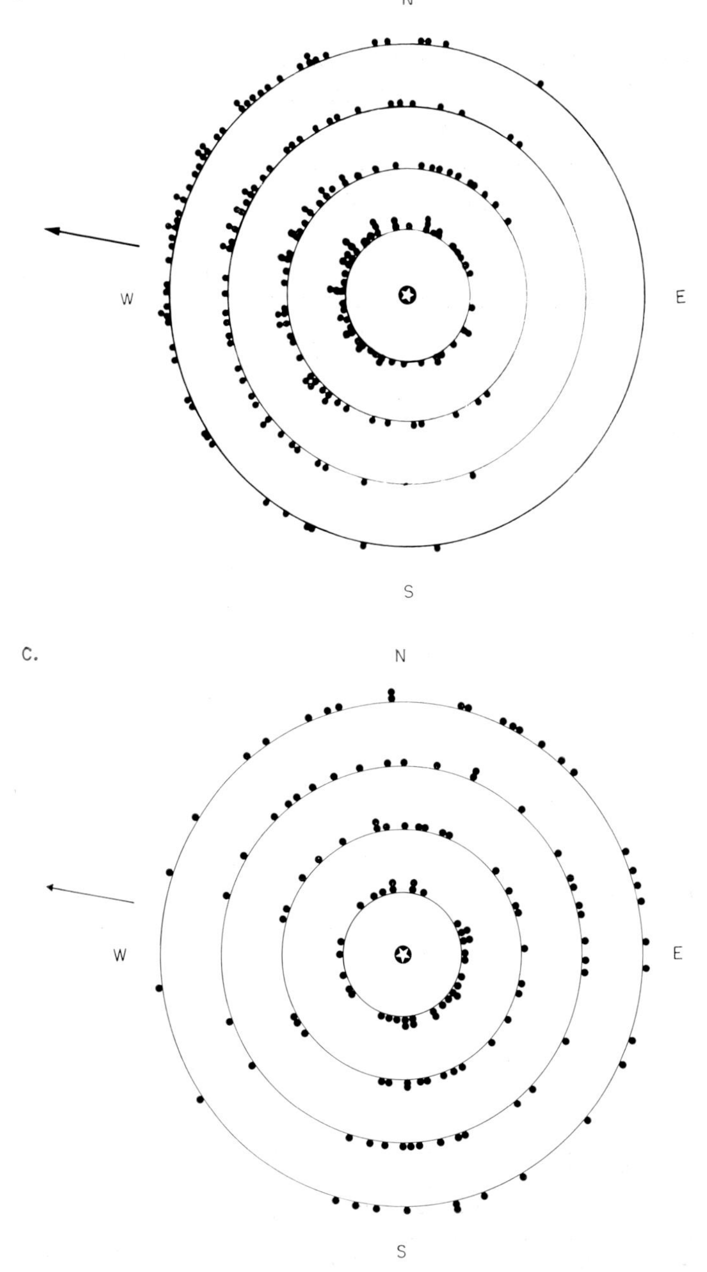

b. Runs in the arena under partially blue sky (half to two-thirds visible), without view of the sun. The tracks are mainly directed towards the west, the direction towards the shore in the former habitat is shown by the arrow. Then experimets were carried out at different times of day.

c. Runs in the arena under a sky wholly covered with clouds. The animals are disoriented.

These experiments show that *Ligia italica* orients astromenotactically to the polarization pattern of the sky.

M

Jacobs-Jessen 1959); bumble-bees (*Bombus hypnorum*, Jacobs-Jessen 1959), dung beetle (*Geotrupes*, Birukow 1953), water-strider (*Velia currens*, Birukow 1956) and other insects. Also crustacea and spiders orient with respect to polarized light (for example the crustaceans: *Daphnia*, Jander & Waterman 1960; *Idothea*, Pardi 1962, 1963, *Talitrus*, Pardi & Papi 1952, 1953; and the spiders *Arctosa*, Papi 1955, 1959, *Agelena*, Görner 1958, 1962; for review see Waterman 1969).

An example of a crustacean orienting to polarized light is shown in Figs. 12.1a, b, c. The Isopod *Ligia italica* lives on the mediterranean shores. *Ligia* always runs back to the shore if it is taken away and achieves this by orientating astromenotactically. With sun and blue sky visible (Fig. 12.1a) their paths are exactly directed towards the shore. In cloudy conditions when only a small part of sky is visible ($< 1/3$) the runs are still well oriented. However, when the sky is completely covered with clouds the runs are disoriented.

These preliminary experiments were repeated under better controlled conditions in Darmstadt (Germany). A circular arena (diam. 2 m) was placed on the flat roof of the laboratory building. The arena was surrounded by a white band which hid any landmarks on the horizon. The animals were placed in the centre of the arena facing in different directions and the runs were quantified as shown in Fig. 12.1 a, b and c. When at least 1/4 of the sky is visible the tracks are mainly directed towards the west. The coastline of their former habitat runs approximately from north to south. However, when the whole sky is covered with clouds the runs are evenly distributed in all directions. (Menzel unpublished experiments.)

Bees direct their dances with respect to a polarization pattern but only when the polarized light is in the UV. (v. Frisch 1967, v. Helversen & Edrich 1974). Similarly the desert ant *Cataglyphis bicolor* orients only to polarized light of short wavelengths (<410 nm, Duelli & Wehner 1973). It is important to note that although the skylight throughout the spectrum is polarized these animals use only the short wavelengths. It was found that atmospheric disturbances affect the long wavelength light 10 times more than UV light (v. Frisch 1967). The greater stability of the UV light may be an important reason for long wavelengths not being used for polarization detection. Furthermore after Rayleigh's law the intensity of scattered light is inversely proportional to the fourth power of the wavelength ($I \sim \frac{1}{\lambda^4}$). Therefore direct sunlight appears redder than scattered light. This is clearly seen when one considers the chromatic distribution of sky light at different distances from the sun (Plate 12.2). The columns in Plate 12.2 show the relative stimulation received by the 3 known colour receptors of the bee eye (Autrum & v. Zwehl 1964, Menzel unpublished), when the bee looks at different points in the sky. It is clear that the long wavelength receptors are more

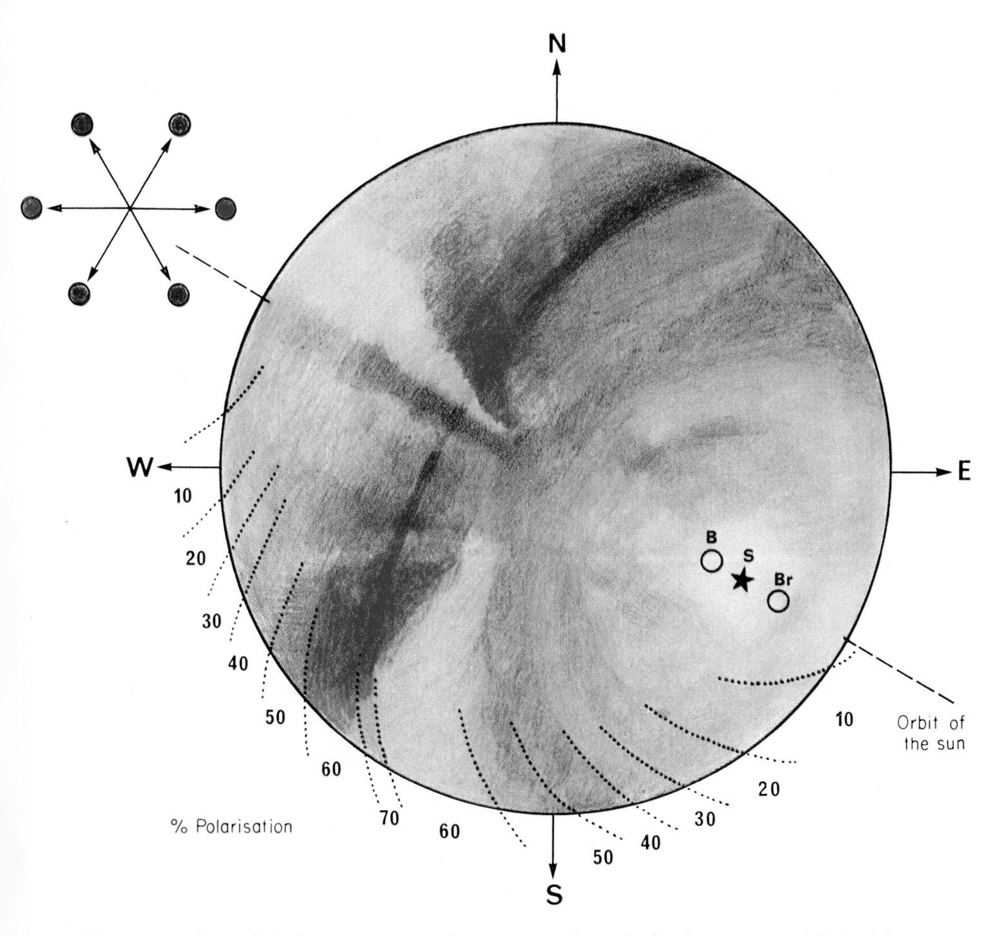

Plate 12.1. (opposite) An attempt to demonstrate the polarization pattern of the blue sky by using different colours for different directions of the E-vector. The three main E-vector directions 210/30°, 90°/270° and 150°/330° relative to the orbit of the sun are represented by three colours: blue, green and red (see upper left corner); other E-vector angles are given by intermediate colours. The amount of light polarized is illustrated by the colours' saturation. The light is not polarized when it comes directly from the sun (★, here 20° above horizon), from Babinet's point (B, 20° above the sun), Brewster's point (Br, 20° below the sun), the point directly opposite the sun (not seen here) and Arago's point (A, 20° above the opposite point). Polarization is maximal at 90° to the sun (70% of the light, see values along the lower side) and decreases in both directions. See text for further explanation.

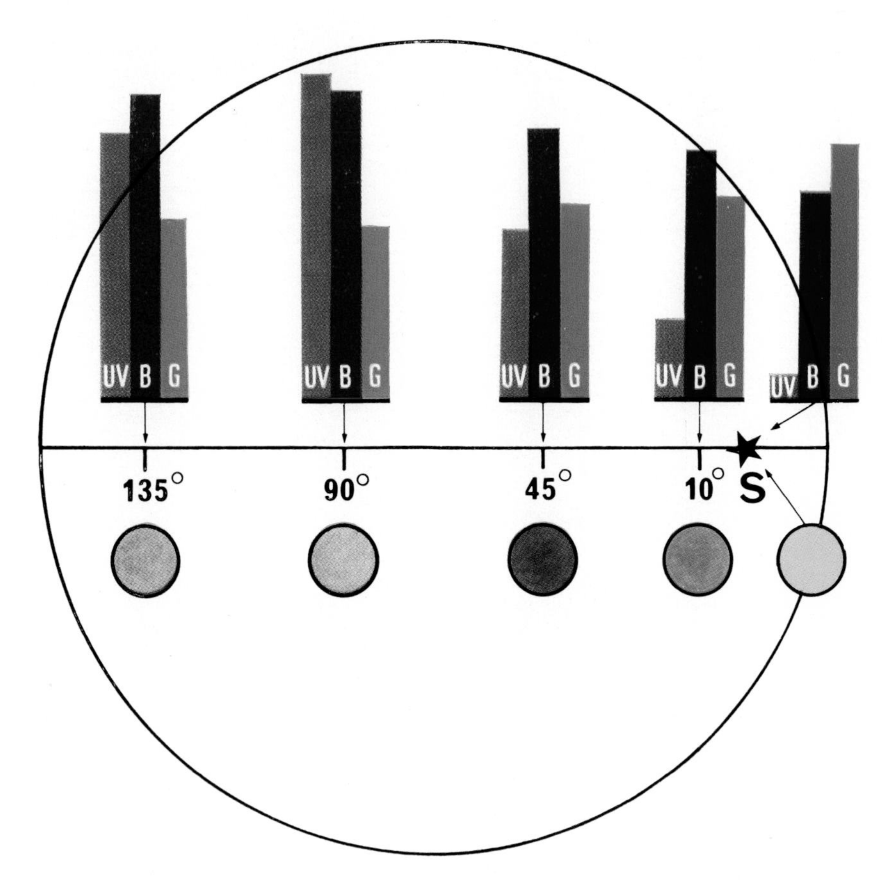

Plate 12.2. Demonstration of the spectral distribution of light coming from different parts of the blue sky. The diagram is based on measurements by Hess (1939), cited in Rosenberg (1966) and were carried out along the sun's orbit with the sun 20° above horizon. Spectral energy is normalized to 500 nm for each point of measurement.

strongly stimulated by the light near the sun, which is weakly polarized. However, at 90° from the sun, where polarization is maximum (Fig. 12.1), the UV receptors clearly dominate. *Therefore UV receptors are most reliable as polarization detectors, both because UV light is least influenced by atmospheric disturbances and because the area of maximal polarization has the maximal relative intensity of UV light.*

Functional basis of polarized light detection

For an animal to perceive the plane of linearly polarized light (the E-vector) it is necessary that the photo-receptors have polarization sensitivity (PS). The structure containing the photopigment molecules in the insect eye (e.g. the rhabdom of the ant eye in Fig. 12.2) consists of sectors of microvilli

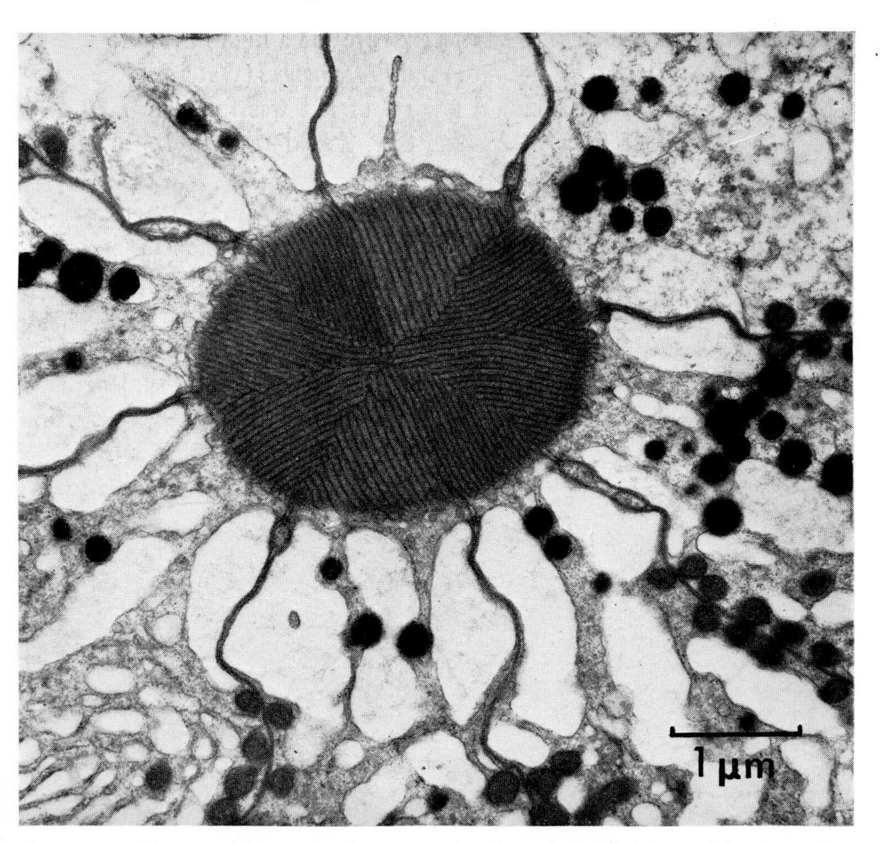

Figure 12.2. Electronmicrograph of a cross section through the rhabdom of the Australian bulldog ant, *Myrmecia gulosa*. The rhabdom is composed of the rhabdomeres of the 8 retinula cells. Four rhabdomeres contain microvilli lying in the same direction, the four others have their microvilli at ±60° to this group. The black dots within the retinula cells are screening pigment granules which move close to the rhabdom during light adaptation.

arranged in different directions. In the example given here there are three groups of microvilli directions arranged at 60° to each other. Obviously one can postulate that these three microvilli directions are the basis for the three assumed receptor types in Plate 12.1. Indeed microspectrophotometry (MSP) both, in the fly (Langer & Thorell 1966, Kirschfeld 1969) and the crayfish (Waterman *et al.* 1969, Waterman & Fernandez 1970) has shown that the absorption of polarized light is higher when the E-vector lies parallel to the long axis of the microvilli. (An exception is receptor 7 in the fly ommatidium, which absorbs most strongly when the E-vector lies perpendicular to the microvilli, Kirschfeld 1969). However, these absorption changes have been found to be small, the dichroic ratio is only 1:2 and this result has been taken to indicate a random distribution of photopigment molecules on the microvillimembranes (Moody & Parriss 1961).

The recordings from single receptors in the almost intact eye (unlike the eye treated for MSP measurement) have revealed a high PS, up to 1:12, in the crustaceans *Carcinus* (Shaw 1969) and *Procamburus* (Waterman & Fernandez 1970). In the fly, although a low PS of 1:2 was found in most cases (Burkhardt & Wendler 1960, Autrum & v. Zwehl 1972), a PS of 1:5 was found occasionally. The occurrence of cells with a high PS suggests that the low values of dichroic absorption ratios found using MSP are due to artefacts arising during preparation and measurement (see discussion in Goldsmith 1975).

Dichroic absorption is obviously a prerequisite for PS in receptors, but other factors add to and are maybe more important than the initial dichroic ratio. The gross morphology of the light absorbing structure (e.g. length and diameter of the rhabdom) and its fine structure (e.g. microvilli orientation, number of microvilli directions) have a strong influence on the retinula cells' PS (Snyder 1973). Also electrical coupling between retinula cells can alter the PS (Snyder *et al.* 1973). The example given below demonstrates how important these factors are in determining the PS of each receptor cell.

Example: The bee retina

The first model for the mechanism of polarization detection was designed for the bee, the first animal found to orient towards polarized light. Von Frisch (for summary v. Frisch 1967) devised a model in which the starlike arrangement of the 8 retinula cells produced 8 polarization sensitive inputs. Later he adapted his model to Goldsmith's (1962) finding that the microvilli are arranged only in two directions. The new model consisted of a 'four branched polarizer' in each ommatidium.

This model requires that all 8 long retinula cells in each ommatidium are polarization sensitive. However, to my great surprise, my intracellular recordings from single retinula cells in the worker bee revealed that these cells were either not or only slightly sensitive to the direction of the E-vector (Fig. 12.3a). The intensity, spectral and polarization sensitivities were

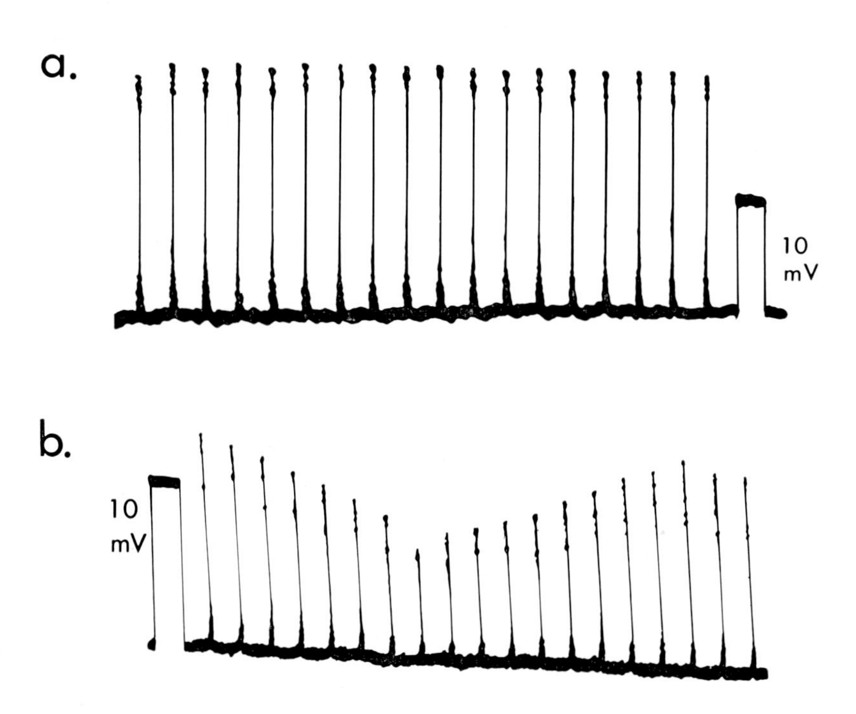

Figure 12.3. Intracellular recorded response of retinula cells in the worker bee eye to light flashes through a polarization filter. The polarization filter has moved by 10° between each light flash.

a. An example typical of most of the retinula cells which show no PS.

b. An example of a retinula cell which is highly sensitive to the rotation of the E-vector. This cell was recorded in the proximal third of the ommatidium and had a spectral sensitivity function with a maximum at 348 nm and no sensitivity above 420 nm.

measured in over 250 cells in the eye of the worker bee. Fig. 12.4 shows the polarization sensitivity from all cells which maintained a stable resting potential for at least 30 minutes. The green cells which one finds most frequently have either no PS (group 1) or a very small PS (group 2). The blue receptors too, as in the drone eye (compare Shaw 1969), have only a small PS. There are two types of UV-receptors, those with a low PS and some with a high PS. Those UV receptors with the low PS respond not only to UV light, but also to light of longer wavelength ($>$ 420 nm). On the other hand the group with the high PS is much less sensitive to longer wavelengths (below 5% relative sensitivity for $\lambda >$ 420 nm). These UV cells were only found when the electrode was positioned in the proximal third of the ommatidium near the basement membrane.

The bee's ommatidium contains 8 long and one short proximal retinula cell (Fig. 12.5). This short cell, the 9th cell, appears in the basal quarter of the

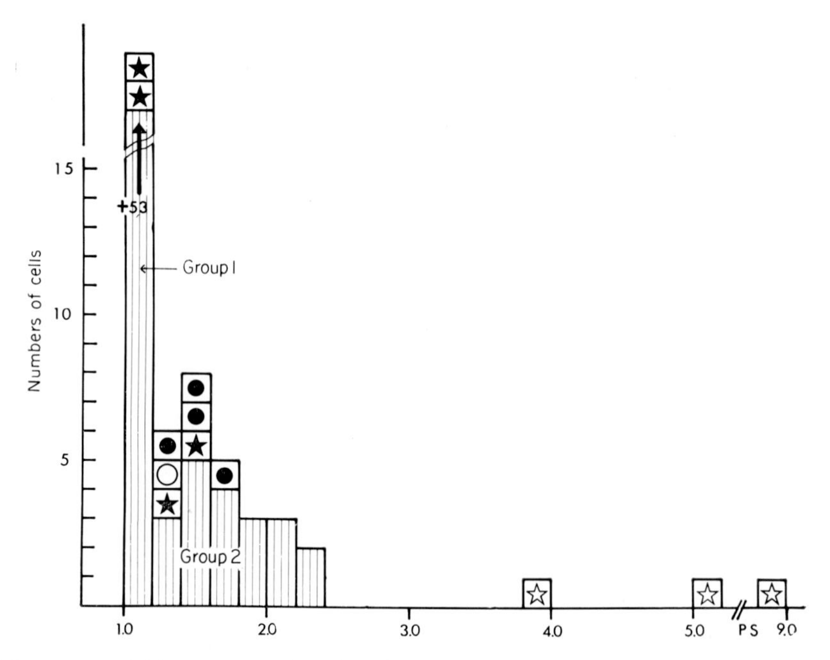

Figure 12.4. A histogram showing the PS values (abscissa) of retinula cells in the worker bee eye. The vertically striped bars give the green sensitive cells, the black dots indicate blue sensitive drone-bee cells, the bar with a circle represents a blue sensitive cell of a worker bee eye. Those marked with black stars are UV sensitive worker bee cells, which had considerable sensitivity to long wavelength (>5% relative spectral sensitivity above 420 nm), and the white stars mark UV cells (worker bee), which were recorded in the proximal third of the ommatidium and had less than 5% relative spectral sensitivity to longer wavelengths (>420 nm).

ommatidium and it has a total length of about 50–80 μm. It replaces the 7th cell which withdraws from the rhabdom as an axon. When the 9th cell appears the two directions of microvilli within the rhabdom are not changed. We know from Gribakin's (1969) spectral adaptation experiments that cell Nos. 1 and 4 (his group 1) are UV cells. It is of great importance for the PS of the 9th cell that the microvilli of these 2 UV cells are orientated perpendicular to those of the 9th cell. The arrangement of the colour receptors in the bee ommatidium given in Fig. 12.5 was deduced from a comparison of Gribakin's data of the distal part of the ommatidium with our own electron miscroscope study of the proximal part, and is based on the assumption that the ommatidium does not twist around its long axis. Meanwhile we have found that ommatidia in hymenopteran eyes do twist (see also Grundler 1974). The total twist angle between the distal end of the ommatidium and the appearance of the 9th cell is about 90°. In addition we found that the 9th cell does not replace retinula cell No. 7 (as shown in Fig. 12.5) but one of the two UV cells (No. 1 or 4). Therefore, the 9th cell microvilli are oriented perpendicular to the distal part of the UV cells. This means that our general conclusion on the crossed polarizer effect of the UV cells and the 9th cell holds with this different arrangement of the colour receptors.

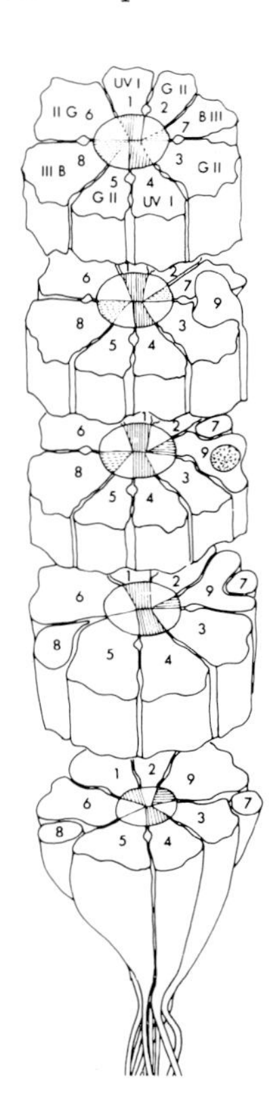

Figure. 12.5. A schematic drawing of the proximal third of the worker bee ommatidium. The ommatidium contains 8 long retinula cells (Nos. 1–8) and a proximal 9th cell. When the 9th cell appears it replaces No. 7. No. 8 withdraws from the rhabdom some microns more proximal than No. 7. Also shown is the distribution of the colour receptor types in the worker bee eye, based on the findings of Gribakin (1969, 1972). UV, B and G indicate the UV, blue and green receptors. The roman numerals (I, II and III) give Gribakin's grouping of the retinula cells. The microvilli direction is shown for the UV cells and the 9th cell. Note that the 9th cell has its microvilli perpendicular to those of the long UV cells.

We interpret our electrophysiological results to mean that only the 9th cell is a polarization detector. Theoretical considerations show that the 9th cell is especially suited for polarization detection:
(1) Because the rhabdom is short, the absorption is smaller, and the smaller the absorption the smaller the influence of the self absorption which reduces the PS;
(2) Above the 9th cell the two long UV cells work as polarization filters in the UV and thereby raise the PS of the 9th cell.

It is astonishing that the 8 long retinula cells have so little PS as the parallel filter effect in the fused rhabdom (Snyder *et al.* 1973) should raise the PS in spite of the high absorption. The low PS of the 8 long cells may have several causes:

(1) The dichroic absorption is much smaller in these 8 cells than in the 9th cell;

(2) the retinula cells are electrically coupled to each other;

(3) the microvilli of each retinula cell may not lie in a constant direction throughout the length of the rhabdom as the ommatidium may be twisted.

We can say nothing about the first possibility as there is no information on the dichroic absorption of individual rhabdomeres in the bee eye. It is unlikely that the 9th cell has a different dichroic absorption from the 8 long cells (see Snyder *et al.* 1973 for discussion). Electrical coupling between the retinula cells is most probable. Simultaneous recordings from 2 cells in the drone eye by Shaw (1969b) clearly demonstrated a coupling. We find that cells which show a low PS also have secondary peaks in their spectral sensitivity curves. These secondary peaks are thought to be caused by electrical coupling between the different colour receptor cells (Menzel & Snyder 1974, Menzel unpubl.). The third possibility (twisting of the rhabdom) has been recently verified in the bee (Grundler 1974). The ommatidia are very regularly arranged immediately under the crystalline cone but the arrangement becomes less regular in the proximal portion of the ommatidia (Fig. 12.5). This twisting and the electrical coupling would both reduce the PS of the long cells.

The separation of function between the 8 long cells and the 9th cell solves the problem of the ambiguity which would occur if all cells simultaneously code intensity, wavelength and polarization direction. Also the development of a specialized polarization detector reduces the complexity of the neuronal wiring in the optic lobe compared with that required if all cells were polarization sensitive.

One PS cell in an ommatidium is not sufficient to provide unambiguous determination of the polarization plane. This failure can be overcome by the interaction of neighbouring 9th cells, either through simultaneous or successive interaction. In both cases it is necessary that the neighbouring 9th cells have different orientations of their microvilli. In order to varify this assumption we have examined the orientation of the 9th cell's microvilli in different parts of the eye (Fig. 12.6). The results show clearly that the ommatidial pattern and especially the directions of the microvilli differs in the various eye regions. The two eyes are mirror images of each other with respect to the orientation of their ommatidia and the 9th cell. We know from v. Frisch's experiments that the dorsal part of the eye is most important for polaro-menotactic orientation, and only in this eye region do the 9th cells show clear groupings of microvilli orientations. The two groups lie at 120° to each other. Furthermore the probability of two neighbouring 9th cells having their microvilli at

120° to each other is much higher than for pairs with 2 ommatidia between them.

These results do not distinguish between the two possibilities of either simultaneous or successive polarization detection. The structural require-

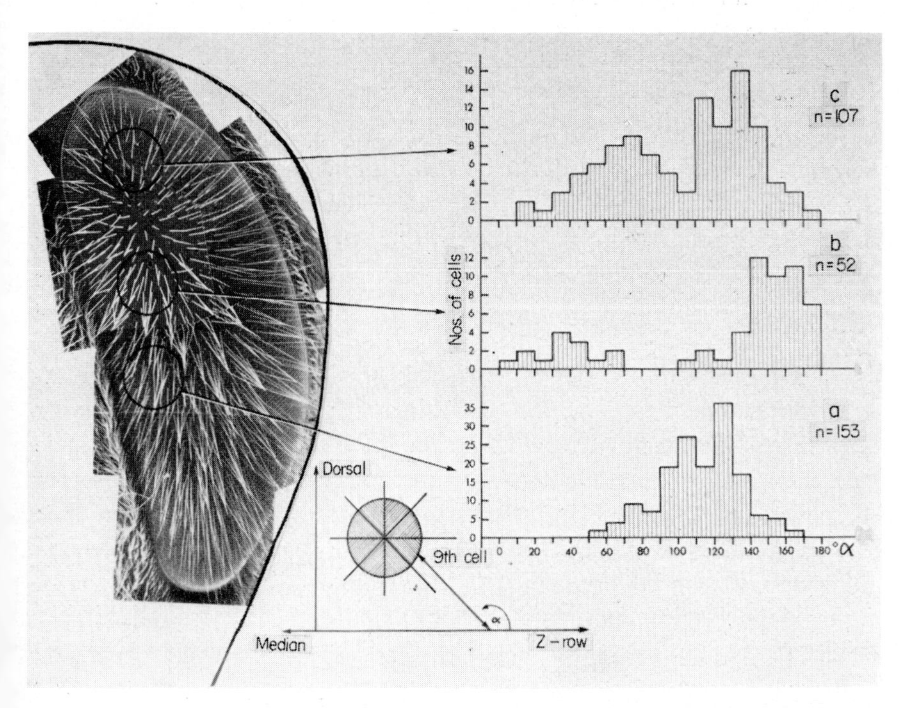

Figure 12.6. The distribution of orientations of the 9th cell microvilli in different areas of the left eye of the worker bee. The scanning micrograph shows the areas from which the cross sections for the electron microscope were taken. The inset gives the orientation of the eye relative to the head axis and defines α, the angle between the Z-row of the ommatidia and the long axis of the microvilli. The 3 histograms on the right show that the mediofrontal eye part contains only one population of 9th cell microvilli directions whereas the mediodorsal and dorsal eye part exhibit two populatoins of microvilli directions.

ments of a simultanous detection method are certainly fullfilled as the receptive fields of neighbouring ommatidia in the dorsal eye region show a broad overlap. (Baumgärtner 1928, Portillo 1936), Laughlin & Horridge 1971). The neuronal mechanisms whereby groups of neighbouring 9th cells can simultaneously measure the direction of the E-vector appear simpler than a method of successive comparison. Behavioural experiments may be able to solve this question.

Conclusion: Structure of rhabdom polarization sensitivity

An eye composed of many receptors is able to deliver information which increases in detail with narrower acceptance angles, higher temporal resolution, narrow spectral sensitivity and strong polarization sensitivity. Thus a fine receptor mosaic made up of different cell types allows the visual world to be analysed with respect to several parameters. The number of quanta reaching the receptors is strongly reduced with the sharpening of the spatial, temporal, spectral and E-vector sensitivities. Therefore the absolute sensitivity must be increased. The increase in sensitivity is of extreme importance in insects with apposition eyes as the aperture of the ommatidia is aproximately 20 μm. This is much smaller than the aperture in superposition eyes (ca 500 μm) or in lens eyes (some mm's).

The mechanisms for increasing absolute sensitivity are a high concentration of photopigment, the dense packing of the photopigment carrying membranes into microvilli, the formation of the rhabdom into a light guide and the length of the rhabdom. Asymmetric photopigment molecules can be packed most tightly when their axes lie parallel to one another. Also the most efficient way to pack membranes into a confined space, while still allowing diffusion of metabolites, is in a hexagonal lattice. Absorption of light is most efficient if the dichroic axis of the pigment molecules is at right angles to the light path (Wald *et al.* 1963). If the rhabdom contains several directions of microvilli it is a more efficient light gatherer than if it contains only one direction of microvilli (Shaw 1969).

As a consequence of the strategy to increase absolute sensitivity the rhabdomeres become polarization sensitive. This PS is maintained in the fused rhabdom by means of optical coupling. Similarly optical coupling sharpens the spectral sensitivities if rhabdomeres with different photopigments are combined in a light guide structure. (Snyder *et al.* 1973), *The great functional advantage of the fusion of several rhabdomeres with different photopigments and different microvilli orientations is that a high absolute sensitivity can be maintained simultaneously with high spectral and polarization sensitivity.*

It is not then astonishing that the PS is destroyed by electrical coupling between cells whose microvilli lie in different directions and twisting of the rhabdom. Possibly partial electrical coupling is a consequence of the dense packing of membranes in a fused rhabdom. However, electrical coupling has also several functional advantages. PS depends on the wavelength of the incident light, and this effect is very pronounced (Menzel 1975). There could be an ambiguity between differences in receptor outputs due to polarization and colour. Although this could be corrected by integration in the optic ganglion it might be considerably easier for the system to abolish the PS by electrical coupling and twisting and leave the detection of polarized light to a single specialized cell.

Snyder (1973) has shown that the tiered rhabdom of crustaceans is a special adaptation to maximize polarization sensitivity. The functional consequence of these mechanisms is a high PS at low intensities in the crustacean type of rhabdom. Since each receptor is highly sensitive to polarized light the rhabdom has high PS over a broad wavelength region if more than one colour receptor type exists in each ommatidium. The presence of more than one colour receptor type per ommatidium has been verified by Eguchi *et al.* (1973) who found two violet and five yellow receptors in each ommatidium of *Procamburus*. Underwater light is strongly polarized in a definite pattern related to the sun's position (Waterman 1972). Because of the strong attenuation of light with depth and the chromatic light climate under water there are distinct adaptive advantages in having a high PS at low intensities throughout a broad spectral region. In contrast, as insects use the polarization pattern of the blue sky light, intensity is high enough and PS at low intensity is of less importance. Their fused rhabdom seems to have evolved to combine high sensitivity with fine grain colour vision, while polarization sensitivity is restricted to a special receptor type and exists only in a small spectral range.

Acknowledgement

I would like to thank my friend K. Gräser for his help with the experiments with *Ligia*, my wife and Ms I. Sewering for technical help, and Dr. J. Kien for her assistance with the translation.

References

Autrum H. & Zwehl V. v. (1962) Die Sehzellen der Insekten als Analysatoren für polarisiertes Licht. *Z. vergl. Physiol.* 46, 1–7.

Autrum H. & Zwehl V. v. (1964) Die spektrale Empfindlichkeit einzelner Sehzellen des Bienenauges. *Z. vergl. Physiol.* 48, 357–84.

Baumgärtner H. (1928) Der Formensinn und die Sehschärfe der Bienen. *Z. vergl. Physiol.* 7, 56–143.

Birukow G. (1953) Menotaxis im polarisierten Licht bei *Geotrupes silvaticus*. *Naturwiss.* 40, 611–12.

Birukow G. (1956) Lichkompaßorientierung beim Wasserlaäufer *Velia currens* (Heteroptera) am Tage und zur Nachtzeit. *Z. Tierphysiol.* 13, 463–84.

BURKHARDT D. & WENDLER L. (1960) Ein direkter Beweis für die Fähigkeit, einzelner Sehzellen des Insektenauges, die Schwingungsrichtung polarisierten Lichtes zu analysieren. *Z. vergl. Physiol.* 43, 687–92.

CARTHY J.D. (1951) The orientation of 2 allied species of British ant. *Behaviour* 3, 275–318.

DUELLI P. & WEHNER R. (1973) The spectral sensitivity of polarised light orientation in *Cataglyphis bicolor* (Formicidae, Hymenoptera). *J. comp. Physiol.* 86, 37–53.

FRISCH K. von (1967) *The dance language and orientation of bees.* Cambridge, Harvard University Press.

GÖRNER P. (1958) Die optische und kinästhetische Orientierung der Trichterspinne *Agelena labyrinthica. Z. vergl. Physiol.* 41, 111–53.

GÖRNER P. (1962) Die Orientierung der Trichterspinne nach polarisiertem Licht. *Z. vergl. Physiol.* 45, 307–14.

GOLDSMITH T.H. (1962) Fine structure of the retinula in the compound eye of the honey bee. *J. Cell. Biol.* 14, 489–94.

GOLDSMITH T.H.(1975) The polarisation sensitivity–dichroic absorption paradox in arthropod photoreceptors. In: *Photoreceptor Optics*, Eds. A.W. Snyder and R. Menzel, Springer Verlag, Heidelberg, New York.

GRIBAKIN F.G. (1969) Cellular basis of colour vision in the honey bee. *Nature* (*Lond.*) 223, 639–41.

GRIBAKIN F.G. (1972) The distribution of the long wave photoreceptors in the compound eye of the honey bee as revealed by selective osmic staining. *Vision Res.* 12, 1225–30.

GRUNDLER O.J. (1974) Elektronenmikroskopische Untersuchungen am Auge der Honigbiene (*Apis mellifica*). *Cytobiologie* 9, 203–20.

HELVERSEN O.v. & EDRICH W. (1974) Der Polarisationsempfänger im Bienenauge: ein Ultraviolettrezeptor. *J. comp. Physiol.* 94, 33–48.

JACOBS-JESSEN U.F. (1959) Zur Orientierung der Hummel und einiger anderer Hymenopteren. *Z. vergl. Physiol.* 41, 597–641.

JANDER R. (1957) Die optische Richtungsorientierung der Roten Waldameise (*Formica rufa*). *Z. vergl. Physiol.* 40, 162–238.

JANDER R. & WATERMAN T.H. (1960) Sensory discrimination between polarized light and light intensity patterns by arthropods. *J. Cell. Comp. Physiol.* 56, 137–59.

KIRSCHFELD K. (1969) Absorption properties of photopigments in single rods, cones and rhabdomeres. In: *Processing of optical data by organisms and by machines.* Academic Press.

KIRSCHFELD K. & FRANCESCHINI N. (1969) Ein Mechanismus zur Steuerung des Lichtflusses in den Rhabdomeren des Komplexauges von Musca. *Kybernetik* 6, 13–22.

LANGER H. & THORELL B. (1966) Microspectrophotometry of single rhabdomeres in the insect eye. *Exp. Cell. Res.* 41, 673–6.

LAUGHLIN S.B. & HORRIDGE G.A. (1971) Angular sensitivity of the retinular cells of dark adapted worker bee. *Z. vergl. Physiol.* 74, 329–35.

MENZEL R. (1975) Polarisation sensitivity in insect eyes with fused rhabdoms. In: *Photoreceptor Optics*, Eds. A.W. Snyder and R. Menzel, Springer Verlag, Heidelberg, New York.

MENZEL R. & KNAUT R. (1973) Pigment movement during light and chromatic adaptation in the retinula cells of *Formica polyctena. J. Comp. Physiol.* 86, 125–38.

MENZEL R. & SNYDER A.W. (1974) Polarised light detection in the bee, *Apis mellifera. J. Comp. Physiol.* 88, 247–70.

MOODY M.F. & PARRISS J.R. (1961) The discrimination of polarised light by Octopus. A behavioural and morphological study. *Z. vergl. Physiol.* 44, 268–91.

PAPI F. (1955) Astronomische Orientierung bei der Wolfsspinne *Arctosa perita*. *Z. vergl. Physiol.* **37**, 230–3.
PAPI F. (1959) Sull'orientamento astronomico in specie del gen. *Arctosa*. *Z. vergl. Physiol.* **41**, 481–9.
PARDI L. (1962/1963) Orientamento astronomico vero in un isopodo marino: *Idothea baltica*. Monitore zoologico italiano 70/71, 491–5.
PARDI L. & PAPI F. (1952) Die Sonne als Kompaß bei *Talitrus saltator* (Amphipoda, Talitridae). Naturwiss. **39**, 262–3.
PARDI L. & PAPI F. (1953) Ricerche sull'orientamento di *Talitrus saltator*. *Z. vergl. Physiol.* **35**, 459–89.
PORTILLO J.del (1936) Beziehungen zwischen den Öffnungswinkeln der Ommatidien, Krümmung und Gestalt der Insektenaugen und ihre funktionellen Aufgabe. *Z. vergl. Physiol.* **23**, 100–45.
ROZENBERG G.V. (1966) *Twilight. A study in atmospheric optics.* Plenum Press, N.Y.
SHAW S.R. (1969a) Sense cell structure and interspecies comparisons of polarised light absorption in arthropod compound eyes. *Vision Res.* **9**, 1031–41.
SHAW S.R. (1969b) Interreceptor coupling in ommatidia of drone bee and locust compound eyes. *Vision Res.* **9**, 999–1029.
SNYDER A.W. (1973) Polarisation sensitivity of individual retinula cells. *J. comp. Physiol.* **83**, 331–60.
SNYDER A.W., MENZEL R. & LAUGHLIN S.B. (1973) Structure and function of the fused rhabdom. *J. comp. Physiol.* **87**, 99–135.
SNYDER A.W. & SAMMUT R.A. (1973) Direction of E for maximum response of a retinula cell. *J. comp. Physiol.* **85**, 37–45.
WALD G. & BROWN P.K. & GIBBONS I.R. (1963) The problem of visual excitation. *J. opt. soc. Amer.* **53**, 20–35.
WATERMAN T.H., FERNANDEZ H.R. & GOLDSMITH T.H. (1969) Dichroism of photosensitive pigment in rhabdoms of the crayfish **Orconectes.** *J. Gen. Physiol.* **54**, 415–32.
WATERMAN T.H. & FERNANDEZ H.R. (1970) E-vector and wavelength discrimination by retinula cells of the crayfish Procamburus. *Z. vergl. Physiol.* 68, 154–74.
WATERMAN T.H. (1973) Responses to polarised light: animals. In: *Biology Data Book*. 2. Edition Vol II, Ed. by P.L. Altman and D.S. Dittmer, pp. 1722–89, Federation Americ. Soc. Exper. Biol. Bethesda, Maryland.

Questions and discussion

In answer to a question from the Chairman Dr Menzel dismissed any importance of the differential reflection of polarized light from the surface of the cornea on the basis that the polarization sensitivity in different ommatidia differs in its plane of maximal response and any effect of dichroic activity in the lens system can have no influence.

13 Natural polarized light and *e*-vector discrimination by vertebrates[1]

TALBOT H. WATERMAN *610 Kline Biology Tower, Department of Biology, Yale University, New Haven, Connecticut 06520, U.S.A.*

Partially polarized light is ubiquitous in nature.[2] The polarization is almost entirely plane and arises by Rayleigh and Mie scattering in air or water as well as by reflection from surfaces. The degree of polarization is often high. For instance the zenith blue sky at dawn or dusk may be 90% polarized and clear oceanic water near the surface may reach 60%. At Brewster's angle the reflected light may be 100% polarized with the *e*-vector parallel to the plane of the reflecting surface.

For reflected light, polarization depends on the orientation and nature of the reflecting surface. No biological significance for perceiving the *e*-vector of such light is obvious. On the other hand, a visual system which could suppress it might reduce glare significantly and improve the perception of underwater objects for a viewer in air. Both these adavantages are well known from the use of polaroid glasses by man. Comparable effects have been proposed for some *e*-vector sensitive insect eyes but no demonstration of such actual function has yet been made.

For scattered light, however, the first order effects both in the atmosphere and under water depend on the sun's position in the sky (Fig. 13.1). Since animals in several of the higher major groups (annelids, molluscs, arthropods and vertebrates) have been experimentally proven capable of using the sun as a compass, polarized scattered sunlight can provide an extension of this direction finding cue. The sun's disk subtends only about one half degree of the spherical irradiance pattern. The sky on the other hand fills up to a full partially polarized hemisphere. Underwater irradiance including its polarization pattern will occupy the whole sphere around any

[1] The author's research is being supported by grants from the U.S. Public Health Service (EY 00405) and the Research Committee of the National Geographic Society.

[2] More detailed references supporting these introductory generalizations are cited in other recent reviews (Waterman 1972, 1974b, Coulson 1974, Tyler 1974).

point in the water sufficiently above the limit of sunlight penetration (Waterman 1974b).

Figure 13.1. Sky polarization patterns photographed with a fisheye lens at three different times of day. In A-C a Polaroid KN36 linear filter with its transmission axis parallel to the sun's bearing was placed over the camera lens. The characteristic dark band in the sky marks the most strongly polarized region of the sky. It lies 90° away from the sun and is perpendicular to the sun's direction; the *e*-vector in the band is also perpendicular to the sun's direction. Seen even with such a single channel analyser the changes in polarization pattern are dramatic. A. Polarization pattern at 9:45 am; the dark band perpendicular to the sun runs approximately NE–SW across the sky. B. At 6:30 pm this band had shifted about 90° to a NW–SE direction. C. At local noon (1:15 pm EDT) the band was horizontal and EW in direction. D. Another visualization of the sky's *e*-vector in the band of maximum polarization at local noon. Here a tangential analyser described more fully in Fig. 13.3 demonstrates a high degree of sky polarization (which determines the visibility of the dark and light sector pattern) and a horizontal *e*-vector parallel to the midline of the dark sectors.

Rhabdom-bearing eyes and polarized light

In view of the widespread occurrence and characteristic patterns of natural polarized light, one might well have predicted that animals should be able to perceive this optical parameter. Yet its discovery about 25 years ago seemed rather a surprise. Von Frisch's famous behavioural experiments with the honeybee (1948) were the first demonstration that any animal other than man can perceive the *e*-vector orientation of plane polarized light. Furthermore his work proved that the bee can effect menotactic (or compass-like) orientation from the natural polarization of the blue sky, even from a localized 15° patch. Thus sky polarization can in fact be used by animals as an extension of the sun compass (review, von Frisch 1965).

Von Frisch and his students assumed on the basis of their experimental controls that the underlying polarized light sensitivity (PLS) depended on a dichroic analyser in the compound eye. However, despite massive experimental data favouring this conclusion, sweeping counterclaims were made by some others that the analyser concerned must be extraocular and dependent only on differential scattering or reflection in the environment. While these claims were largely unsupported by biological experiments they did focus attention on the importance of controlling reflection and scattering artifacts in any such studies.

Intensity patterns of this kind are widespread both in the field and in the laboratory. They constantly challenge the conclusion that behavioural responses in polarized light are really polarotaxes rather than phototaxes. They also haunt the apparatus for physiological tests where reflection from electrodes and other components, polarization of light by the filament and envelope of a light source or differential scattering in a fluid medium are almost unavoidable.

Nevertheless the original polarized light compass reaction of the honeybee has been proven unequivocally to be a polarotaxis and the *e*-vector analyser to be located in the retina. Meanwhile a large number of other animals have been found to respond to *e*-vector direction; a recent tabulation lists 107 species (Waterman 1973) mostly arthropods (91 distributed among 8 arachnids, 24 crustaceans and 59 insects) but also including 9 molluscs and 7 vertebrates (in addition to man).

Of these more than 100 species already studied, 36 have been shown to be capable of polarotactic light compass reactions in the field. No direct evidence for such menotaxis has yet been obtained for animals underwater. Furthermore there are only two cases where both polarotaxis and the related visual physiology have been studied in the same species: the spider *Arctosa variana* (Magni *et al.* 1962, 1965, Papi & Tongiorgi 1963) and the honeybee *Apis mellifera* (Autrum & Stumpf 1950, Stockhammer 1959, von Frisch 1965, Shaw 1969). Clearly available evidence is still rather sparse and incoherent. Thus the orientation behavioural data relate to one set of species while the physiological data (measured as electrophysiological, microspectrophotometric, optomotor, etc., responses) relate to an almost entirely different set of species.

Nevertheless the primary receptor mechanism in crustaceans, insects and molluscs has been determined from fine structural, microspectrophotometric and electrophysiological studies. In these types the visual pigment lies in the rhabdom which is the light sensitive organelle. Within this structure the photoreceptor membrane is made up of regularly arranged microvilli lying perpendicular to the optic axis. These are fingerlike processes projecting from the axial surfaces of the receptor cells. Normally all microvilli from a single cell are closely parallel (but careful measurements have not been made). Most often there are two orthogonal directions for the microvilli contributed

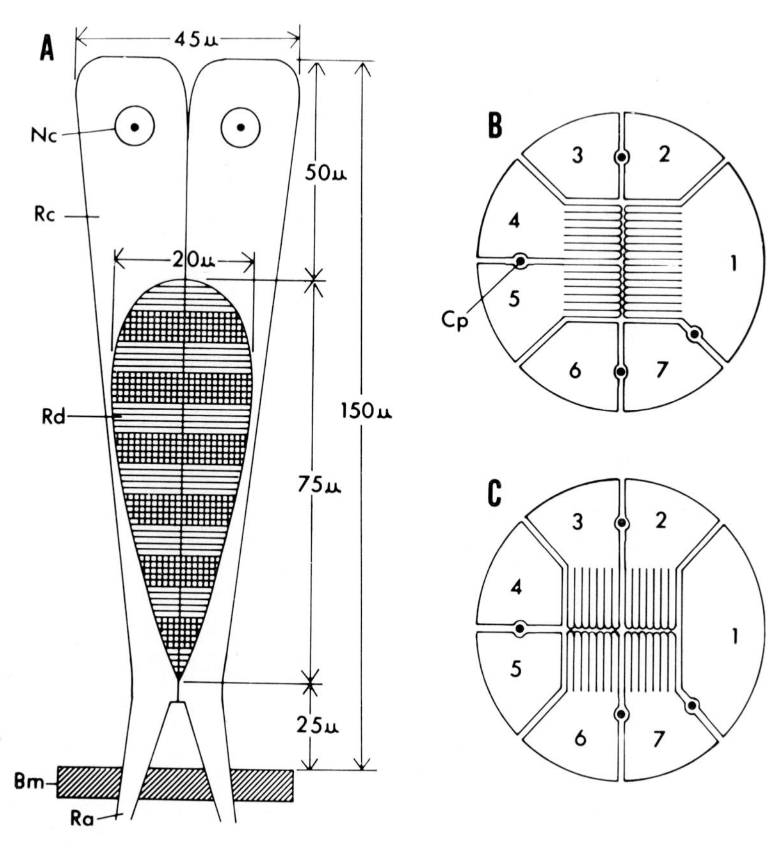

Figure 13.2. Diagram of the receptor unit (retinula) of a decapod crustacean compound eye typified by the crayfish. Such a cluster of seven regular photoreceptor cells lies beneath the dioptric elements of each facet of the eye. Light is absorbed by the rhabdom which contains the visual pigment and comprises regularly oriented microvilli. These are arranged in alternating layers originating from different sets of retinular cells. Since the microvilli are dichroic and absorb preferentially parallel to their long axis retinular cells 1, 4 and 5 are more sensitive to horizontal *e*-vector, R2, 3, 6 and 7 to vertical. A. Axial view; light enters from above. The actual number of layers of microvilli is about twice the number diagrammed. Typical adult dimensions are given. B. and C. Transverse sections through adjacent layers of microvilli. Bm, basement membrane; Cp, proximal process of crystalline cone stalk; Nc, nucleus of retinular cell, Ra, retinular cell axon; Rc, retinular cell body Rd, rhabdom. From Eguchi *et al.* 1973.

by the 4 (cephalopod) to 8 or 9 (typical decapod crustacean and insect) cells which form the rhabdom (Fig. 13.2).

The rhabdom microvilli are dichroic and in almost all known cases absorb more light parallel to their long axis than transverse to it. Their dichroism as shown by microspectrophotometric measurements depends directly on the visual pigment rhodopsin and hence on the orientation of its chromophores within or on the receptor membrane system (Hagins & Liebman 1963, Langer 1965, Waterman *et al.* 1969).

Even though the primary principles involved in *e*-vector discrimination are reasonably clear for rhabdom bearing eyes, more recent experiments and theories suggest that we still have much to learn about the specific manner in which *e*-vector information is ultimately processed in various arthropod visual systems (McCann & Arnett 1972, Smola & Gemperlein 1972, Eguchi & Waterman 1973, Eguchi *et al.* 1973, Muller 1973, Skrzipek & Skrzipek 1973, 1974, Menzel & Snyder 1974, Meyer-Rochow 1974, Waterman, 1974a).

The situation for those animals whose eyes lack rhabdoms is even less settled. Actually we are just beginning to learn that responses to polarized light are not uncommon in vertebrates in addition to man. Furthermore our understanding of the analyser mechanism involved has not yet moved beyond a quite preliminary stage. A review of the current status of this field particularly in relation to our own research on fishes is the main objective of this review.

Vertebrate eyes and polarized light

To begin with, the specific fine structural geometrical details of rhabdoms which are crucial to their effective dichroism are lacking in vertebrate outer segments. Indeed the latter have been known for some time to be isotropic to axial light (Schmidt 1935). They are, however, dichroic to obliquely incident light, strongly so to rays perpendicular to the normal optical axis. Similarly there is no evidence in vertebrates that dioptric elements like the normal cornea or lens are dichroic.

Nevertheless there are increasing numbers of recent reports of *e*-vector sensitivity in vertebrates, mainly in fishes[3] (Forward & Waterman 1973, Kleerekoper *et al.* 1973) but also in an amphibian (Adler & Taylor 1973, Taylor & Adler 1973) and a bird (Kreithen & Keeton 1974). Of course in man Haidinger's brushes have been known since 1844 and several other polarized light related phenomena in human vision have been reported although they have not been much studied.

[3] Earlier reports of fish *e*-vector sensitivity are summarized in Waterman 1972, pp. 445–7.

To begin with Haidinger's brushes these comprise a faint foveal entoptic image evoked by polarized light in the normal eye focused at infinity (Haidinger 1844). With linearly polarized light it appears as a small cross-like figure having the two arms parallel to the *e*-vector electric blue in colour and the two perpendicular ones dirty yellow (Fig. 13.3). The brushes fade from

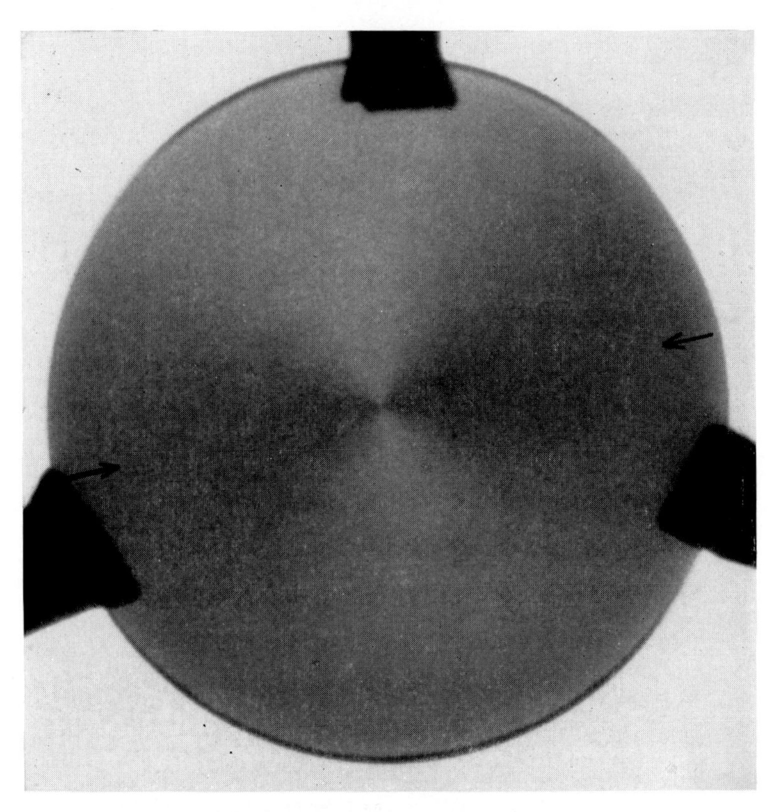

Figure 13.3. Image produced by a tangential analyser when traversed by partially polarized sky light as in Fig. 13.1D. Dichroic elements in the analyser are lined up so that their major transmission axes are oriented in a polar pattern being everywhere tangential to radii from the centre. Plane polarized light is therefore transmitted maximally at 90° to its *e*-vector (shown by arrows) and minimally at 0°. The resultant dark and light sector pattern has strong similarities to the *e*-vector sensitivity distribution we have observed in the goldfish optic tectum. Similarities and differences between this phenomenon in the fish eye and Boehm's and Haidinger's brushes in man are discussed in the text.

sight in a few seconds when the eye is fixated. Reports on the colours as well as the image's vividness vary a lot among observers. Without a little training with a polarizer most people never see Haidinger's brushes. Yet after some experience they are relatively easy to see with the unaided eye in the clear blue sky especially at dawn or dusk.

The generally accepted hypothesis of mechanism is that proposed long ago by Helmholz; it depends on the dichroism and regular arrangement of the chromophores in the macular pigment on the retinal surface. Since this is only known to occur in primates it can hardly be the basis of *e*-vector perception in other vertebrates. However, three other entoptic images for man have been reported in polarized light (LeGrand 1936, 1937, Boehm 1940).

Boehm's brushes

One of these, Boehm's brushes, although rarely mentioned in the literature seems particularly interesting for us. It is a parafoveal entoptic image apparently due to intraocular light scattering and only visible when the *e*-vector being perceived is rotating fairly rapidly (at about 360°/sec). This image is also describable as a cross in which two arms perpendicular to the *e*-vector are light and the orthogonal ones dark (Fig. 13.3).

This can be readily observed using a small bright spot of plane polarized light with its *e*-vector revolving once a second and fixated about 15–20° outside the fovea. At lower speeds of rotation contrast decreases to zero and at higher speeds the light sectors appear more like the arms of a spiral nebula and become fainter and shorter. As will be seen below, our work on the goldfish suggests that it must see two light sectors perpendicular to dark sectors coincident with *e*-vector direction.

This would suggest that something comparable to Boehm's brushes occurs quite generally in vertebrate eyes. If so *e*-vector perception could be widespread throughout this major group. In the goldfish, however, the contrast between sectors remains strong even with a fixed *e*-vector. Also its PLS is strong peripherally and weak centrally, unlike Boehm's brushes. These relations are discussed in detail below.

Turning now to a summary of our recent experiments on *e*-vector perception in teleosts that work falls into two major areas: (1) behavioural experiments on spontaneous orientation in polarized light and (2) electrophysiological (and some optical) measurements designed to identify the analyser mechanism. Both kinds of research are being actively continued but the behavioural work so far completed has mainly been published; the electrophysiological work has been published only briefly in one abstract and recently in detail in two technical papers (Waterman & Aoki 1974, Waterman & Hashimoto 1974).

Polarotaxis in fish

Our behavioural experiments have been carried out in the field on *Zenarchop-*

terus a tropical marine hemiramphid common in Palau, Western Caroline Islands (Fig. 13.4); they have also been done in the laboratory on a fresh-water relative *Dermogenys*. For *Zenarchopterus* an extensive set of experiments were done underwater using SCUBA to execute the work (Waterman & Forward 1970, 1972).

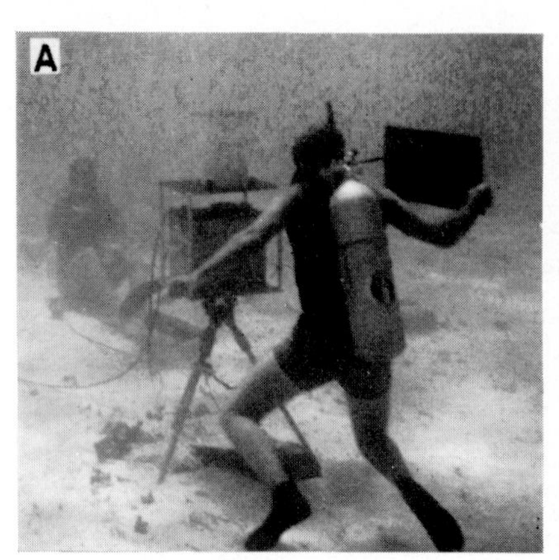

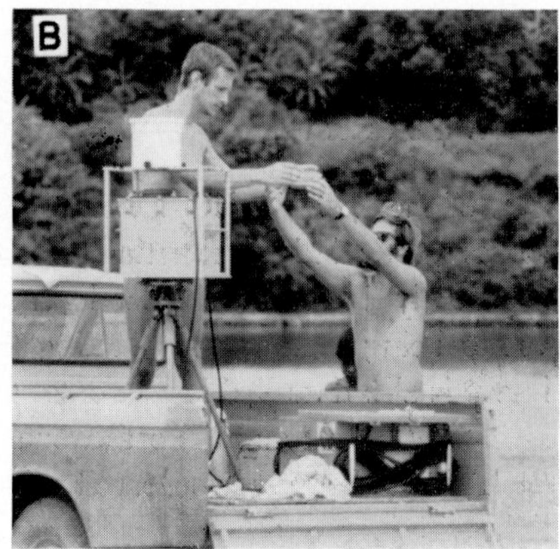

Figure 13.4. Experimental conditions used to measure spontaneous orientation of the teleost *Zenarchopterus* to an imposed polarization pattern in the field. A. Setup underwater at about 5 m depth. A camera looking upward photographed the fish's azimuth orientation under a variety of optic conditions. The fish was contained in a covered lucite vessel and could see the whole upper hemisphere in the water including the sun and sky through the critical angle. B. Similar study of the fish orienting at the water surface of the uncovered experimental vessel. The orientation observed under these two conditions is shown in Fig. 13.5.

Spontaneous orientation of juvenile specimens enclosed in a covered cylindrical plexiglas experimental vessel and responding to a randomized sequence of optical conditions were photographed from below. The conditions included six different orientations (at 30° intervals) of an imposed overhead plane polarization. The sun, sky and water were visible above the fish but white screens shielded the lower hemisphere to cut off any view of the bottom, divers, etc.

In relation to the imposed *e*-vector a significant preference was shown for directions perpendicular to the plane of polarization (Fig. 13.5A). For all 36 fish tested the degree of orientation involved was not very high, being 8% (defined as $o_{\perp} = (d_{\perp} - d_{\parallel})/(d_{\perp} + d_{\parallel})$ where d is the observed count in directions perpendicular (⊥) and parallel (∥) to the *e*-vector). But 15 fish selected from these for moderate turning rates and absence of clouds over the sun during any part of their run showed $o_{\perp} = 17\%$. Both preferences are statistically significant.

Since *Zenarchopterus* spends a considerable part of its time swimming a the water surface a second set of field experiments was done with the experimental vessel open to the air (Forward *et al.* 1972). As before the upper hemisphere in this case including sun and sky was visible to the fish but no landmarks or the observers. Under these conditions *e*-vector induced preferential orientation was parallel to the plane and the degree of orientation was 28% (Fig. 13.5B).

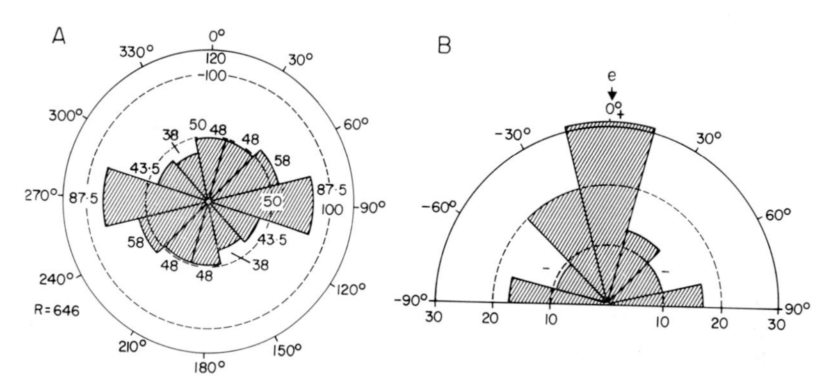

Figure 13.5. Spontaneous orientation of *Zenarchopterus* to imposed *e*-vector in the field. A. Response underwater in a covered vessel showing significant preference for headings perpendicular to the *e*-vector at 0–180° (Fig. 13.4A). From Waterman & Forward 1972. B. Response at the water surface in an open vessel showing even stronger orientation preference this time parallel to the *e*-vector (Fig. 13·4B). From Forward *et. al.* 1973. Parallel preference to the polarization plane was also shown in laboratory experiments by goldfish (Fig. 13.6) and by *Dermogenys* a close relative of *Zenarchopterus* (Fig. 13.7).

Similar preferential orientation parallel to the *e*-vector ($o_{\parallel} = 26\%$) was found for the goldfish *Carassius auratus* by Kleerekoper *et al.* (1973) in laboratory experiments (Fig. 13.6). Here the circular experimental tank was much larger (2 m in diameter) and the fish's spontaenous entries into a series of 16 22·5° peripheral compartments were summed relative to *e*-vector orientation. The preferred direction parallel to the polarization plane was similar to that of *Zenarchopterus* at the water surface.

Polarotaxis vs. Phototaxis

In our field experiments as well as those of Kleerekoper *et al.* on the goldfish in the laboratory no sure statement can be made on the specific cue used by the animal. The preferential orientation observed in polarized light could be due either to polarotactic responses derived from the perceived *e*-vector itself or to phototactic responses arising from light intensity patterns caused by differential scattering of the polarized light. We know for instance that

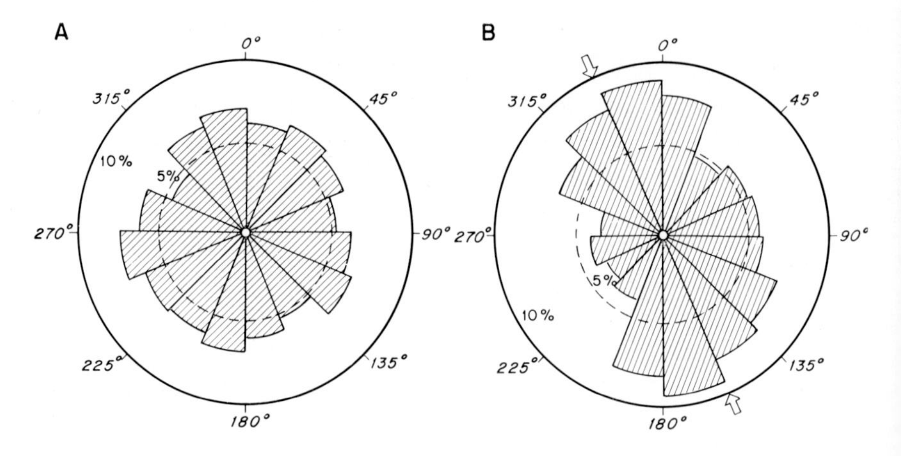

Figure 13.6. Spontaneous azimuth orientation of *Carassius auratus* measured during a period of three days as per cent entries into a circular array of 16 22·5° radial compartments in a 2 m diameter experimental area. A. Overhead light (1 m in diameter) not polarized. Orientation random. B. Overhead light plane polarized in the direction shown by the broad arrows. Orientation significantly favoured directions approximately parallel to the *e*-vector. Redrawn from Kleerekoper *et al.* 1973.

scattered light intensities in directions perpendicular to a light beam can be 4·8X greater perpendicular to the *e*-vector than parallel to it in distilled water; this ratio was 3·6 in a filtered sample of seawater (Waterman 1960).

If viewing conditions are such that these intensity differences can be discriminated the fishes could respond to them without necessarily being able to perceive the polarization plane directly. However, in favourable cases and with appropriate precautions one may determine whether the observed behaviour is in fact a polarotaxis or a phototaxis (Jander & Waterman 1960). Thus our laboratory experiments on *Dermogenys* indicate that its preference for orientation parallel to the *e*-vector cannot reasonably be a phototaxis and hence must depend on *e*-vector perception *per se* (Forward & Waterman 1973).

To establish this we compared orientation under two conditions: (1) In a vertical beam of plane polarized light with a uniform white screen surrounding the experimental vessel. (Note that by using a white surround rather than a black surround the Weber fraction [$\triangle$ I/I] which is critical for detecting the intensity pattern will be decreased as long as the differential scattering $\triangle$I [determined by the medium] remains the same [Miller & Benedek 1973, pp. 40–52]). (2) In a vertical beam of unpolarized light with a black and white surround divided into alternating quadrants crudely simulating strong differential scatter.

With the uniform white surround and a vertical plane polarized beam preference was weak ($o_{\parallel}$ = 4%) but significant in the direction parallel to the *e*-vector (Fig. 13.7). As we have seen that is the direction of minimum horizontal scatter and hence corresponds to the dark sectors of the differentially scattered light intensity pattern. But under the second condition

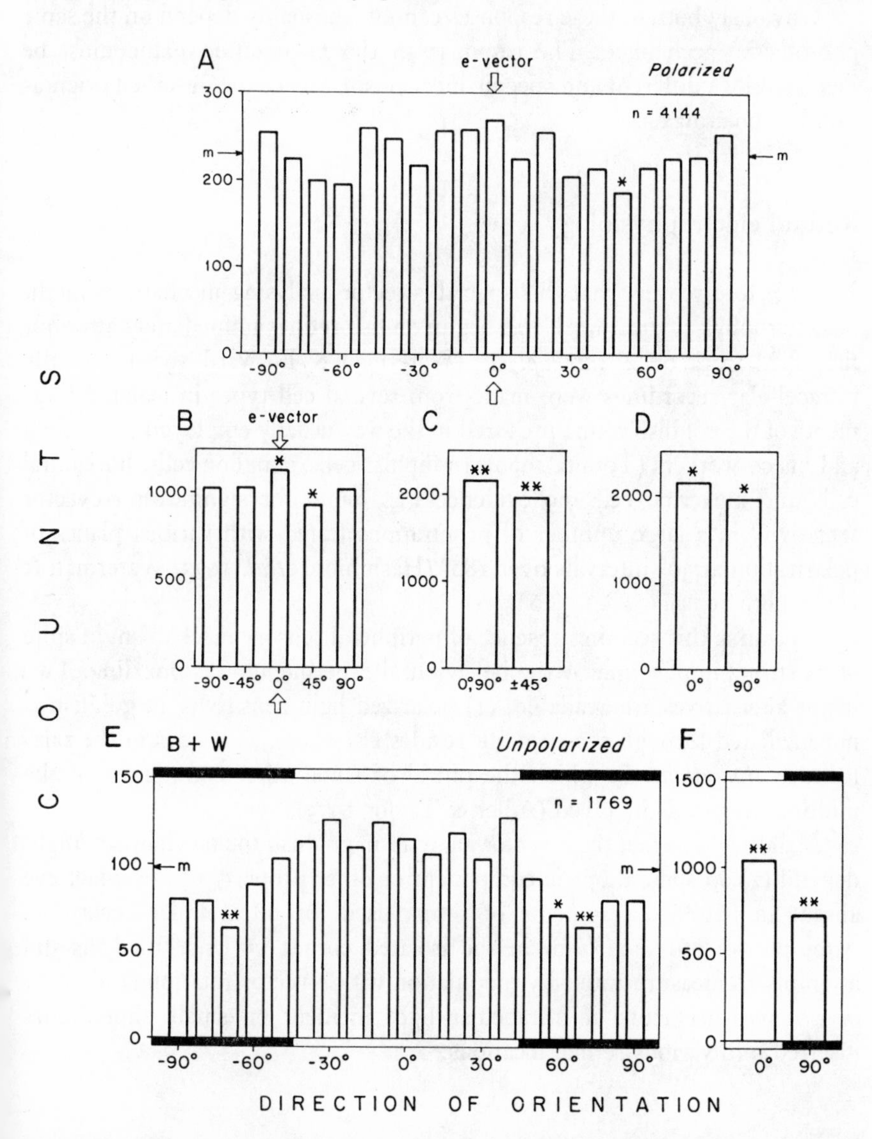

Figure 13.7. Comparison of polarotaxis (A–D) and phototaxis (E, F) in the spontaneous orientation of the teleost *Dermogenys*. A-D. Overhead light plane polarized at 0°, surrounding screen uniform white. Significant orientation parallel to the *e*-vector which lies in the directions of minimum intensity for horizontally scattered light. E, F. Overhead light not polarized, surrounding screen divided into alternating black and white quadrants. Strong orientation favouring the directions of maximum horizontal light intensity. Modified from Forward & Waterman 1973.

Dermogenys shows a clear positive phototactic response (Forward & Waterman 1973) by preferentially orienting toward the white sectors ($o_W = 18\%$) (Fig. 13.8).

Obviously both of these responses cannot reasonably depend on the same phototactic mechanism. The response to the polarization plane must be mediated by a different and specific mechanism. Hence the observed orientation is a polarotaxis.

Retinal electrophysiology

In order to discover what this special *e*-vector analysing mechanism might be electrophysiological work was begun on fish retinas immediately after our first field work on *Zenarchopterus* (Waterman & Forward 1970). Initially intracellular recordings were made from several cell types in isolated fragments of the goldfish retina prepared in the way usually employed by Tomita and his co-workers (Tomita 1968). In bipolar cells, ganglion cells, horizontal cells and amacrine cells no evidence was found for significant *e*-evector sensitivity in a large number of penetrations tested with various planes of polarization at 30° intervals over 180° (Hashimoto *et al.* 1973, Waterman & Hashimoto 1974).

Of course this seeming absence of peripheral sensory mechanism in spite of its strong implication by the behavioural experiments was puzzling. Two major alternatives are available: (1) polarized light sensitivity in goldfish is not mediated through the eyes but resides elsewhere, as it does in the salamander *Ambystoma* in which the pineal or some other component of the midbrain region is involved (Adler & Taylor 1973).

If the eye is in fact the *e*-vector analyser in goldfish the mechanism might depend (2) on some dioptric component or other property of the intact eye absent in flattened fragments of retina used for intracellular recording. Attempts to find dichroism in the isolated cornea or lens failed, as did attempts to measure intensity modulation with *e*-vector rotation studied by *in situ* measurements of intraocular light incident on small photodiodes inserted into various retinal locations.

Optic tectum electrophysiology

Finally we decided to test units in the optic tectum for their sensitivity to *e*-vector direction. Despite our failure to find any peripheral signs of sensitivity practically every tectal unit we recorded showed some differential responses to the orientation of the stimulus polarization plane (Waterman &

Hashimoto 1974). So far we have not attempted to identify the cells from which we were recording. Instead any units from which we could pick up good spike responses were explored for evidence of *e*-vector discrimination in the visual system. Their specific identity and their role in processing retinal input for this parameter remain for further study.

Before summarizing these results the methodology will be briefly described (Fig. 13.8). Adult comet goldfish 15–20 cm in length were used,

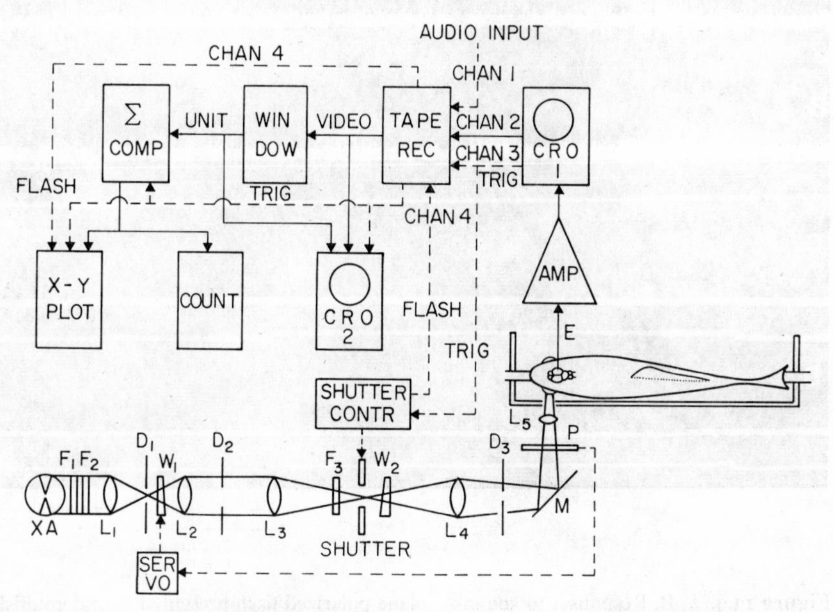

Figure 13.8. Experimental setup used to study single units in the goldfish optic tectum responding to retinal stimulation with plane polarized light. See text for general description. The light source is a xenon arc controlled by filters (F), lenses (L), diaphragms (D), neutral wedges (W), a mirror (M) and a plane polarizer (P). The spike responses of single tectal units (Fig. 13.9.) are ultimately displayed on an X-Y printout as sums of 8–10 individual flash responses. Waterman & Hashimoto 1974.

anesthetized initially with MS 222 and relaxed throughout with tubocurarine-Forced ventilation was maintained with aerated water. Light flashes 500 m. sec in duration and comprising a large beam focused on and filling the pupil were presented to one eye and unit activity was recorded extracellularly with tapered tungsten electrodes (0·5–1·0 μm tip diameter) inserted into the contralateral optic tectum.

The flashes were plane polarized with a Polaroid NK36 filter; the *e*-vector direction was changed in random order in 30° steps from 0° to 1°; zero reference for the plane of polarization was chosen to correspond with the anterior pole of the anterior-posterior axis; 90° was dorsal. Overall stimulus intensity was adjusted to compensate automatically for intensity changes caused inadvertently by some stray polarization in the light source, the mirror or other components. The reference light level was $1 \cdot 5 \times 10^{10}$ quanta/$mm^{-2}sec^{-1}$ and could be attenuated with neutral filters over a range of 4 or more log units.

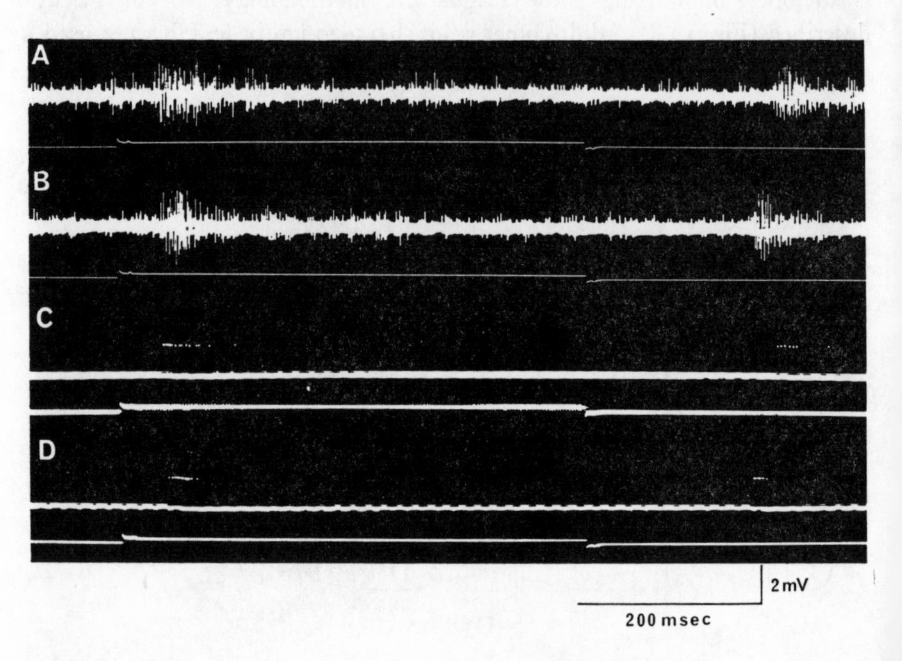

Figure 13.9. A, B. Responses to 500 msec plane polarized flash presented to the goldfish eye and recorded extracellularly in the heterolateral optic tectum. Observed on CRO 1 of Fig. 13.8. Maximum responses were found with the *e*-vector oriented at 0° (parallel to the longitudinal body axis) (A), the minimum occurred at 90°, parallel to the dorso-ventral axis (B). C, D. Responses of one of the units included in the records above selected by an appropriate window. Observed on CRO 2 of Fig. 13.8. The unit is obviously an on-off element with the same *e*-vector sensitivity as in the overall responses A, B. From Waterman & Hashimoto 1974.

Spike responses from the tectum were amplified, monitored on a CRO and recorded in one channel of a tape recorder. Trigger signals for synchronizing the stimulus and recorder time bases were put into a second channel while a third provided an audio record of protocol details. Shutter timing was recorded in channel 4. Typically the responses to 8–10 flashes were taken under each experimental condition.

In analysing the data the response of the best unit in the primary record of several usually picked up by the electrode (Fig. 13.9A) was isolated with a 'window' (Fig. 13.9B) and its responses to 8 or more similar flashes were summated with a Fabritek signal averager. These totals were graphed with an X–Y plotter (Fig. 13.10) and spike counts of the total response or its components were made electronically.

A survey of the characteristics of about 60 tectal units in the first experimental series yields the following main facts. All or nearly all of the units recorded showed some significant *e*-vector sensitivity; in many this orientation discrimination was substantial. In general the number of spikes in the response varied sinusoidally with *e*-vector rotation and this function had a period of 180° (Fig. 13.11).

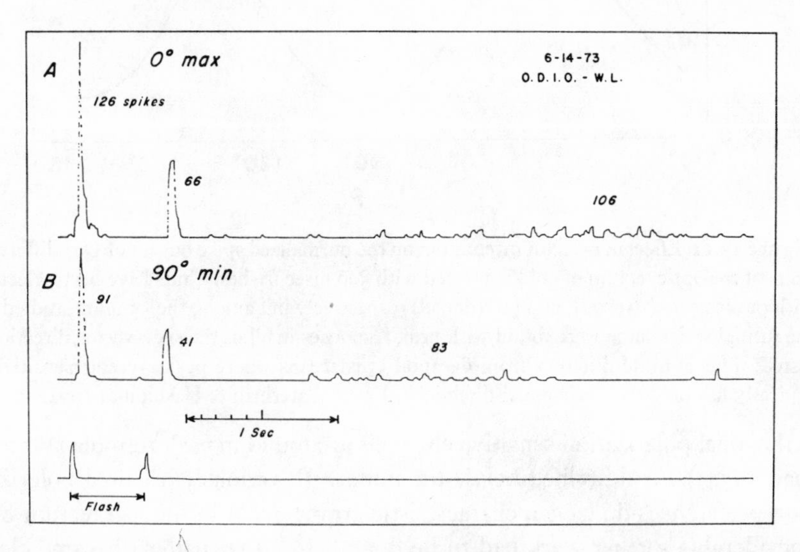

Figure 13.10. X–Y printout of the summed spike responses to eight 500 msec flashes by the same single goldfish tectal unit shown in Fig. 13.9C, D. In addition to the on-off responses there are delayed off responses. Both the former are significantly greater at 0°; the delayed off appears greater in A but the χ^2 is not significant at the 5% level. The delay of the delayed off appears longer at 0°. In this unit then excitatory effects of the optimum plane of polarization on the on and off components may be shared by the delayed off response. The latter appears to be suppressed for a longer period (inhibited) by the 'maximum' *e*-vector. From Waterman & Hashimoto 1974.

Different units had maximal responses in some particular directions for instance 0° and 180° (anterior-posterior) with a minimum at 90° and 270°. Others had maxima at 90° and 270° (dorsal-ventral) and minima at 0° and 180°. Review of the *e*-vector directions evoking maximal responses shows that all tested directions (0°, 30°, 60°, 90°, 120° and 150°) had cells responding maximally to one of these and minimally to the orthogonal direction.

Obviously this situation is quite different from the typical case in many decapod crustaceans and insects, as well as in cephalopods, where only two

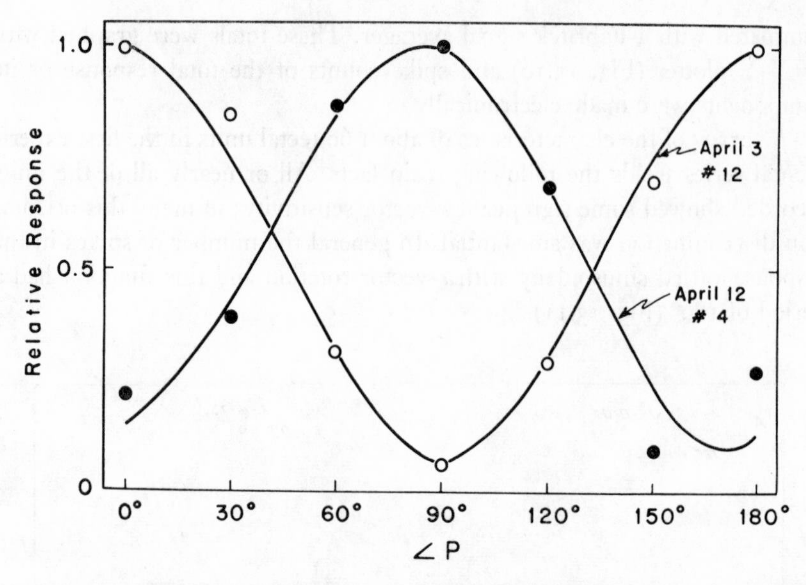

Figure 13.11. Effect of *e*-vector orientation on the normalized spike output of two different units of the optic tectum of goldfish tested with 500 msec flashes. Units have been selected with peaks near 0° (rostral) and 90° (dorsal) respectively but among the 55 units studied in the intitial series some were found with peak responses in all of the six *e*-vector directions tested. This is quite different from decapod crustaceans where peak *e*-vector sensitivity typically has only two orthogonal directions. From Waterman & Hashimoto 1974.

orthogonal polarization sensitive channels are found in each retinula (Waterman 1974a). As described below the units with variously oriented polarization preferences do have a characteristic arrangement in the fish tectum but considerable further work had to be done before that pattern became clear to us.

In the first series of cells studied (Waterman & Hashimoto 1974) no evidence was found for localization of *e*-vector discrimination either at some particular depth in the tectum or in any restricted area. Thus in all locations tested, where spike responses were recorded to the stimulating flash, there was evidence for significant *e*-vector sensitivity. We will see below, however, that there is a basic areal pattern of such sensitivity but again this did not emerge until our later experimental series.

Tectal unit types

Looking more closely at the original data shows that beneath the generalities reported above there are a number of interesting details. For example, although all were differentially affected by *e*-vector orientation 47 tectal units from 28 different goldfish had rather distinct characteristics as follows.

Under the conditions of test five responded with just 'on' responses, one gave stronger on than 'off' elements, 31s howed approximately balanced on-off discharges, while seven had stronger off than on components and four were pure off. Thus 81% of the units tested showed an on-off type response.

In some cells the on component had just an initial transient peak but in others this was followed by a sustained plateau response. The sustained response could also occur without obvious transients. Likewise the off element often consisted of two components, an early brief typical off response followed by a long delayed (500 msec or more) sustained discharge. The latter in turn sometimes showed several identifiable subcomponents. In some instances units with marked delayed off response showed remarkably high sensitivity to *e*-vector orientation.

The reason for mentioning these different response patterns here is that various components of the total discharge pattern reacted differently to *e*-vector orientation. Thus both on and off elements sometimes showed the same *e*-vector maxima and minima. Yet commonly the maxima for on and off differed from one another by 90° (Fig. 13.12). In such a case the most spikes in the on response would be at say 90° (where the off response was at its minimum) and the fewest in the on at 0° (where the off spikes were most numerous).

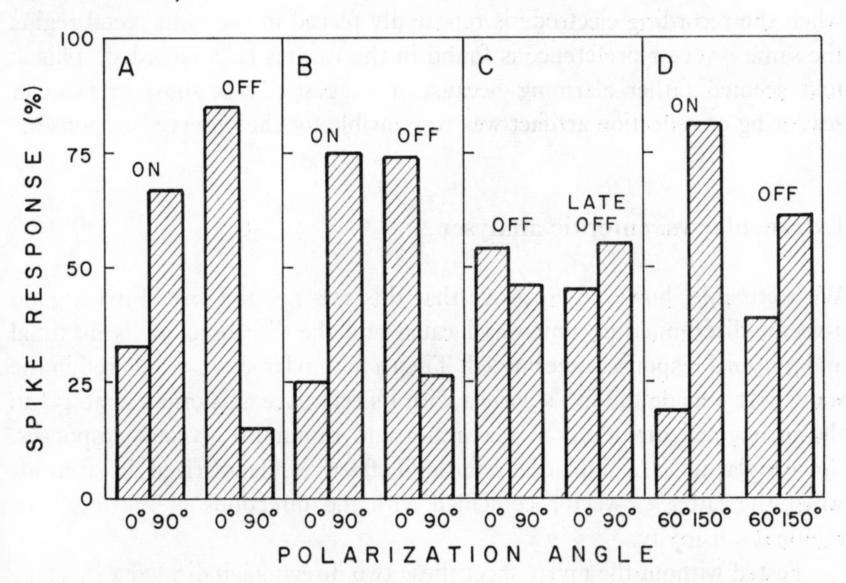

Figure 13.12. Some variations in *e*-vector response of selected goldfish tectal units. The maximum and minimum are often reversed for the on and off components of the response (A, B) but they may be the same (C). The effects of *e*-vector direction on components may be symmetrical (B, C) or not (A, D). Other components than the on, off may be involved (C). The responses are plotted as normalized spike counts in the components concerned for the four units. Data from Waterman & Hashimoto 1974.

A somewhat analogous situation was observed in colour (C-) cells which showed differential wavelength responses which we tested in the green (533 nm) and red (633 nm). Ten such cells were studied in all, five of them were red on, green off units and five were green on, red off units. Four showed orthogonal antagonistic preferences for *e*-vector directions. But in two units both the red and green sensitive responses were modulated in parallel by rotating the plane of polarization. The remaining four units did not demonstrate consistent differential effects of *e*-vector orientation.

Tectal location and PLS

Further insight into the possible mechanism involved in the tectal cells' PLS described above was provided by a subsequent series of experiments (Waterman & Aoki 1974). In these *e*-vector discrimination was studied systematically as a function of tectal cell location and hence retinal position of the sensory units activated (Fig. 13.13C). Fortunately good maps of retinal projection onto the tectum are well known in the goldfish (Jacobson & Gaze 1964, Schwassmann & Kruger 1965).

Our interest in this kind of study was provoked by the observation that when the recording electrode is repeatedly placed in the same tectal region the same *e*-vector preference is found in the various cells recorded. This at first seemed rather alarming because it suggested that some extraocular scattering or reflection artifact was responsible for the observed responses.

Extraocular vs. entoptic analyser

We were able, however, to prove that this was not the case. First a good *e*-vector discriminating unit was located and the directions of its maximal and minimal responses determined. Then a flat mylar sheet was placed in the water just outside the fish's cornea with its reference axis oriented at 45° to the orthogonal directions of maximum and minimum *e*-vector responses. Since mylar acts as a high order retarder it effectively depolarized the stimulus when the latter's *e*-vector coincided with the directions of maximum or minimal sensitivity.

Tested without the mylar sheet these two directions indicated a substantial PLS but with this depolarizer in place the responses to orthogonal *e*-vector orientations were not significantly different. Thus the depolarizer next to the cornea abolished PLS but could not have done so by affecting any extraocular intensity patterns due to differential reflection or scattering of the polarized stimulus by elements outside the eye. Hence we can conclude that the fish's *e*-vector analyser is indeed ocular as we had been assuming.

But we were still faced with the curious fact that *e*-vector preference seemed coupled to the position of the recording electrode in the tectum. This perhaps contradicted the results of earlier experiments in which maximum *e*-vector response in different cells occurred in various planes of polarization

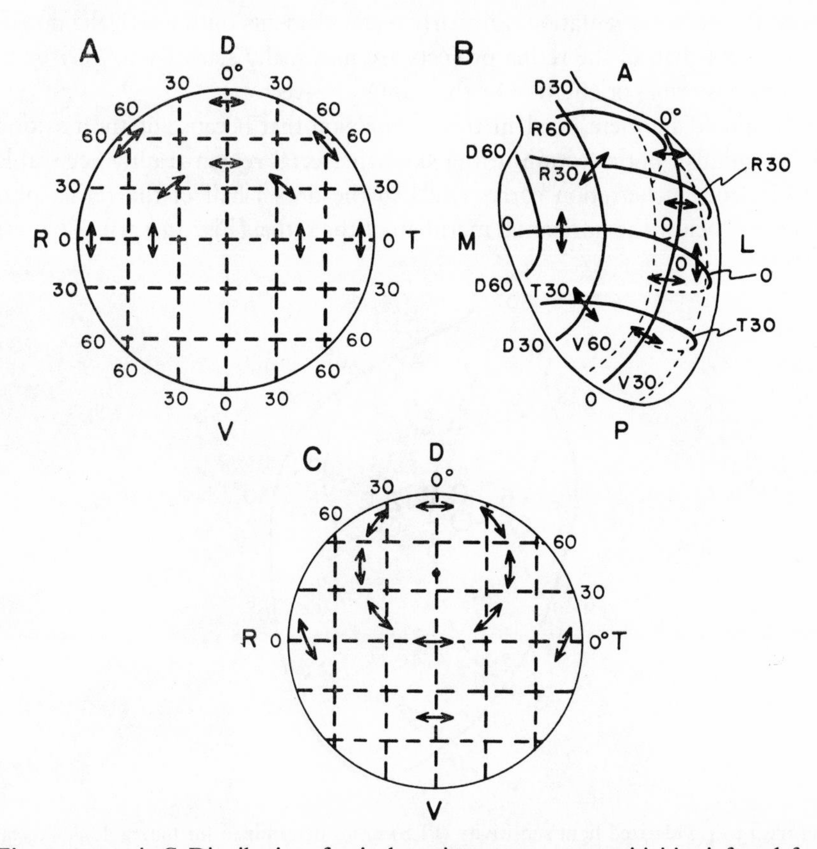

Figure 13.13. A, C. Distribution of retinal maximum *e*-vector sensitivities inferred from single unit responses recorded in the optic tectum. B. Double headed arrows show diagrammatic distribution of maximally effective *e*-vector directions superimposed on a tectal projection map determined by Schwassmann and Kruger. In A and B the axis of the stimulating beam coincided with the eye's optic axis (0, 0 in the projection). In C the retinal pattern was derived from tectal measurements where the stimulus axis was directed upward toward a point mid-dorsal 45° above the eye's optic axis. From Waterman & Aoki 1974.

around 180°. This apparent anomaly was resolved when *e*-vector orientation for maximum response was correlated specifically with the retinal coordinates of the tectal region recorded.

Pattern of *e*-vector discrimination

A relatively simple systematic relationship exists between these two factors. Thus tectal units along the projection line corresponding to the horizontal meridian of the retina are always maximally sensitive to vertical (dorso-ventral) *e*-vector orientation. Similarly tectal elements onto which the dorso-ventral meridian of the retina projects are maximally sensitive to horizontal (antero-posterior) orientation of the stimulus *e*-vector.

It should be mentioned in this connection that because of the position of the goldfish optic tectum in the skull the tectal region readily accessible for electrode penetration corresponds to the dorsal half of the retina plus about 15° below the horizontal meridian of the retina (Fig. 13.13B). The rest

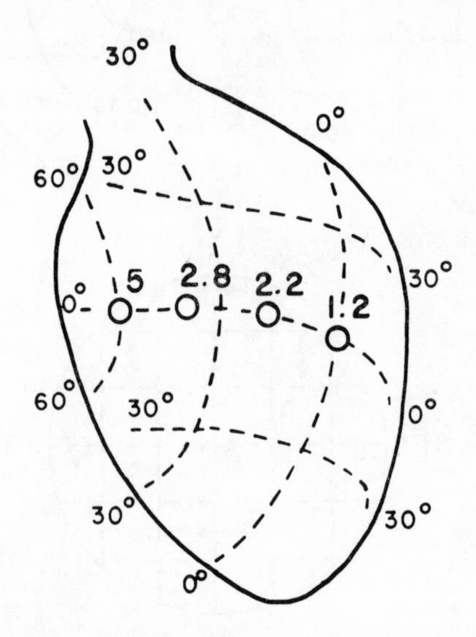

Figure 13.14. Polarized light sensitivity (PLS) ratios determined for four goldfish tectal units located along the projection of the mid-dorsal retinal radius. Sensitivity to *e*-vector direction was minimal near the centre of the retina (0, 0 on the tectal coordinates) and increased progressively towards the periphery. As usual for a plane polarized stimulus directed along the eye's optic axis all four units showed their maximum response to 0° (horizontal) *e*-vector (Fig. 13.13B). From Waterman & Aoki 1974.

of the projection area for the ventral half of the retina curves down and under the accessible part. Some exploratory forays into this lower region have yielded results like those for the dorsal area but we have not yet made detailed or systematic recordings.

In areas between the dorso-ventral and horizontal meridians of the retina maximum responses occur at oblique angles intermediate to those found along the vertical and horizontal coordinates. When all these data are plotted onto the retina it appears that maximum responses to the plane of polarization are everywhere tangential to a polar coordinate system centred on the eye's optic axis (Figs. 13.13A & B). This implies that two entoptic light sectors perpendicular to the stimulus *e*-vector should be visible to the fish in a pattern resembling stationary Boehm's brushes but having substantially greater angular extent. No signs of discrete channels for particular polarization planes were evident.

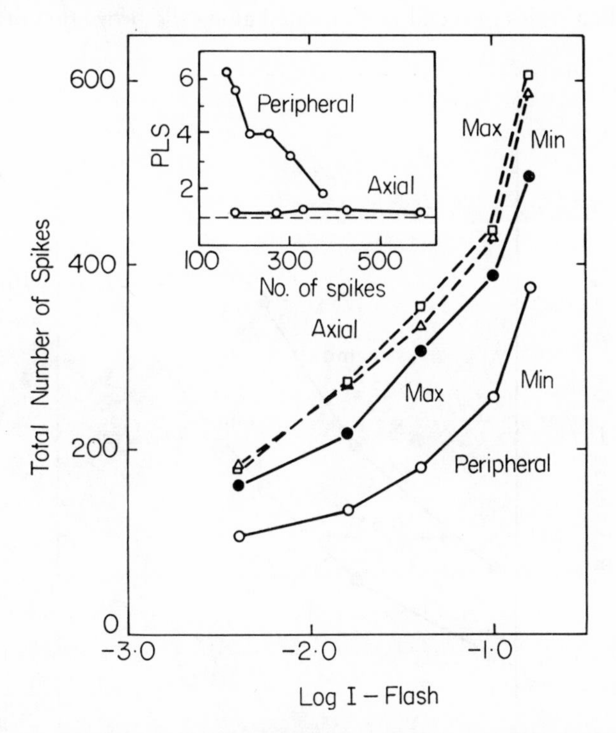

Figure 13.15. Comparison of PLS in axial and peripheral retinal units in the goldfish (inferred from tectal responses). The intensity response (IR) curves for the axial unit show that it responded somewhat more strongly at all intensities than the peripheral one. Measured from intensity differences required to produce equal responses PLS was small for the axial element at all intensities. As shown by the insert the sensitivity ratio of the peripheral element reached 6 at low response levels but decreased progressively at higher response levels evoked by increasing stimulus intensities. From Waterman & Aoki 1974.

Sensitivity pattern

In addition to this directional discrimination pattern our data show that there is a circular distribution of the degree of PLS with a centrally located minimum and peripheral maxima surrounding it in all directions. With axial illumination used in all the experiments so far described PLS is minimal or negligible at the intersection of the horizontal and vertical meridians of the retina. Indeed one would anticipate that zero *e*-vector discrimination must occur in the centre of the retina where the horizontal preference along the vertical meridian and the vertical preference of the horizontal meridian cross over.

In addition PLS increases steeply as one moves radially out from the polar coordinate origin to locations in the retina 30–60° away from the optic axis. Thus in a series of tectal units located along the projection of the mid-

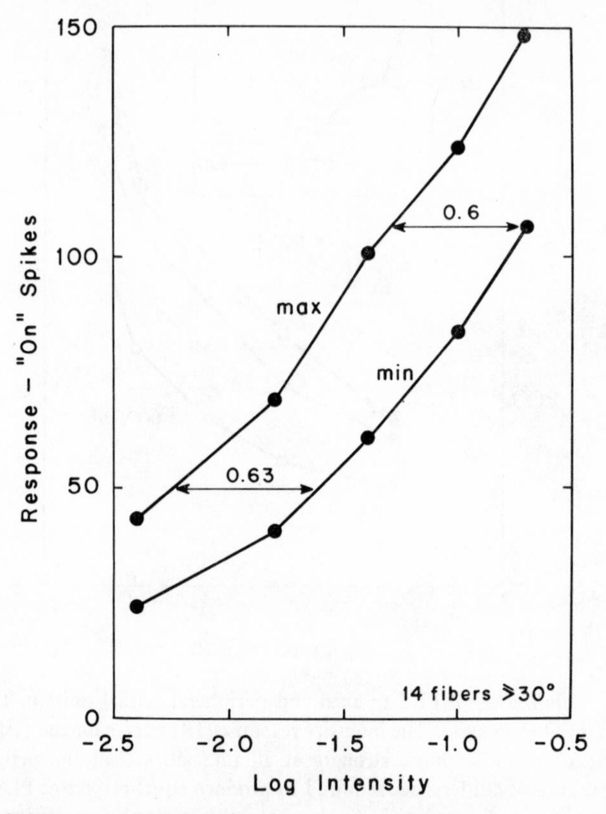

Figure 13.16. Average intensity response curves for 14 goldfish tectal units whose location corresponded to peripheral retinal locations 30° or more from the centre. Here the sensitivity ratios for the on spike responses were about the same over the intensity range tested and equalled about 4. From Waterman & Aoki 1974.

dorsal radius of the retina the one nearest the centre had the lowest PLS ratio, 1·2. The value for units at increasing angular distances from this were 2·2 at 20° dorsal, 2·8 at 40° and finally 5 at 60° (Fig. 13.14). Ordinarily PLS ratios observed peripherally in other preparations reached values between 6 and 7 (Figs. 13.15 & 13.16). Occasionally, however, units of much greater PLS have been recorded. These have not yet been related to the above *e*-

vector response pattern which is the one usually found. As mentioned above very high PLS's have been observed mainly when a strong delayed response was present which usually occurred at high light intensities.

So far we have found little evidence that overall PLS in the goldfish is dependent on wavelength or on the state of light and dark adaptation. Thus IR curves for a tectal unit studied in strongly dark adapted condition are closely superimposable for stimuli at 460 nm, 540 nm and 620 nm (Fig. 13.17). PLS is about the same over the 2–3 log units of intensity tested. Note also

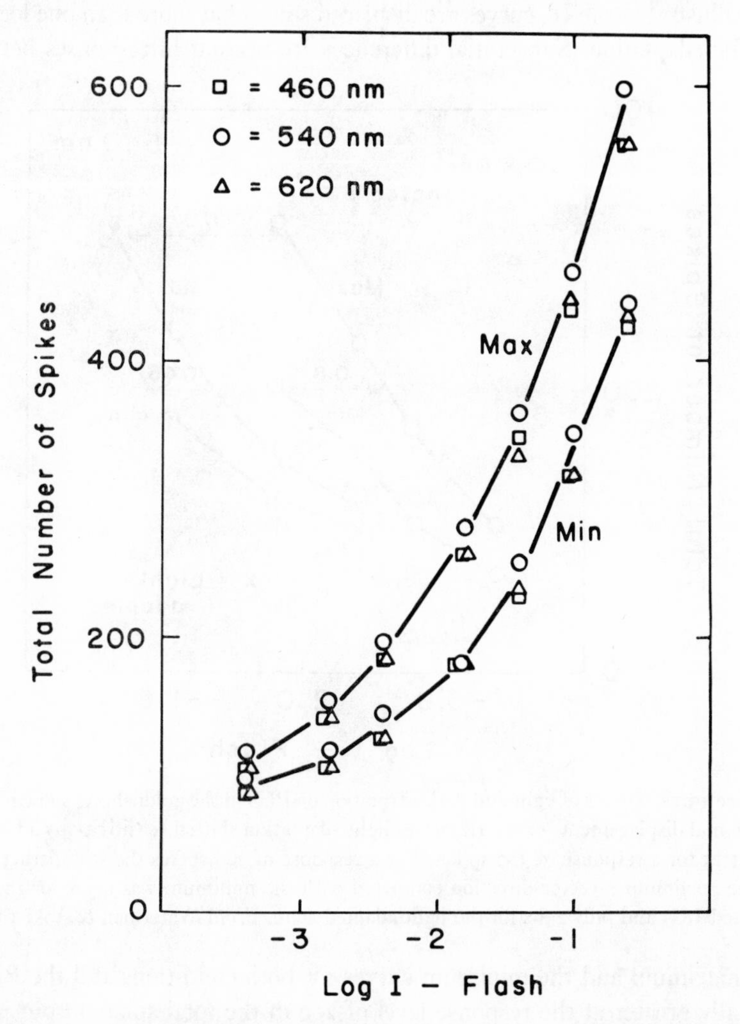

Figure 13.17. Lack of specific wavelength effects on PLS in a goldfish optic tectal unit. Intensity response curves for blue, green and red narrow band stimuli closely superimpose at both maximum and minimum *e*-vector directions. The PLS ratio is relatively stable over the intensity range tested. This unit was studied in a strongly dark adapted state. From Waterman & Aoki 1974.

that PLS is clearly present in the dark adapted eye which means that Snyder's hypothesis (1973) suggested to account for the behavioural polarotaxis we reported for *Zenarchopterus* (Waterman & Forward 1970, 1972) and dependent on a high degree of light adaptation would not appear to be valid for the goldfish.

Further evidence that *e*-vector discrimination occurs in both the light and dark adapted state can be found in many of our other experiments where the same unit was studied under both conditions (e.g. Fig. 13.18). In the case illustrated the IR curves are displaced somewhat more than one log unit by the adaptation. Substantial differences are present in responses between

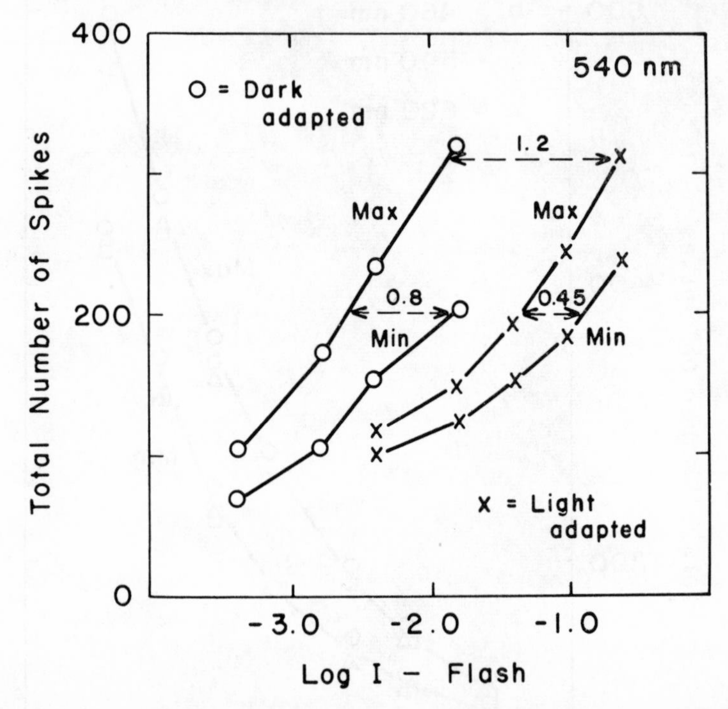

Figure 13.18. Effects of light and dark adaptation on PLS in the goldfish. As shown by the horizontal displacements of the IR curves light adaptation shifted sensitivity by a factor of about 15 for a response of 300 spikes. For a response of 200 spikes the stimulating effect of the maximum *e*-vector direction compared with the minimum was 6·3 x for the dark adapted IR's and only 2·8 x for the light adapted state. From Waterman & Aoki 1974.

the maximum and the minimum curves for both conditions and the PLS is actually greater at the response level of 200 in the total spike output in the dark adapted as compared to the light adapted state.

Whatever the mechanism of the analyser involved in the goldfish eye the PLS pattern we have generally observed could be due to different inherent properties of the central compared with the peripheral parts of the retina. Or it could be imposed by the particular geometry of our stimulating beam whose axis was directed primarily onto the central retina. In fact by altering the direction of our stimulus we have found that the pattern is actually correlated with stimulus direction.

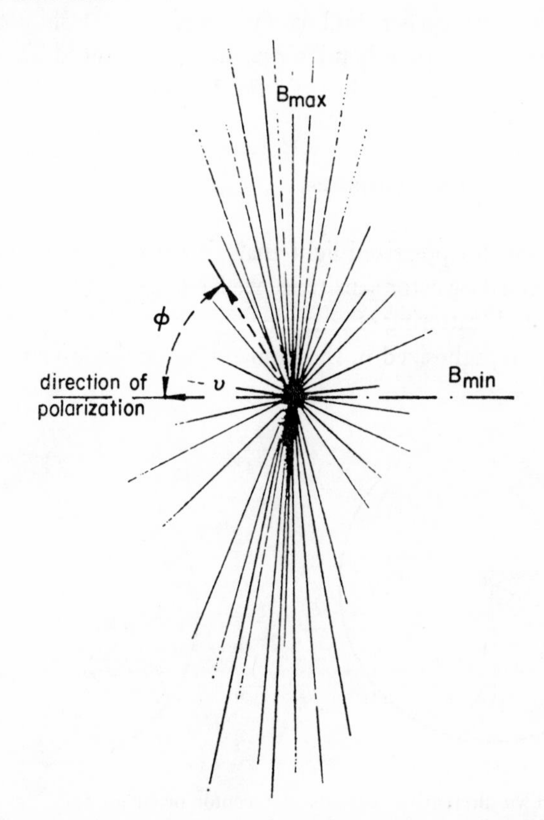

Figure 13.19. Diagram of Boehm's brushes, an entoptic image visible in the human eye when a small bright plane polarized light is viewed parafoveally. It can only be seen when the *e*-vector is rotating and optimally at a rate around 1 Hz. Two opposite sectors perpendicular to the polarization plane appear light and the rest of the field is dark in sectors parallel to the *e*-vector. There are interesting similarities and differences between Boehm's brushes and our results in the goldfish. Both are peripheral entoptic images with light sectors perpendicular to the *e*-vector. Both are nonchromatic. In contrast Haidinger's brushes are coloured and foveal in location. From Vos & Bouman 1964.

To test this we shifted the test beam mid-dorsally so that its direction made 45° and 60° angles with the optic axis of eye. When this was done the PLS pattern showed equivalent shifts (Fig. 13.13C). The tangential pattern of maximum sensitivity directions now centred around the respective stimulus centres and not the eye centre as it did in all experiments with the beam direction axial (Figs. 13.13A & B).

Similarly the degree of *e*-vector sensitivity also moved with the beam direction. Near the beam axis (45° or 60° to the retina's centre) PLS was minimum and increased in all radial directions from the retinal position of the stimulus axis. This proves that the pattern is not inherent in specific central or peripheral location of the retinal receptors concerned. This is consistent with our earlier finding (Waterman & Hashimoto 1974) that retinal cells from the peripheral retina did not show different properties *vis à vis e*-vector orientation than did fundic units.

Possible analyser mechanisms

Two mechanisms for polarized light analysis are suggested by the tangential pattern of maximal *e*-vector sensitivity we have observed. One would depend on the differential scattering of polarized light by ocular elements (Fig. 13.19) similar to that hypothesized in the human eye for Boehm's brushes (Boehm

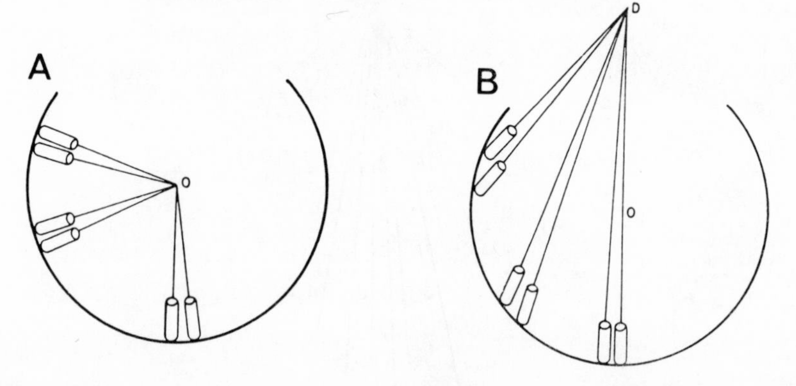

Figure 13.20. Two alternative patterns of receptor orientation in the camera eye. A. Optic axes of receptors are radial so that the photoreceptors are looking at the centre of the eyeball. B. Optic axes of receptors show differential graded orientation so that they are looking at a point or plane just outside the eye in its optical axis. All eyes studied so far in this respect show the second pattern (g.d.o.). From Laties & Enoch 1971.

1940, Vos & Bouman 1964). The other would depend on the dichroism of the outer segments of the photoreceptor cells to oblique and transverse illumination.

If differential scattering in the retina and elsewhere in the eye is responsible for *e*-vector discrimination, maximal responses should occur perpendicular to the *e*-vector as they do. But on the basis of measurements in the human eye the intensity of scattered light should be maximum close to the stimulus axis and fall off rather steeply with distance from this (Miller & Benedek 1973). Hence *e*-vector discrimination due to differential scatter should diminish with distance from the stimulus axis, becoming negligible beyond 10–20°. Instead we have found the reverse pattern with PLS increasing from a minimum axially to a peripheral maximum as much as 50° to 60° from the retinal centre.

On the other hand this aspect of the centrifugally augmenting pattern is compatible with the second of the two hypotheses above, namely that oblique incidence of the light into receptor outer segments is involved in the analyser. For example, if the outer segments of the fish's rods and cones are strictly radially oriented in the nearly spherical eyeball, such an *e*-vector sensitivity pattern might be expected (Fig. 13.20A).

Obviously all light entering the camera eye emanates as from some axial point (or plane) in front of the organ. Thus retinal units in the fundus will receive axial or near axial illumination if their axes are radially oriented. However, radially arranged receptor elements will receive increasingly oblique illuminations as their location becomes more peripheral (Laties & Enoch 1971).

Pari passu one would expect PLS to increase to a maximum for transverse illumination determined by the orientation of visual pigment chromophores. Measured dichroic ratios for transverse absorption have been reported with maximum values between 4 and 6 (e.g., Wald *et al.* 1962, Liebman 1962) but recent figures for the goldfish are less: rods 1–2:1; cones 2–3:1 (Hárosi & MacNichol 1974). In addition if such a radial receptor geometry actually exists the overall pattern of PLS should show maximal responses in sectors perpendicular to the *e*-vector as we have observed in the goldfish optic tectum.

Conflicting evidence

However, our experiments with oblique light appear inconsistent with a mechanism dependent on radial outer segment orientation in the peripheral retina. If such a receptor geometry were present one would expect at 45–60° from the retinal centre that outer segments oblique to direct rays striking them would show *e*-vector sensitivity. Hence the axial region of an oblique beam should have significant *e*-vector sensitivity.

Also since the radial receptor pattern has its origin at the centre of the eyeball and the stimulating light originates considerably peripheral to this centre the maximum *e*-vector sensitivity of units around an oblique beam axis should show a distorted asymmetrical pattern. However, neither of these expectations is fulfilled by our data.

More direct evidence against a radial outer segment geometry is provided by the fact that all vertebrates eyes so far studied show differential orientation of the photoreceptors (Fig. 13.20B). This graduated differential orientation (g.d.o.) is marked in the periphery by tilt away from radial directions so that the receptor elements there also receive axial rather than oblique primary illumination (Laties & Enoch 1971). Indeed these authors state that g.d.o. is present in the goldfish although no data have been published.

If then the peripheral as well as central receptor units in the goldfish are all oriented so that they effectively receive axial primary illumination they would be isotropic for such light. But if some scattering mechanism causes stray light to enter the receptor outer segments obliquely or transversely to a degree which increases with angular distance from the axis of the stimulating beam, then our data so far might be accounted for. Some high sensitivity ratios which greatly exceed the known dichroic ratios for vertebrate outer segments cited above might seem to be an exception. However, a somewhat similar anomaly has been found in arthropod eyes where microspectrophotometrically determined dichroic ratios are two or less. Yet electrophysiologically measured sensitivities can reach 10 or more.

Conclusions and summary

Growing evidence indicates that a variety of vertebrates, especially fishes, have an intraocular *e*-vector analyser. Since receptor outer segments are isotropic to light traversing their optic axis and since intracellular recordings from isolated retinal fragments yield no sensitivity to the plane of polarization, the perceptive mechanism must be quite different from that in rhabdom-bearing eyes. There the dichroism of the visual pigment coupled with the fine structural geometry of the receptor membrane provides differential axial absorption of plane polarized light by individual retinular cells. The direction of preferential absorption by a single retinular cell is fixed by the orientation of the microvilli in its rhabdomere. This has been established by selective adaptation, electrophysiological recordings and direct, microspectrophotometry.

Our work on the goldfish optic tectum supports the previous evidence that vertebrate retinal cells do not detect *e*-vector orientation of axially absorbed light. Nevertheless most if not all tectal cells recorded by us show PLS under the proper conditions. Analysis of the PLS patterns involved shows that a large entoptic image must be formed on the retina in linearly polarized light. This image comprises two opposite dark sectors parallel to the *e*-vector and two lighter sectors perpendicular to it.

Unlike Haidinger's brushes which are small ($< 5°$) coloured and foveal in location, this image is large ($> 50°$) achromatic and weak or absent axially. It is centred around the axis of the light stimulus rather than the eye's optic axis. Although it thus bears some resemblance to Boehm's brushes in human vision the latter are only visible while the *e*-vector is rotating about once a second and when fixation is parafoveal.

The PLS pattern in the fish retina reflects tangentially oriented preferential absorption around the stimulus axis. The preferred direction for an individual retinal cell changes with its location relative to the stimulus axis. Mechanisms dependent on selective intraocular scattering or differential oblique absorption by receptor cell outer segments would seem adequate to account for the known facts. But in our present state of knowledge it seems too early to press hard for the definitive mechanism.

References

Adler K. & Taylor D.H. (1973) Extraocular perception of polarized light by orienting salamanders. *J. Comp. Physiol.* 87, 203–12.

Autrum H. & Stumpf H. (1950) Das Bienenauge als Analysator für polarisiertes Licht. *Z. Naturforsch.* 5b, 116–22.

Boehm G. (1940) Über ein neues entoptisches Phänomen im polarisierten Licht. 'Periphere' Polarisationsbüschel. *Acta Ophthalmol.* 18, 143–69.

Coulson K.L. (1974) The polarization of light in the environment. In *Planets, Stars and Nebulae Studied with Photopolarimetry* (T. Gehrels, ed.), pp. 444–71. University of Arizona Press, Tucson.

Eguchi E. & Waterman T.H. (1973) Orthogonal microvillus pattern in the eighth rhabdomere of the rock crab *Grapsus. Z. Zellforsch.* 137, 145–57.

Eguchi E., Waterman T.H. & Akiyama J. (1973) Localization of the violet and yellow receptor cells in the crayfish retinula. *J. Gen. Physiol.* 62, 355–74.

Forward R. B., Jr., Horch K.W. & Waterman T.H. (1972) Visual orientation at the water surface by the teleost *Zenarchopterus. Biol. Bull.* 143, 112–26.

Forward R.B., Jr. & Waterman T.H. (1973) Evidence for *e*-vector and light intensity pattern discrimination by the teleost *Dermogenys. J. Comp. Physiol.* 87, 189–202.

Frisch K. von (1948) Gelöste und ungelöste Rätsel der Bienensprache. *Naturwiss.* 35, 38–43.

Frisch K. von (1965) *Tanzsprache und Orientierung der Bienen.* 578 pp. Springer-Verlag. Berlin, Heidelberg, New York.

Hagins W.A. & Liebman P.A. (1963) The relationship between photochemical and electrical processes in living squid photoreceptors. Abstracts of the Biophysical Society 7th Annual Meeting, New York, N.Y. ME 6.

Haidinger W. (1844) Ueber das direckte Erkennen des polarisierten Lichts und der Lage der Polarisationsbene. *Ann. Physik. Chemie* 63, 29–39.

Hashimoto H., Aoki K. & Waterman T.H. (1973) Discrimination of *e*-vector direction by single units of the goldfish optic tectum. (Abstr.) *Amer. Zool.* 13, 1305.

Hárosi F.I. & MacNichol E.F., Jr. (1974) Visual pigments in goldfish cones. Spectral properties and dichroism. *J. Gen. Physiol.* 63, 279–304.

Jacobson M. & Gaze R.M. (1964) Types of visual response from single units in the optic tectum and optic nerve of the goldfish. *Quart. J. Exp. Physiol.* 49, 199–209.

Jander R. & Waterman T.H. (1960) Sensory discrimination between polarized light and light intensity patterns by arthropods. *J. Cell. Comp. Physiol.* 56, 137–60.

KLEEREKOPER H., MATIS J.H., TIMMS A.M. & GENSLER P. (1973) Locomotor response of the goldfish to polarized light and its *e*-vector. *J. Comp. Physiol.* 86, 27–36.

KREITHEN M.L. & KEETON W.T. (1974) Detection of polarized light by the homing pigeon, *Columba livia*. *J. Comp. Physiol.* 89, 83–92.

LANGER H. (1965) Nachweis dichroitischer Absorption des Sehfarbstoffes in den Rhabdomeren des Insektenauges. *Z. Vergl. Physiol.* **51**, 258–63.

LATIES A.M. & ENOCH J.M. (1971) An analysis of retinal receptor orientation. I. Angular relationship of neighbouring photoreceptors. *Investig. Opthal.* **10**, 69–77.

LeGRAND Y. (1936) Sur deux propriétés des sources de lumière polarisée. *Comp. Rend. Acad. Sci. Paris* **202**, 939–41.

LeGRAND Y. (1937) Recherches sur la diffusion de la lumière dans l'oeil humain. *Rev. d'Optique* 16, (6–7), 201–14.

LIEBMAN P.A. (1962) *In situ* microspectrophotometric studies on the pigments of single retinal rods. *Biophys. J.* **2**, 161–78.

MAGNI F., PAPI F., SAVELY H.E. & TONGIORGI P. (1962) Electroretinographic responses to polarized light in the wolf-spider *Arctosa variana* C.L. Kock. *Experientia*, 18, 511.

MAGNI F., PAPI F., SAVELY H.E. & TONGIORGI P. (1965) Research on the structure and physiology of the eyes of a lycosid spider. III. Electroretinographic responses to polarized light. *Arch. Ital. Biol.* **103**, 146–58.

McCANN G.D. & ARNETT D.W. (1972) Spectral and polarization sensitivity of the dipteran visual system. *J. Gen. Physiol.* 59, 534–58.

MENZEL R. & SNYDER A.W. (1974) Polarized light detection in the bee, *Apis mellifera*. *J. Comp. Physiol.* 88, 247–70.

MEYER-ROCHOW V.B. (1974) Fine structural changes in dark-light adaptation in relation to unit studies of an insect compound eye with a crustacean-like rhabdom. *J. Insect Physiol.* **20**, 573–89.

MILLER D. & BENEDEK G. (1973) Intraocular Light Scattering. 121 pp. Thomas, Springfield, Illinois.

MULLER K.J. (1973) Photoreceptors in the crayfish compound eye: electrical interactions between cells as related to polarized light sensitivity. *J. Physiol.* **232**, 573–95.

PAPI F. & TONGIORGI P. (1963) Innate and learned components in the astronomical orientation of wolf spiders. *Ergeb. Biol.* **26**, 259–80.

SCHMIDT W.J. (1935) Doppelbrechung, Dichroismus und Feinbau des Aussengliedes der Sehzellen vom Frosch. *Z. Zellforsch.* **22**, 485–522.

SCHWASSMANN H.O. & KRUGER L. (1965) Organization of the visual projection upon the optic tectum of some freshwater fish. *J. Comp. Neurol.* **124**, 113–26.

SHAW S.R. (1969) Interreceptor coupling in ommatidia of drone honeybee and locust compound eyes. *Vision Res.* **9**, 999–1029.

SKRZIPEK K.-H. & SKRZIPEK H. (1973) Die Anordnung der Ommatidien in der Retina der Biene (*Apis mellifica* L.) *Z. Zellforsch.* **139**, 567–82.

SKRZIPEK K.-H. & SKRZIPEK H. (1974) The ninth retinula cell in the ommatidium of the worker honey bee (*Apis mellifica* L.) *Z. Zellforsch.* **147**, 589–93.

SMOLA U. & GEMPERLEIN R. (1972) Übertragungseigenschaften der Sehzelle der Schmeissfliege *Calliphora erythrocephala*. 2. Retina—Lamina ganglionaris. *J. Comp. Physiol.* 79, 363–92.

SNYDER A.W. (1973) How fish detect polarized light. *Investig. Ophthal.* **12**, 78–9.

STOCKHAMMER K. (1959) Die Orientierung nach der Schwingungsrichtung linear polarisierten Lichtes und ihre sinnesphysiologischen Grundlagen. *Ergeb. Biol.* **21**, 23–56.

TAYLOR D.H. & ADLER K. (1973) Spatial orientation by salamanders using plane-polarized light. *Science* 181, 285–7.

TOMITA T. (1968) Electrical response of single photoreceptors. *Proc. IEEE* 56, 1015–23.

TYLER J.E. (1974) Heuristic arguments for the pattern of polarization in deep ocean water. In *Planets, Stars and Nebulae Studied with Photopolarimetry* (T. Gehrels, ed.) pp. 434–43. University of Arizona Press, Tucson.

VOS J.J. & BOUMAN M.A. (1964) Contribution of the retina to entoptic scatter. *J. Opt. Soc. Amer.* 54, 95–100.

WALD G., BROWN P.K. & GIBBONS I.R. (1962) Visual excitation: a chemo-anatomical study. *Symp. Soc. Exp. Biol.* 16, 32–57.

WATERMAN T.H. (1960) Interaction of polarized light and turbidity in the orientation of *Daphnia* and *Mysidium*. *Z. Vergl. Physiol.* 43, 149–72.

WATERMAN T.H. (1972) Visual direction finding by fishes. In *Animal Orientation and Navigation* (S.R. Galler, K. Schmidt-Koenig, G.J. Jacobs and R.E. Belleville, eds.), pp. 437–56. National Aeronautics and Space Administration, Washington, D.C.

WATERMAN T.H. (1973) Responses to polarized light: Animals. In *Biology Data Book*, vol. II, 2nd edition (P.L. Altman and D.S. Dittmer, eds.), pp. 1272–89. Federation of American Societies for Experimental Biology, Bethesda, Maryland.

WATERMAN T.H. (1974a) Polarimeters in animals. In: 'Planets, Stars and Nebulae Studied with Photopolarimetry' (T. Gehrels, ed.), pp. 472–94. University of Arizona Press, Tucson, Arizona.

WATERMAN T.H. (1974b) Underwater light and the orientation of animals. In *Optical Aspects of Oceanography* (N.G. Jerlov and E. Steeman Nielsen, eds.), pp. 415–43. Academic Press, London (in press).

WATERMAN T.H. & AOKI K. (1974) *E*-vector sensitivity patterns in the goldfish optic tectum. *J. comp. Physiol.* 95, 13–27.

WATERMAN T.H., FERNÁNDEZ H.R. & GOLDSMITH T.H. (1969) Dichroism of photosensitive pigment in rhabdoms of the crayfish *Orconectes*. *J. Gen. Physiol.* 54, 415–32.

WATERMAN T.H. & FORWARD R.B., Jr. (1970) Field evidence for polarized light sensitivity in the fish *Zenarchopterus*. *Nature* 228, 85–7.

WATERMAN T.H. & FORWARD R.B., Jr. (1972) Field demonstration of polarotaxis in the fish *Zenarchopterus*. *J. Exp. Zool.* 180, 33–54.

WATERMAN T.H. & HASHIMOTO H. (1974) *E*-vector discrimination by the goldfish optic tectum. *J. comp. Physiol.* 95, 1–12.

Questions and discussion

During the discussion Professor Waterman pointed out that it was still necessary to demonstrate unequivocally that the natural pattern of polarization means something to the fish. In answer to a question from Dr Lythgoe he pointed out that the field experiments were performed in shallow water where reflection from the broken coral and sand bottom would somewhat reduce polarization but not upset its pattern.

14 The role of daylength in regulating the breeding seasons and distribution of wildfowl

R. K. MURTON & JANET KEAR, *N.E.R.C. Monks Wood Experimental Station, Huntingdon and The Wildfowl Trust, Slimbridge, Gloucester*

Introduction

Most birds inhabit environments in which the food supply alters seasonally, and often drastically. Each individual must take full advantage of favourable ecological conditions to reproduce, for selection will favour those that leave the most progeny; but they must also moult, and in some cases migrate. In order to ensure the optimum budgeting of seasonal resources, conflicting energy-demanding functions need to be spaced, and this spacing requires that birds have either an extremely accurate endogenous clock, which they do not, or else rely on an external timer. The most reliable of these is the seasonal change in daylength, and photoperiodic responses have, therefore, become the important proximate factors initiating nesting (Emmé 1960 Farner & Follett 1966, Farner 1970, Farner & Lewis 1971, Lofts & Murton 1968, Wolfson 1970). These responses are adapted to suit the ultimate ecological factors that determine when breeding is best attempted, a topic reviewed by Lack (1968), Lofts & Murton (1968) and Immelmann (1971). An adaptive feature of avian photoperiodism is the phenomenon of refractoriness, which is marked by the spontaneous regression of the pituitary-gonad axis following continued exposure to 'long' stimulatory days. Refractoriness can only be broken if the subject experiences a period of 'short' days whereupon it regains the capacity to respond to stimulatory photoperiods. A species from temperate latitudes, such as the Mallard *Anas platyrhynchos*, breeds in the spring, whereupon refractoriness to long summer days develops (Lofts & Coombs 1965), thereby inhibiting breeding at a time when environmental conditions are unfavourable yet stimulatory daylengths are experienced.

There is considerable interest in the evolutionary and adaptive aspects of photoperiodism, of which refractoriness is an example, and we have found that many waterfowl species in the collection of the Wildfowl Trust come into breeding condition, judged by the date of egg-laying, under a characteristic photoperiod (Kear 1966, Murton & Kear 1973a, b, Kear & Murton 1973). Moreover, when the photoperiod at egg-laying is related to the mid-latitude of the natural breeding range (from Delacour 1954–1964), straight-line regressions result if closely related (same genus) species are considered, while different genera occupy other regression lines. We have argued that this is because closely related species differ by only slight modifications in their photo-response mechanism, perhaps threshold differences, while more fundamental adjustments are involved at generic level. We explore the topic further in this contribution, in which the photoresponses of the rather 'primitive' whistling ducks of the genus *Dendrocygna* are compared with those of the more 'advanced' *Anas* dabbling ducks.

Methods

The wildfowl collection at Slimbridge, Gloucester (lat. 51° 44'N) is kept mostly in open outdoor paddocks, which contain a pond and reasonably natural vegetation. The number of birds per group varies with paddock size, species and their availability, so that a few of the very rare or difficult species are unrepresented. A regular, balanced diet is provided so that the only seasonal variables affecting the species are daylength, temperature and vegetational cycles which are much the same for the whole collection. The birds form pairs, if they are not already held as such, and most species lay normal clutches of eggs in nests which they construct in the shelter of vegetation, or in the special boxes and burrows provided. A record is kept of the date each year that the first egg is laid by each species. For some kinds, records of several pairs extend over 20 years and an accurate median laying date for first clutches is available. However, it was established for swans, geese, shelducks and sheldgeese that the variability in laying date from year to year is slight and that, even when few records are available, a reliable datum is usually obtainable.

In Anatidae, and birds in general, the female lays when she receives appropriate courtship from a male at the peak of his reproductive cycle, and provided that she is not inhibited by a non-stimulatory daylength or inappropriate environmental conditions. Sometimes eggs may be laid spontaneously, but the appearance of a full clutch in a properly constructed nest marks a precise phase in the breeding cycle.

Once egg laying begins, it is not easy to distinguish layings by other birds or repeats and so the pattern of total egg production is unavailable. The date of laying the last clutches can be calculated because records are kept of most hatching dates.

Results

With the exception of the Magpie Goose *Anseranas semipalmata*, which is allocated to the monotypic sub-family Anseranatinae, the eight species of whistling ducks of the genus *Dendrocygna* show more primitive features than any other living members of the Anatidae. These features, apart from anatomy, include: a tropical and sedentary distribution (Fig. 14.1); a single

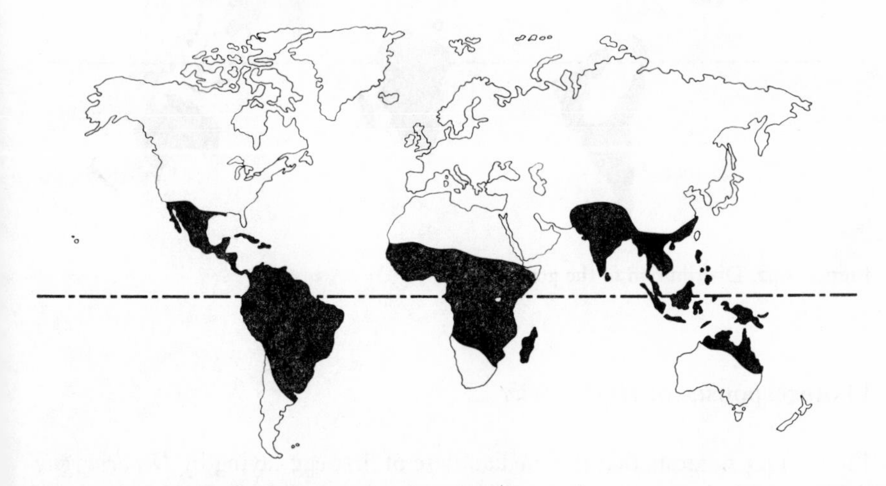

Figure 14.1. Distribution of the genus *Dendrocygna*.

annual moult; a monomorphism and a long-standing pair bond; mutual displays and allopreening; copulation while standing at the edge of water; and the participation of both sexes in nest building, incubation and the defence of eggs and young (Kear 1970). The birds are generally small in size: seven species range on average from 500 to 840 g in body weight, and the eighth, the Cuban, averages 1,150 g.

In contrast to *Dendrocygna*, many *Anas* species show advanced features:- the genus has a predominantly holarctic distribution (Fig. 14.2); there are two moults of body feathers a year; the sexual bond is typically seasonal, especially in migratory forms; the males tend to have bright distinctive plumage and conspicuous displays, serving for species isolation, while the female is often brown and inconspicuous to reduce predation risks when incubating (sedentary, tropical species are, however, often monomorphic); copulation occurs during swimming; and the female takes sole charge of nest building, incubation and usually of the brood (Kear 1970). *Anas* species range in body weight from the Hottentot Teal at 225 g to the Bronzewing at

1,500 g, but the majority of the 50 odd forms are at the heavier end of this scale.

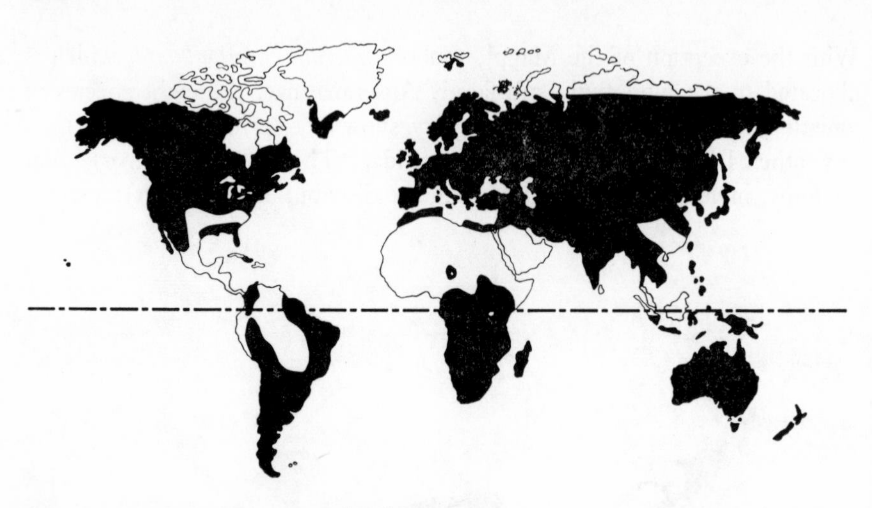

Figure 14.2. Distribution of the genus *Anas*.

Photoresponses of *Dendrocygna*

Figure 14.3 suggests that the median date of first egg-laying by *Dendrocygna* ducks at Slimbridge has been slightly correlated with the mid-latitude of their natural breeding range. The correlation is probably spurious for it depends on the reading obtained for the Australian Plumed Whistling Duck *D.eytoni* and disappears if the earliest laying dates recorded, as distinct from medians, are considered ($r_7 = -0{\cdot}071$). Most of the *Dendrocygna* ducks inhabit low latitudes where seasonal changes in daylength can be of little consequence in timing the onset of the breeding season. The Plumed Whistling Duck (mid-lat. 25°S) certainly cannot depend on a photoperiod of 18-hr to initiate breeding, for such a long daylength is never experienced in the natural range. In fact, climatic conditions at Slimbridge possibly inhibit some of the smaller, more delicate species from their potential of laying eggs during a greater part of the year, for all whistling ducks are vulnerable to cold and, for example, easily get frost-bitten toes in winter. In the cold winter of 1962–3, many died of apparent cold-stress in spite of being well fed (Beer 1964, but see also Delacour 1954, p. 34). Even in high summer, some of the species are reluctant to breed at Slimbridge, so that for the Indian *D.javanica* (mid-lat. 12°N), the Spotted *D.guttata* (mid-lat. 1°S) and the Plumed Whistling Ducks there are breeding records covering only three seasons. In none of these years was nesting attempted before late May, yet second and repeat clutches were laid until early autumn under daylengths of 16·8-hr and 15·7-hr in the case of *javanica* and *guttata* respectively.

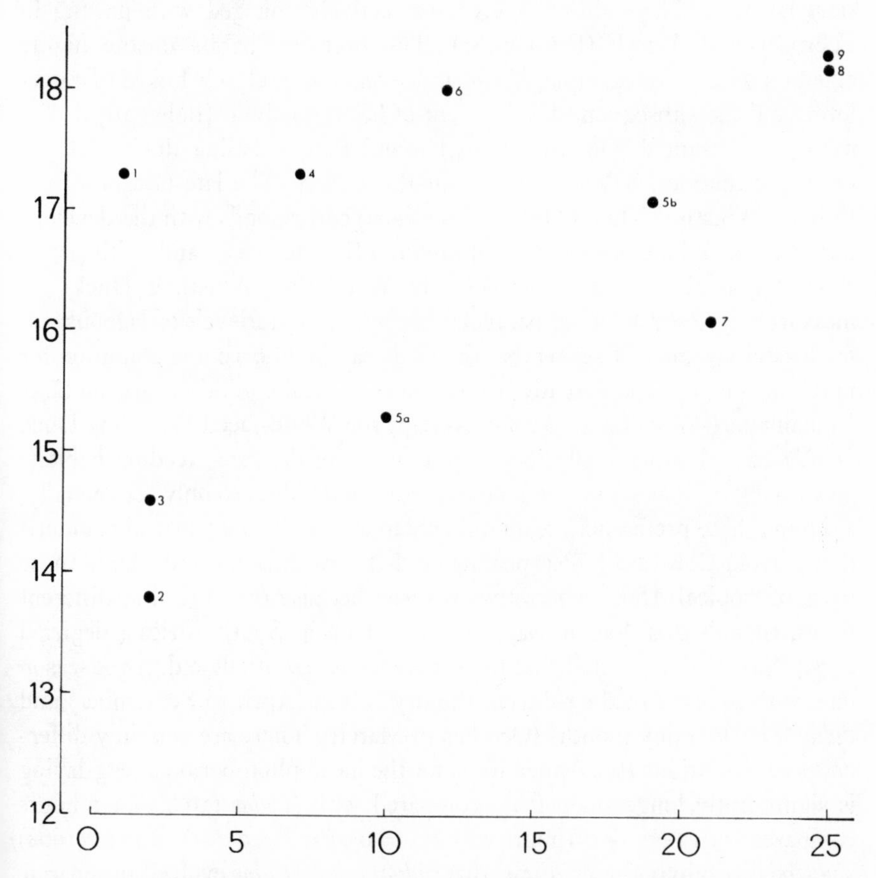

Figure 14.3. Median photoperiod under which first clutches of eggs are laid by *Dendrocygna* ducks at Slimbridge in relation to the mid-latitude of their natural breeding range. Numbers refer to the species as listed in Appendix 1. $r_7 = 0{\cdot}519$; n.s. Number 9 is *Anseranas semipalmata*, not included in the regression.

The potential breeding season of these equatorial and tropical species when exposed to a north-temperate seasonal daylight cycle is probably best illustrated by the geographically widespread Fulvous Whistling Duck *D.bicolor* (mid-lat. of breeding range 2°N) for which there are breeding records covering 16 years. The median date for first eggs at Slimbridge is 28 March but laying has begun as early as 24 February. More than one brood may be reared, so that clutches have been produced until early August.

Many *Dendrocygna* species in the wild, as Bolen (1973) pointed out, do not have nesting peaks; instead only a small part of the population begins laying at any time. Lavery (1967) reported that clutches of the Wandering

Whistling Duck *D.arcuata* in Australia are started as early as mid-December and that nesting extends until early April. Plumed Whistling Ducks may start laying in November but are most actively engaged with nesting in February and March (Frith 1967). The breeding habits of the highly aquatic species—for example, *bicolor* and *arcuata*—are closely linked to water-levels and the subsequent development of nesting habitat (Bolen 1973). For instance, Cottam & Glazener (1959) found that whistling ducks delayed nesting if the local habitat was not suitable earlier. The late-nesting of the Fulvous Whistling Ducks (July in Louisiana) corresponds with the development of rice and the nesting cover it provides (Lynch 1943); and Frith (1967) showed that the breeding activities of Wandering Whistling Ducks, as measured by testis volume, paralleled changes in waterlevels of lagoons and freshwater swamps. It seems that the birds can be in breeding condition for much of the year and actually nest when ecological conditions are suitable. In many parts of its range, *bicolor* overlaps the White-faced Whistling Duck *D.viduata* and superficially they appear to share the same feeding habitat: *viduata* regularly nests in tree holes whereas *bicolor* does so only occasionally, although these preferences seem unlikely to explain how ecological competition is avoided. Ruwet (1964) postulated that competition is reduced in those parts of tropical Africa where they co-exist because they breed at different times, though this does not appear to be true in South Africa (Siegfried 1973). Siegfried suggested that the difference is essentially a dry/wet season one, with *bicolor* breeding during the dry season (April to September) and *viduata* in the rainy months (October to March). There are certainly differences at Slimbridge (see Appendix 1) for the mean photoperiod at egg-laying is significantly longer in *viduata* compared with *bicolor* ($16{\cdot}8 \pm 1{\cdot}1$ hours compared with $13{\cdot}7 \pm 1{\cdot}6$ hours with $t_{24} = 6{\cdot}016$; $P < 0{\cdot}001$). Ruwet (1964) suggested two hypotheses. First, that *viduata* and *bicolor* evolved in separate geographical areas, without developing any notable ecological and ethological divergences. On coming into contact, the species avoided competition by evolving their ecologically isolated breeding cycles. Alternatively, he proposed that the two forms had their centres of distribution to the north and south of the equator, and after colonizing the area of the other form in the opposite hemisphere, each species retained the annual breeding cycle of its original hemisphere. This second explanation does not seem quite correct, although judged on the basis of their photoresponses the two species did evolve at different latitudes (see pp. 351, 352).

The earliest the East Indian Wandering Whistling Duck *D.a.arcuata* has nested at Slimbridge is 19 March (13·3-hr photoperiod), while in California (35°N) they breed for almost the entire year (Sibley 1940). The southern race of the Red-billed Whistling Duck *D.autumnalis discolor* (10°S) has begun breeding on a 12·3-hr photoperiod at Slimbridge and the northern form *D.a.autumnalis* (19°N) with a daylength of 13·4-hr. In the wild in Texas, birds began laying on an average date of 5 May (in a three-year study) and nesting lasted on average 109 days thereafter (Bolen 1973). Finally, the Cuban Whistling Duck *D.arborea* (the largest in body size) from lat. 21°N, confirms the ability of the group to begin breeding with a short daylength by having nested on 25 February (11·8-hr photoperiod).

If not inhibited by cold, the whistling ducks tend to have long egg-laying seasons at Slimbridge which extend on either side of the summer solstice, but breeding is clearly not attempted during the season of very short winter photoperiods (Fig. 14.4). A similar pattern applies to the Magpie Goose for,

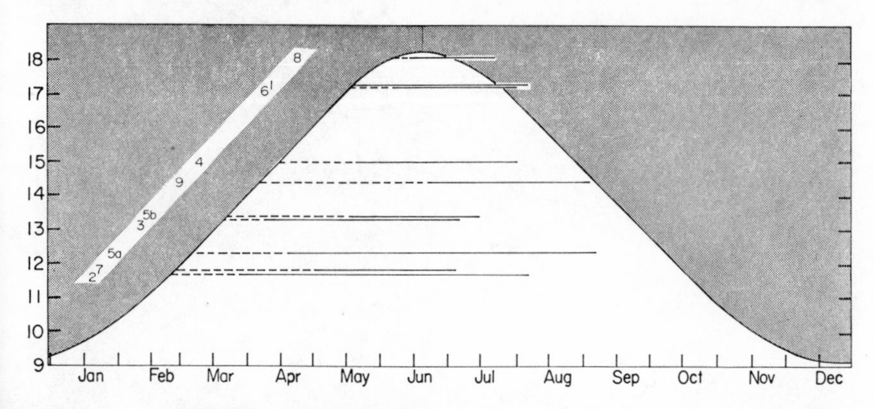

Figure 14.4. Egg-laying season of *Dendrocygna* species and *Anseranas* at Slimbridge in relation to local changes in daylength. The *earliest* recorded date of egg-laying is plotted against the appropriate photoperiod and date and extended as a dotted line to the *median* date for first clutches. A solid line then extends to the date of laying of the latest recorded clutch. Key to numbers given in Appendix 1.

although the median date for first eggs is 22 June, it has begun as early as 6 April continuing, sometimes with four clutches in a season, until 4 September (Figs. 14.3 & 14.4) (Kear 1973). In the wild in tropical Australia it is also an opportunist breeder, relying on the rains to provide a suitable nesting habitat (Frith 1967).

Photoresponses of *Anas*

The photoperiod needed to initiate egg-laying in the *Anas* ducks at Slimbridge is clearly related to the mid-latitude of the natural breeding range (Fig. 14.5). The plots in Fig. 14.5 do indicate a significant correlation ($r_{47} = 0{\cdot}327$; $P < 0{\cdot}02$) although it is evident that some anomalous results are depicted in the upper left portion of the graph. In fact, the very small

Hottentot Teal *Anas punctata* (No. 4) and the Red-billed Pintail *Anas erythrorhynchos* (No. 5) are African species which are probably affected by climate at Slimbridge. The former is often kept indoors in the Tropical

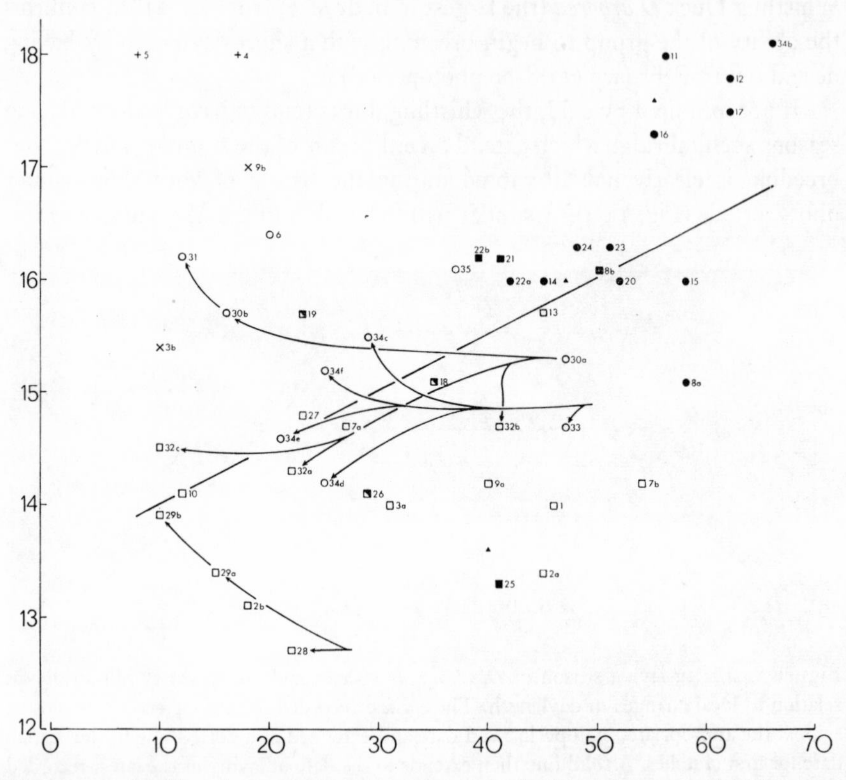

Figure 14.5. Median photoperiod under which first clutches of eggs are laid by *Anas* ducks at Slimbridge in relation to the mid-latitude of their natural breeding range. Numbers refer to the species as listed in Appendix 1. Circles denote northern hemisphere species and squares southern hemisphere forms. Open symbols indicate monomorphic species, solid symbols seasonally dimorphic ones and half-solid squares represent three permanently dimorphic species. Crosses refer to two high altitude forms and + signs to two equatorial species which are affected by the Slimbridge climate; these four species are omitted from the regression analysis which gives:

$r_{43} = 0{\cdot}559$; $P < 0{\cdot}001$

$y = 13{\cdot}5 + 0{\cdot}05x$

Arrows connect some closely related species in which those inhabiting 'low' latitudes have been derived from 'higher' latitudes; see text. The median date for egg-laying by wild Mallard populations in N. America is indicated by small triangles, the data being derived from Bent (1925).

House where, with some extra light in winter, it has nested in January, June, July, October and December. In Lincolnshire (53° 30'N), Hottentots kept outside have laid first eggs on 30 January, 27 February and 29 November

(R. Dawson, pers. com.). In Zambia, nests of the Red-billed Pintail have been recorded from December until May (Benson *et al.* 1971). The Puna Teal *Anas versicolor puna* (No. 36) from lat. 10°S is presumably a high altitude derivative of the Versicolor Teal *A.v.versicolor* (No. 3a) of South America (mid-lat. 31°S) which is closely related to the Hottentot Teal. We have already established that montane species may derive from stocks having distribution centres at high latitudes (Murton & Kear 1973a). This results in their photo-response thresholds being set high relative to latitude, thus causing the onset of breeding to be delayed. Since spring arrives late in montane zones relative to sea-level, these photoresponses are obviously adaptive. Perhaps more interesting is the possibility that a suitable breeding response is more easily evolved at a higher latitude. This is also shown by the Sharp-winged Teal *A.flavirostris oxyptera* (9b) which inhabits the highland plateau of Peru, Bolivia and Chile (mid-lat. 18°S) and which is a sub-species of the Chilean Teal *A.f.flavirostris* (9a) whose mid-lat. is 40°S.

The regression line calculated for Fig. 14.5 omits the anomalous species just discussed in the preceding paragraph. It also omits the Mallard *Anas platyrhynchos* because the population at Slimbridge, as elsewhere in Britain, has been affected by domesticated strains. The regression is highly significant ($r_{43} = 0{\cdot}001$) but accounts for only 31% of the variability in the data; considerably less variability was found in the case of the swans and true geese where regressions accounted for 78% and 76% of the data. We suggest that this is because these phylogenetically older lines have been more vigorously selected. Conversely, the *Anas* ducks represent a modern and successful evolutionary line in terms of numbers of species and individuals and many species that have evolved at a particular latitude zone are in process of radiating to new geographical zones (see below). But first we must consider the remarkable fact that those species emanating from higher latitudes require a longer photoperiod to initiate breeding at Slimbridge than do those from lower latitudes. The requirement of a species-specific photoperiod has been retained even though many of the birds have been bred for several generations in captivity. The responses appear to be generally adaptive for, with increase in latitude, spring arrives later and the northern tundras become suitable breeding places only in late May or early June. Migrants must not be stimulated to fly north too early in the season.

Now it is well established that photo-responsive bird species with regressed gonads can be stimulated into premature breeding condition by exposing them to unnaturally long photoperiods; the longer the photoperiods, the more rapid is the rate of gonad recrudescence and the quicker is a functional gonad state achieved (Wolfson 1959). Moreover, so far as most temperate species are concerned, long daylengths reduce the period during which a functional gonad condition is maintained before spontaneous regression into a refractory state occurs. That is, rapid stimulation from a resting condition is achieved with the cost of a shortened breeding season. The converse applies if a given species is exposed to shorter than normal daylengths. Therefore, as the emergent ducks radiated north to colonize habitats created during inter- and post-glacial epochs, their reproductive apparatus would have been stimulated into breeding condition earlier in the year. The various species appear to have compensated by an adjustment to their photo-response thresholds.

When the breeding season (mean date of first eggs—date of last recorded laying) of each *Anas* species is plotted according to daylength and date at Slimbridge, it becomes evident that the genus can be reasonably divided into two broad categories (Figs. 14.6 & 14.7). There are a few species for which

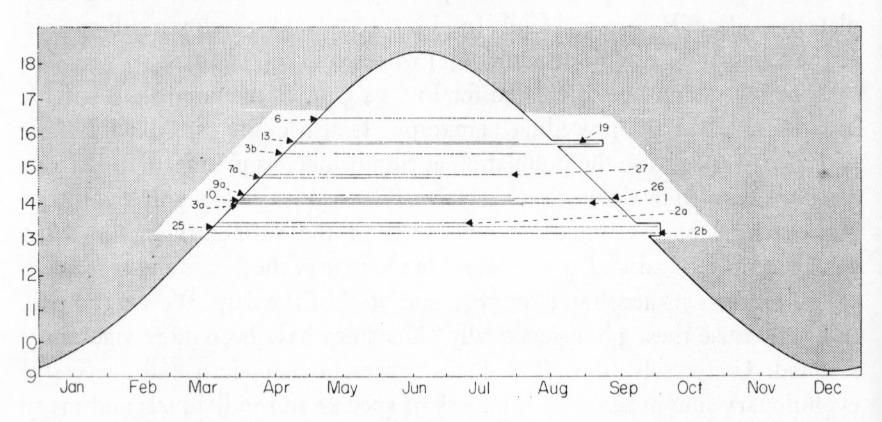

Figure 14.6. Egg-laying of some *Anas* species (Type A) at Slimbridge in relation to local changes in daylength. Numbers refer to the species listed in Appendix 1. The median date of first clutches is plotted against the appropriate photoperiod and a solid line extends to the date on which the latest laid clutch was recorded, that is, the lines are the number of days over which egg-laying was recorded (y). The length of the breeding season (y) is related to the photoperiod at first laying (x) as:

$y = 424{\cdot}7 - 20{\cdot}4x$

$r_{14} = -0{\cdot}743;\ P < 0{\cdot}001$

more complete data would be helpful but the trends are clear. One group is characterized by having a long potential breeding season which is more or less symmetrically positioned in relation to the summer solstice (Type A). The Cape Teal *Anas capensis* is a good example. It comes from South Africa (mid-lat. 12°S) and on average begins breeding at Slimbridge in early April, when the daylength is 14·1-hr. It continues nesting until late August when the photoperiod is 14·0-hr, apparently ceasing to breed when the photoperiod falls below the stimulatory threshold. The Chiloe Wigeon *A.sibilatrix* is another excellent example of Type A breeding. A pair kept at Harpenden

(51° 30'N) hatched 37 ducklings from three successive clutches produced during the summer of 1972, highlighting the capacity to be multi-brooded (W. M. H. Williams, pers. com.).

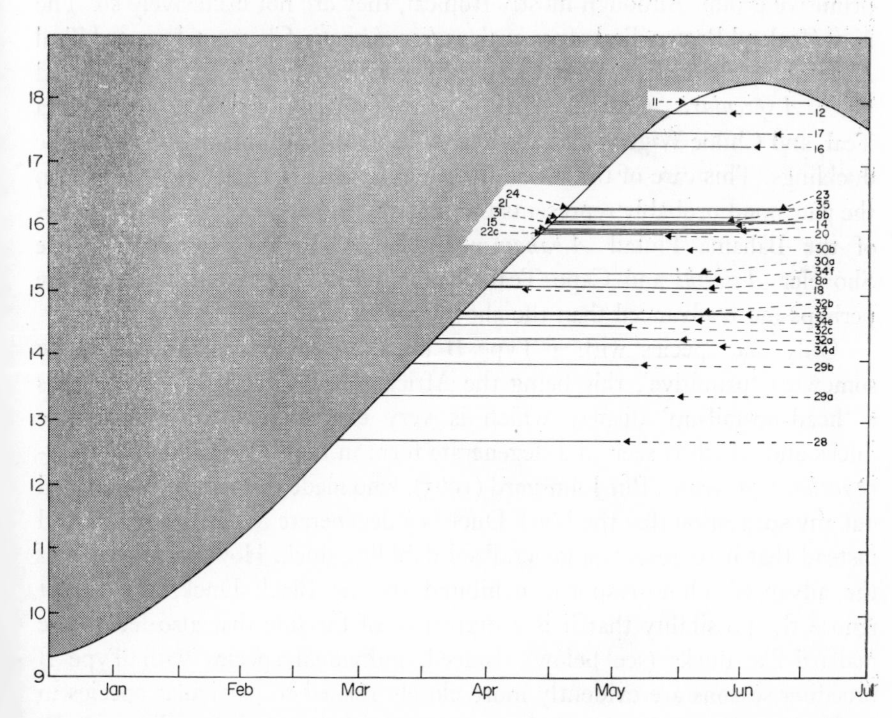

Figure 14.7. Egg-laying season of some *Anas* species (Type B) at Slimbridge in relation to local changes in daylength. Conventions as for Fig. 14.6. The length of the breeding season in days (y) is negatively correlated with the photoperiod (x) needed to initiate breeding:-

$r_{26} = -0{\cdot}667$; $P < 0{\cdot}001$

$y = 178{\cdot}9 - 8{\cdot}5x$

The second *Anas* group (Fig. 14.7) have distinctly short laying seasons and spontaneously cease breeding when daylengths are still stimulatory (Type B). They appear to develop characteristic refractory periods of the kind experimentally demonstrated in the Mallard *Anas platyrhynchos* (Lofts & Coombs 1965). There is a tendency for the breeding season to become shorter in those species which need the longer photoperiod to begin breeding (see Figs. 14.6 & 14.7).

Adaptive radiation in *Anas*

The species plotted in Figure 14.6 have, on plumage and behavioural criteria, a number of primitive features in comparison with other members of the genus (Kear 1970) and it is interesting that they have photoresponses which resemble the pattern seen in *Dendrocygna* (Fig. 14.4) which is probably a primitive group. Although mostly tropical, they are not exclusively so. The New Zealand Brown Teal *Anas aucklandica chlorotis*, Chestnut-breasted Teal *A.castanea*, Australian Grey Teal *A.gibberifrons gracilis*, Bronze-winged Duck *A.specularis*, Crested Duck *A.specularioides*, Versicolor Teal, Puna Teal, and Chiloe Wigeon all have long pair bonds, and the male defends the ducklings. This care of the brood does not seem a secondary feature and so the pair bond probably represents the original primary condition. The males of the Bahama Pintail *A.bahamensis*, Chilean Pintail *A.georgica*, Cape Shoveler *A.smithi* and Cape Teal do not defend their young and so are perhaps more advanced than the above species.

Only one species with a Type B breeding pattern is thought to be somewhat 'primitive', this being the African Black Duck *A.sparsa*. It has a 'head-up-tail-up' display which is very characteristic of mallard-like ducks and which is seen in a degenerate form in the Laysan Duck *A.platyrhynchos laysanensis*. But Johnsgard (1965), who made this comparison, ruled out any suggestion that the Black Duck is a degenerate Mallard and claimed instead that it represents a generalized dabbling duck. However, in view of the advanced photo-response exhibited by the Black Duck, we cannot ignore the possibility that it is a derivative of the line that also led to the mallard-like ducks (see below). Indeed, individual species with Type B breeding seasons are evidently more closely related to particular species in group A than to each other and a series of parallel evolutionary lines can be traced in the genus.

The *Anas* ducks can be assembled into species groups or super-species which are evidently very closely related and which in many cases represent lines which are adapted to specialized feeding niches. Details are given in Fig. 14.8 for the majority of species considered in this paper. The affinities and possible lines of evolution depend to a large extent on behavioural and anatomical criteria presented by Johnsgard (1965).

The green-winged teals are relatively unspecialized small dabbling ducks. The monomorphic Chilean Teal, which has a Type A breeding season, forms a super-species with the dimorphic Green-winged Teal *A.crecca*, which has a Type B breeding pattern. The Cape Teal of Africa seems more primitive than the Chilean Teal and is probably an example of an earlier teal stock which was presumably once distributed in both Old and New Worlds. The austral-teal are related and probably arose early in the history of the group. The wigeon have become specialist grazers with a short stout bill of goose-like proportions. They are linked via the Gadwall *A.strepera* to the green-winged teals. The pintails are long necked dabbling ducks which can reach

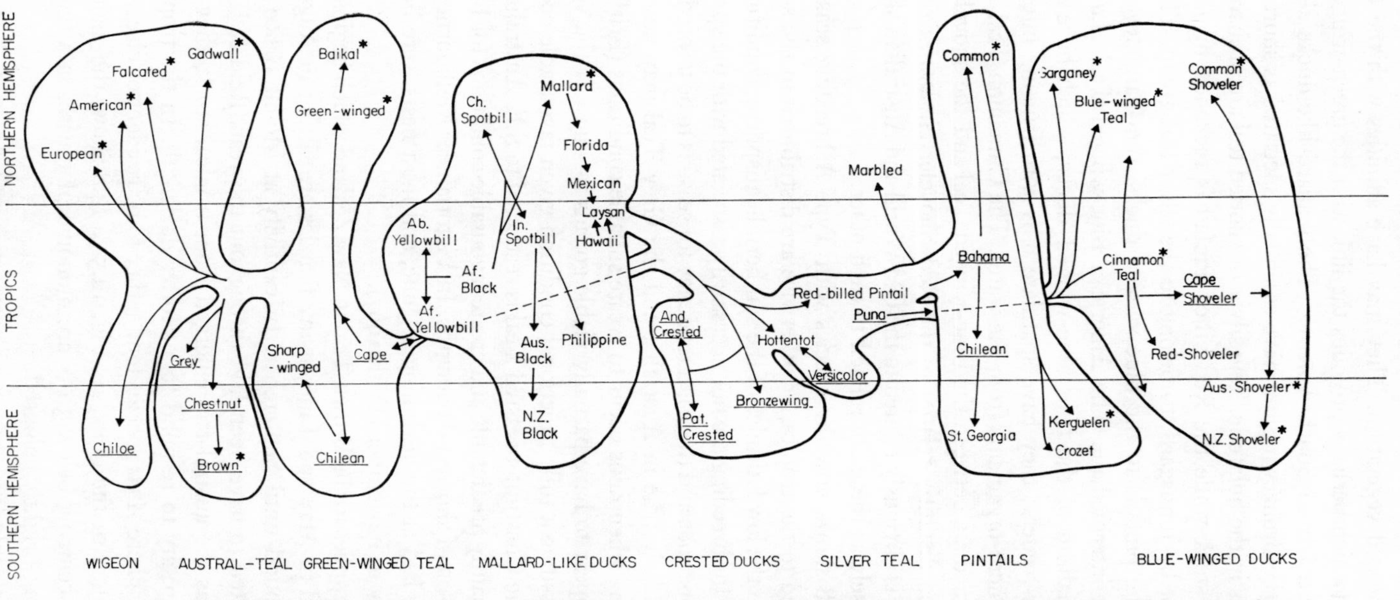

Figure 14.8. Supposed evolutionary relationships in the genus *Anas*. The species and sub-species are plotted in approximate relation to the mid-latitude of their natural breeding range. Closely related-species groups and super-species are enclosed and arrows within and between groups indicate affinities. Those species which are underlined have Type A breeding cycles (see Fig. 14.6) while sexually dimorphic ones are marked by an asterisk.

down to submerged vegetation. They may have affinities with the Bronze-winged Duck (which perhaps provides the link with the green-winged teal) and have radiated from South America. The mallard-like ducks also have affinities with the Bronzewing but have essentially speciated as short-necked dabbling ducks in the holarctic. The silver or spotted teal are probably close to the pintail stock while the Red-billed Pintail is reckoned by Delacour (1956), but not by Johnsgard, to belong to the silver teal group. The silver teal do link the pintails to the blue-winged ducks, the Garganey *A.querquedula*, being intermediate. The American Blue-winged Teal *A.discors* is the least specialized of the true blue-winged ducks, which have become specialist filter-feeders; they have spatulate bills with fringing filter plates for straining micro-organisms from the water. The Cinnamon Teal *A.cyanoptera* is intermediate between the Blue-winged Teal and the shovelers and may be not too distantly related to the Red Shoveler *A.platalea* of South America. This is obviously related to the Old World and Australian shovelers but not so closely as these are related to each other.

Figure 14.8 shows how the species with Type A breeding seasons are mostly confined to the tropics, or otherwise are distributed in the southern hemisphere. In the low latitudes of the southern hemisphere, conditions that are favourable for breeding are not necessarily associated with the spring and summer photoperiods and there is a tendency for seasons to be unpredictable. This is particularly true in Australia, and the Grey Teal may nest at any time, depending when rains occur to create suitable conditions (Frith 1967). Individuals appear to have partially viable gonads for most of the year and they rapidly assume a fully functional condition given favourable ecological conditions. The Chestnut-breasted Teal is restricted to S.E. Australia where the rains are mainly predictable and create favourable conditions for breeding during the season of shortest daylengths. In Victoria near Melbourne (38°S), first eggs were laid at the beginning of June, and most nests were begun in the period July to November (Frith 1967).

The same applies to the Brown Teal in New Zealand which begins breeding in the wild in May and June; many females are sitting in August and clutches are found until September. In captivity at Mount Bruce (41°S), eggs are reported to have been laid throughout the year (Reid & Roderick 1973). Thus, as the austral-teal emerged from tropical stock, they had to acquire the capacity to respond to a shorter daylength. In the tropics and southern hemisphere *Anas* species tend either to be resident or else sporadic nomads moving with the rains. As Fig. 14.5 & 14.8 show, the majority of these species occurring below 38°N are monomorphic with, in a few cases, both sexes being brightly coloured.

The high latitude land masses of northern North America, Europe and Asia provide conditions not found in the southern hemisphere in that a regular food supply becomes available each spring or summer. Inter- and post-glacial events have created conditions that favour migrants able to take advantage of the short arctic summer. Courtship occurs in the contra-nuptial quarters so that pairs are already formed on arrival on the breeding grounds. Since many closely related species share the same wintering grounds, species isolation during courtship is important. The tendency for bright colours to emerge in the Anatidae assumes its full expression in these species. But bright colours would make females conspicuous when forced to nest in open situations in treeless terrain. Accordingly, these north-temperate species are dimorphic, as Figs. 14.5 & 14.8 demonstrate, for the females retain the inconspicuous juvenile dress as adults. This plumage neotony is probably controlled by hormone secretions, as has been proved in the Mallard (Walton 1937, Caridroit 1938). Ostrogenic steroids inhibit the male-type plumage from developing and females remain dull coloured unless ovariectomized, whereupon they assume the male type plumage. Males acquire an eclipse or female-like dull feathering during the spring moult for this coincides with high plasma steroid titres. However, the bright plumage is assumed again with the post-nuptial moult for this occurs during the refractory period when steroid titres are minimal.

The Chestnut-breasted Teal, Cape Shoveler and Red Shoveler are unique in being permanently dimorphic. Conceivably this represents a condition which has evolved from the hormonally controlled condition typical of north-temperate forms. The Type B kind of photo-response of the Red Shoveler suggests a derivation from Cinnamon Teal-like ancestors. The New Zealand Brown Teal does have an eclipse plumage, but the mechanism for its control has not been studied experimentally. It is tempting, however, to suggest that it has arisen from northern hemisphere ancestors of the green-winged teal or perhaps even Mallard stocks for there are some similarities to both. Since the Chestnut-breasted Teal is permanently dimorphic it seems unlikely that a seasonally dimorphic form should arise from it, and the relationships of the Brown Teal are still uncertain. Johnsgard's (1965) view that the monomorphic Grey Teal has reached Australia by island hopping from the East Indies is probably correct so that the dimorphic *castanea* and monomorphic *gibberifrons* have only recently come into secondary contact. A repeated evolutionary trend in *Anas* has been for forms speciating in the north-temperate zone to invade and colonize the southern hemisphere as we now show.

Species that have evolved photo-responses which are adaptive at mid to high latitudes in the northern hemisphere may be able to adapt to a similar latitude in the south. The Kerguelen Pintail *A.acuta eatoni* and Crozet Pintail *A.a.drygalskii* resident at 50°S in the South Indian Ocean are monomorphic descendents from the type species of the northern hemisphere perhaps from migrants which did not return north to breed. As they experience a slightly less extreme seasonal daylength cycle, and have an extended breeding season, Lofts & Murton (1968) suggested that the prime reason for the loss of bright plumage in the male resulted from steroid secretion occurring for a longer period. The mallards illustrate this very well. The Laysan Duck *A.platyrhynchos laysanensis* (25°N), Hawaiian Duck *A.p. wyvilliana* (21°N), Florida Duck *A.p.fulvigula* (29°N), and Mexican Duck *A.p. diazi* (30°N) are probably all recent derivatives of the Mallard which have colonized lower latitudes and lost their dimorphism. In fact, the first two occasionally assume some of the bright feathers of the drake Mallard.

The above trends may have affected earlier mallard lines (see Fig. 14.8.) to give the North American Black Duck *A.rubripes*, a forest form of the Mallard, and Meller's Duck *A.melleri* of Madagascar. During this earlier evolutionary history of the mallards, the stock was represented in the eastern part of the range by the Chinese Spotbill *A.poecilorhyncha zonorhyncha*. The Australian Black Duck and New Zealand Black Duck and Pelew Island Duck are here regarded as races of *A.superciliosa* although Johnsgard (1965) treats them as races of *A.poecilorhyncha*. They are certainly close and doubtless derive from a northern stock, as does the Philippine Duck *A.luzonica* which forms a super-species with *superciliosa*. The African Yellowbill Ducks *A.undulata* are more distinctive and probably represent a still earlier offshoot from the mallard-line. Like the perhaps phylogenetically older African Black Ducks *A.sparsa*, they have Type B breeding seasons which on the evidence presented here suggests a derivation from more temperate forms. These inter-relationships between Mallard-like ducks have been incorporated into Fig. 14.5 as arrows to indicate range changes. It does seem as if some of the variability in the regression has resulted from species moving back from high to low latitudes so that their photo-response mechanisms are set at a 'too-high' threshold being adapted to photoperiods with more seasonal variation.

Discussion

Photoperiodism in birds is currently explained in terms of the Bünning (1960, 1963) hypothesis. The theory was initially formulated to account for diurnal rhythms in plants and it invokes the concept of a circadian rhythm of light sensitivity. It has been made explicit, so far as photoperiodic time measurement is concerned, by the Princeton model of Pittendrigh (1966) and Pittendrigh & Minis (1964, 1971). The daily cycle of day and night acts as a controlling oscillation (hence the term *Zeitgeber* given by Aschoff (1965)) to entrain an endogenous oscillation in the animal. The peak, or more strictly that part of the oscillation above some threshold, is reckoned by Pittendrigh & Minis to be the photo-inducible phase. In the Princeton coincidence model, the daily light-dark cycle entrains the oscillation thereby determining where the light sensitive phase will be positioned relative to the *Zeitgeber* so that, if the light sensitive phase occurs during the daytime, induction can occur. There is much evidence to implicate a circadian rhythm mechanism of light sensitivity in birds (Hamner 1963, Eskin 1971). Although light does have an important modifying inductive role, its importance in this respect is not absolute for several species have been shown to assume breeding condition in constant dark.

In birds the photoperiodic regulation of reproduction and other seasonal functions essentially resolves into the integration of pituitary hormone release via neuroendocrine connections in the hypothalamus and feed-back inhibition or stimulation mediated through gonadal hormone secretion (Lofts *et al.* 1970, Follett 1973). It is now well established that avian gonadotrophin release depends on a circadian rhythm mechanism and that specific hormones, e.g. follicle stimulating hormone (FSH) and luteinizing hormone (LH), are released at different times (Murton *et al.* 1969, Murton *et al.* 1970, Follett & Davies 1975). The photo-inducible phase is evidently compounded from a light-sensitive releasing mechanism for LH and FSH and perhaps other hormones. We do not know whether the oscillations controlling the rhythms of LH and FSH release are coupled to a common light-sensitive pacemaker or whether they are independently controlled; they need not themselves have the same frequencies (see below).

In the House Sparrow *Passer domesticus* there is evidence that with seasonal changes in daylength the phase-angles assumed by the rhythm of LH and FSH alter to different degrees so that the phasing of FSH release alters relative to LH (Murton 1974). It has been established that the temporal relationship of corticotrophin and prolactin determine whether or not a response can occur in the pigeon crop sac (Meier *et al.* 1971) and similar temporal synergisms appear to be important in determining the nature of the gonad response to gonadotrophin stimulation (Meier & Dusseau 1973). It has been suggested that refractoriness occurs when the temporal phasing of LH and FSH secretion prevent the rehabilitation of the interstitial elements of the gonad (Murton 1974, Murton & Westwood 1974).

Two features apparent in the adaptive radiation of the photo-response mechanism of waterfowl need to be explained in terms of biological oscillation theory. First, there appears to be a threshold effect, that is, species differ in the total daily photoperiod at which overt indications of reproductive capacity are noted. Second, refractoriness to long daylengths is noted in some species but not in others. The apparently more primitive species of *Anas* appear not to manifest a refractory period. We assume that they have mostly evolved below latitude 30° where the daily photoperiod, including twilight, would be within the range 11 to 15-hr. The secretion rhythms of LH and FSH must occur within this range and assume a stable phase-angle in spite of alteration of *Zeitgeber* strength. That is, a pacemaker(s) must have properties, perhaps the frequency of its free-running oscillation, that render it, and any coupled systems, resistant to phase-shifting. Evidently, these species can tolerate quite extreme temperate light cycles (LD 9:15—LD 18:6, see Fig. 14.6) without disruption of their oscillator system.

As early *Anas* species moved into higher latitudes they might need to respond to lower or higher photo-thresholds depending on the environment. Thus in temperate S.E. Australia, breeding conditions tend to coincide with the shortest daylengths so that selection should favour a capacity to breed under a photoperiod of LD 10:14 or even LD 9:15. The Chestnut-breasted and Brown Teal are perhaps examples. An alternative requirement might be for breeding to coincide with the season of longest days, in which case a raising of the photo-response threshold is needed. The Chiloe Wigeon from 45°S in South America seems to be an example. None of these species appears to develop refractoriness when exposed to an extremely long daylength, suggesting that their oscillation systems are stable to phase-shifting. It is conceivable that adaptation to latitude has involved the re-phasing of the pacemaker controlling FSH and LH secretion relative to the *Zeitgeber* so that all coupled systems have been shifted as a unit, thereby allowing them to maintain constant phase-angles relative to each other.

Species radiating to north temperate latitudes have experienced ecological conditions favouring spring rather than summer breeding. It is possible that their threshold of response could be set in the way already suggested, that is, oscillations responsible for gonadotrophin secretion would be re-phased as a unit relative to the *Zeitgeber*. But an additional mechanism would be required to regulate refractoriness and, following the line of reasoning presented earlier in the discussion, it is to be wondered whether this could be achieved by a frequency alteration in the coupled oscillators. With change in *Zeitgeber* strength, changes in frequency and phasing of any controlled pacemakers would be expected. If, in addition, the frequencies of any oscillations coupled to this pacemaker had changed to a variable extent relative to each other, a new range of phase-relationships might be possible. We are impressed by the systematic order in which refractoriness develops in the species depicted in Fig. 14.7 but with present knowledge of biological oscillator systems are unable to offer an explicit explanation. The problems of formulating realistic hypotheses may be judged from the recent excellent mathematical analysis of biological oscillators by Pavlidis (1973).

It is probable that the waterfowl cycles discussed in this paper are manifestations of long-term endogenous rhythms, for such have been claimed in other species (Lofts 1964, Gwinner 1971, Schwab 1971). However, it is not yet clear whether these long-term, and sometimes circannual, rhythms are compounded from circadian periodicities involving endogenous feedback, or whether oscillations with frequencies of several months exist. Comparison of the breeding responses of closely related waterfowl pose interesting problems so far as photo periodic phenomena are concerned and there is clearly scope for an extensive experimental programme.

References

Aschoff J. (Ed.) (1965) *Circadian Clocks*. North-Holland Publ., Amsterdam.

Beer J.V. (1964) Wildfowl mortality in the Slimbridge collection during the winters of 1961–62 and 1962–63. *Wildfowl Trust Ann. Rep.* 15, 50–6.

Benson C.W., Brooke R.K., Dowsett R.J. & Irwin M.P.S. (1971) *The Birds of Zambia*. Collins, London.

Bent A.C. (1925) Life Histories of North American Wild Fowl. *U.S. National Museum Bull.* 130, Vol. 2.

Bolen E.G. (1973) Breeding whistling ducks *Dendrocygna* spp. in captivity. *Int. Zoo Yb.* 13, 32–8

Bünning E. (1960) Circadian rhythms and time-measurement in photoperiodism. *Cold Spring Harb. Symp. quant. Biol.* 25, 249–56.

Bünning E. (1963) *Die Physiologische Uhr*. Springer-Verlag, Berlin.

Caridroit F. (1938) Recherches expérimentales sur les rapports entre testicules, plumage d'éclipse et mues chez le Canard sauvage. *Trav. Stn. zool. Wimereux* 13, 47–67.

Cottam C. & Glazener W.C. (1959) Late nesting of water birds in South Texas. *Trans. N. Am. Wildl. Conf.* 24, 382–95.

Delacour J. (1954, 1956, 1959, 1964) *Waterfowl of the World*, Vols. 1–4. Country Life, London.

Emmé A.M. (1960) Photoperiodic response in reproduction. *Russ. Rev. Biol.*, Aug. 1960: 223–40 (translated from *Adv. Contemp. Biol.* 49: March–April 1960).

Eskin A. (1971) Some properties of the system controlling the circadian activity rhythm of sparrows. pp. 55–78 in *Biochronometry*, ed. M. Menaker. Friday Harbor Symp., Washington, D.C., 1969.

Farner D.S. (1970) Daylength as environmental information in the control of reproduction of birds. *Colloq. Int. Cent. Nat. Rech. Sci.* 172, 71–91.

Farner D.S. & Follett B.K. (1966) Light and other environmental factors affecting avian reproduction. *J. Anim. Sci.* 25 Suppl., 90–115.

Farner D.S. & Lewis R.A. (1971) Photoperiodism and reproductive cycles in birds. *Photophysiology* 6, 325–70.

Follett B.K. (1973) The neuroendocrine regulation of gonadotropin secretion in avian reproduction. In *Breeding Biology of Birds*, Ed. D.S. Farner. Natl. Acad. Sci., Washington, D.C.

Follett B. & Davies D.T. (1975) The neuro endocrine control of reproduction. *Symp. Zool. Soc. Lond.* 35 (in press).

Frith H.J. (1967) *Waterfowl in Australia*. Angus and Robertson, Sydney.

Gwinner E. (1971) A comparative study of circannual rhythms in Warblers. In *Biochronometry*, Ed. M. Menaker. Natl. Acad. Sci., Washington, D.C.

IMMELMANN K. (1971) Ecological aspects of periodic reproduction. In *Avian Biology*, Ed. D.S. Farner & J.R. King. Vol. I Academic Press, New York & London.

JOHNSGARD P.A. (1965) *Handbook of Waterfowl Behavior*. Cornell University Press, Ithaca.

KEAR J. (1966) The food of geese. *Int. Zoo Yb.* 6, 96–103.

KEAR J. (1970) The adaptive radiation of parental care in waterfowl. In *Social Behaviour in Birds and Mammals*, Ed. J.H. Crook. Academic Press, London.

KEAR J. (1973) The Magpie Goose *Anseranas semipalmata* in captivity. *Int. Zoo Yb.* 13, 28–32.

KEAR J. & MURTON R.K. (1973) The systematic status of the Cape Barren Goose as judged by its photo-responses. *Wildfowl* 24, 141–3.

LACK D. (1968) *Ecological Adaptations for Breeding in Birds*. Methuen, London.

LACK D. (1971) *Ecological Isolation in Birds*. Blackwell, Oxford & Edinburgh.

LAVERY H.J. (1967) Whistling ducks in Queensland. *Advis. Leafl. Div. Plant Indust. Queensl.* No. 917.

LOFTS B. (1964) Evidence of an autonomous reproductive rhythm in an equatorial bird (*Quelea quelea*). *Nature, Lond.* 201, 523–4.

LOFTS B. & COOMBS C.J.F. (1965) Photoperiodism and the testicular refractory period in the mallard. *J. Zool., Lond.* 146, 44–54.

LOFTS B., FOLLETT B.K. & MURTON R.K. (1970) Temporal changes in the pituitary-gonadal axis. *Mem. Soc. Endocrinol.* 18, 546–75.

LOFTS B. & MURTON R.K. (1968) Photoperiodic and physiological adaptations regulating avian breeding cycles and their ecological significance. *J. Zool., Lond.* 155, 327–94.

LYNCH J.J. (1943) Fulvous tree duck in Louisiana. *Auk* 60, 100–2.

MEIER A.H., JOHN T.M. & JOSEPH M.M. (1971) Cortiscosterone and the circadian pigeon crop sac response to prolactin. *Comp. Biochem. Physiol.* 40, 459–66.

MEIER A.H. & DUSSEAU J.W. (1973) Daily entrainment of the photoinducible phases for photostimulation of the reproductive system in the sparrows, *Zonotrichia albicollis* and *Passer domesticus*. *Biology of Reproduction* 8, 400–10.

MURTON R.K. (1974) Ecological adaptation in avian reproductive physiology. *Sym. Zool. Soc. Lond.* In press.

MURTON R.K., BAGSHAWE K.D. & LOFTS B. (1969) The circadian basis of specific gonadotrophin release in relation to avian spermatogenesis. *J. Endocr.* 45, 311–12.

MURTON R.K. & KEAR J. (1973a) The nature and evolution of the photoperiodic control of reproduction in wildfowl of the family Anatidae. *J. Reprod. Fert., Suppl.* 19, 67–84.

MURTON R.K. & KEAR J. (1973b) The influence of daylight in the breeding of diving ducks. *Int. Zoo Yb.* 13, 19–23.

MURTON R.K., LOFTS B. & WESTWOOD N.J. (1970) The circadian basis of photoperiodically controlled spermatogenesis in the Greenfinch *Chloris chloris*. *J. Zool., Lond.* 161, 125–36.

MURTON R.K. & WESTWOOD N.J. (1974) An investigation of photorefractoriness in the house sparrow by artificial photoperiods. *Ibis*.

PAVLIDIS T. (1973) *Biological Oscillators: Their Mathematical Analysis*. Academic Press, London & New York.

PITTENDRIGH C.S. (1966) The circadian oscillation in *Drosophila pseudoobscura* pupae; a model for the photoperiodic clock. *Z.Pflanzenphysiol.* 54, 275–307.

PITTENDRIGH C.S. & MINIS D.H. (1964) The entrainment of circadian oscillations by light and their role as photoperiodic clocks. *Am. Nat.* 98, 261–94.

PITTENDRIGH C.S. & MINIS D.H. (1971) The photoperiodic time measurement in *Pecti-*

nophora gossypiella and its relation to the circadian system in that species. In *Biochronometry*, Ed. M. Menaker. Natl. Acad. Sci., Washington, D.C.

REID B. & RODERICK C. (1973) New Zealand scaup *Aythya novae-seelandiae* and Brown teal *Anas aucklandica chlorotis* in captivity. *Int. Zoo Yb.* 13, 12–15.

RUWET J.C. (1964) Quelques exemples de séparation écologique d'espéces congènériques de l'avifaune Katangaise. *Le Gerfaut* 54, 159–66.

SCHWAB R.G. (1971) Circannian testicular periodicity in the European starling in the absence of photoperiodic change. In *Biochronometry*, Ed. M. Menaker. Natl. Acad. Sci., Washington, D.C.

SIBLEY C.L. (1940) Breeding various tree ducks in America. *Avic. Mag.* 5th series V, No. 6, 155–8.

SIEGFRIED W.R. (1973) Morphology and ecology of the southern African whistling ducks (*Dendrocygna*). *Auk* 90, 198–201.

WALTON A. (1937) On the eclipse plumage of the Mallard (*Anas platyrhynchos platyrhynchos*). *J. exp. Biol.* 14, 440–7.

WOLFSON A. (1959) Role of light and darkness in the regulation of spring migration and reproductive cycles in birds. In *Photoperiodism and related phenomena in plants and Animals*. Ed. R.B. Withrow. American Association for the Advancement of Science, Washington, D.C.

WOLFSON A. (1970) Light and darkness and circadian rhythms in the regulation of annual reproductive cycles in birds. *Colloq. Int. Cent. Nat. Rech. Sci.* 172, 93–119.

Questions and discussion

In answer to a question on technique Dr Kear made clear that all eggs were removed on laying and incubated artificially, each egg being replaced by a dummy. The date of laying and other factors were recorded by four assistants constantly combing the laying sites.

Appendix Key to *Dendrocygna* and *Anas* species considered in text and in figures

	Mid. lat. of natural breeding range	Photoperiod at median date of first egg-laying	Photoperiod when last eggs laid italic figs. means that laying occurs over the longest day
Dendrocygna			
1. Spotted Whistling Duck *D. guttata*	1°S	17·3 (17·3)*	*15·6*
2. Fulvous Whistling Duck *D. bicolor*	2°N	13·8 (11·7)	*16·5*
3. East Indian Wandering Whistling Duck *D. a. arcuata*	2°S	14·6 (13·3)	*18·0*
4. White-faced Whistling Duck *D. viduata*	7°S	17·3 (15·0)	*16·7*
5a. Southern Red-billed Whistling Duck *D. autumnalis discolor*	10°S	15·3 (12·3)	*14·3*
5b. Northern Red-billed Whistling Duck *D. a. autumnalis*	19°N	17·1 (13·4)	*17·7*
6. Indian Whistling Duck *D. javanica*	12°N	18·0 (17·3)	*16·7*
7. Cuban Whistling Duck *D. arborea*	21°N	16·1 (11·8)	*18.1*
8. Plumed Whistling Duck *D. eytoni*	25°S	18'2 (18.1)	*17·4*
Anseranas			
9. Magpie Goose *A. semipalmata*	25°S	18·3 (14·4)	*14·5*
Anas			
1. Bronzewing *A. specularis*	46°S	14·0	*14·7*
2a. Patagonian Crested Duck *A. s. specularioides*	45°S	13·4	*17·9*
2b. Andan Crested Duck *A. s. alticola*	18°S	13·1	*12·8*
3a. Spotted Teal *A. v. versicolor*	31°S	14·0	*16·7*

3b.	Puna Teal *A. v. puna*	10°S	15·4	*16·5*
4.	Hottentot Teal *A. punctata*	17°S	18·0	*12·5*
5.	Red-billed Pintail *A. erythrorhyncha*	8°S	18·0	*17·9*
6.	Bahama Pintail *A. b. bahamensis*	20°N	16·4	*16·5*
7.a	Chilean Pintail *A. georgica spinicauda*	27°S	14·7	18·2
7b.	St. Georgia Pintail *A. g. georgica*	54°S	14·2	–
8a.	Pintail *A. a. acuta*	58°N	15·1	18·1
8b.	Kerguelen Pintail *A. a. eatoni*	50°S	16·1	18·2
9a.	Chilean Teal *A. f. flavirostris*	40°S	14·2	*16·9*
9b.	Sharp-winged Teal *A. f. oxyptera*	18°S	17·0	–
10.	Cape Teal *A. capensis*	12°S	14·1	*14·0*
11.	Green-winged Teal *A. crecca*	56°N	18·0	18·1
12.	Baikal Teal *A. formosa*	62°N	17·8	18·1
13.	Chiloe Wigeon *A. sibilatrix*	45°S	15·7	*16·7*
14.	Gadwell *A. s. strepera*	45°N	16·0	18·2
15.	European Wigeon *A. penelope*	58°N	16.0	18·3
16.	American Wigeon *A. americana*	55°N	17·3	18·3
17.	Falcated Teal *A. falcata*	62°N	17·5	18·3
18.	Red Shoveler *A. platalea*	35°S	15·1	15·7
19.	Cape Shoveler *A. smithi*	23°S	15·7	*14·3*
20.	Common Shoveler *A. clypeata*	55°N	16·0	17·6
21.	New Zealand Shoveler *A. rhynchotis variegata*	41°S	16·2	18·3
22a.	Northern Cinnamon Teal *A. cyanoptera septentrionalium*	42°N	16·0	18·1
22b.	Argentine Cinnamon Teal *A. c. cyanoptera*	39°S	16·2	–

23.	Garganey *A. querquedula*	51°N	16·3	18·2
24.	Blue-winged Teal *A. discors orphna*	48°N	16·3	18·2
25.	Brown Teal *A. auklandica chlorotis*	41°S	13·3	*12·7*
26.	Chestnut-breasted Teal *A. castanea*	29°S	14·1	*16·8*
27.	Grey Teal *A. gibberifrons gracilis*	23°S	14·8	*17·0*
28.	African Black Duck *A. s. sparsa*	22°S	12·7	17·2
29a.	African Yellowbill *A. undulata undulata*	15°S	13·4	17·8
29b.	Abyssinian Yellowbill *A. u. ruppelli*	10°N	13·9	17·4
30a.	Chinese Spotbill *A. peocilorhycha zonorhyncha*	47°N	15·3	18·1
30b.	Indian Spotbill *A. p. poecilorhycha*	16°N	15·7	17·9
31.	Philippine Duck *A. luzonica*	12°N	16·2	18·2
32a.	Australian Black Duck *A. superciliosa rogersi*	22°S	14·3	17·8
32b.	New Zealand Black Duck *A. s. superciliosa*	41°S	14·7	18·2
32c.	Pelew Island Duck *A. s. pelewensis*	10°S	14·5	17·2
33.	North American Black Duck *A. rubripes*	47°N	14·7	18·2
34a.	Mallard *A. p. platyrhynchos*	49°N	11·8	18·2
34b.	Greenland Mallard *A. p. conboschas*	66°N	18·1	–
34c.	Florida Duck *A. p. fulvigula*	29°N	15·5	–
34d.	Mexican Duck *A. p. diazi*	25°N	14·2	18·0
34e.	Hawaiian Duck *A. p. wyvilliana*	21°N	14·6	17·9
34f.	Laysan Teal *A. p. laysanensis*	25°N	15·2	18·1
35.	Marbled Teal *A. angustirostris*	37°N	16·1	18·1

* Data in brackets refer to photoperiod of first egg-laying. All photoperiods are given as sunrise to sunset plus civil twilight.

15 Climate, topography and germination

I. H. RORISON and F. SUTTON *Unit of Comparative Plant Ecology (Natural Environment Research Council) Department of Botany, University of Sheffield*

Introduction

Consideration of solar radiation as an ecological factor is difficult in isolation. Its effect on plant growth and distribution involves a chain of reactions and, for a given situation, it may be necessary to determine the limiting link in the chain. Is it light intensity or quality, soil temperature, or soil moisture, and what influence has topography or soil fertility?

The ultimate source of heat energy for the soil is solar radiation and the thermal pattern it produces is one of the primary factors controlling seedling establishment, plant growth and distribution. Its influence is not only direct through temperature regulation of plant processes but also indirect through its influence on moisture availability, soil development, mineralization and, therefore, nutrient availability.

The soil acts as a heat reservoir with a continually changing temperature at any point in the profile. Its temperature will vary in amplitude according to season, topography, soil texture (Monteith 1973), colour (Watts, Chapter 17 in this volume) and moisture, and, of course, depth.

Aspect obviously has a marked influence on the amount of radiation a surface will receive. In Britain south-facing slopes receive considerably more radiation per annum than do north-facing slopes, with east-facing and west-facing slopes in an intermediate position. Similarly, annual radiation received at the top of a north-slope[1] may not differ greatly from the bottom of a south-slope in a deep valley. In the Derbyshire dales there are bigger heat losses and gains on the south-slope compared with the north-slope not only because of the greater amount of solar radiation received but also because the soil is drier and the vegetation cover less complete.

Soil temperature and its effects on plant growth has been the subject of innumerable papers over the last hundred years from Haberlandt (1874) to Geiger (1961) but the ecological literature is singularly lacking in references

[1] Used in this paper to mean a north-*facing* slope following the current German use of *Nordhang*.

to experimental work directly concerned with the relationships between solar radiation, soil temperature and plant growth. There are many references to plant distribution in relation to aspect and a particularly well-documented case was given by Ludwig *et al.* (1957). They made sequential sowings of maize varieties on pyramids of soil of slope 22·5°, 11·25° and 6·25°, with faces to north, south, east and west. They also maintained water levels on all slopes at approximately field capacity. For early emergence and good establishment, the advantages of south- and, to a lesser extent, west-slopes contrasted with the disadvantages of north- and east-slopes. The importance of angle of slope changed with season and the rate of emergence varied inversely with soil temperature. An additional aim of their work was to identify criteria for selection of sites to extend the northern limits for the cultivation of maize in England.

Such limits are also of importance in the distribution of naturally occurring species (Jarvis 1963) but again reports of experimental work are few. Pigott (1958) drew attention to the prevalence of *Polemonium caeruleum* L. on steep, damp north-slopes in Derbyshire, which are about the southern limit of its British distribution. Later (1968) he presented experimental evidence relating the limits of distribution of the southern species *Cirsium acaulon* (L.) Scop. to climatic factors. This species is close to its northern limits in Derbyshire and is almost entirely confined to south- to southwest-slopes of the dales, where with summer daily maximum air temperatures of +20°C it produces ripe fruit. In addition, the survival of certain winter annual species of south-slopes is related to high soil temperatures and moisture tensions to a degree which still requires experimental clarification (Ratcliffe 1961).

In order to measure something of the effects that these various environmental factors have on plant growth and distribution we embarked on a detailed study of the germination and seedling establishment of selected species, the measurement of seasonal nutrient availability and of climatic variations in one such dale.

Materials and methods

The sites

In order to simplify the system to manageable proportions and to make the most of the equipment available, three points were selected across Lathkilldale, a valley running E–W through the limestone plateau.

The two sites on the south- and north-slopes lie halfway up the slopes at *c.* 230 m OD (700 ft). The slope is *c.* 30° and the soil, overlying stabilized limestone scree, is nowhere more than 15 cm deep. Since texture influences the thermal characteristics of a soil, the relevant data are given in Table 15.1. There is typical development according to aspect with the north-slope having more organic matter (and greater water holding capacity), greater depth and lower pH than the south-slope (Bridges 1960, Franzmeier *et al.* 1969). On the plateau the brown podzolic soil has an A_0 layer consisting mainly of mor humus. The A_1 layer is also highly humified and its inorganic components are largely non-calcareous, whereas those of the rendzina soils on the slopes are mainly limestone fragments. Both the percentage and the textural quality of organic matter varies from profile to profile. In the plateau A_1 horizon and in north-slope soil it is largely amorphous; in the south-slope soil it occurs as a fluffy comminution and in the plateau A_0 horizon only a small proportion is fully decomposed. These factors are all reflected in some thermal conductivity measurements (after de Vries 1952) made when the soils were at field capacity moisture level.

Table 15.1. Physico-chemical characteristics of the soils at the three experimental sites.

Percentage by weight consisting of	Plateau A_0	Plateau A_1	North-slope A	South-slope A
Organic matter (H_2O_2)	76	41	21	14
Carbonate (HCl)	1	3	7	21
Coarse sand	1	4	12	4
Fine sand	5	22	35	40
Silt	5	14	14	9
Clay	12	16	10	8
Total percent recovered	101 ± 0.7	100 ± 1.4	98 ± 3.0	95 ± 2.3
Soil depth in cm	<20		<14	<5
pH (0·1 N $CaCl_2$)	3·0–3·4		6·3–7·1	6·6–7·5
Thermal conductivity at field capacity $W\ m^{-1}\ °C^{-1}$	1·39	2·06	2·06	1·78

At each of the three dale sites 3 replicate 10×1 m plots were laid out and each divided into ten 1 m squares, each of which was again divided into sixteen 25×25 cm squares. Each replicate was separated by a 1 m pathway and the whole layout was designed to provide replicated sampling for a number of years. Resistance blocks and thermistors were positioned to monitor soil moisture and soil and air temperatures respectively.

Since the investigation was concerned mainly with seedling emergence and nutrient availability in relatively shallow soils, measurements of temperature were made 2 cm below and 2 cm above the soil surface.

More recently a solarimeter system was set up to make continuous recordings of total and diffuse radiation. It is sited on the plateau to the north of the dale in a field with an acceptably clear horizon.

Data collection systems

(a) Solar radiation

Two solarimeters[1] are mounted on top of a metal cabinet[2] (Plate 15.1), the length of which runs north to south. From a southerly viewpoint, the 'diffuse' solarimeter is mounted behind and a little above the 'total' solarimeter, the latter being mounted, as supplied, on a raised platform and the former,

Plate 15.1. The solarimeter system (for details see text).

[1] CM5 Solarimeters (Kipp & Zonen).
[2] Botany/Zoology Workshop, Sheffield University.

having been removed from the manufacturers' base, fixed to a specially-cast framework[1] which includes an adjustable shade ring.

The cabinet is divided into two compartments, one of which houses the two supply batteries[2] and the other, thermally insulated, contains the measuring equipment described in the Appendix.

A time-sharing circuit,[3] with a switching cycle of approximately four minutes, ensures that each solarimeter in turn is switched, through a common digital integrator,[4] to an associated digital counter.[5] A battery-driven clock[6] initiates an hourly print-out of total counts simultaneously on both counters.

Calibration was carried out *in situ* using a standard solarimeter, by Dr. M. Unsworth of the Nottingham University School of Agriculture, Sutton Bonington.

Little maintenance is required beyond changing the supply batteries, collecting the printed data strips and adjusting the shade ring at weekly intervals. The solarimeter domes are cleaned weekly and, less frequently, the levelling of each solarimeter is checked and the desiccant changed.

From weekly data collections, computer cards are punched, each with the date, daily 'diffuse' count and daily 'total' count. Calibration and shade-ring corrections are carried out using a FORTRAN programme. A print-out is obtained of the corrected values and also a histogram of the daily means of total, diffuse and direct radiation in MJ m^{-2} d^{-1}. For reasons of economy in computer time, the programme is normally run at two-monthly intervals.

(b) Soil and air temperatures

Modified D-Mac 10-channel Limpet Loggers (Sutton and Rorison 1970) are in continuous use, measuring soil- and air-temperatures on north- and south-slopes and recently on a plateau site in Lathkilldale (approximately 1 km from the solarimeter equipment). The sensors used are matched bead thermistors. The data are recorded at half-hourly intervals on domestic magnetic tape. The tapes are collected weekly and translated (Sutton and Rorison 1972) to 8-hole paper tape. This tape is fed into the computer with a programme which corrects the data, labels each line with the date, time and a code number which indicates whether or not detectable errors are present. In this form, the lines are written on magnetic tape, to be stored for future

[1] University of Nottingham Workshops, Sutton Bonington.
[2] Dryfit PC, 12v, 20 amp-h (F.W.O. Bauch Ltd).
[3] NERC, Sheffield (see Appendix).
[4] TS100A Digital Integrator (Time Electronics Ltd).
[5] PL103 Printing Impulse Counters (Sodeco).
[6] Modified 12v car clock (Smith's) (see Appendix).

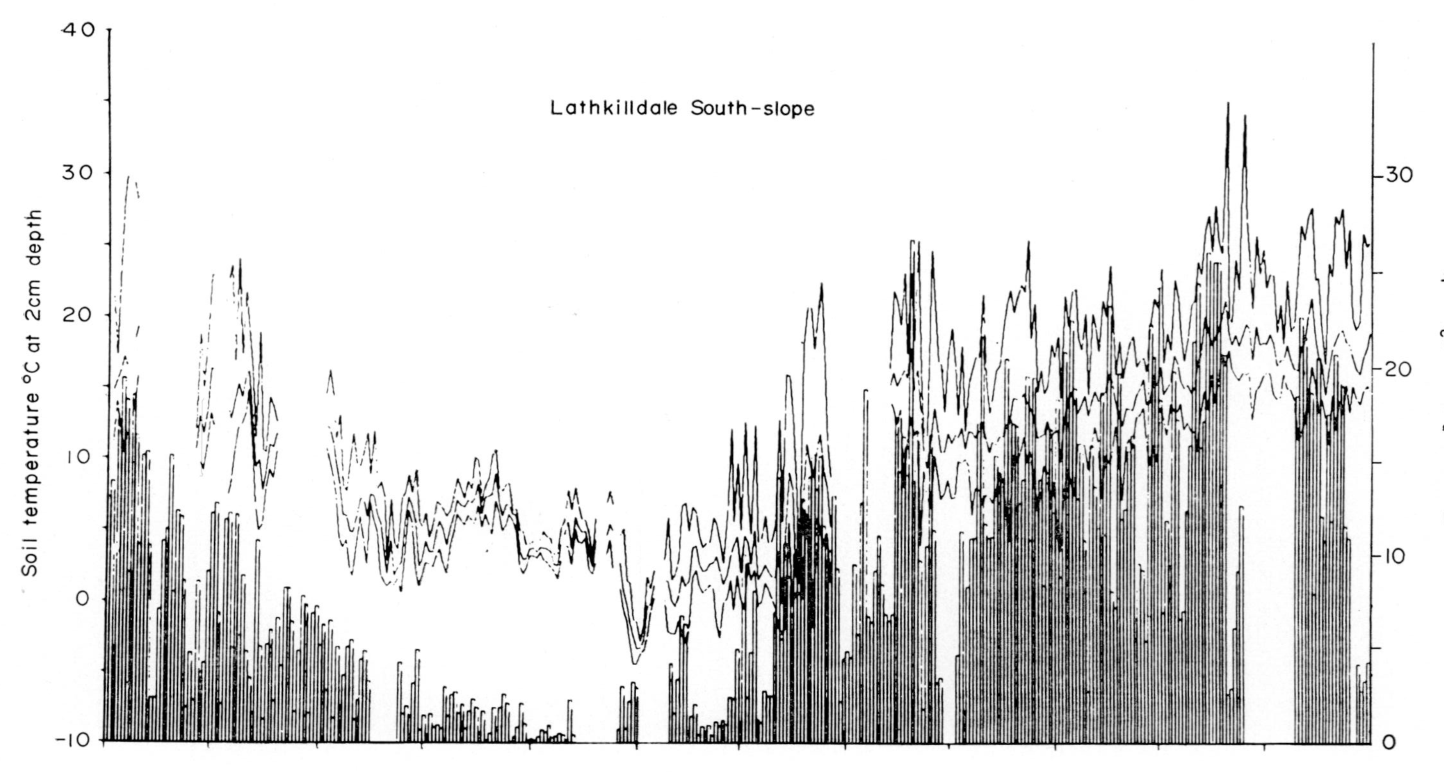
Lathkilldale South-slope
Soil temperature °C at 2cm depth
40
30
20
10
0
-10
Total radⁿ (MJ m⁻² day⁻¹)
30
20
10
0

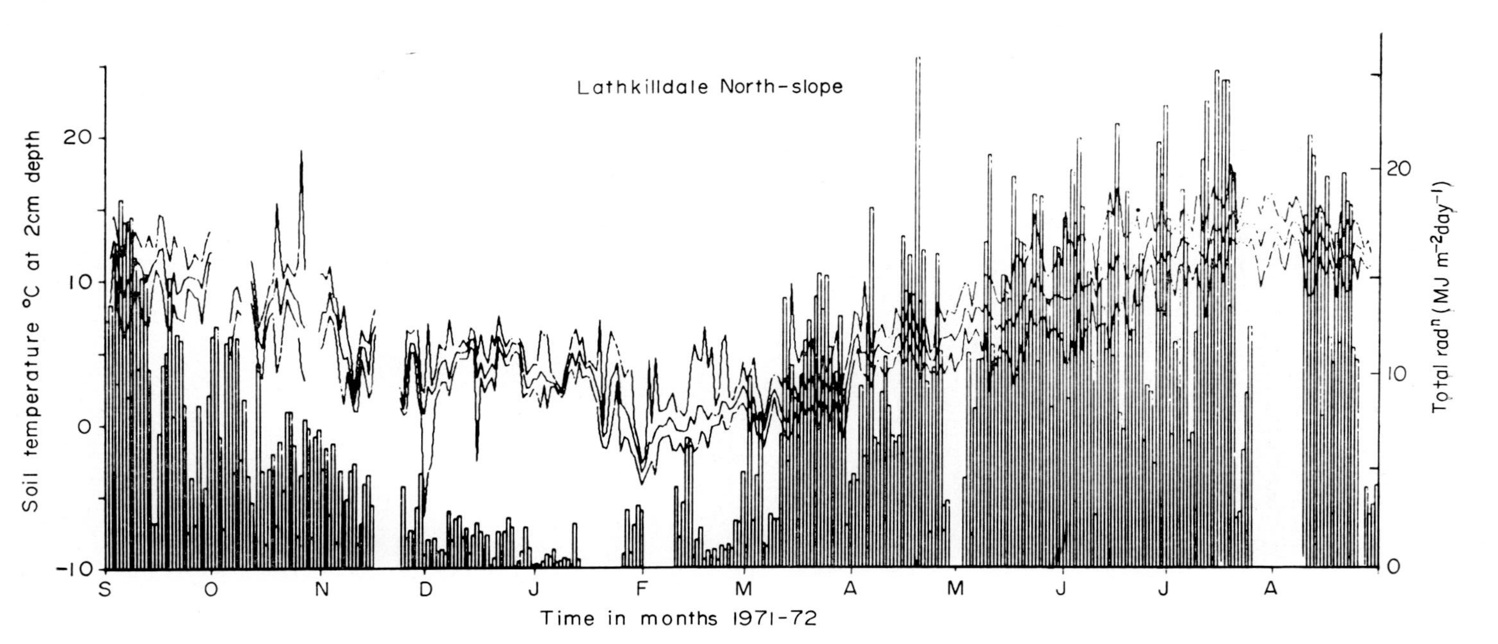

Figure 15.1. Maximum, mean and minimum soil temperatures at 2 cm depth on north-and south-slopes (continuous lines) compared with daily solar radiation as received by a horizontal surface with an unobstructed horizon (histogram). Computer drawing.

analysis, each tape being capable of holding approximately two and a half years' data from one logger. A graph of weekly temperature fluctuations is also obtained.

A comprehensive FORTRAN programme, still being developed, allows the processing of data related to any specified period. Diurnal fluctuations of soil and/or air temperature may be plotted. Maximum, minimum and mean temperatures over fixed or specified irregular intervals may be plotted together with, if desired, superimposed histograms of solar radiation. A paper tape output, punched with the plotted values, may also be obtained. Some of the forms of presentation are illustrated in this paper.

(c) Soil moisture

Calibrated miniature gypsum resistance blocks (Lloyd 1968) buried to a depth of 2–5 cm were read at weekly intervals by means of a Bouyoucos Bridge.

Germination experiments

Sufficient ripe seed of each species to be tested was collected from specific sites to allow for a sequence of sowings at five-week intervals throughout one year. It was stored in loosely stoppered bottles at *c.* 20°C in the laboratory and its viability tested at intervals. At each field site five 25 × 25 cm sub-plots were chosen at random from among the major plot replicates and at each planting time 20 seeds were inserted along a shallow slit in the soil at about the depth of their own size from the surface. The surface vegetation had been previously removed for analysis and any underground organs likely to regenerate were removed. Counts of emerging seedlings were made at weekly intervals throughout the year.

Results

Patterns of solar radiation and soil temperature

Figure 15.1 shows daily total solar radiation plotted as histograms from 1.9.71 to 31.8.72 with daily maximum, mean and minimum soil temperatures at a depth of 2 cm superimposed on the same time scale.

The general trend is for soil temperatures to follow the solar radiation pattern with greater daily fluctuations of temperature on the south-slope than on the north-slope. A closer examination is needed to distinguish precise relationships.

A simple regression analysis was carried out to test for linear, exponential and hyperbolic fit. It allotted maximum significance to the linear regression with a decreasing degree of correlation between radiation and maximum, mean and minimum soil temperature. A polynomial regression analysis revealed the same trend and confirmed that the linear regression gave the best line of fit for radiation versus maximum temperature. Figure 15.2 shows the resulting scatter diagrams with the calculated regression line drawn in. The slope of the line for the south-slope is approximately three times that for the north-slope. Taken overall there is on the north-slope a 1°C rise in maximum soil temperature for every 3 MJ m^{-2} d^{-1} of solar radiation and on the south-slope a 1°C rise for every 1 MJ m^{-2} d^{-1}. The daily mean and minimum soil temperatures follow the same general pattern but the correlations are weaker as may be seen from the r values given in Table 15.2.

Table 15.2. Correlation coefficients (r) between soil temperature and solar radiation for the period 1.9.71 until 31.8.72. All the r values are significant at $P<0.001$. Missing values occur due to failure of either the temperature or radiation recording equipment.

	South	North
T_{max}	0·8013	0·5921
T_{mean}	0·6896	0·5347
T_{min}	0·5278	0·4679
Degrees of freedom	258	297

The scatter of points in Fig. 15.2 reflects the general relationship between total radiation and temperature throughout the year. This may be seen in more detail from an inspection of two ten-day periods, one from the south-slope when the relationship is very close, and the other from the north-slope when no such relationship exists. On the south-slope, 5–15 October 1971 (Fig. 15.3a), soil moisture was at approximately field capacity, total radiation ranged from 1·8 to over 10·8 MJ m^{-2} d^{-1} and mean soil temperature was *c.* 15°C. Inspection of the computer-drawn curves suggests that there is a strong relationship between daily total solar radiation and daily maximum soil temperature. Mean and minimum temperatures appear to be less immediately affected. Correlation coefficients confirm that there is a close linear relationship between total radiation and T_{max} ($r = 0{\cdot}84^{***}$) but not between total radiation and T_{mean} ($r = 0{\cdot}25$) or T_{min} ($r = 0{\cdot}13$). There is a suggestion of a lag response from T_{mean} to T_{min} which is much more precisely

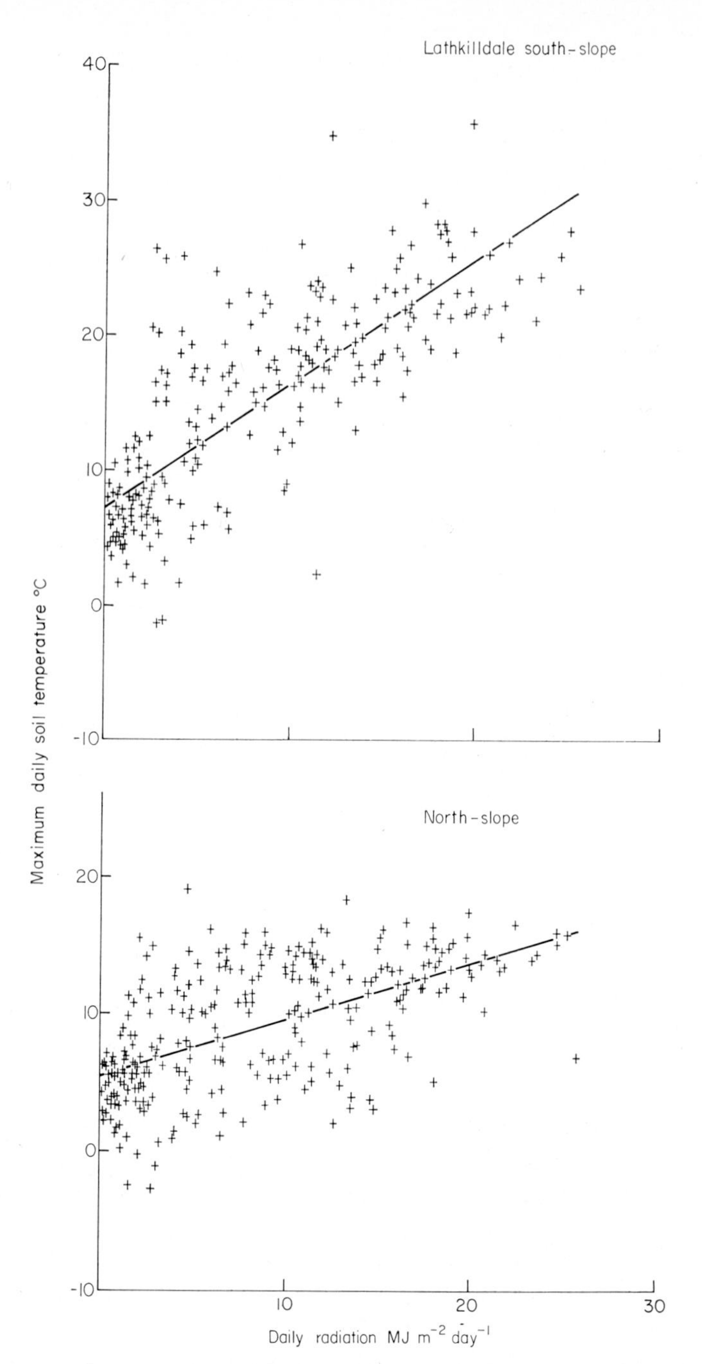

Figure 15.2. Scatter diagrams with linear regression lines fitted to show correlations between maximum daily soil temperatures and daily total radiation for the period 1.9.71 to 31.8.72.

South-slope: $T_{max} = 7{\cdot}207 + 0{\cdot}033$ R (95% limits $\pm 0{\cdot}003$)
Correlation coefficient $= 0{\cdot}801$ ($P < 0{\cdot}001$)

North-slope: $T_{max} = 5{\cdot}480 + 0{\cdot}015$ R (95% limits $\pm 0{\cdot}002$)
Correlation coefficient $= 0{\cdot}594$ ($P < 0{\cdot}001$)

demonstrated in other periods. When a 24 h lag is allowed for response in soil temperature, correlations relating to daily minimum temperatures are only marginally improved.

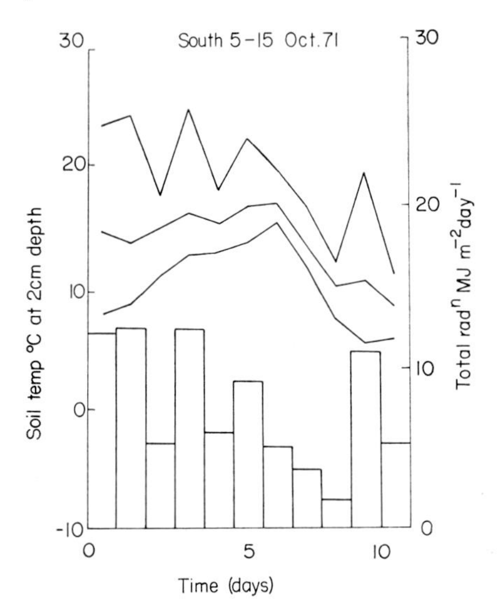

Figure 15.3a. Soil temperature and solar radiation data for the period 5–15.10.71 (details as in Fig. 15.1).

Figure 15.3b shows that on the north-slope between 1–12 March 1972 there was no measurable response in soil temperature to a fluctuation between 0·7 and 10·8 MJ m^{-2} d^{-1} in solar radiation. Mean soil temperature was *c.* 0°C and soil moisture was at field capacity. At such a low soil temperature fluctuation in solar radiation therefore has no detectable effect even at 2 cm depth.

Climatic extremes involving aspect differences are also important when studying plant germination and survival. These can occur in both summer and winter.

Figure 15.4 is a computer plot of mean soil and air temperatures on opposite dale slopes in midsummer. Days of intermittent cloud and sunshine are represented by multiple air peaks (e.g. day 12); completely overcast days by relative smooth peaks of small variation (day 21) and days with steady sunshine by extended smooth peaks (day 24).

This is best seen from the south-slope where maximum diurnal fluctuations in both air and soil are in the order of 30°C about a mean of 20°C. On the north-slope diurnal fluctuations are around 10–15°C, with a marked difference between soil and air—due in part to the much moister conditions prevailing on the north-slope at the time and also to the more complete vegetative cover.

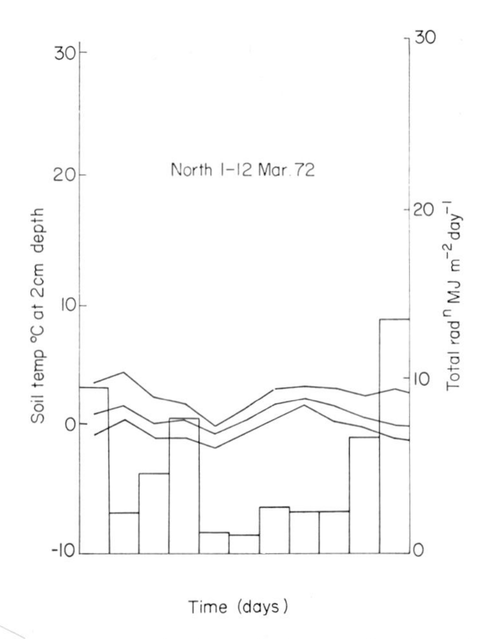

Figure 15.3b. Soil temperature and solar radiation data for the period 1–10.3.72 (details as in Fig. 15.1).

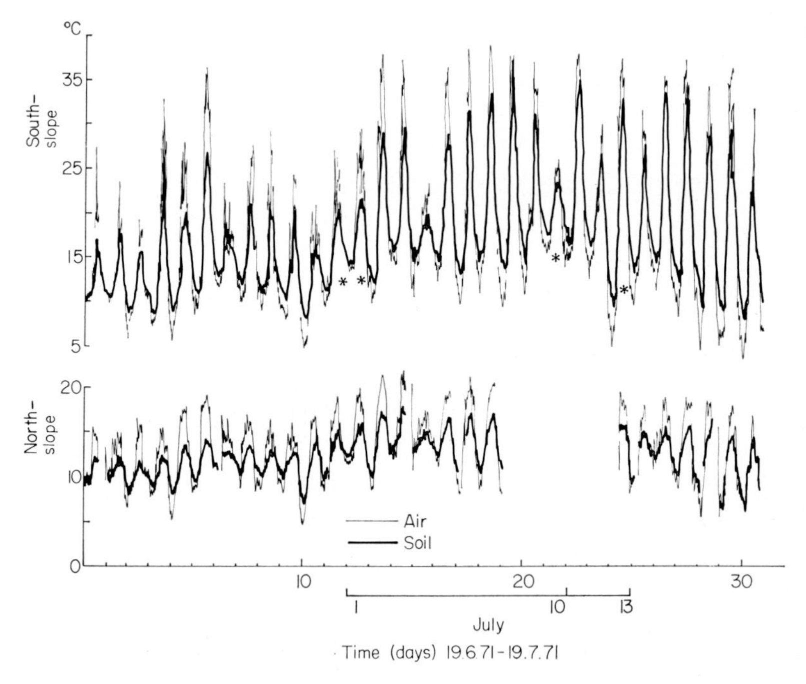

Figure 15.4. Temperatures 2 cm above and 2 cm below ground level for the period 19.6.71–19.7.71.

The second example (Fig. 15.5) relates to midwinter, starting with a mild period and passing through a cold spell of frost and snow. Apart from sunny spells, air temperatures are lower than soil temperatures. The effect of snow lying is particularly interesting. Snow fell from 22 December for seven days, gradually covering the air probes 2 cm above the ground. From 29 December onwards it was cold, clear and bright—snow melted on the south-slope only and exposed the air probes. Air temperatures fluctuated from −7·5° to +10°C and the soil was frozen and thawed daily for six days, while the north-slope lay insulated under a layer of snow.

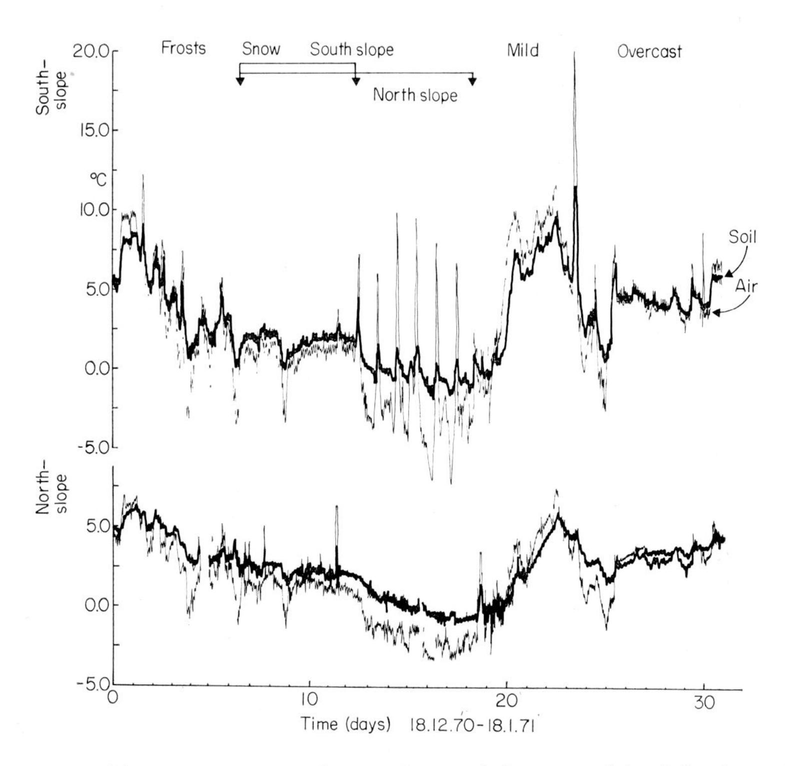

Figure 15.5. Temperatures 2 cm above and 2 cm below ground level for the period 18.12.70–18.1.71. Duration of snow cover is indicated.

Again, when a mild spell arrived (6 January) air temperatures on the south-slope reached +20°C, but were no higher than +7·5°C on the north-slope. In a final overcast spell (12 January 1971) soil and air temperatures were remarkably similar on both slopes.

The extremes of climate that occur tend to make the south-slope a less hospitable place for plant growth in both winter and summer.

In winter, freezing and thawing may increase mineralization in the soil but also cause heaving and solifluction and subsequent disturbance to plants which on the north-slope are unmoved and are often insulated from extremely low temperatures by a layer of snow.

In summer, the shallow soils of the south-slope quickly dry out so that there is little to slow the flow of heat in sunny weather and surface soil temperatures fluctuate widely.

Vegetation of south-slopes tends therefore to be more stunted, open and xeromorphic than that of the north-slope and a comparison of standing crop data (Sutton and Rorison 1970) confirms its relatively poor productivity.

Soil temperature and seedling emergence

Emergence characteristics

Preliminary experiments showed that whereas some species, e.g. *Rumex acetosa*, will germinate to some degree at almost any time of year, others such as *Scabiosa columbaria* L. are much more restricted. Apart from showing sensitivity to both climate and soil reaction, *S. columbaria* was chosen because its seeds have no dormancy and their germination is discontinuous. It was judged therefore to be a useful indicator of the range of factors operating at the three experimental sites.

The times taken to emerge were very similar on the north- and south-slopes and on the plateau. They were shortest in June and July and longest in March and October. No emergence was recorded between November and February. Since soil moisture levels were adequate throughout the winter, emergence data were examined in relation to soil temperature.

Figure 15.6 shows a computer trace of maximum, mean and minimum temperatures with an indication along the base-line of seedling emergence. Those sown at the end of March appeared at approximately the same time as the early May sowing and temperature readings suggested that a daily mean of 8–10°C was required to initiate germination.

Results of germination tests on thermogradient bars in the laboratory (Rorison 1973) supported the field results. The lowest steady temperature at which 50% germination was achieved was *c.* 9°C and the highest *c.* 34°C. When diurnal fluctuations of 5 and 10°C were applied the highest temperature tolerated remained at around 34°C and the lowest *mean* temperature remained at 9°C. In other words, *S. columbaria* could germinate while spending at least 12 hours at *c.* 4°C as long as the mean temperature was *c.* 9°C.

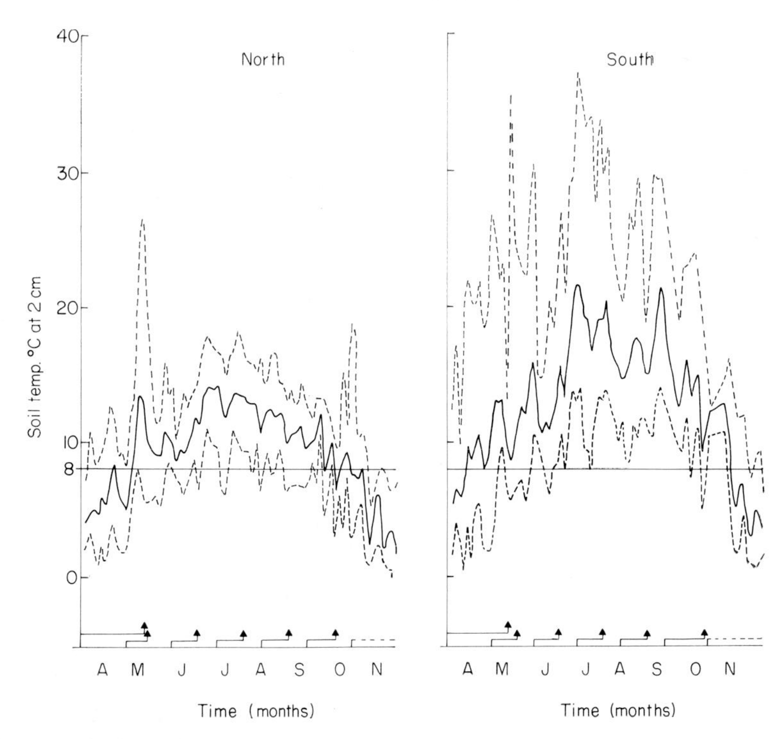

Figure 15.6. Time taken for start of emergence of sequential sowings of *Scabiosa* (arrowed lines at base) compared with three-day maximum, mean and minimum temperatures on north- and south-slopes. April to November 1971.

Survival patterns

Rate of seedling emergence was inversely proportional to the time taken to emerge (Fig. 15.6). The most complete emergence was from seed sown in early June which appeared within 21 days of sowing. The lowest percentage emergence was early and late in the season.

Figure 15.7a shows the pattern for the south-slope with low survival of Nos 1 and 5 and very low survival of No. 6. The soil was dry and open and the vegetation included *Festuca ovina*, *Helianthemum chamaecistus*, and *Arrhenatherum elatius*.

On the north-slope the soil was moister and the plant cover was denser and less xeromorphic, including *Succisa pratensis*, *Centaurea nigra*, and *Festuca ovina*. The survival pattern was similar to that on the south-slope; the seedlings were more vigorous although the total number emerging was smaller.

On the plateau the pattern of emergence and survival was very different from those on the slopes (Fig. 15.7c). Part of the reason lies in the acidic

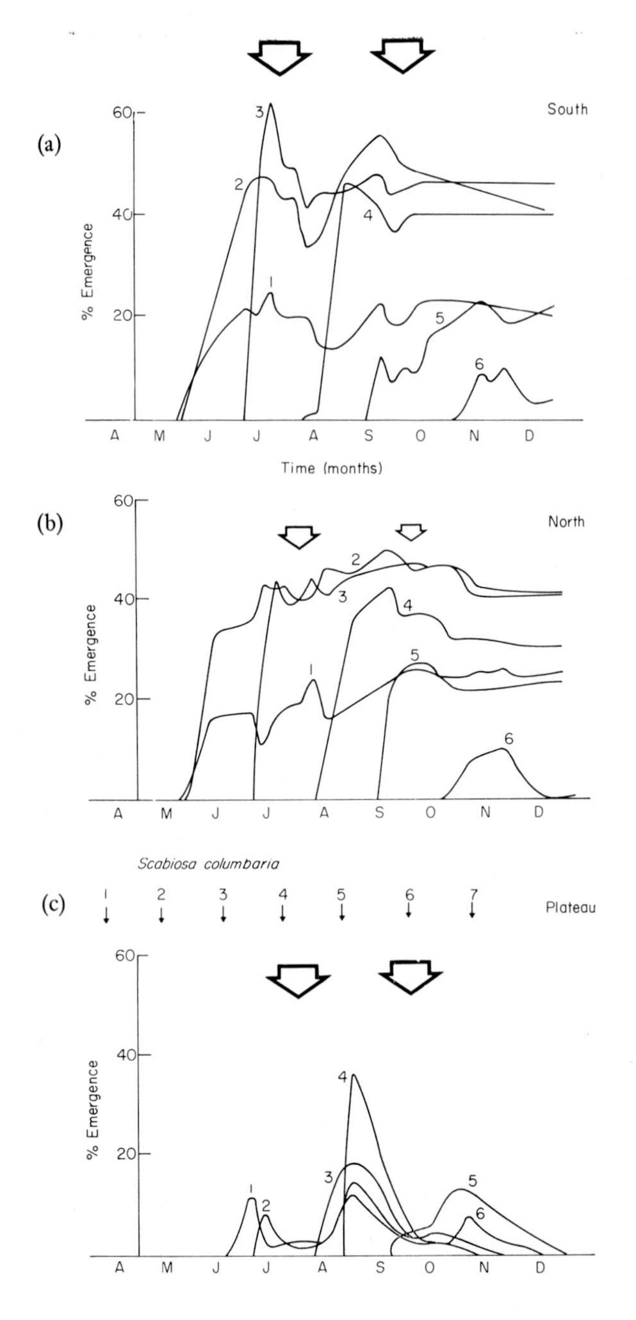

Figure 15.7a. *Scabiosa columbaria*. Seedling emergence on the south-slope.

Downward-pointing arrows numbered 1–7 indicate sowing dates. The two open arrows indicate two periods of soil moisture stress. The six numbered curves indicate seedling emergence and subsequent survival throughout the season from May until December. The seventh sowing failed to emerge.

Figure 15.7b. *Scabiosa columbaria*. Seedling emergence on the north-slope.

Figure 15.7c. *Scabiosa columbaria*. Seedling emergence on the plateau.

easily-dried surface of the treatment plots. The vegetation comprised *Deschampsia flexuosa* and *Festuca ovina* with occasional *Galium saxatile*. Only a few seedlings emerged and with few exceptions these failed to unfold their cotyledons.

Three years later (March 1974), *S. columbaria* survives only on the south-slope and the squares are still not invaded by the surrounding vegetation. All the squares are re-invaded on the north-slope and none of the seedlings remain. On the plateau, no seedlings survived the first winter and the dense turf has closed in again.

Soil moisture and soil reaction

The irregularities in the three sets of survival curves cannot be attributed to temperature differences at the three sites for these were all within the known tolerance limits for the species. On the slopes the losses incurred in July and, to a lesser extent, in September, coincide with two periods of high soil moisture tension (Pope 1973). These caused the deaths of recently-germinated seedlings and were increasingly severe from the north-slope (2 atm) through the south-slope (>15 atm) and on the mor layer of the plateau where radicles were particularly short. The dramatic reduction in seedling numbers on the plateau was due to both dryness and soil acidity—the latter preventing any root penetration into the A_1 layer where the native vegetation was rooted and which suffered only slight moisture tension (0·2 atm) even when the A_0 layer was at >15 atm.

Soil fertility

As a simple, integrated measure of fertility, four species were grown on the soils from the three sites for six weeks in a controlled environment. They were *Deschampsia flexuosa*, which occurred only on the plateau, *Festuca ovina* from all three sites, *Arrhenatherum elatius* from the south-slope and *Centaurea nigra* from the north-slope.

To distinguish sensitivity to acidity factors, the four species were also grown in plateau soil, the pH of which had been raised from 3·9 to 5·5 by the addition of $CaCO_3$.

The species responded to the soils very much as might be expected from their natural distribution (Table 15.3). They all grew as well on the north- as on the south-slope, suggesting equal fertility. *Festuca ovina* and *Arrhenatherum elatius* grew best on the plateau + $CaCO_3$, which implied that this soil was potentially the most fertile.

Table 15.3. Dry weight of seedlings (mg) after 6 weeks growth in a controlled environment.

Dry wt mg/plant	*Deschampsia flexuosa*	*Festuca ovina*	*Arrhenatherum elatius*	*Centaurea nigra*
Plateau (+ $CaCO_3$)	14·5	15·9	57·9	40·5
Plateau	14·7	6·1	14·7	3·3
North-slope	3·8	9·0	34·3	41·5
South-slope	4·3	7·3	29·7	42·2
Least significant difference (P = 5%)	3·1	1·6	5·4	6·3

The relative fertility of the soils may be judged from the response of *A. elatius* to potting compost in which it achieved 459·7 mg mean dry wt per plant in six weeks.

Discussion

This has been an exploratory examination of the relationship between solar radiation measured on a flat surface and the nearby soil temperatures of a plateau and of a north- and a south-slope. The analyses of one year's data show a highly significant linear relationship between the two, with the daily maximum temperature being closest to the total daily solar radiation. This is a much closer relationship than would be supposed from a comparison of monthly means (Carson 1961, Pope 1973). But while it may be a good general guide there are times when such relationships are considerably modified and, as such times may be important to the growth or germination of plants, they must be treated individually. Even the highest correlation coefficients quoted on p. 10 do not suggest a complete relationship between daily maximum soil temperature and daily total radiation unaffected by any other factor. Other components for which correction might be made are topographical obstruction (Pope and Lloyd, Chapter 16), variable air mass (e.g. fog) and wind pattern, and the thermal properties of each soil, all of which are complex within a dale system.

The levels of total radiation that were measured and related to soil temperature were not growth-limiting. Both the existing vegetation and early seedlings established in bare patches were more productive on the north- than the south-slope. Subsequent elimination of seedlings from the north-slope, which was completed by the winter of 1973, further attested to the vigour of the resident species. This vigour is most likely to reflect the better moisture levels maintained in the north-slope soils.

Soil temperature as such was never too high to inhibit germination even on the south-slope and it is probable that the major limitation to growth at the highest temperatures recorded is the associated soil moisture stress. *Scabiosa columbaria* and other tap-rooted species have root systems 40 cm into the limestone crevices (Salisbury 1952) and so established plants would rarely suffer droughting. Their growth however may be limited by high shoot temperatures (Watts, Chapter 17 in this volume; Clark, Chapter 18) which could build up in sunny weather. Critical measurements are needed.

Mean daily soil temperature was more often below 5°C on the north- than on the south-slope. Since this did not result in lower productivity, the better moisture conditions during most of the growing season must make up for any deficiency in soil temperature or direct solar radiation.

S. columbaria seedlings from all the 1971 sowings have survived on the south-slope and flowered in 1974 for the first time. The open nature of the vegetation and the apparently slow rate of growth of the resident species make the possibility of inter-specific competition above ground unlikely for several more years. On this slope there are a number of species, including *Scabiosa*, which are adapted to relatively warm, dry and infertile conditions. The north-slope, which differs in being moist and cool, and the plateau, with intermediate physical conditions and highly acidic reaction, support other distinct floras. An exception which occurs at all three sites is *Festuca ovina*.

Despite differences in the amount of mineralization that occurs, aspect has little effect on soil fertility as measured by the bioassay using four of the local species (Table 15.3). Regular analyses for available mineral nutrients (Rorison and Gupta, in preparation) over several years have also failed to show any marked differences between the contents of soils of the north- and south-slopes. They are, however, very shallow infertile soils in which nutrient levels are low at all times.

It is concluded that the major effect of solar radiation on plant growth in these dale sites is through evaporation influencing soil moisture. Soil temperature, fertility, and light intensity play a lesser role. For germination and seedling establishment, minimum soil temperature and soil reaction are critical factors.

To be more precise requires more experimentation, particularly into growth response, and this is currently proceeding. Refinements in calculating relationships between solar radiation and soil temperature which take account of wind and variable air mass have still to be applied.

Acknowledgements

We are most grateful to Professor J. L. Monteith and Dr. M. H. Unsworth for advice and practical help during the setting up of the solarimeter system, to our colleagues in the Unit who were involved in data collection and handling, particularly Mrs. R. Spencer, P. L. Gupta and Dr. R. Hunt, and to Mr. K. Bocock (Institute of Terrestrial Ecology, Merlewood) for preliminary computer analyses.

Appendix

Switching circuits for a self-contained solarimeter system

Details of the time-sharing circuit are given in Fig. 15.8. The integrated timer circuit, A, is connected as a free-running multivibrator producing a negative going pulse, the width and recurrence frequency of which is determined by the external network, R_1, R_2 and C_1. The values shown give a pulse width of approx. 15 sec. recurring at approx. 2 min. The leading edge of each pulse triggers the bistable circuit, C, via the steering circuit, B, providing at the output a symmetrical rectangular wave of twice the timer frequency and amplitude 6 volts. This output is fed to a compound emitter-follower circuit, Tr1 and Tr2, which, during the positive half-cycle, amplifies the current sufficiently to energize the four reed coils, L1 to L4. These in turn operate the four changeover reed switches, S1 to S4. The value of the resistor, R6, should be selected, on test, to produce a maximum of 6 volts at the emitter of Tr2. Thus the integrator is alternately switched to measure and record Total and Diffuse radiation, each for a period of 2 mins.

The timing contacts, Km, Kf, were cut from thin phosphor-bronze sheet and bent to the shape shown. The trailing contact, Km, was soldered to the minute hand of the clock, Kf being fixed to, but insulated from, the clock face in a position such that the contacts closed on the hour. The moving contact was manipulated so that it exerted a minimum of pressure during its travel over the fixed contact. An insulated lead was brought out from Kf, whilst connection to Km was made through the body of the clock. The wiping action of the contacts has a self-cleaning effect but it was thought advisable to apply a little contact lubricant before replacing the clock in its case. The original equipment has been in operation for more than two years without need of further adjustment or cleaning of the clock contacts.

The capacitor, C4, is charged to +12v through the resistor, R7, and, on closure of the clock contacts, the reed relay, L5, S5, is energized, allowing C4 to discharge via R8, into the base of Tr3. Relay RL1 is energized by the amplified emitter current and its contacts, K1, initiate the printout of both the Diffuse and Total counters. The relatively low-resistance discharge path

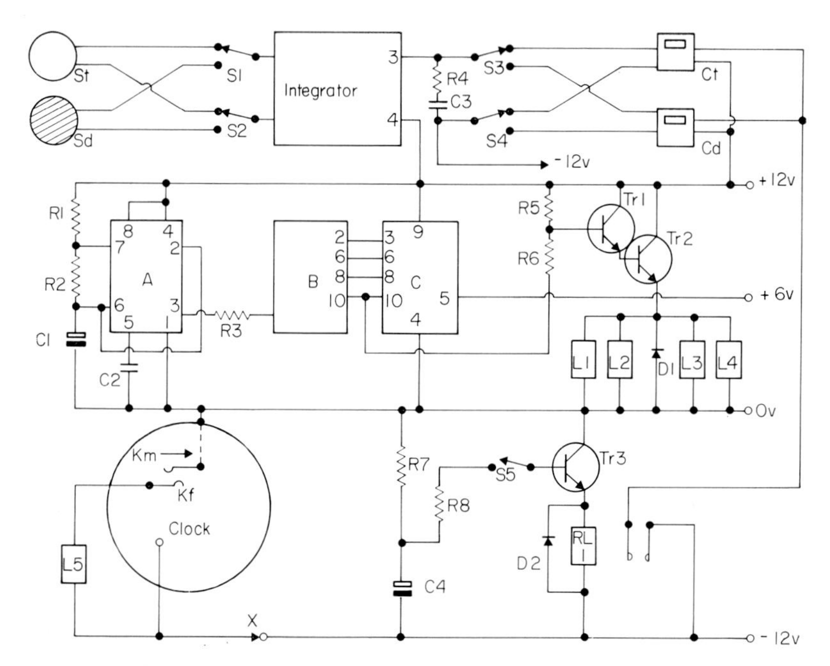

Figure 15.8. Time-sharing and printout initiation circuits. The connections required for the Digital Integrator and the Printing Impulse Counters are clearly described in the manufacturers' literature.

Key to components:
*A Timer A; B OMY102; C OMY121; Cd Diffuse counter; Ct Total counter; Sd Diffuse solarimeter; St Total solarimeter; Tr1–3 BC109; D1, 2 1N4148; R1 4·7M; R2 820K; R3 2·2K; R4 47; R5 1M; R6 See text; R7 10M; R8 100K; C1 25μF; C2 0·01μF; C3 2μF; C4 10μF; *L1–4 Reed coil 1; *L5 Reed coil 2; *S1–4 13-RSR-A; *S5 4-RSR-A; *RL1 Relay 912.
* Supplied by RS Components Ltd.

Other components may be obtained from local suppliers. Connection is shown for a positive-earth clock; if, however, a negative-earth instrument is used, connection X should be to + 12v.

prevents C4 from recharging until the clock contacts open, when a period of approx. 1 minute must elapse before the circuit is ready for further operation. This prevents unwanted printout due to possible irregularity in operation of the clock contacts.

References

BRIDGES E.M. (1961) Aspect and Time in Soil Formation. *Agriculture* 68, 358–63.

CARSON J.E. (1961) Soil temperature and weather conditions. Argonne National Laboratory Report ANL 6470.

DE VRIES D.A. (1952) A non-stationary method for determining thermal conductivity of soil in situ. *Soil Sci.* 73, 83–9.

FRANZMEIER D.P., PEDERSON E.J., LONGWELL T.F., BYRNE J.G. & LOSCHE C.K. (1969) Properties of some soils in the Cumberland plateau as related to slope aspect and position. *Soil Sci. Soc. Amer. Proc.* 33, 755–61.

GEIGER R. (1961) *Das Klima der bodennahen Luftschicht.* 4th ed. Braunschweig. (Other editions include translation, *The climate near the ground*, Cambridge (Mass.), 1965.)

HABERLANDT F. (1874) Die oberen und unteren Temperaturgrenzen für die Keimung der wichtigeren landwirtschaftlichen Sämereien. *Landw. Vers. Sta.* 17, 104–16.

JARVIS M.S. (1963) A comparison between the water relations of species with contrasting types of geographical distribution in the British Isles. *The Water Relations of Plants.* (Ed. by A.J. Rutter and F.H. Whitehead.) Blackwell, Oxford, pp. 289–312.

LLOYD P.S. (1968) A miniature gypsum resistance block for measuring soil moisture tensions. *The Measurement of Environmental Factors in Terrestrial Ecology* (Ed. by R.M. Wadsworth). Blackwell, Oxford, p. 273.

LUDWIG J.W., BUNTING E.S. & HARPER J.L. (1957) The influence of environment on seed and seedling mortality. III. The influence of aspect on maize germination. *J. Ecol.* 45, 205–24.

MONTEITH J.L. (1973) *Principles of Environmental Physics.* Edward Arnold, London.

PIGOTT C.D. (1958) Biological Flora of the British Isles: *Polemonium caeruleum* L. *J. Ecol.* 46, 507–26.

PIGOTTT C.D. (1968) Biological Flora of the British Isles: *Cirsium acaulon* (L.) Scop. *J. Ecol.* 56, 597–612.

POPE D.J. (1973) A study of plant growth in relation to topography in the Derbyshire dales. Ph.D. thesis, University of Sheffield.

RATCLIFFE D. (1961) Adaptation to habitat in a group of annual plants. *J. Ecol.* 49, 187–203.

RORISON I.H. (1973) Seed ecology—present and future. *Seed Ecology* (ed. by W. Heydecker). Butterworth, London, pp. 497–519.

SALISBURY E.J. (1952) *Downs and Dunes: their plant life and its environment.* G. Bell & Sons Ltd., London.

SUTTON F. & RORISON I.H. (1970) The modification of a data logger for the recording of temperatures in the field, using thermistor sensors. *J. appl. Ecol.* 7, 321–9.

SUTTON F. & RORISON I.H. (1972) An automatic data-translation system, using a simple adaptation to link a D-Mac portable translator and a Solartron data logger. *J. appl. Ecol.* 9, 112–6.

Questions and discussion

Weather and soil temperature

Professor J. L. Monteith drew attention to the relatively weak annual correlations between daily solar radiation and daily maximum soil temperature (Fig. 15.2). He attributed the scatter to variations of air mass—to incursions of air coming from different regions and in consequence having different characteristics. Some information existed on the frequency and temperature distributions of air masses. It was possible for different air masses to invert the usual correlation between radiation input and temperature: for instance, cloudy moist air masses could be warm relative to the time of year, while relatively cold Arctic air could produce long periods of sunshine.

He also pointed out that differences in soil moisture content from year to year, especially in early summer, could be influenced by rainfall distribution at least as strongly as by differences in radiation through the year. Dr. Rorison added that water stress was relatively infrequent in the Derbyshire climate.

Soil heat balance and rates of drying

Professor C. D. Pigott attributed part of the scatter in the relation between soil temperature and daily radiation to differences in the rate at which the surface soil dried out. On south-slopes the surface was liable to dry on the first dry day after rain, forming a hard crust or a fluff and leading to a rapid decline in evaporative cooling. On north-slopes the soil might remain moist for 3–4 days after rain. The resulting pattern of differences in energy balance was further complicated by differences in soil texture and plant cover on the two aspects. (See pp. 363, 375, and Table 15.1.)

Survival of seedlings of Scabiosa columbaria *germinating at different times*

In response to questions by Dr. G. C. Evans, Dr. Rorison pointed out that the highest rate of survival to the end of the first season, as a proportion of the number of seeds sown, was shown by the seedlings germinating in May and July (Fig. 15.7). Evans suggested that in some species seedlings that had germinated early in the season would be at an advantage because they could then resist soil moisture stress better; Rorison showed that such an advantage existed in *Scabiosa* but was offset by the much lower germination rate of the earlier-sown seed.

Rorison said that in this species the optimum conditions for germination were not synchronized with the time of seed fall. In the *natural* environment the best chance that seeds had of surviving was to germinate immediately after seed fall in September and October, even though the germination rate at that time of year was poor, because the survival of seed in the ground over the winter was virtually nil.

16 Hemispherical photography, topography and plant distribution

D. J. POPE[1] and P. S. LLOYD[2] (*N.E.R.C. Unit of Comparative Plant Ecology, University of Sheffield*)

Introduction

The influence of topography on the distribution of plants has long been recognized. Similarly the effect of topography on the radiation properties of receiving surfaces has been well recognized and the subject of numerous calculations (Dirmhirn 1964, Geiger 1966). Ecologists in northern latitudes know slopes of northern aspect to be relatively cool and moist, while slopes of southern aspect are relatively warm and dry. The interpretation of plant distribution in relation to topography has often been left at that level of analysis, but some studies have involved detailed measurement and correlation (e.g. Lundqvist 1968) and a few (e.g. Pigott 1968) have given strong evidence of how, in the case of particular species, the observed distribution is controlled. In the last example, *Cirsium acaulon* (L.) Scop. was studied in Derbyshire, where it is near the northern limit of its distribution in Britain and where it is largely confined to south- or south-west-facing slopes. It was demonstrated that temperature differences between shaded and unshaded flowers were sufficient to prevent the proper formation of seed in the capitula of shaded plants. The inference is that slopes of northern aspect are not sufficiently warm to allow the regeneration of the species.

Such direct effects of the differential radiation properties of slopes of different aspect may be reinforced if there is additional topographic obstruction of the horizon. In this paper a method is described for recording and calculating the distribution of radiation on slopes of differing aspect, taking quantitative account of the effect of topographic obstruction. The calculation is used to work out the instantaneous distribution of radiation in relation to solar altitude and to compute the pattern of daily and cumulative radiation totals. These provide the figures for a site radiation index, which is used to interpret the distribution of plant species.

The distinction between direct and diffuse solar radiation is an important

[1] Present address: Middlesex Polytechnic, Queensway, Enfield.

[2] Later at the departments of Botany and Forestry, University of Oxford. The Editors greatly regret to announce the untimely death of Dr. Lloyd while this paper was in proof.

one in the context of topography. Input of direct radiation to sites is strongly influenced by the site topography, because of the directional nature of the radiation. Input of diffuse radiation is much less dependent on topography, because it is less strongly directional. Indeed under the conditions of a uniformly overcast sky, diffuse radiation is non-directional and the proportion of diffuse radiation received on a sloping surface (relative to that received on a horizontal surface) will be related to the proportion of the sky visible from the site. Because of the variable and unpredictable nature of diffuse radiation, a simple index of the radiation properties of sites is best based on direct radiation alone. Attempts to allow for diffuse radiation (e.g. Garnier & Ohmura 1970) have not met with great success. A further cogent reason for focussing attention on direct radiation only is that the intensity of energy input is potentially much greater than from diffuse radiation. Although on an annual basis as much as 60% of the total radiation may be from diffuse sources (Hughes 1966), extreme conditions are brought about by direct radiation rather than diffuse. The short term heating and drying effects of intense direct radiation are likely to be of much greater ecological significance than an equivalent quantity of diffuse radiation received over a longer period. Development of a site radiation index has thus been limited to consideration of direct radiation only.

Early calculations of the radiation properties of sites of differing slope and aspect were of general applicability and did not attempt to make allowance for topographic peculiarities of individual sites, which may seriously deplete the input of direct radiation. Garnett (1937) included estimates of sunshine duration in her map-based calculations of insolation in alpine topography. More recently, Williams *et al.* (1972) have used a geometric solution for topographic obstruction, also involving altitude measurements from contour maps. Garnier & Ohmura (1968, 1970) have developed a method of calculating radiation input to sites which includes an unspecified technique for allowing for skyline obstructions. Turner (1966) has published a detailed analysis of the radiation properties of a small area (6·86 ha) in mountainous terrain involving more than 2700 site measurements of the skyline, using the horizontoscope of Tonne (1958). This is a parabolic convex mirror in which the reflected horizon can be viewed against solar track diagrams or other constructed grids. We here use a technique for allowing for topographic obstruction based on hemispherical photography which was pioneered by Hill (1924) and developed for use in woodlands by Evans & Coombe (1959) and Anderson (1964). This initial site characterization is followed by a calculation of radiation properties using the method of Garnier & Ohmura (1968).

The methods described were developed as part of a study of the influence of topography on the distribution of plants in grassland communities in the Peak District of Derbyshire (Pope 1973).

Site characterization

The technique involves taking a hemispherical photograph vertically upwards at each site to record the angular height and position of any topographic obstruction of the horizon. The superimposition of calculated solar track diagrams enables the potential duration of direct radiation to be measured for any day of the year. It is therefore possible to characterize the light climate of each site in a single brief visit. Unlike studies in deciduous woodlands (Anderson 1964) it is not necessary to photograph grassland sites more than once, since the interest is in light reaching the community and not the fate of light within the community.

Photographic method

The camera used was a Nikon F 35 mm fitted with a Nikon 7·5 mm f5·6 Fish-eye lens, which has an acceptance angle of nearly 180° and an approximately equiangular projection. Measurements of the characteristics of another lens of this model are given on p. 555.

The lens has a fixed focus (from 70 cm to ∞) and has built-in filters. Preliminary trials showed that a light green filter (×0) gave good results. Exposures were based on the need for a clear image of the horizon and a Weston Mark IV exposure meter was used to obtain an appropriate value. A panchromatic film was used. Exposure was increased two times for the filter and also increased two stops for the lens.

The photographs produced are circular and are enlarged to a standardised diameter (24 cm). They were developed to give a good image of the horizon, ignoring detail of vegetation, clouds or other features. Similar lenses have been used previously for environmental studies by Hill (1924), Evans & Coombe (1959), Anderson (1964) and Madgwick & Brumfield (1969).

In the field the camera was set up on an aluminium platform approximately 30 cm square. One edge of the platform rested against the soil surface; two independently extendable legs on the opposite side enabled levelling to be carried out using an accurate spirit level. The camera was positioned in a rivetted aluminium assembly such that when the platform was horizontal the lens pointed vertically upwards. The platform design enabled the photograph to be taken within 10 cm of the ground surface so that the apparent horizon was virtually that seen from the level of the grassland vegetation. A marker post was placed approximately 1 metre from the platform in the direction of magnetic North, carrying a card numbering the site. Each negative was thus numbered and could be subsequently orientated to true North.

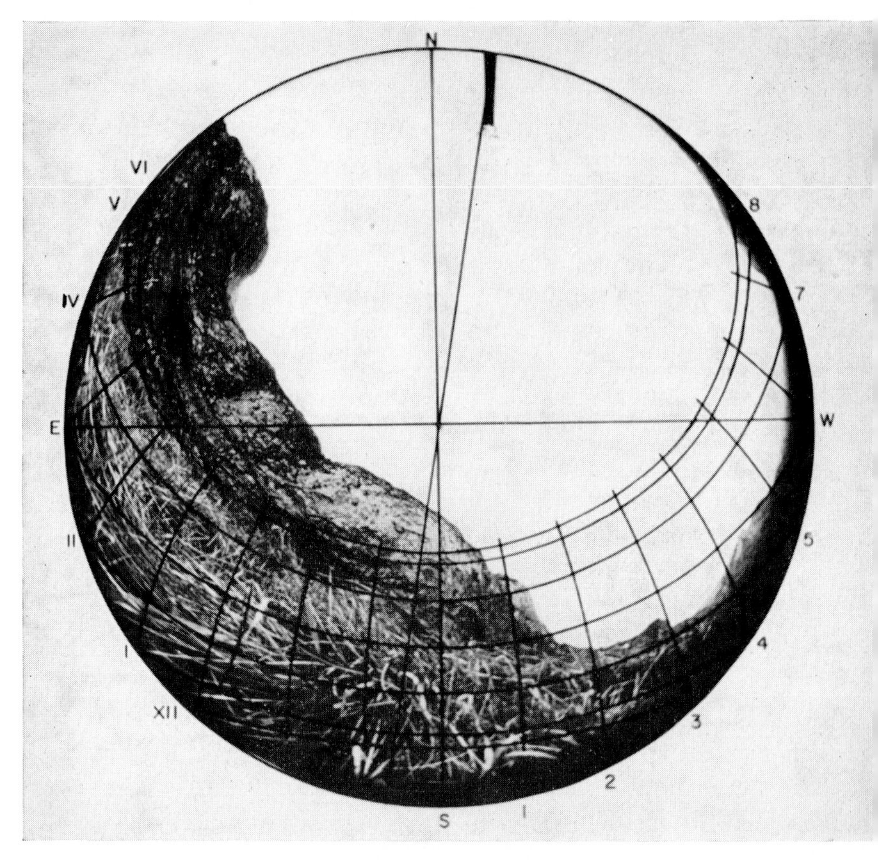

Plate 16.1. Hemispherical photographs of two north-facing sites, with superimposed solar

Adjustments for exposure were made before placing the camera on the platform and the photograph was taken by time delay to enable the operator to retreat to an inconspicuous position.

Photographs were taken at each of 199 sites in the Derbyshire dales and were interpreted using transparent overlays marked with solar tracks appropriate for 53°15′ N (the latitude of the Wye Valley, Derbyshire). Tracks were calculated by the method of Evans & Coombe (1959) using their table of mean declination. The seven tracks are correct respectively for 22 June, 22 May and July, 22 April and August, 21 March and 23 September, 21 February and 22 October, 22 January and 21 November, and 22 December.

Using the overlays, which were constructed to fit the size of the photographs (24 cm), the effect of topography on the apparent time of local sunrise and sunset could be read for each of the solar tracks, representing 12 days in the year. These data were used in the subsequent calculation of the insolation

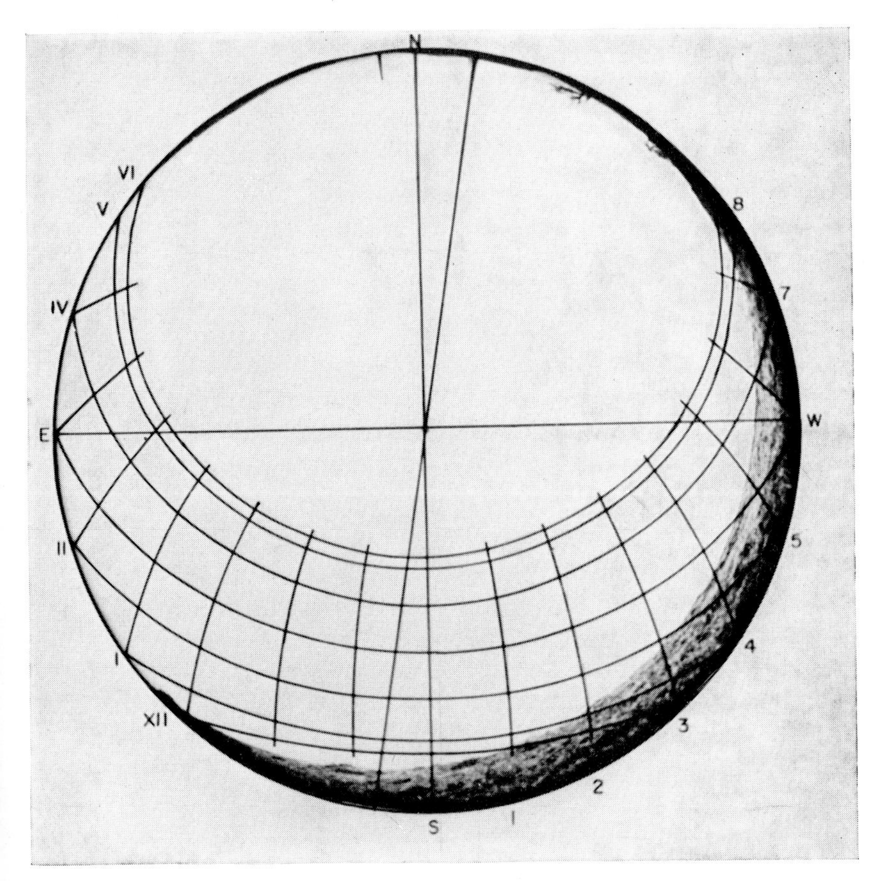

tracks: (a) Winnats Pass. Aspect 350°, slope 37°. (b) Monsal Dale. Aspect 56°, slope 20°.

properties of each site. (Examples of photographs with superimposed overlays are shown in Plates 16.1 and 16.2.)

Calculation of direct solar radiation

Calculations are based on a simplified law of transmission (Haltiner & Martin 1957):

$$I_m = I_0 . p^m \qquad \text{Equation (1)}$$

where I_m is the direct radiation incident at the earth's surface, I_0 is the solar constant, p is the atmospheric transmissivity and m is the optical air mass (i.e. the path length through the atmosphere based on $m = 1{\cdot}0$ for the vertical distance from the earth's surface to the top of the atmosphere). According to recent estimates the value of I_0 is approximately 2·00 cal

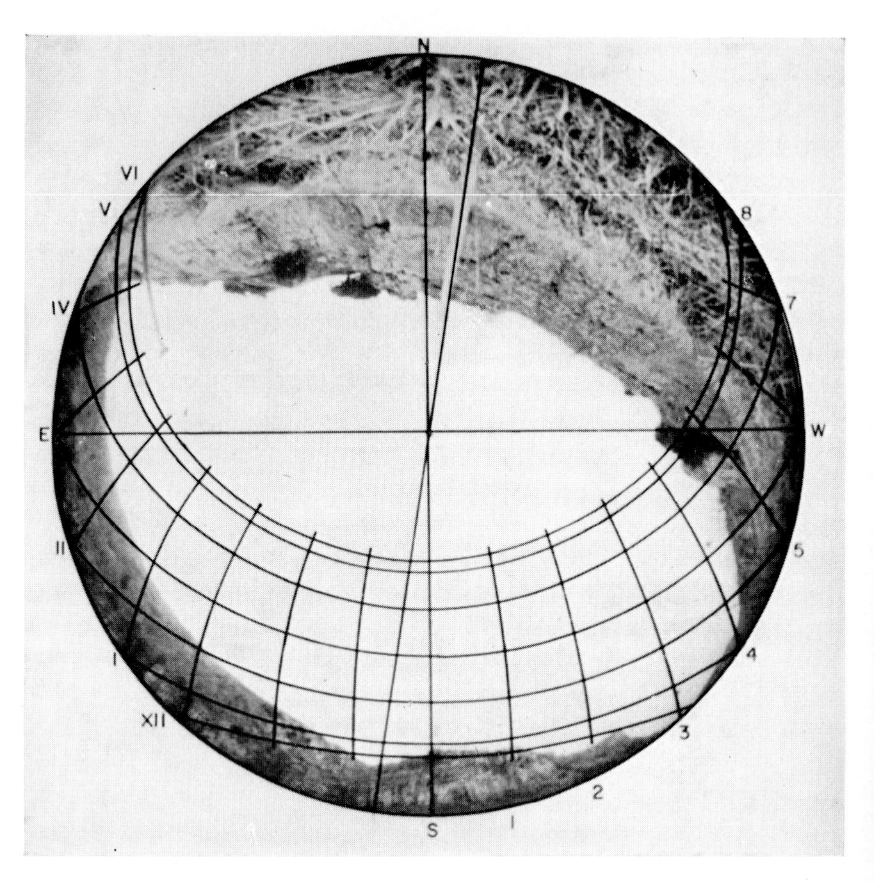

Plate 16.2. Hemispherical photographs of two south-facing sites, with superimposed solar

cm^{-2} min^{-1} ($13{\cdot}96 \times 10^2$ W m^{-2}). The transmissivity p is that factor, usually expressed as a decimal, by which light is reduced as it passes through air of a depth equal to the vertical height of the atmosphere. It is a measure of atmospheric absorption. By Beer's Law transmission is reduced exponentially by a linear increase in path length of optical air mass m.

Garnier & Ohmura (1968) proposed a correlation factor for slope and aspect that can be used to calculate the attenuation of incident direct radiation at any slope and aspect at any latitude:

$$I_s = I_0 . p^m . \cos(\mathbf{X} \wedge \mathbf{S}) \qquad \text{Equation (2)}$$

where I_s is the direct radiation incident on the slope, **X** is a unit coordinate vector normal to the slope, **S** is a unit coordinate vector expressing the height and position of the sun and $\wedge$ is a symbol expressing the angle between **X** and **S**.

The equation given for the calculation of $\cos X \wedge S$ is:

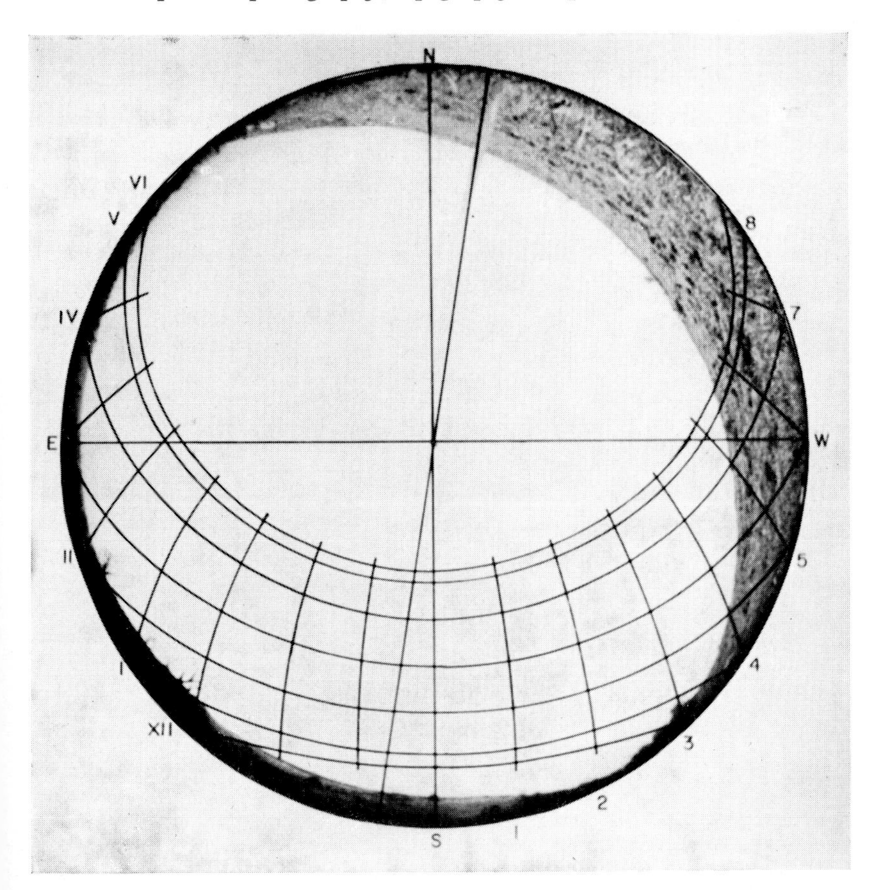

tracks: (a) Wye Dale. Aspect 198°, slope 38°. (b) Blackwell Dale. Aspect 147°, slope 28°.

$$\cos \mathbf{X} \wedge \mathbf{S} = [(\sin \phi \cos H)(-\cos A \sin Z) - \sin H (\sin A \sin Z) + (\cos \phi \cos H) \cos Z] \cos \delta + [\cos \phi (\cos A \sin Z) + \sin \phi \cos Z] \sin \delta \qquad \text{Equation (3)}$$

where ϕ is the latitude, H is the hour angle measured from solar noon positively toward west, A is the azimuth (aspect) of the site, Z is the zenith angle (slope) of the site and δ is the sun's declination.

These equations have been used to calculate the direct radiation potentially incident on each site. Equation (2) gives the value of radiation reaching a slope per minute. The daily total of direct radiation potentially received by the site can be obtained by calculating the value of I_s for each minute between sunrise and sunset and totalling the values. However, Garnier and Ohmura showed that a calculation for each 20 minute period, treated as a mean value for the period, gives the same total (to within 5%) as the longer calculation. The shorter method has been adopted here.

A computer program (written in FORTRAN IV) was devised to calculate the potential direct irradiance of any site using Garnier & Ohmura's correction factor and at the same time allowing for the effects of topographic obstruction of the horizon. For convenience the program was written to allow for up to two periods of direct radiation during the day although in principle there is no reason why allowance should not be made for any number. In practice, in the sample of 199 sites for which photographs were available, only one was found in which more than two periods of direct sunlight were possible. As this was due to a momentary occlusion of the sun behind a peak, followed by a second occlusion during which sunlight could not reach the site for a much longer period, the calculation was made on the basis of two periods of sunlight ignoring the shorter occlusion (thereby incurring an over-estimate of only *c.* 2%).

The major steps in the computation are shown in Fig. 16.1. The following parameters are incorporated in the program and can be altered as appropriate:

Latitude is set by a single card. For present purposes a value of 53° 15′ N has been adopted. The full latitudinal range of the study area was 53° 11′ N to 53° 21′ N;
declination is listed in the programme from a table given by Evans & Coombe (1959);
transmissivity is also set by a single card;
optical air mass is listed in the programme from a table given by List (1951).

The data cards for each site comprise a series of numbers: the site number, then the aspect and slope in radians, and then seven groups of four numbers which are the data taken from the site photograph by means of the seven solar tracks. The numbers are the times, in units, of sunrise and sunset for the first and second periods of direct sunlight. The time units are 20-minute periods numbered from 0300h. Since in the majority of sites there is only a single period of sunshine, the second pair of numbers is usually zero.

The optical air mass was read from tabulated values (List 1951) appropriate to the solar altitude as calculated by the equation given by Evans & Coombe (1959). Daily totals of direct radiation, calculated by summing the 20-min. values of I_s (Equation 2), are printed out together with the site number, slope, aspect and transmissivity value used. The whole procedure is repeated for each of the seven data sets corresponding to the solar tracks. In order to simplify the use of the results, a site radiation index has been devised by reducing the seven daily totals to a single annual value (by summing the estimates appropriate for each of the twelve months). Alternative indices may be obtained by summing the values appropriate for any desired period.

The daily values calculated by the computer programme represent the maximum amount of direct radiation potentially received by a site on a cloudless day, subject to the value adopted for the transmissivity. The yearly or seasonal indices are thus purely nominal, but serve to order the sites in respect of insolation properties.

Of the parameters used in the calculation, latitude is a constant; aspect

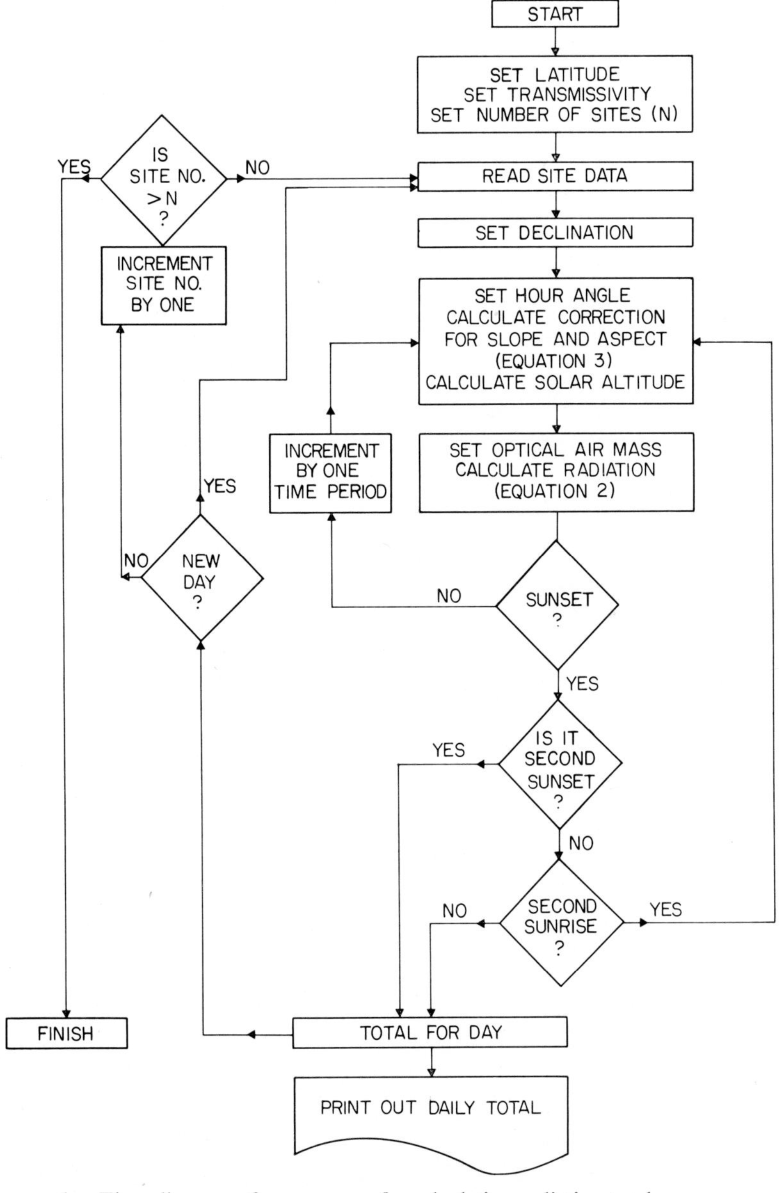

Figure 16.1. Flow diagram of programme for calculating radiation totals.

and slope, sunrise and sunset are features of each site; the hour angle is a function of time of day; the solar constant is treated as constant (though subject to seasonal fluctuation of $\pm 3\%$); and both declination and optical air mass can be read from tables. The only parameter not easily found is transmissivity. The choice of transmissivity values requires further discussion.

The transmissivity factor expresses the depletion of direct radiation as it passes through the atmosphere. It is therefore the sum of several different processes including Rayleigh scattering by molecules, scattering by dust, and absorption by gases such as ozone and water vapour. It is possible to calculate this depletion using Linke's formula (cf. Stagg 1950), but it is also possible to justify the assumption of a particular value. Firstly, Linke's formula involves a number of additional assumptions; secondly, as Garnier & Ohmura (1968) state, it is not worthwhile to calculate for transmissivities less than 0·5 since by then diffuse radiation is contributing a high percentage of the total radiation. Haltiner & Martin (1957) give a diagram of radiation calculated by Haurwitz using a transmissivity of 0·7 but they also suggest figures for values of 0·6–0·9.

The relationship between values of I_s obtained with different values of transmissivity p is determined by p^m (Equation 2), where m varies diurnally and seasonally. The minimum (noon) values taken by m at the solstices at latitude 53° 15′ N are 1·15 (June) and 4·22 (December). Data for 38 dummy sites are plotted in Fig. 16.2 for two values of p and three declinations. These

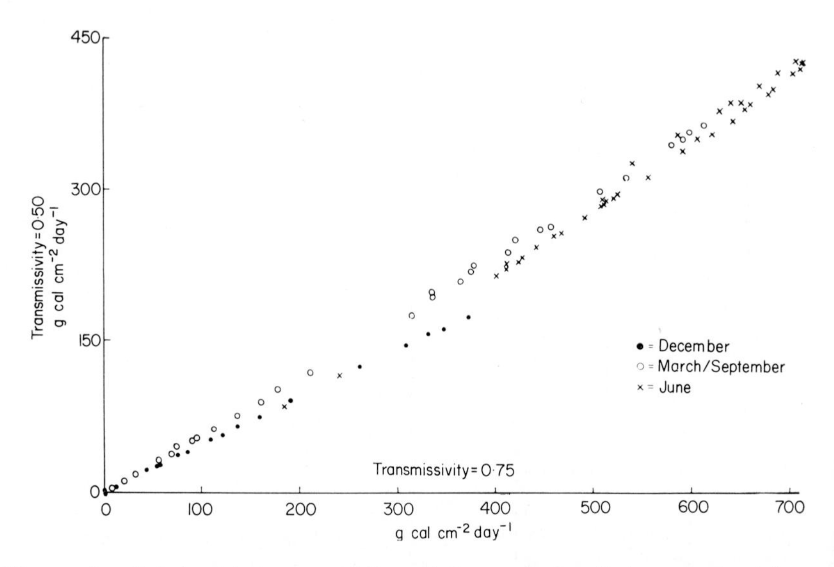

Figure 16.2. Relationship between daily radiation totals for 38 sites calculated for two values of atmospheric transmissivity at three solar declinations. For clarity zero values are omitted.

'site' data were contrived to simulate a range of slopes and aspects with maximum insolation period in each case. Scatter between sites at any one declination is due to the effects of slope and aspect in modifying the angle of incidence and in limiting the duration of insolation. The form of the relationship is seen to be roughly linear over much of its range and, since the object is a comparative estimate of potential radiation, it seems reasonable to regard any value of p between 0·5 and 0·85 as acceptable. Subsequent calculations have been based on a value of $p = 0{\cdot}70$ except where specified.

Radiation and topography

The influence of slope and aspect on the irradiance of a site can be examined by reference to the surface of a dome. For most ecological purposes a maximum slope value of 60° is realistic and at this stage no account is taken of obstruction of the horizon save that due to the slope alone.

The instantaneous distribution of direct radiation on surfaces of different slope and aspect is shown in Fig. 16.3 relative to that received on a surface normal to the sun's rays. The data have been obtained by calculating cosine depletion of radiation from Equation 3, assuming the sun to be in the south ($H = O$), for declinations appropriate to 22 December, 21 March, and 22 June at latitudes 50° and 60° N. At the lowest solar altitudes, a large sector of the northern quadrant does not receive direct radiation and the intensity of radiation increases towards the steeper south-facing slopes. At the highest solar altitudes, almost all parts of the dome are insolated; but the intensity on steep north-facing slopes is very much less than that on shallow south-facing slopes, and at latitude 60° N. north-facing slopes steeper than $53\frac{1}{2}$° are still in shadow. (If the sun were overhead, as at noon in the tropics, a horizontal surface would receive direct radiation at full intensity and all slopes irrespective of aspect would receive proportionately less radiation according to the cosine of the angle of slope).

The same instantaneous pattern of irradiation is to be found for the same solar altitude at any time of year, provided that the sun be regarded as being in the south and the polargraph orientated appropriately. Figure 16.3 thus provides a guide to the relative intensity of radiation and of heating for the most extreme conditions at latitudes 50° and 60° N and for two less extreme conditions. These latitudes span the range encountered in the British Isles. Cosine depletion is of course valid for all surfaces and angles of incidence.

Similar diagrams in Fig. 16.4 show the daily totals of radiation potentially received on the same three days for latitude 53° 15′ N. The figures have been calculated from Equation 2 using dummy data giving maximum daylength for each site. As before, the distribution of radiation in December is very

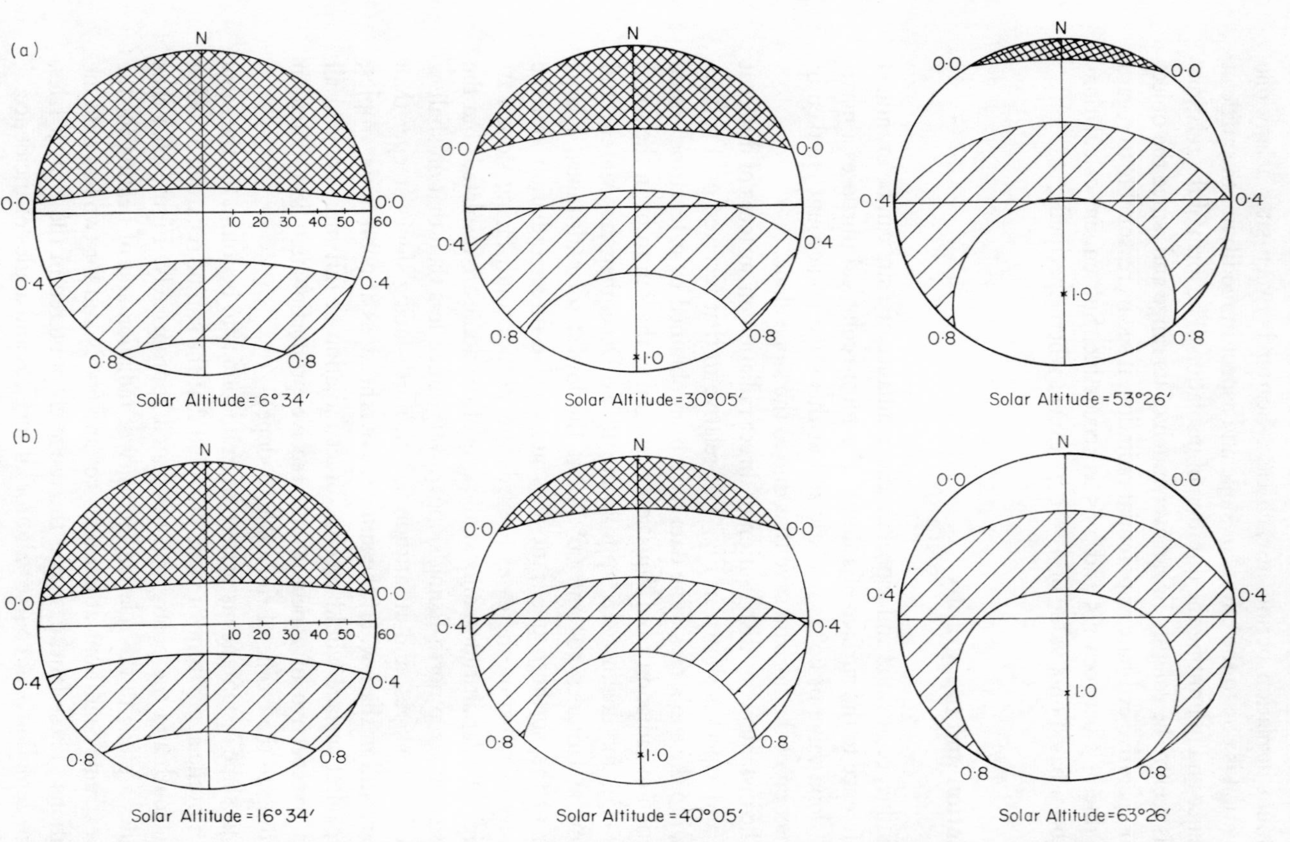

Figure 16.3. Relative intensity of direct radiation at noon on slopes of up to 60° and all aspects for three solar altitudes appropriate to the solstices and equinoxes at (a) 60°N and (b) 50°N. In each polargraph slope is read concentrically

uneven and increases from the sunless northern sector to a value of over 300 cal cm^{-2} day^{-1} on a 60° south-facing slope. In March the sector without direct radiation is smaller but still present and the zone of maximum radiation is now on a south-facing slope slightly less than normal to the sun's rays at noon. In June the whole dome receives direct radiation, a wide range of slopes and aspects receive similar levels and the maximum is at about 12° on a south-facing slope.

The radiation contours for the daily totals are similar to the contours for instantaneous intensity, but the distribution of radiation is rather more uniform. Figure 16.4 indicates the relative intensity of direct radiation potentially received on different sites on a short term basis around the days illustrated.

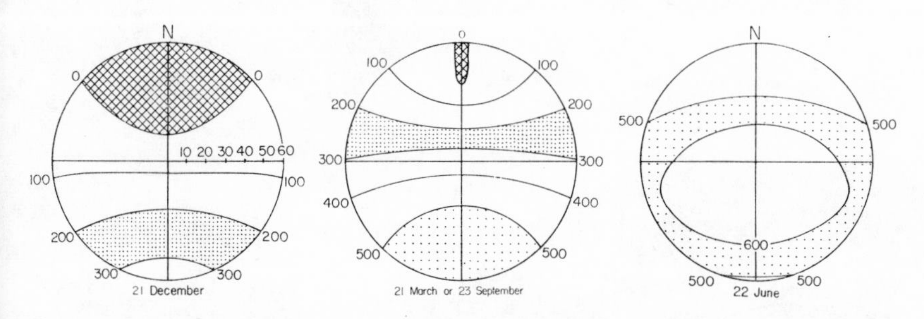

Figure 16.4. Daily totals of direct radiation potentially received by sites of all aspects and slopes up to 60° on three days at latitude 53° 15′ N (g cal cm^{-2}day^{-1}).

Cumulative totals of direct radiation throughout the year provide a guide to the relative insolation of different sites and to the probable rates at which they would heat up as spring and summer advance, if other factors such as vegetation cover, albedo or soil moisture were constant and uniform. Such cumulative totals are shown in Fig. 16.5 for December to June and for the year. The figures are the sums of one daily calculation per month and are thus nominal values. It will be seen that the developing pattern of the contours lags behind the daily pattern (Fig. 16.4) and never reaches the relative degree of uniformity of the daily value for June. Based on radiation values, these contours are symmetrical about the north-south axis. In practice it is likely that under uniform conditions, the zone of maximum heating will be displaced towards the south-west because ground and plant surfaces will be drier after noon than before noon (Geiger 1969).

The distributions of radiation on the three time scales—instantaneous, daily and cumulative—are respectively most relevant to different ecological situations. Two recurrent features of the distributions should be noted:

firstly the increase in received radiation southwards along the north-south axis and secondly the equivalence of east- and west-facing slopes.

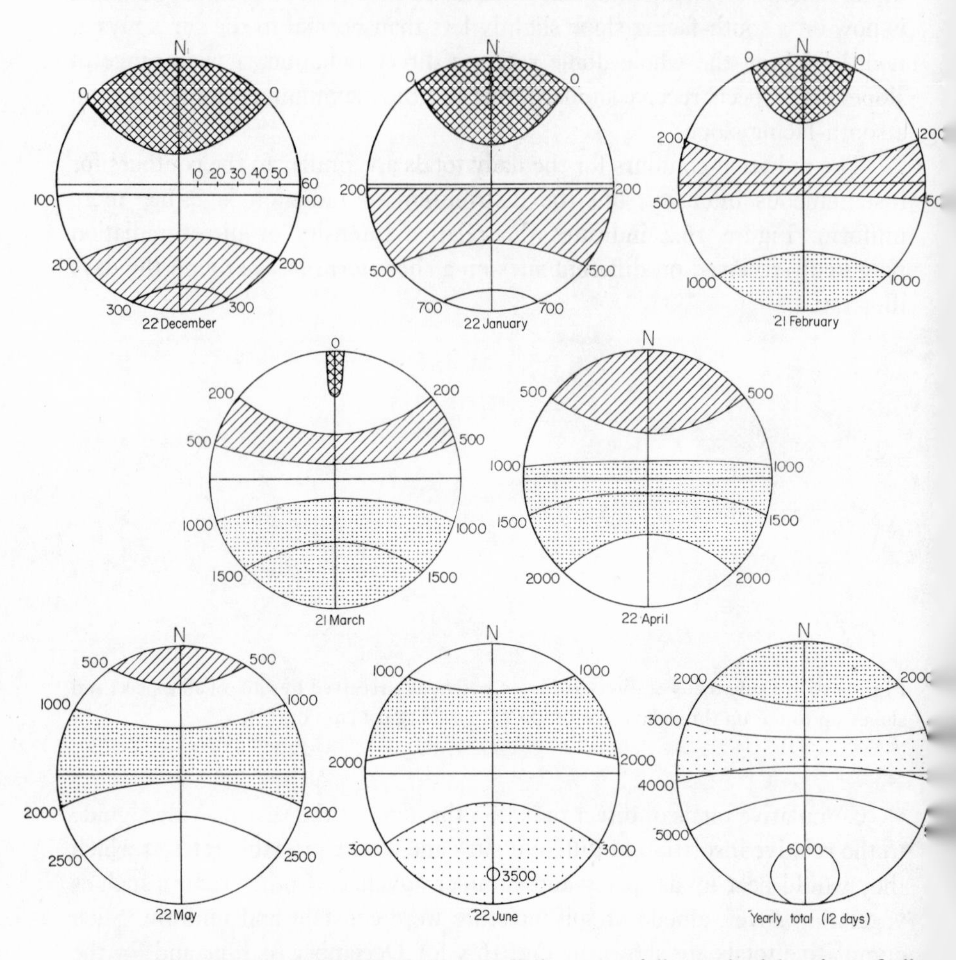

Figure 16.5. Cumulative totals of direct radiation potentially received by sites of all aspects and slopes up to 60°, based on one daily value per month. (Latitude 53° 15′ N) (cal. cm^{-2}).

The influence of slope on the north-south axis is illustrated in Fig. 16.6, using the annual cumulative values for radiation as above. It is emphasized that these are dependent only on slope and aspect; no additional topographic obstruction is involved. From the steepest north-facing slope depicted, there is a progressive increase in received radiation through the horizontal to a maximum between 45° and 50° on a south-facing slope. Further increase in slope results in progressively lower totals.

The equivalence of east and west aspects is also shown in Fig. 16.6, and it appears that the yearly total of radiation is virtually independent of slope, at least over much of the range. However, the yearly figures mask considerable seasonal fluctuations in which the relationship with slope undergoes a

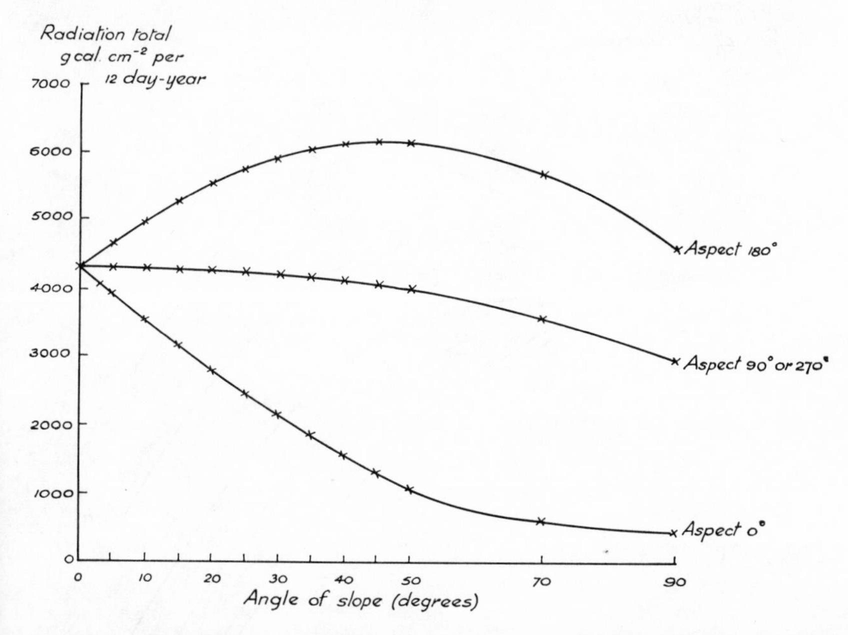

Figure 16.6. Influence of slope and aspect on the yearly total of direct radiation potentially received at Latitude 53° 15′ N.

complete reversal (Fig. 16.7). During the winter, when the sun's declination is negative and the solar altitude always low, received radiation increases with increasing slope up to about 50°. During the summer, when the sun's declination is positive, there is a negative relationship with increasing slope. At no time is the actual difference very great, but in December a 50° slope experiences an increase of 13% over the value for a horizontal surface. Such differences may be of particular significance to species like winter annual plants, but this point has not been examined further.

Topography and plant distribution

Examination of hemispherical photographs taken at a variety of sites in Derbyshire shows that some measure of topographic obstruction of the horizon is normal and that, in areas of high relief, severe obstruction is frequent. In one narrow, steep-sided valley (Chee Dale) all eight sites investigated experience severe topographic obstruction. Selected photographs illustrate some extremes (Plates 16.1 and 16.2). (Note that since the photographs are taken vertically upwards east and west are reversed.) Plate 16.1

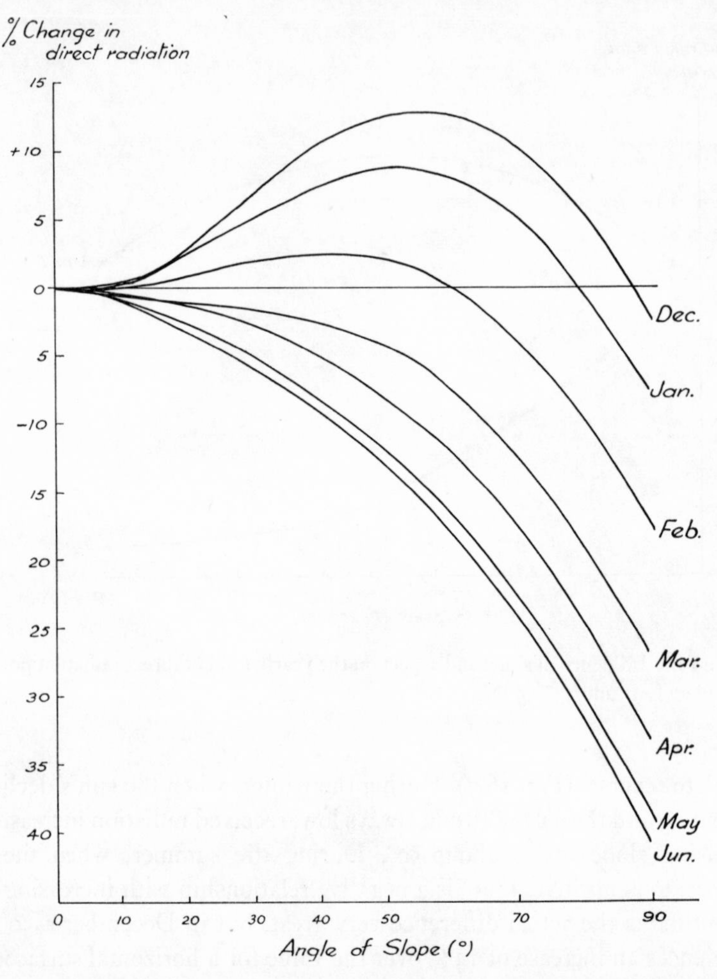

Figure 16.7. Seasonal variation in the influence of slope on the daily total of direct radiation on aspects 90° or 270°, relative to that potentially received by a horizontal surface at Latitude 53° 15′ N.

shows two north-facing sites, one of which does not receive direct radiation before midday on any day of the year and receives no direct radiation at all between October and February. The other site receives almost full duration of sunshine throughout the year despite its aspect. Plate 16.2 shows two south-facing sites. In this case the presence of surrounding cliffs (Plate 16.2a) does not prevent the site receiving a good proportion of the maximum potential direct radiation at any season.

The influence of topographic obstruction on site radiation properties has been assessed by calculating the annual value of the site radiation index for each of 199 sites visited in Derbyshire. These values were plotted on a polargraph in the appropriate position for each site. Contours, drawn at intervals of 1000 cal cm^{-2} per 12-day year and based on data for dummy, unobstructed sites, were superimposed on the polargraph and it was seen that a large number of sites had lower indices than expected from the position of the site on the polargraph relative to the contours. All such sites experience topographic obstruction, as can be verified by reference to the site photographs. Sites with a reduction of at least 500 cal cm^{-2} are regarded as subject to severe obstruction and are marked on Fig. 16.8, with the actual value of the

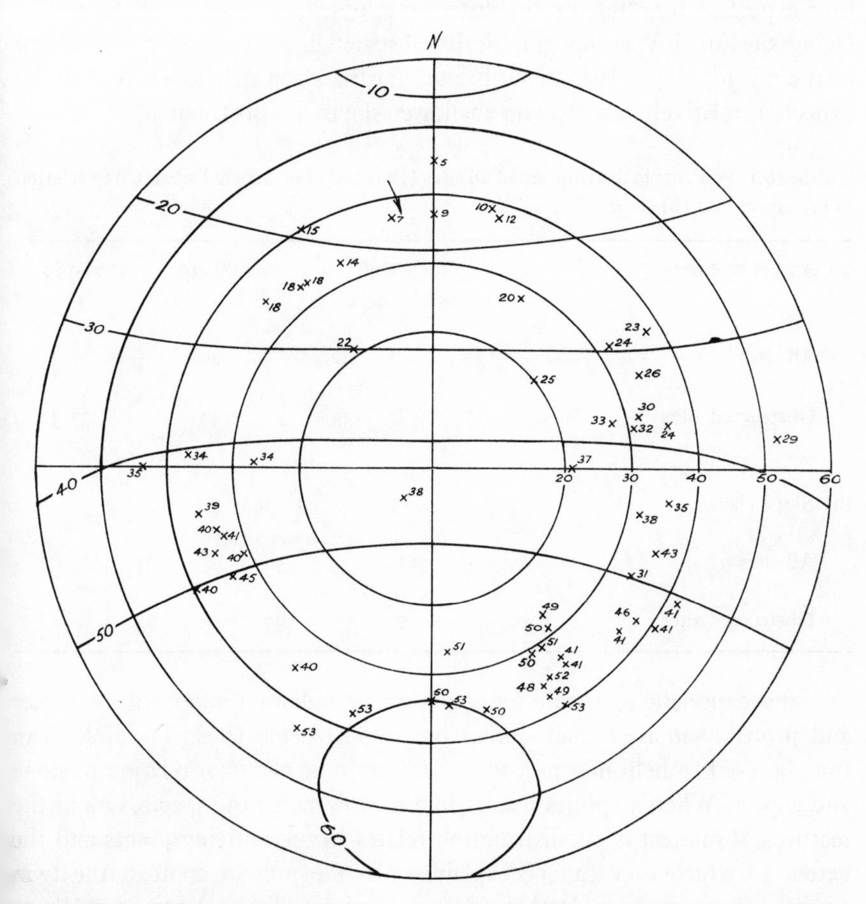

Figure 16.8. Distribution in relation to slope and aspect of potential direct radiation (10^2 cal. cm^{-2} per 12 day-year). The position of topographically obstructed sites is indicated, with the appropriate radiation total. The site illustrated in Plate 16.1a is arrowed.

site radiation index. WIN 13 (Plate 16.1a) is indicated in Fig. 16.8 and is subject to a reduction of over 60% of the potential index value. Even on south-facing slopes reductions of over 20% are recorded. In relation to the actual values of the site radiation index, a reduction of 500 cal cm^{-2} represents a proportionate reduction of direct irradiation of approximately 8% on steep south-facing slopes, 12% on horizontal ground and east- or west-facing slopes and up to 50% on steep north-facing slopes. There is thus a more restrictive definition of obstructed sites on north-facing slopes, which is justified by the greater relative importance of diffuse light on these slopes and by the presumed tolerance of or adaptation to more mesic conditions by plants normally encountered on such slopes. The distribution of the obstructed sites in relation to that of all sites surveyed is summarized in Table 16.1 in terms of slope and aspect. Representation is fairly even in the four aspect sectors but, by the criterion adopted, rather few of the north-facing sites qualify as topographically obstructed. The majority of sites fall in the 30–39° slope class, being located on the steep dale sides. As may be expected, relatively few sites on shallower slopes are obstructed.

Table 16.1. Percentage distribution of all sites (199) and of obstructed sites (58) in relation to (a) aspect and (b) slope.

(a) Aspect sectors:	316°–360° + 0°–45°	46°–90° + 271°–315°	91°–135° + 226°–315°	136°–180° + 181°–225°	
All sites	32	22	20	26	%
Obstructed sites	17	33	42	33	%

(b) Slope classes:	0–29°	30–39°	40°+	
All sites	27	58	15	%
Obstructed sites	17	69	14	%

Sites experiencing severe reduction of the radiation index will be cooler and probably moister than similar but unobstructed sites. The index can thus be used to help interpret the distribution of plants in relation to slope and aspect. When a species occurs in a narrow range of aspects, one of the features of interest of its distribution relates to the outlying points and the extent to which they can be explained. In the present context, the twin hypotheses are proposed: that, in their outlying sites, plants of northern aspects will be particularly associated with obstructing topography and plants of southern aspects will be less frequent in such places.

Within the study area, *Hieracium pilosella* L. and *Thymus drucei* Ronn. are both more common on south- than on north-facing slopes, with a very high statistical significance (Fig. 16.9). Within the southern aspect sector,

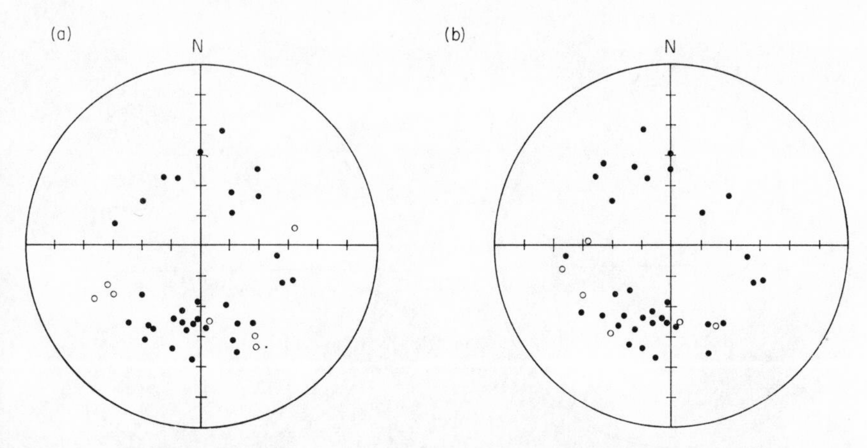

Figure 16.9. Polargraphs showing the presence of species in obstructed (○) and unobstructed (●) sites. (a) *Hieracium pilosella*; (b) *Thymus drucei.*

both species are recorded less frequently than expected from obstructed sites (Fisher's exact test: $P = 0{\cdot}016$ in each case). Neither species is recorded from obstructed sites in the northern sector, though because of the small number of records statistical significance is not attained. Similar distributional patterns are shown by *Plantago lanceolata* L. (which avoids obstructed sites in the northern, $P = 0{\cdot}039$, as well as in the southern sector) and by *Lotus corniculatus* L. All these species were shown to exhibit similar distributions in relation to aspect over a wider geographical area by Grime & Lloyd (1973). The data support the hypothesis that such species are less common in obstructed sites.

Several species of northern aspects were recorded so infrequently and over such a narrow range of aspect that they cannot be used to test the hypothesis that they are more common in obstructed sites. *Oxalis acetosella* L. was recorded only once outside the northern sector: the site in question had an aspect of 314° and was severely obstructed. More wide-ranging species include *Mercurialis perennis* L. (Fig. 16.10a), which shows a positive association with obstructed sites in the northern sector ($P = 0{\cdot}022$). Outside that sector, five of the eight sites are severely obstructed. *Helictotrichon pubescens* (Huds.) Pilger is similarly centred on northern aspects (Fig. 16.10b) and, in this case, obstructed sites are favoured in the southern sector ($P = 0{\cdot}032$). These species with rather restricted aspect distributions, as detected in the present work, tend to support the hypothesis that plants of northern slopes will be more frequent than expected in obstructed sites on other aspects.

Other, more frequent, species show distributions which are not so clear. *Arrhenatherum elatius* (L.) J. & C. Presl (Fig. 16.11a) is significantly biased towards northern aspects, where it is particularly frequent in obstructed

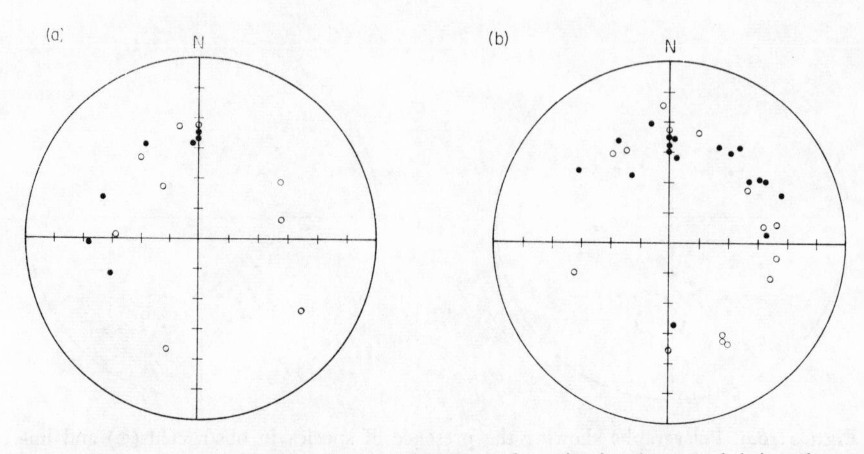

Figure 16.10. Polargraphs showing the presence of species in obstructed (○) and unobstructed (●) sites. (a) *Mercurialis perennis*; (b) *Helictotrichon pubescens.*

sites ($P = 0{\cdot}027$). It is also more frequent in obstructed sites on southern aspects, though not significantly so ($P = 0{\cdot}058$). This evidence of an aspect preference was not detected by Grime & Lloyd (1973), either on the Carboniferous Limestone or over a wider area. The bias for obstructed sites may be a response to edaphic conditions rather than to radiation, since *A. elatius* is often found in the Derbyshire dales growing in gullies between limestone crags and in their outwash fans, where the soil contains an admixture of non-calcareous matter derived from the plateau. *Viola riviniana* Rchb. (Fig. 16.11b) and *Potentilla erecta* (L.) Räusch. both show aspect distributions biased to the north and yet are less frequent than expected in obstructed sites on northern aspects. There is no evidence of site selection on southern aspects. It appears that obstructed sites with northern aspect are too extreme for these species, either because of some site characteristic or because of the nature (perhaps height) of the associated vegetation. A few species which show no aspect bias nevertheless show some relationship with obstructed sites. Thus *Briza media* L. avoids obstructed sites strongly on southern aspects ($P = 0{\cdot}004$) and to a lesser extent on northern aspects ($P = 0{\cdot}054$). This distribution appears more puzzling in the light of the previous demonstration that *B. media* has significantly higher within-quadrat frequency on northern aspects (Grime & Lloyd 1973).

The site radiation index proves suitable for identifying plant distributions associated with reduced radiation, but it cannot detect sites where the radiation climate is enhanced, as by a backing of south-facing cliffs. In such situations thermophilous or xerophytic species should constitute a particularly high proportion of the vegetation. In the absence of a suitable direct

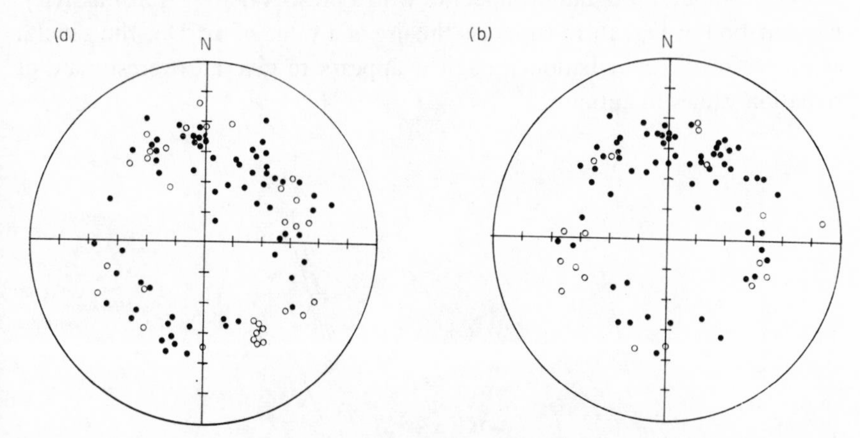

Figure 16.11. Polargraphs showing the presence of species in obstructed (○) and unobstructed (●) sites. (a) *Arrhenatherum elatius*; (b) *Viola riviniana*.

measure of radiation, the hemispherical photographs can be used to indicate which sites may fall into this category. In the present work nine such sites were identified. They were matched with other sites of similar slope and aspect but without the surrounding cliffs, but no consistent floristic differences could be found between the two categories. There are, of course, other variables which have not been taken into account.

Discussion

In the preceding section it has been assumed that the site radiation index is a valid measure of radiation properties of a site. Justification for that assumption is now presented.

Since 1971, total and diffuse radiation have been recorded near Lathkill Dale. Direct radiation figures derived by difference from these records are compared with equivalent data calculated for a horizontal surface by means of the computer programme (Fig. 16.12). Two sets of calculated data are shown, for transmissivities of 0·70 and 0·75. The actual measurements of daily direct radiation represent the highest daily value recorded for each month over a period of two years. In most cases these fall well below the calculated values, but once correction has been made for the number of hours of sunshine in relation to the potential maximum then the corrected measurements agree closely with the calculated values. Only once in each year (June 1972,

July 1973) is there serious disparity between the figures. This seems to arise when abnormally bad weather results in no high values of radiation during the month.

The computer programme operates with a preset value for transmissivity. Examination of Fig. 16.12 supports the use of a value of 0·70 for the general estimates of direct radiation though it appears to give an overestimate of radiation values in autumn.

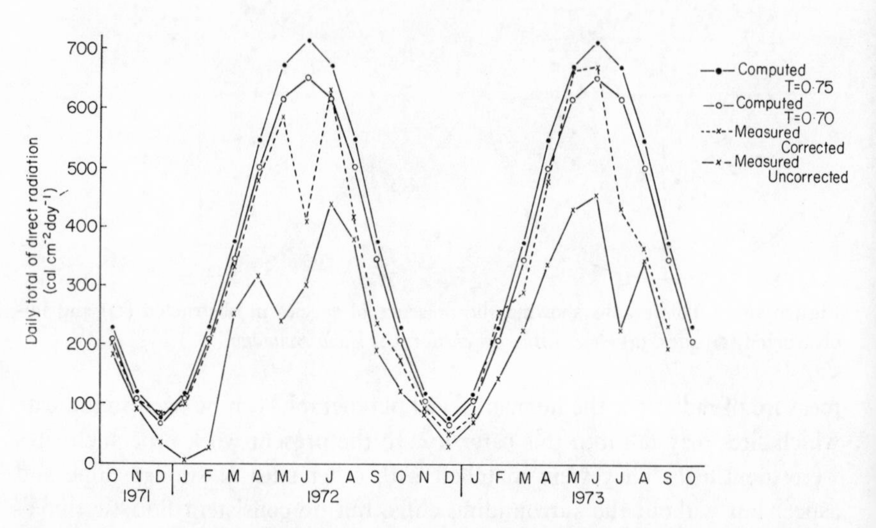

Figure 16.12. Comparison between computed and measured values of direct radiation. Computed values are given for transmissivities of 0·75 (●) and 0·70 (○). The maximum daily value recorded in the field in each month (— × —) has been corrected for duration of bright sunshine (— — × — —). (Field data from Lathkill Dale, courtesy of F. Sutton and I.H. Rorison; sunshine duration from Sheffield, courtesy of Sheffield City Museums).

During the course of the field studies on which this work is based, various attributes of each site were measured, including soil depth, pH, slope and aspect. In addition soil moisture was measured at some of the sites. Correlation coefficients have been calculated for radiation and each other variable (Table 16.2), using the site radiation index. The strong correlation with aspect and the lack of correlation with slope are to be expected in view of the data presented in Fig. 16.6. The strong negative correlation with soil moisture emphasizes the inter-related nature of radiation, topography and soil processes. Clearly aspect and slope measurements together provide good information about the radiation input of most sites, but the permanent record provided by hemispherical photography is in itself a valuable addition to site records and the opportunities that the method presents, of characterizing in detail the radiation properties of any site and of assessing the sig-

Table 16.2. Correlation between radiation and other site factors.

	Soil depth	Soil pH (10 cm)	Slope Angle	Aspect	Soil Moisture (% by weight)
Radiation	−0·13	0·06	−0·11	−0·94 ***	−0·71 ***

Degrees of freedom = 197 for all comparisons except soil moisture (23). Aspect was classified from 1–9 in 20° steps symmetrically from South.

nificance of horizon obstruction, should ensure a wide application in ecological research. Correlation of the site radiation index with quantitative values of species abundance, such as frequency or cover, may well prove more sensitive than the qualitative correlation examined so far.

References

ANDERSON M.C. (1964) Studies of the woodland light climate. I. The photographic computation of light conditions. *J. Ecol.* 52, 27–41.

DIRMHIRN I. (1964) *Das Strahlungsfeld im Lebensraum.* Akad. Verlagsges., Frankfurt-a-M.

EVANS G.C. & COOMBE D.E. (1959) Hemispherical and woodland canopy photography and the light climate. *J. Ecol.* 47, 103–13.

GARNETT A (1937) Insolation and relief. *Publs. Inst. Br. Geogr.* 5, 1–71.

GARNIER B.J. & OHMURA A. (1968) A method of calculating the direct short wave radiation income of slopes. *J. appl. Meteorol.* 7, 796–800.

GARNIER B.J. & OHMURA A. (1970) The evaluation of surface variations in solar radiation income. *Solar Energy* 13, 21–34.

GEIGER R. (1966) *The climate near the ground.* English translation by Scripta Technica Inc. Harvard Univ. Press, Cambridge, Mass.

GEIGER R. (1969) Topoclimates. In H. Flohn (ed.). *World survey of climatology.* Vol. 2, General climatology 2. Elsevier, Amsterdam.

GRIME J.P. & LLOYD P.S. (1973) *An ecological atlas of grassland plants.* Arnold, London.

HALTINER G.J. & MARTIN F.L. (1957) *Dynamical and physical meteorology.* McGraw-Hill, New York.

HILL R. (1924) A lens for whole sky photographs. *Quart. J. R. Met. Soc.* 50, 227–35.

HUGHES A.P. (1966) The importance of light compared with other factors affecting plant growth. In R. Bainbridge, G.C. Evans & O. Rackham (eds). *Light as an ecological factor.* Blackwell, Oxford.

LIST R.J. (1951) *Smithsonian meteorological tables.* 6th edn. Smithsonian Institute Washington, D.C.

LUNDQVIST J. (1968) Plant cover and environment of steep hillsides in Pite Lappmark. *Acta phytogeogr. Suec.* 53, 1–153.

MADGWICK H.A.I. & BRUMFIELD G.L. (1969) The use of hemispherical photographs to assess light climate in the forest. *J. Ecol.* 57, 537–42.

PIGOTT C.D. (1968) Biological flora of the British Isles. *Cirsium acaulon* (L.) Scop. *J. Ecol.* 56, 597–612.

POPE D.J. (1973) *A study of plant growth in relation to topography in the Derbyshire Dales.* Ph.D. Thesis, Sheffield University.

STAGG J.M. (1950) Solar radiation at Kew Observatory. *Geophys. Mem., Lond.* 11 (86), 1–37.

TONNE F. (1958) Optographic computation of insolation duration and insolation energy. In E. Carpenter (ed.), *Trans. Conf. Use of Solar Energy*, Vol. 1. Univ. Arizona Press, Tucson.

TURNER H. (1966) Die globale Hangbestrahlung als Standortsfaktor bei Aufforstungen in der subalpinen Stufe. *Mitt. schweiz. Anst. forstl. VersWes.* 42, 111–68.

WILLIAMS L.D., BARRY R.G. & ANDREW J.T. (1972) Application of computed global radiation for areas of high relief. *J. appl. Meteorol.* 11, 526–33.

Questions and discussion

The possibility was discussed of extending the method to include diffuse radiation by using hemispherical photographs with overlay grids calculated for particular directional distributions of diffuse skylight. Dr. M. H. Unsworth argued that in some circumstances, such as north-facing slopes in winter, diffuse radiation and the energy balance as a whole were ecologically more important than direct sunlight. Dr. G. C. Evans suggested that the chief obstacle to making such an extension, the variable and unpredictable distribution of diffuse radiation, might not be insuperable in a comparison of slopes and aspects within the same region. If the skylight distribution were known to be the same over the sites being compared, it might not be necessary to specify it exactly; one of the standard distribution patterns might provide an adequate approximation.

17 Soil reflection coefficient and its consequences for soil temperature and plant growth

W. R. WATTS *Department of Botany, University of Aberdeen, Scotland*

Introduction

In studies of the relationship between soil colour and the heating effects of solar radiation, it is usual to measure the fraction of total short-wave radiation (400–3,000 nm) that is reflected from the soil surface. This fraction is often referred to as the *albedo* but, as Monteith points out (1973, p. 61), albedo is a term borrowed from astronomy and is associated with the visible spectrum (400–700 nm), so the wider term *reflection coefficient* is preferable.

Distinction should also be made between the *reflectivity* of a surface $\rho(\lambda)$, which is the fraction of incident radiation reflected at a specific wavelength λ, and *reflection coefficient* which is the average reflectivity over a specific waveband.

In the context of this paper, *soil colour* is taken to refer either to natural soil colour, or to thin layers of surface colorants which do not impede evaporation, or to thick or impervious soil coverings (*mulches*) which alter the reflection coefficient and also reduce heat loss through latent heat of vaporization.

As the reflection coefficient changes with soil colour it alters the radiation and energy balance at the soil surface. This can have a direct or indirect influence on plant growth, depending on the stage of development and/or growth habit.

Radiation and energy balance

The radiation balance for unit area of soil surface can be generalized (following Monteith 1973) as demonstrated below.

$$\begin{array}{ccccc} \text{Balance} & & \text{Gains} & & \text{Losses} \\ \text{Net radiation} & = & \left\{\begin{array}{c}\text{Incident short wave}\\ \text{radiation}\\ + \\ \text{Absorbed long wave}\\ \text{radiation}\end{array}\right\} & - & \left\{\begin{array}{c}\text{Reflected and transmitted}\\ \text{short wave radiation}\\ + \\ \text{Emitted long wave}\\ \text{radiation}\end{array}\right\} \end{array}$$

This paper is concerned entirely with horizontal soil surfaces, for which the net radiation flux per unit area can be written R_n. (The effects of slope are a complication dealt with elsewhere in this volume.) Incident short-wave radiation consists of direct and diffuse radiation from the sun and the atmosphere, S_t, plus sunlight reflected from the rest of the environment, S_e. The total incident short-wave radiation is then S_t+S_e and if the reflection coefficient is ρ, the reflected short-wave flux is $\rho(S_t+S_e)$. Fluxes of long-wave radiation are L_a from the atmosphere, L_e from the rest of the environment. If a surface has an emissivity of ε it will gain $\varepsilon(L_a+L_e)$ from its surroundings, and emit εL_b to its surroundings.

The general equation of radiation balance can then be written as:

$$R_n = (1-\rho)(S_t+S_e)+\varepsilon(L_a+L_e-L_b)$$

The major effect of soil colour in this equation is on the short-wave components since emissivity is effectively constant (0·9–1·0) for the range of naturally occurring soil colours (Table 17.1). Even a white surface such as

Table 17.1. Measured values of emissivity for bare soils.

Sand	red, dry	0·95
	wet	0·98
Gravel		0·91
Limestone-derived soil	light grey	0·92
Loam	dry	0·95
	wet	0·97
Peat	dry	0·97
	wet	0·98

(Stanhill, personal communication)

magnesium carbonate has an emissivity of approximately 0·8 and, as such, has only a relatively small influence on the radiation balance in the daytime.

The relative importance of the short wave component is shown in Fig. 17.1 where the radiation and energy balances of bare soil and soil covered with magnesium carbonate are compared. The heat balance equation may be written as:

$$(S_t+S_e)(1-\rho)-\varepsilon(L_a+L_e-L_b) = R_n = \lambda E \pm K_h \pm G$$

where λE is the latent heat lost by evaporation, K_h is the convective heat exchange between the surface and air, and G is the conductive energy exchange between the surface and the soil.

The effect of the treatment is to reduce the radiation balance at the surface from 70 W m^{-2} to only 10 W m^{-2}, by doubling the reflection coefficient (0·64 compared with 0·31). Long-wave radiation losses are reduced by the lower emissivity of the treated surface and by the decrease of soil surface temperature from 33°C to 28°C.

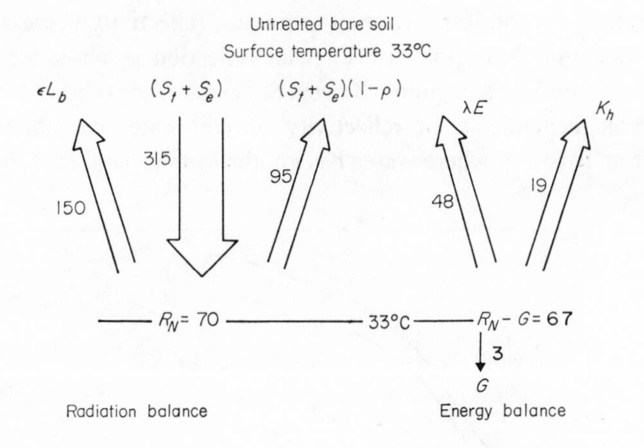

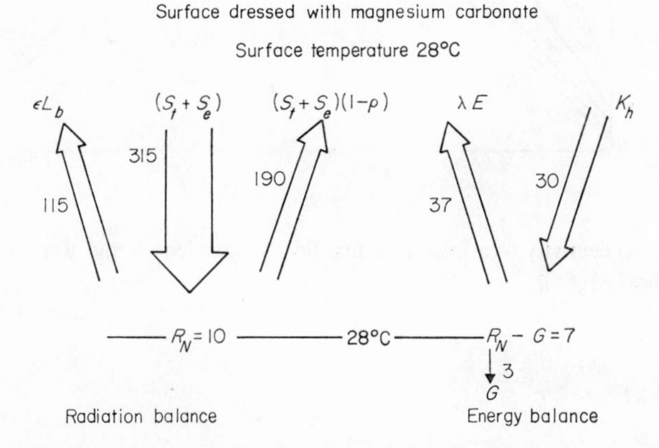

Figure 17.1. Approximate radiation and energy balance of bare soil with and without surface dressing of magnesium carbonate (0·11 mm thick, assumed not to impede evaporation). Mean values for July and August at Gilat. Symbols as explained in text; units in W m^{-2} (after Stanhill 1966).

Seasonal variations of reflection coefficient are usually small as long as the colour of the surface remains constant. The reflection coefficient of sand dunes in Israel had a range of 0·34 to 0·46 for the period March 1961 to March 1962 (Stanhill 1970). Similarly, for a 1 cm high grass surface at Rothamsted, the reflection coefficient remained between 0·24 and 0·26 from April to September 1958 (Monteith 1959).

The main diurnal changes in reflection coefficient occur in the early morning and late evening when the angle of incidence is greater than about 35° (Leonard & Eschner 1968, Stewart 1971). During most of the day the reflection coefficient remains almost constant unless there is a change in soil water content. As soil water content increases, reflectivity decreases, largely because radiation is trapped by internal reflection at air-water interfaces formed by menisci in soil-pores. Bowers & Hanks (1965) showed there was a considerable dependence of reflectivity on soil water content which was strongest at 1950 nm where water has an absorption band (Fig. 17.2).

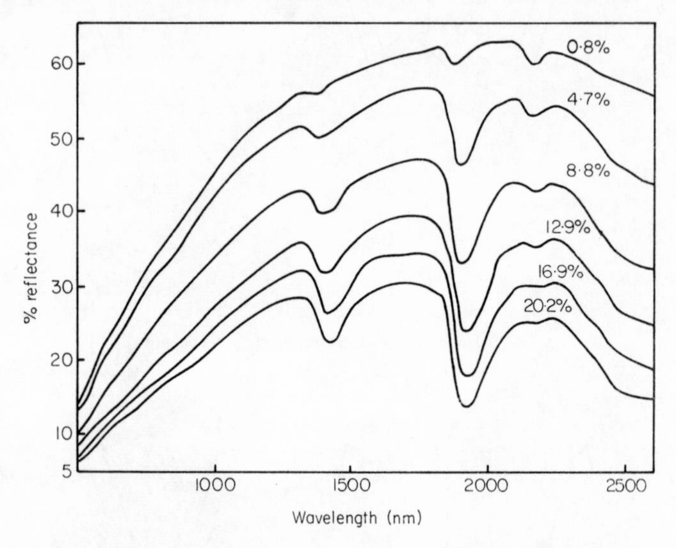

Figure 17.2. Reflectivity of a loam as a function of wavelength and water content (from Bowers & Hanks 1965).

Soil temperature

Having seen how the reflection coefficient alters the radiation balance at the soil surface, we now want to know how this is translated into soil temperature differences, and how it affects plant growth and development. The relationship between soil colour and temperature has been recognized for many years, and Ludwig & Harper (1958) refer to the work of Wollny as far back as 1878. Wollny measured temperature differences between boxes of different coloured soils and summarized his conclusions as follows:

(1) In warm seasons, dark-coloured air-dry soils are usually warmer than light-coloured soils.

(2) The daily fluctuations in temperature are larger in dark- than in light-coloured soils.

(3) The difference between the temperature of the dark- and light-coloured soils is largest at the daily maximum soil temperature and is very slight at the daily minimum.
(4) The differences in temperature between the dark- and the light-coloured soils are reduced with increasing depth and disappear almost completely under reduced insolation.

Quantitative data on the relation between natural soil colour, soil temperature, and plant growth are difficult to find in the literature, but in the last 20–30 years there have been a number of studies dealing with soil colorants and mulches.

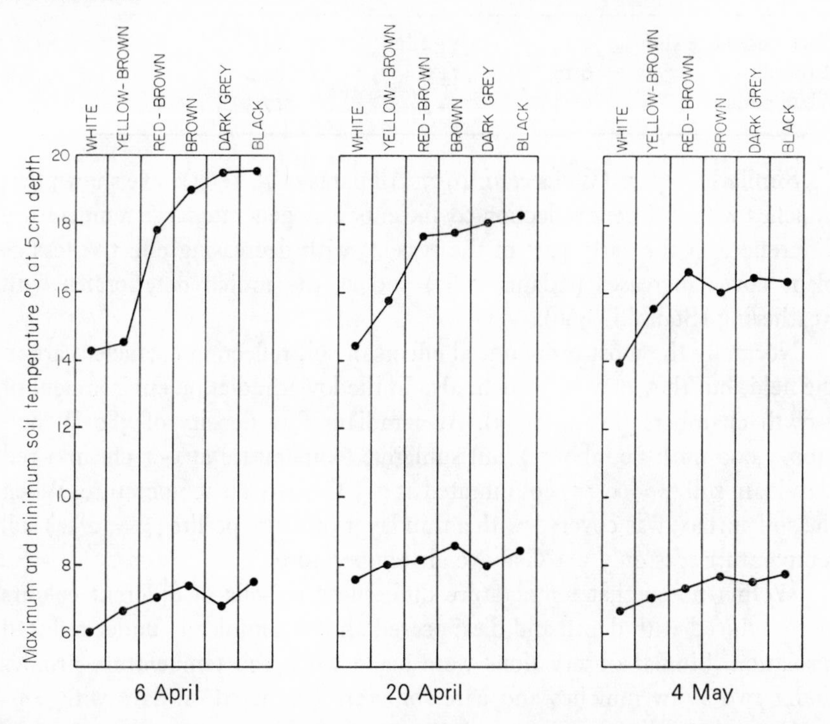

Figure 17.3. Influence of soil surface colour on soil temperature. The values shown are means for three weeks after sowing date (after Ludwig & Harper 1958).

Ludwig & Harper (1958) compared temperature differences in soils with six surface colorants ranging from white to black. The mean maximum and minimum temperatures are shown in Fig. 17.3. The darker-coloured soils had higher maximum temperatures, particularly in the early part of the growing season. The differences in minimum temperatures were much smaller, indicating that at night, when the short-wave component was absent, the soils all reached a similar temperature.

When artificial mulches are applied to soil surfaces, larger temperature differences may be generated, since losses by latent heat of vaporization are also reduced. Watts (1973) used black polythene sheets, perlite (inert, white expanded quartz beads), and a bare reddish loam surface. Large temperature differences were measured on a warm sunny day but these differences were much smaller on a cool cloudy day (Table 17.2).

Table 17.2. Maximum daily soil temperatures under mulch (Sutton Bonington).

Treatment	Reflection coefficient	Temperature (°C) at −5 cm	
		Cool, cloudy	Warm, sunny
Black polythene sheet	0·13	14·2	29·0
Bare soil	0·17	14·0	24·2
White perlite	0·34	12·5	17·5

Similarly, others (Willis *et al.* 1957, Allmaras *et al.* 1964) have shown that mulches with different reflection coefficients may generate large temperature differences in the early part of the season, with decreasing effectiveness as plant cover increases (Adams 1967) and as the mulch deteriorates with weathering (Stanhill 1966).

Normally the most pronounced effects of soil reflection coefficient are in the field, but they may be evident also in the lower radiation environment of growth chambers (Watts 1975). At a radiant flux density of 380 W m^{-2} (400–3,000 nm), i.e. about $\frac{1}{3}$ full sunlight, temperature at −1 cm in a red clay-loam soil ($\rho = 0{\cdot}17$) equilibrated at 7·5°C above air temperature. When the soil surface was covered with a thin layer of white perlite ($\rho = 0{\cdot}34$) soil temperature was only 2·4°C above air temperature.

Wollny noted that temperature differences in soils of different colours were reduced with depth and disappeared almost completely under reduced insolation. Similar observations were made when soil temperature-profiles under two straw mulches and bare soil were compared on days with contrasting solar radiation (Fig. 17.4).

Plant growth

Germination

The effects of soil colours or mulches on soil temperature are usually most pronounced early in the growing season, when plant cover is minimal, and are therefore more likely to influence germination rate and early vegetative growth. In hot climates with high insolation, a high reflection coefficient

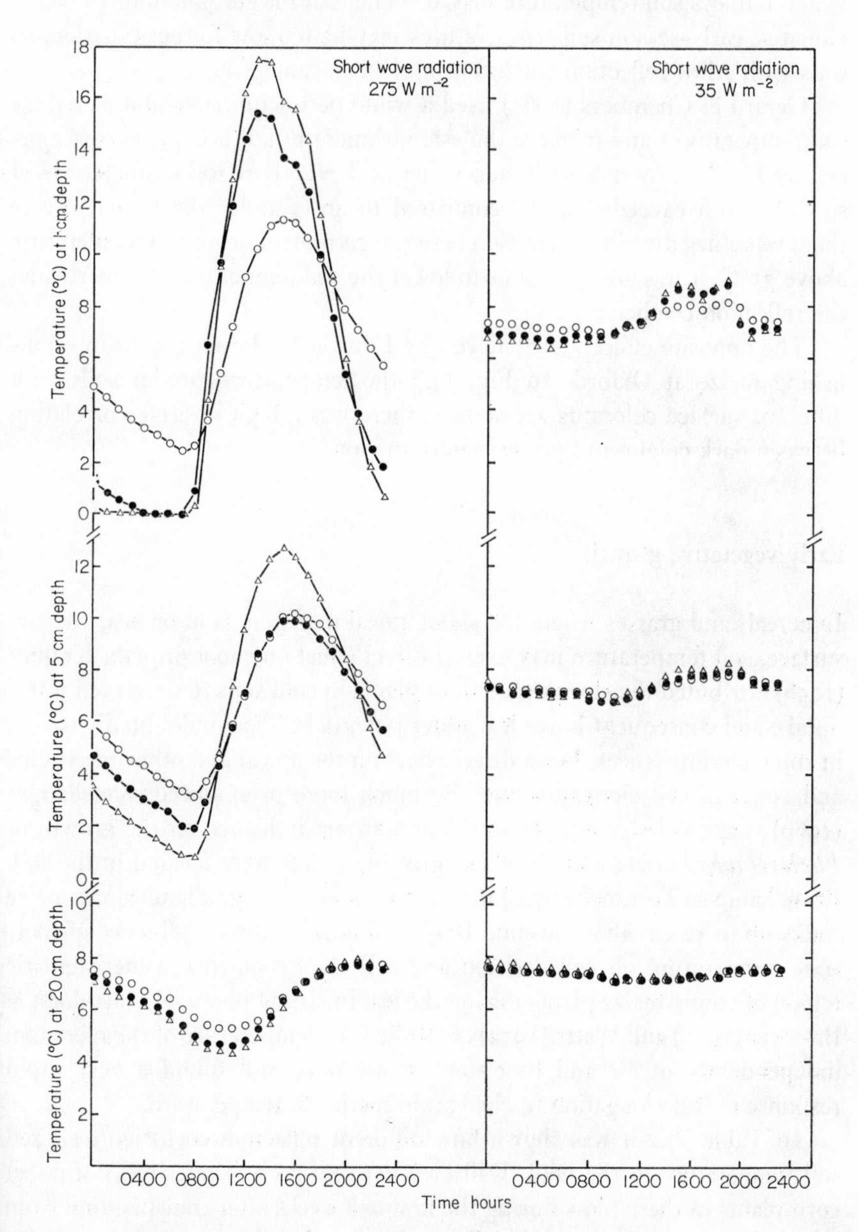

Figure 17.4. Diurnal fluctuations of soil temperature at three depths under three soil surfaces on days with contrasting solar radiation. ○ Heavy straw mulch, $\rho = 0.26$. ● Light straw mulch, $\rho = 0.20$. △ Bare brown soil, $\rho = 0.13$. (R.K.M. Hay, personal communication).

which reduces soil temperature may be beneficial for germination. In cooler climates, early-season soil temperatures may be too low for germination, so soils with a low reflection coefficient may be advantageous.

Gerard & Chambers (1967) used a white petroleum emulsion to reduce soil temperatures and increase the establishment of fall bell-peppers (*Capsicum* sp.) in the Lower Rio Grande Valley of Texas. Bare soil temperatures at sowing depth exceeded 48°C, compared to 40°C under the mulch. Since there was a negative linear relation between germination and soil temperature above 35°C, it was advantageous to lower the soil temperature by increasing the reflection coefficient.

The opposite effect was achieved by Ludwig & Harper (1958) in establishing maize at Oxford. In Fig. 17.3 the temperatures under soils with different surface colorants are shown; there was a high positive correlation between dark colour and per cent germination.

Early vegetative growth

In cereals and grasses where the shoot apical meristem is in or near the soil surface, soil temperature may exert a direct effect on shoot growth. Kramer (1956) attributed the slower growth of plants in cold soils to decreased water uptake and consequent lower leaf water potentials. This undoubtedly occurs in some circumstances, but a direct effect on the apical and other meristems and zones of cell elongation may be much more pronounced. Ketellapper (1960) suggested soil temperature had a strong influence on the growth of *Phalaris tuberosa* because the shoot growing points were located in the soil. Beauchamp & Torrance (1969) measured temperature gradients along maize stalks up to 12 cm above ground level, and found a direct influence of root-zone temperature on cell division and cell elongation in the meristematic region of young maize plants during the leaf initiating phase. Kleinendorst & Brouwer (1970) and Watts (1972) controlled the temperature of the meristem independently of air and root-zone temperature and found a very rapid response of leaf elongation to changes in meristem temperature.

In Table 17.2 it was shown how different reflection coefficients altered soil temperature in the field. Daily relative leaf growth rates (R_A) of sweet corn plants in these plots during the first 6–8 weeks after transplanting from a greenhouse were reasonably well correlated with soil temperature at -5 cm (which is close to the shoot apical meristem and zone of cell elongation) in the high temperature plots, but less so in the cooler perlite plots (Fig. 17.5). It is suggested that soil temperature at -5 cm was a reasonable estimate of shoot meristem temperature until stem elongation commenced, and that this was the main factor controlling *daily* fluctuations of R_a.

Williams and Biddiscombe (1965) reported a strong positive correlation between daily extension growth of temperate grass tillers and temperature of the soil surface, while Willis *et al.* (1957) and Allmaras *et al.* (1964) have shown similar responses by maize and sorghum in the early vegetative stage, before stem elongation commences.

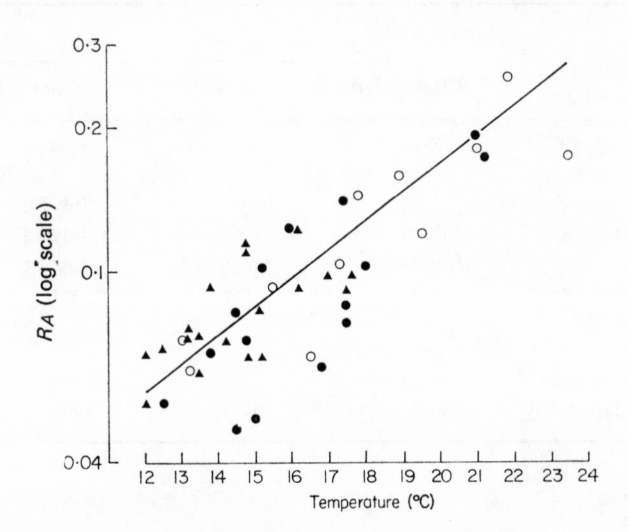

Figure 17.5. Daily relative leaf growth rates (log scale) and mean soil temperature at −5 cm in all treatments. ○ Black polythene sheets. ● Bare red-brown soil. ▲ White perlite.
Fitted regression: $R_A = -1{\cdot}97 \times 10^{0{\cdot}06T}$. Correlation coefficient between T and log R_A = 0·66.

Total crop yields

The response of plants to soil temperatures induced by different reflection coefficients over the course of a growing season are much more complex than in the short term. Long term exposure to low temperatures may result in morphological and anatomical changes in the root system which may restrict water uptake (Brouwer & Hoogland 1964) and nutrient uptake (Neilson & Humphries 1966), and may alter the hormone balance in the plant (Bull 1964, Atkin *et al.* 1973).

In Table 17.3 a number of indices of development and yield of sweet corn are given from the plots with the reflection coefficients in Table 17.2. The results show clearly the benefits of a black polythene mulch at Nottingham, although the higher soil temperatures cannot be attributed entirely to its low reflection coefficient, since mulches also alter the energy balance in other ways previously described. However, the poor growth in the perlite plots *can* be attributed largely to the high reflection coefficient ($\rho = 0{\cdot}34$) and subsequent low soil temperatures throughout the season.

Table 17.3. Development and yield of sweet corn under different mulches (Sutton Bonington).

Index	Treatment		
	Black polythene	Bare soil	White perlite
Mean soil temp. (°C) at −5 cm for June	18·1	16·7	14·8
Tassel emergence date	June 30	July 7	July 10
Anthesis date	July 16	July 19	July 24
Harvest date	August 10	August 14	August 21
Leaf emergence rate (Leaves/week)	1·6	1·4	1·1
Tillers/plant	2·8	2·4	1·1
Number of ears/plant	1·33	1·29	0·92
Weight of ears/plant (g)	156·8	105·9	98·8

Implications and applications

Although the influence of reflection coefficient on soil temperature has been known for many years, little has been done to exploit the potential for increasing crop production in favourable locations or for extending production into less favourable areas. Black polythene has been used to a limited extent in the production of vegetables and soft fruits, primarily for weed control, but the increased soil temperatures have also promoted earlier maturity. Stubble mulch farming was introduced to control soil erosion and conserve soil moisture but, as shown in Fig. 17.4, it may also reduce soil temperature by increasing the reflection coefficient.

Some suggestions have even been made as to the way weather patterns may be changed by altering soil reflection coefficients on a large scale. Wexler (1958) makes some interesting calculations on the effects of blackening the Arctic ice-pack and adjacent snow fields, but has reservations about the logistics of depositing 10^9 tons of carbon black in order to lay down a black layer 0·1 mm thick! However, Black (1963) was much more enthusiastic about applying a black petroleum mulch (he was employed by Esso Research and Engineering Co.) over tens or hundreds of square miles near the shoreline of a sea or large lake in arid areas. He proposed that the differential heating would increase sea breeze circulation, bringing in moist air from over the water, lifting it to its condensation level and bringing cloud formation and subsequent rainfall.

The possibilities of climate control still seem improbable and would have to be carefully assessed. However, in view of recent large increases in costs of crop production, and a consequent need to increase yield per unit area, it may be timely to exploit on a *field scale* the benefits of increasing or decreasing soil temperatures by altering the reflection coefficient.

Acknowledgements

I am grateful to Professor G. Stanhill of the Institute of Soils and Water, The Volcani Centre, Israel, and to Dr. R. K. M. Hay, Crops Research Unit, University of Edinburgh for kindly allowing me to use some of their previously unpublished data.

References

ADAMS J.E. (1967) Effect of mulches and seed bed configuration. I Early season soil temperature and emergence of grain sorghum and corn. *Agron. J.* 59, 595–9.

ALLMARAS R.R., BURROWS W.C. & LARSON W.E. (1964) Early growth of corn as affected by soil temperature. *Soil Sci. Soc. Am. Proc.* 28, 271–5.

ATKIN R.K., BARTON G.E. & ROBINSON D.K. (1973) Effect of root-growing temperature on growth substances in xylem exudate of *Zea mays*. *J. exp. Bot.* 24, 475–87.

BEAUCHAMP E.G. & TORRANCE J.K. (1969) Temperature gradients within young maize stalks as influenced by aerial and root-zone temperatures. *Plant and Soil* 30, 241–51.

BLACK J.F. (1963) Weather control: use of asphalt coatings to tap solar energy. *Science* 139, 226–7.

BOWERS S.A. & HANKS R.J. (1965) Reflection of radiant energy from soils. *Soil Science* 100, 130–8.

BROUWER R. & HOOGLAND A. (1964) Response of bean plants to root temperatures. *Jaarb. I. B. S.*, 23–31.

BULL T.A. (1964) The effects of temperature, variety and age on the response of *Saccharum* spp. to applied gibberellic acid. *Aust. J. Agric. Res.* 15, 77–84.

GERARD C.J. & CHAMBERS G. (1967) Effect of reflective coatings on soil temperatures, soil moisture, and the establishment of fall bell peppers. *Agron. J.* 59, 293–6.

KETELLAPPER H.J. (1960) The effect of soil temperature on the growth of *Phalaris tuberosa* L. *Physiol. Planta* 13, 641–7.

KLEINENDORST A. & BROUWER R. (1970) The effect of temperature of the root medium and of the growing point of the shoot on growth, water content and sugar content of maize leaves. *Neth. J. Agric. Sci.* 18, 140–8.

KRAMER P.J. (1956) Physical and physiological aspects of water absorption. In *Encyclopedia of Plant Physiology*: W. Ruhland (ed.) 3, 124–59, Springer-Verlag Berlin.

LEONARD R.E. & ESCHNER A.R. (1968) Albedo of intercepted snow. *Water Resources Research* 4, 931–5.

LUDWIG J.W. & HARPER J.L. (1958) The influence of the environment on seed and seedling mortality. VIII The influence of soil colour. *J. Ecol.* 46, 381–9.

MONTEITH J.L. (1959) The reflection of short-wave radiation by vegetation. *Quart. J. R. Met. Soc.* 85, 386–92.

MONTEITH J.L. (1973) *Principles of environmental physics.* Edward Arnold, London.

NEILSON K.F. & HUMPHRIES E.C. (1966) Effects of root temperature on plant growth. *Soils and Fertilisers* 29, 1–7.

STANHILL G. (1966) Observations on the reduction of soil temperature. *Agr. Meteorol.* 2, 197–203.

STANHILL G. (1970) Some results of helicopter measurements of the albedo of different land surfaces. *Solar Energy* 13, 59–66.

STEWART J.B. (1971) The albedo of a pine forest. *Quart J. R. Met. Soc.* 97, 561–64.

WATTS W.R. (1972) Leaf extension in *Zea mays* II. Leaf extension in response to independent variation of the temperature of the apical meristem, of the air around the leaves, and of the root-zone. *J. exp. Bot.* 23, 713–21.

WATTS W.R. (1973) Soil temperature and leaf expansion in *Zea mays. Expl. Agric.* 9, 1–8.

WATTS W.R. (1975) Air and soil temperature differences in controlled environments, as a consequence of high radiant flux densities and of day/night temperature changes. *Plant and Soil* 42, 299–303.

WEXLER H. (1958) Modifying weather on a large scale. *Science* 128, 1059–63.

WILLIAMS C.N. & BIDDISCOMBE E.F. (1965) Extension growth of grass tillers in the field. *Aust. J. Agric. Res.* 16, 14–22.

WILLIS W.O., LARSON W.E. & KIRKHAM D. (1957) Corn growth as affected by soil temperature and mulch. *Agron. J.* 49, 323–8.

Questions and discussion

Professor A. J. Willis and Dr. O. Rackham pointed out that certain plants, such as *Pteridium aquilinum*, mulched themselves with their own litter and thereby controlled the reflection coefficient and energy balance of the soil in which they perennated. Some of these plants had seemingly 'inefficient' seed-dispersal mechanisms, dropping their seeds into this deposit.

Dr. I. H. Rorison suggested that, with plants that were relatively insensitive to soil temperature, a white soil covering—magnesium carbonate would often be unsuitable because of its chemical activity—might promote growth by increasing the available light. Dr. Watts, who had not himself investigated the matter, quoted a parallel study by R. Geiger in which the yield of tomatoes grown against a wall painted white, to increase the light, had been enhanced to a greater degree than that of plants against a black-painted wall which raised the air temperature.

The meeting failed to agree on the derivation of the word 'albedo'. Dr. Watts suggested it came from the Latin *albus* 'white'. Professor Monteith claimed it derived from a classic paper[1] by Professor Al Bedo in the Bulletin of Shorter Contributions of the American Astronomical Society. The word was first applied to the visible part of the spectrum but was later used by meteorologists for the whole spectrum.

[1] Bedo A. (1899) The reflectivity of stars and other optical aspects of the celestial bodies. *A.A.S. Bull. Sht. Contrib.*, 40, 37–48.

18 Temperatures of plant communities as measured by pyrometric and other methods

OLIVER RACKHAM *Botany School, University of Cambridge*

Introduction

Physiological ecologists are interested in the temperature of plants either because of its effect on metabolism or because it provides a key to the state of the plant's water and energy balances. Plant temperature is involved in all attempts at a comprehensive study of plant communities or plant parts in relation to the aerial environment. It is also widely used in that much larger number of ecological studies which, because of the complexity of the vegetation or the limited manpower and equipment available, do not aspire to a comprehensive interpretation but seek to extract the maximum evidence from simple characterization of plants and their environment.

Contact instruments—thermometers, thermocouples, thermistors, etc.—enable the temperature of a small area of leaf or the like to be followed for long periods. They are useful in situations, especially in controlled environments, where it is reasonably certain that the site of measurement is representative of the vegetation under study. They are less suitable for a rapid survey of temperature differences within a plant community. Clark's paper in this volume reminds us that parts of the same leaf can differ in temperature by several degrees. Moreover, by making contact with the leaf, they inevitably disturb its environment; this is particularly serious in a cornfield where it is difficult to make sure that the element stays on the leaf without holding the leaf still. In the Appendix is a note of a thermocouple needle designed for the thin leaves of *Impatiens parviflora*.

Aerodynamic methods infer plant temperatures from measurements on the exchange of energy, gases, or momentum between a stand of vegetation and the air above it. The simplest method (Monteith 1973) is to measure wind speed and air temperature as functions of height above the ground, and to determine (by two extrapolations) the temperature at the height corresponding to zero wind speed. The theory of these methods limits their use to extensive uninterrupted stretches of farm or forest crops: natural and semi-natural vegetation is rarely sufficiently simple. And because of the outlay involved in providing and maintaining instruments and in handling data, aerodynamic methods are usually restricted to the more ambitious and comprehensive types of investigation.

This paper is chiefly concerned with *radiometric* methods, depending on the long-wave radiation which all objects emit by virtue of their surface temperatures. Contact, aerodynamic, and radiometric methods measure three different temperatures: the actual temperature of the contact probe, the mean effective temperature at which the crop acts as a source for heat transfer, and the mean effective temperature at which it emits radiation. There will normally be some correlation between these temperatures, but we cannot expect them to agree in detail. The last two temperatures are abstractions; there is no reason why any part of any plant should actually be at either 'mean effective' temperature.

Pyrometric measurements

A *black-body* or *full* radiator, typically a cavity like the stoke-hole of a furnace, emits radiant energy according to the classic *Stefan's Law*

$$W = \sigma T^4$$

where W is the energy emitted by unit surface area in unit time, T is the temperature measured from absolute zero, and σ is $5{\cdot}57 \times 10^{-1}$ W m^{-2} deg C^{-4} $= 8{\cdot}13 \times 10^{-9}$ cal cm^{-2} min^{-1} deg C^{-4}.

Objects like tungsten filaments behave in a more complex way, but fortunately plants and most other natural surfaces act for all practical purposes as black bodies.

The spectral distribution of the radiation emitted by a black body is given by *Planck's Formula*

$$W_\lambda = \frac{C_1 \lambda^{-5}}{\exp (C_2/\lambda T) - 1}$$

where W_λ is the energy emitted, at wavelength λ, per unit wavelength interval from unit surface area in unit time, C_1 is $3{\cdot}74 \times 10^{-5}$ erg cm^{-2} sec^{-1}, and C_2 is $1{\cdot}438$ cm deg. It follows from Planck's formula that the emission spectrum has a maximum at a wavelength, λ_{max}, which is inversely proportional to T (*Wien's Displacement Law*). Tables for calculating black-body radiation and spectra are to be found in *Smithsonian Meteorological Tables* (1951).

The use of emitted radiation to measure surface temperatures is called *pyrometry* (pedants have invented longer words for it, like *pyrgeometry*). For at least a century it has been used as a painless means of measuring furnace temperatures, but only recently has it become possible to buy equipment of various kinds designed for the much feebler radiations emitted at ordinary temperatures. This paper is concerned with a simple hand-held instrument and with an aircraft-mounted line-scan device producing detailed black-and-white pictures. Pictures giving less detail but coded in false colour are the subject of the following paper. Most of these instruments work on the basis of radiant intensity rather than spectral distribution.

The ecological use of pyrometry is helped by the fact that most natural radiant energy comes in two distinct wavebands. The sun has an effective surface temperature of about 6,000°K and emits radiation almost entirely in the *short-wave* region of the spectrum, i.e. in the *ultra-violet* ($\lambda < 0{\cdot}4\ \mu$), *visible* ($0{\cdot}4\ \mu < \lambda < 0{\cdot}7\ \mu$), and (*near-*) *infra-red* ($0{\cdot}7\ \mu < \lambda < 3\ \mu$). Plants and their environment have surface temperatures not far from 300°K and emit almost entirely in the *long-wave* (*far-infra-red*, *terrestrial*, *thermal*) region ($3\ \mu < \lambda < 30\ \mu$). Unless we live on the edge of a volcano, or in other odd environments, we seldom encounter temperatures or radiation spectra in between. The sun's emission can neatly be separated from long-wave radiation by dividing the spectrum at 3 μ, a point at which there is little naturally-occurring radiation and around which, as it happens, most glasses become opaque. Moreover, short-wave and long-wave intensities, thus divided, are of the same order of magnitude. The sun emits 160,000 times as much energy per unit surface area as an object at 300°K, but this is offset by the fact that it occupies only 1/100,000 of the hemisphere. Figure 18.1 and Table 18.1 compare the intensity and spectral properties of the sun's radiation as *received* by surfaces on earth with those of the radiation *emitted* by surfaces at ordinary temperatures. It is instructive to compare this Figure, where the ordinate scale is logarithmic and the abscissae linear, with the corresponding Fig. 1 of Blackwell (1966), where the ordinate scale is linear and the abscissae logarithmic.

The relatively cheap and simple pyrometer used for most of the measurements in this paper was a hand-held battery-operated instrument made by Barnes Instrument Company. Radiation within a 7° field of view enters through a thin polythene window and is focused by a concave mirror on to the detector. According to the makers the latter is a thermopile, spectral sensitivity being limited by filters to the 6·9–20 μ waveband. By pointing the instrument during bright sunshine at a diffusely-reflecting concrete surface of high thermal capacity, and then shading the field of view for a few seconds, I have not been able to detect any sensitivity to short-wave radiation; but most of the short-wave radiation that enters the device must somehow be dissipated inside it and presumably contributes to its instability. The zero setting was found to drift unpredictably, probably because of temperature changes inside the instrument, and it was necessary to look every minute or so at a reference surface (which the makers provide, though a blackened tin can weighted to float in a bucket of water will do). Two direct-reading scales are provided, one being coarse or 'absolute' (–10°–+60°C in 2° steps), the

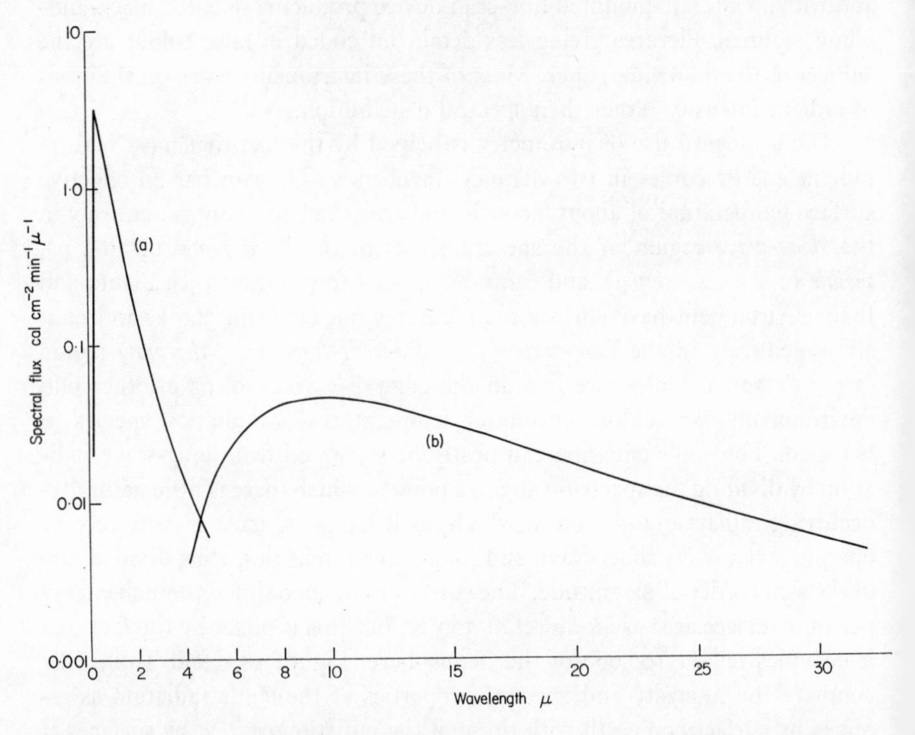

Figure 18.1. Spectral composition (a) of solar radiation as received by a surface at right angles to the sun's rays at the top of the earth's atmosphere; (b) of black-body radiation as emitted by a surface at 27°C (300°K). From *Smithsonian Meteorological Tables*; solar constant taken as 1·94 cal cm^{-2} min^{-1}.

other being fine or 'differential' (10° range in ½° steps). No recorder outlet is provided and I have not attempted to contrive one, partly because of the need for constant rezeroing, and partly because the method is more suited to rapid survey over an area than to prolonged recording at one site. The scale is non-linear, though the non-linearity inherent in Stefan's Law is partly offset by the fact that the 6·9–20 μ waveband includes only about ⅔ of black-body emission at 300°K, a fraction which itself varies with temperature in a curvilinear manner.

There follow two applications of this instrument to the measurement of surface temperature ranges within a heterogeneous stand of vegetation. For this purpose its portability enables many sites to be covered rapidly with a single instrument, while its field of view covers an area large enough to be representative of some particular element in the mosaic of species and microenvironments. In both applications, periods of stable weather were chosen; rain, broken cloud, etc. were avoided. This is a substantial but necessary restriction on the measurements: the interaction between variable vegetation and changing weather would demand a separate and much more comprehensive study with continuous recording at a large number of sites.

Table 18.1.

		Total intensity cal cm^{-2} min^{-1}	KW m^{-2}	λ_{max}, μ	Percentage in the long-wave range ($\lambda > 3\ \mu$)
Maximum sun and sky radiation received by horizontal surface at sea-level in Britain (cloudless day, mid-June)		1·3	0·91	0·48	$< 1\frac{1}{2}$ (depends on atmospheric absorption)
Radiation emitted by full radiator at:	0°C	0·45	0·31	10·6	99·998
	10°C	0·52	0·36	10·2	99·996
	20°C	0·69	0·48	9·9	99·994
	30°C	0·78	0·54	9·6	99·990
	40°C	0·89	0·62	9·3	99·98

The first application is a simple one in which the vegetation is a uniform monoculture and its heterogeneity arises from manipulation of soil moisture in experimental plots. The second application is to woodlands in which variation arises from structure and species composition.

Surface temperature and water deficit in barley

The vegetation in this case consisted of a field of spring barley, *Hordeum vulgare* 'Proctor', cultivated and sown with normal agricultural machinery. Treatments were applied to plots, each measuring 1·8 × 2·3 m, assembled in one small part of the field but separated by substantial untreated baulks. Four treatments were used (Rackham 1972a):

(i) Control.

(ii) Irrigation, applied by trickle nozzles to the soil surface at intervals of a week or more in quantities sufficient to keep the soil moisture deficit oscillating between two predetermined limits.

(iii) Plastic rainwater gutters laid between the rows of crop to intercept a proportion of the rainfall (normally about 40% in the part of the growing season with which we are concerned).

(iv) 'Leaky gutters' having slots cut in them to return the intercepted rainwater immediately to the soil and thus to allow for the effects of gutters on the crop other than by withholding moisture.

Treatments were replicated in what amounted to a randomized block design. In one of the two years (1971) a second variety, 'Zephyr', was added as a further factor.

Surface temperature measurements formed part of an extensive study of the physiology of the barley crop in relation to drought. The experimental field was equipped with a wide range of micrometeorological and related instruments, two of which chiefly concern us here. Aerodynamic profile equipment enabled the temperature and transpiration rate of the field as a whole to be estimated from hour to hour. Some of the treatment plots were fitted with neutron probe access tubes which were visited from week to week to provide a long-term record of the changes in soil moisture and transpiration associated with particular treatments (Grant 1970). These instruments were run by Mr. D. R. Grant and the staff of the then Meteorological Office Research Unit, Cambridge, to whom I am indebted for the following figures of soil moisture and aerodynamic crop temperature.

Crop surface temperatures were usually measured on the 'differential' scale of the Barnes instrument. It was held so as to look at the crop at a high angle of altitude, normally either east or west (preliminary trials indicated that it did not greatly matter in which azimuth direction the crop was looked at relative to the sun and wind). Because its response time is only 2 sec. it was affected by puffs of wind acting on the crop, and practice was needed in reading it. Each measurement on a plot was immediately preceded or followed by a reading from an adjacent area of untreated field; measurements relative to untreated field were sometimes found to be more self-consistent than absolute readings, presumably because the disturbing effects of soil and weather variation were partially eliminated. No measurements were taken when the crop was young or senescent, because of incomplete ground cover.

The work was done in 1970 and 1971. The 1971 data were subjected to a 4-way analysis of variance (observations × replications × treatments × varieties).

Sensitivity

The 'differential' scale can be read to 0·1°. In 1971 the smallest difference significant at the 5% probability level was 0·12°, but that was on 292 residual degrees of freedom. The standard error of a difference of two single observations can seldom be less than 0·5° and in gusty weather probably exceeds 1·0°.

Comparison with aerodynamic method

In 1971 there were 20 occasions (in ten pairs about ½ hour apart) on which it was possible to compare radiometric surface temperature with air temperature extrapolated to zero wind. The results are shown in Fig. 18.2. The two

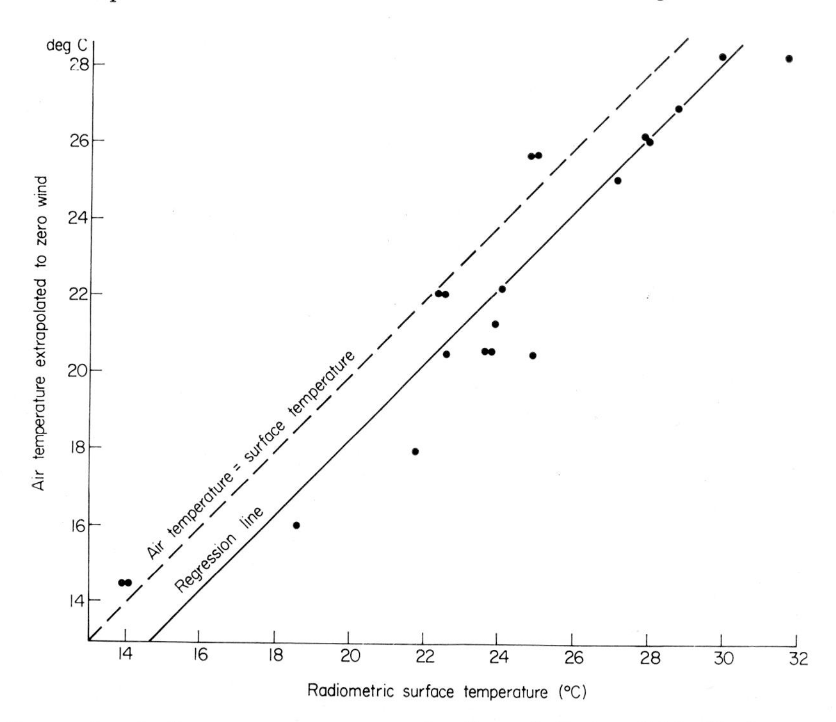

Figure 18.2. Surface temperature, measured radiometrically in a field of barley, compared with crop temperature measured aerodynamically by extrapolating the air temperature profile to zero wind. May–July 1971. Each point results from comparing the mean of 20 radiometric observations, spread over a period of about 20 minutes, with an hourly mean aerodynamic temperature.

temperatures vary in parallel—the regression coefficient is 1·00—but the radiometric temperature is an average of 1·7° higher.*

Surface temperature and soil moisture

It was consistently found that the wetter the treatment the cooler the plot.

* The 1970 figures gave less consistency between radiometric and aerodynamic temperatures, but this may result from defects in the *absolute* calibration of the reference temperature standard.

In 1970 there was a naturally dry June and in consequence irrigation had a bigger effect than the gutter treatment on soil moisture. Irrigated plants were consistently cooler than the other treatments. The difference—present throughout the irrigation cycle—was largest on highly-evaporative days, reaching 7° on one hot afternoon; but irrigated plots were still 0·9° cooler than control on a particularly cold overcast morning. Rain fell around the end of June and 'leaky gutters' plots thereafter became wetter than 'gutters', while some of the 'gutters' plots were irrigated from then onwards as an additional treatment. Temperature differences then appeared to correspond with these differences in soil moisture.

In 1971 there was a wetter June and accordingly gutters produced a bigger soil moisture difference. With weather generally less evaporative than in 1970, the 'gutters' crop was warmer by a maximum of 2°. Irrigation produced a smaller but still significant opposite effect.

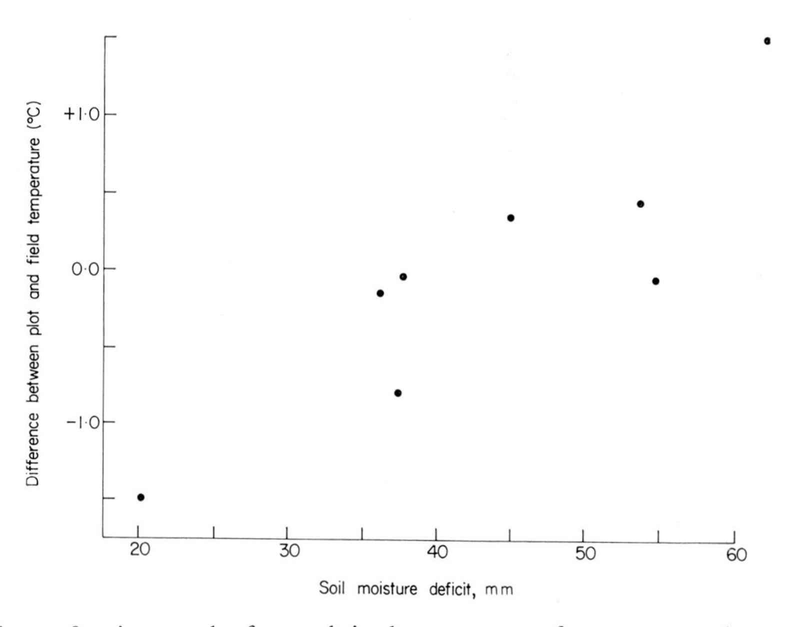

Figure 18.3. An example of a correlation between crop surface temperature (measured radiometrically) and soil moisture deficit (determined by the neutron probe). 'Proctor' barley, 2 June 1971. The points represent measurements on eight plots subjected to four moisture treatments. Plot temperatures are given relative to areas of untreated crop immediately adjacent. Correlation coefficient on this occasion=0·89.

These differences must arise mainly from differences in transpiration. Neutron probe and lysimeter measurements (Grant (1970) and later work as yet unpublished) have shown that transpiration in barley is sensitive to differences of soil moisture content: irrigated plots evaporate substantially faster than control, and provided there has been rain 'leaky gutters' plots evaporate faster than 'gutters'.

Figure 18.3 gives an example of the correlations which were sometimes obtained between soil moisture deficit (by neutron scattering) and radiometric crop temperature. A linear relation was obtained only for soil moisture deficits below about 70 mm; at higher deficits crop temperatures could range over several degrees without being correlated with soil moisture deficit. Soil moisture deficit, when high, is a poor measure of the availability of water to barley, which does not depend merely on soil moisture tension—this soil has a highly non-linear relation between moisture content and tension—but also on the vertical distribution of water in the soil profile and probably on sources of horizontal variation such as worm channels. Passioura (1972) has shown that viscous resistance to flow in the xylem is substantial in cereals: for this reason free water deep in the soil, even if roots have reached it, is less readily available than water near the surface.

The quantitative relations between the radiometric surface temperatures and evaporation rates of treatment plots await further analysis.

Other aspects of surface temperature

Table 18.2 summarizes the results of the analysis of variance on the 1971 measurements. Attention is drawn to the following points:
(i) There is a significant variety effect, Zephyr being warmer than Proctor. There is some indication that treatments have less effect on the temperature of Proctor than they have on Zephyr. This is interesting in relation to Clark's and Hadfield's papers in this volume. Proctor has floppy leaves which bend over while Zephyr is more erect in habit.
(ii) There is a significant block (replication) effect. Even in a supposedly uniform crop of barley, some parts of the field are a fraction of a degree warmer than others, due presumably to soil variation.

Temperatures in woods

This study forms part of an investigation of the history of woodlands and their management in Eastern England. For at least 700 years the usual method of managing woods in most of lowland Britain was the coppice-with-standards system. The majority of the trees and shrubs (the *underwood*) are felled at intervals of 3 to 25 years and allowed to grow again from the stumps (*stools*) or by root suckers. Other trees (the *standards*) are allowed to stand for several coppice rotations and form a scattered upper storey. This process results in great complexity of structure: a wood consists (if large enough) of a number of *panels*, areas at different phases of the underwood cycle, each of which contains several microclimatic niches. Figure 18.4 illustrates some

Table 18.2. Summary of surface temperatures in the 1971 barley experiment.

Effect	Actual temperature in °C				Temperature in °C above or below adjacent field			
Variety	Proctor	23·63		P < 0·001	Proctor	+0·06		P = 0·006
	Zephyr	24·24		LSD = 0·12	Zephyr	+0·22		LSD = 0·12
Treatment	Gutters	24·88		P < 0·001	Gutters	+1·03		P = 0·001
	Control	23·72		LSD = 0·16	Control	−0·07		LSD = 0·17
	Leaky	23·83			Leaky	−0·03		
	Irrigation	23·32			Irrigation	−0.43		
Variety × treatment		Proc.	Zeph.			Proc.	Zeph.	
	Gutt.	24·56	25·20	P=0·05	Gutt.	+0·86	+1·20	P=0·04
	Cont.	23·44	24·00		Cont.	−0·06	−0·10	
	Leaky	23·41	24·25		Leaky	−0·14	+0·19	
	Irri.	23·13	23·51		Irri.	−0·45	−0·41	
Time of observation	P < 0·001, LSD = 0·27				P < 0·001, LSD = 0·28			
Block	P < 0·001, LSD = 0·18				P = 0·007, LSD = 0·19			
Time of obs. × variety	P < 0·001				P < 0·001			
Time of obs. × treatment	P < 0·001				P < 0·001			
Time of obs. × block	P < 0·001				P = 0·04			
Time of obs. × var. × trt.	P < 0·001				P < 0·001			

P: level of statistical probability. LSD = least significant difference in °C at P = 0·05.

of these. Such attention as has been paid to the microclimatic effects of management has mainly been concerned with the effects of changing shade on photosynthesis and growth of the ground vegetation. Here we are concerned with the heating effects of the sun's radiation in relation to woodland structure.

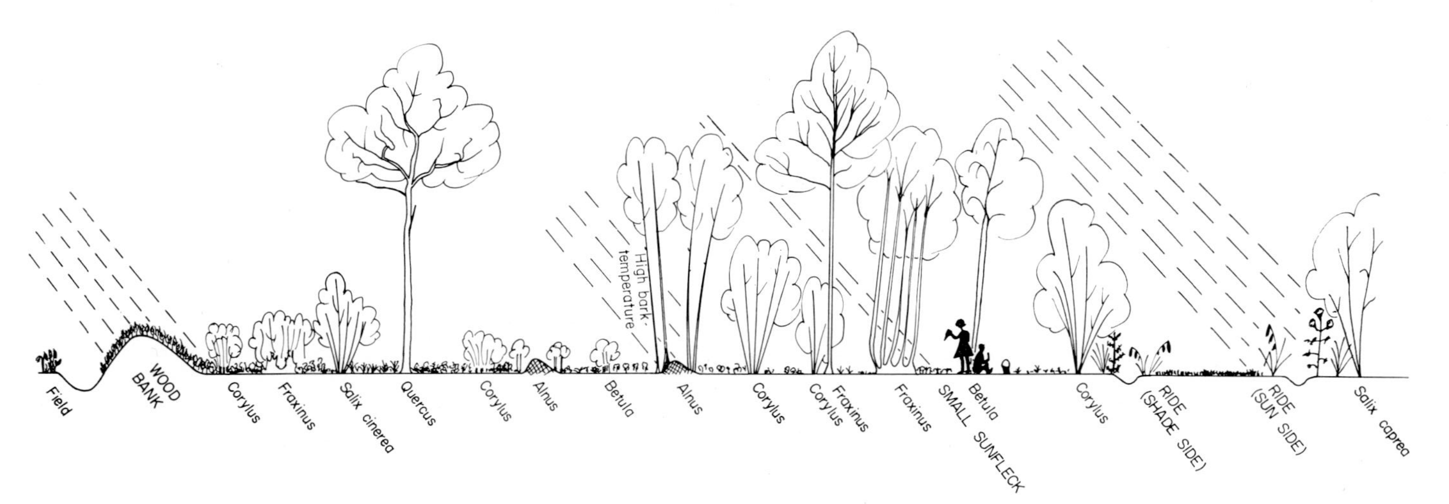

Figure 18.4. Structure of the Bradfield Woods. This profile sketch (idealized and compressed) shows, left to right, the wood bank; a panel of underwood one year after felling, with standard oak tree; a panel of 'old' underwood (about 15 years' growth) with standard ash tree; and a ride. Sun is at top left and diagonal broken lines indicate its rays.

Some preliminary observations

Some years ago, observations were made on the energy balance of *Impatiens parviflora* in Madingley Wood, Cambridgeshire. *Impatiens* is an annual plant serving as a physiological model for the native ground vegetation of these woods. The site was a long-neglected stand of *Ulmus minor*, *Quercus robur*, *Fraxinus excelsior*, etc. Its light environment has been studied extensively by Anderson (1964), Coombe (1966) and others; we are concerned here with the summer aspect. On the woodland floor there is a general background of diffuse light, supplemented at infrequent and irregular intervals, during periods of sunshine, by direct sunlight in the form of *sunflecks* when the sun gets in line with a gap in the canopy. Although infrequent, sunflecks contribute a large proportion of the total light. Their occurrence varies according to the woodland structure and in particular to the number and size of canopy gaps as seen from the ground. The intensity and duration of any particular sunfleck depends on the time of day and of year at which it arises, and on the size of the canopy gap which generates it. Since the sun's disk subtends an angle of $\frac{1}{2}°$ at the earth's surface, a canopy gap of at least this apparent size is needed to transmit full direct sunlight. Most sunflecks are produced by small canopy gaps, subtending angles of less than about 2°. The actual intensity of a sunfleck produced by such a small gap depends partly on the fraction of the sun's disk that can be seen, but may be markedly reduced by atmospheric haze or thin cloud, which (even if almost invisible) scatter a proportion of the direct radiation so that it appears to come from the sky close to the sun rather than from the sun's disk itself. Wind also reduces the intensity of small sunflecks through the obstruction of the canopy gap by waving foliage.

In these early observations on *Impatiens*, leaf temperature excess—the difference between leaf and air temperature—was measured with a thermocouple needle and recorded by hand at 10 sec. intervals. During periods of diffuse light, i.e. for the whole time when the sun is obscured by cloud and for most of the time when it is shining, *Impatiens* leaves are very close to air temperature. Short-term averages ranged from 3·1° below air temperature to 0·5° above, with a general mean of −0·3°. Transpiration resistance is low in this plant and for most of the time in shade it loses more energy in evaporating water than it gains from radiation absorbed.

Figure 18.5 shows the effect of sunflecks of different intensities on the temperature of *Impatiens* leaves. During an intense sunfleck, when the leaf is suddenly exposed to full sunlight for several minutes, its temperature may rise several degrees, the maximum observed being 9° above air temperature.*

* This is a minimum estimate. The thermocouple needle gives the correct sign to leaf temperature excess but tends to underestimate its magnitude, partly because the air tem-

21.1

18.1a

18.1b

facing p. 435

Such high leaf temperatures are very sensitive to wind, because the effects of puffs of wind on the transfer of sensible and latent heat reinforce each other (they oppose each other when the leaf is cooler than the air). Figure 18.6 shows that the amount by which the leaf temperature rises—averaged over the duration of the sunfleck—depends mainly on the change in incident radiation. A rise in temperature of 1° is produced by the addition of 0·054 cal cm^{-2} min^{-1} (= 37 W m^{-2}) of direct radiation.* The wide variations which were observed between one sunfleck and another in air temperature and humidity, wind, leaf attitude, stomatal aperture, etc. create the scatter observed in Fig. 18.6 but do not destroy the linear relation.

The frequency distribution of leaf temperatures in *Impatiens* is highly skewed: temperatures much higher than the normal range occur on the infrequent occasion of an intense sunfleck. A sunfleck of the highest possible intensity is very infrequent: it can only occur within a few hours of noon and within a few weeks of the summer solstice, and requires that an appropriate canopy gap should be in line with the solar track, that the sun should be shining, the atmosphere clear and without much wind. In these circumstances incident radiation would rise by nearly half the solar constant (p. 426), giving a rise of leaf temperature several degrees greater than any arising in these observations.

This raises the question of whether damaging leaf temperatures might be produced when an intense sunfleck happens to coincide with high humidity and low wind. Such a combination of circumstances would be rare and would probably not be detected in the ordinary course of microclimatic recording at one or a few sites. This may be the explanation for the leaf necrosis, strongly resembling the heat damage reported by Hadfield in *Camellia* in this volume, which occurs widely in *Mercurialis perennis* in the Cambridgeshire woods in some years. Plate 18.1 illustrates an example of this observed near the south edge of Hayley Wood in 1966. A study of the canopy gaps shown on hemispherical photographs (Evans & Coombe 1959) taken in the area showing this damage in relation to solar tracks suggest that it was produced by a very large sunfleck occurring on a June afternoon that year; the shadows thrown by tree-trunks appeared as areas of unaffected plants.

perature junction is only a few cm from the plant, and partly because of heat conduction along the needle (see Appendix).

* This parameter happens to coincide with the advective heat transfer coefficient of 0·05 cal cm^{-2} min^{-1} deg^{-1} estimated for *Impatiens* plants grown in a similar range of wind speeds in controlled-environment cabinets (Rackham 1965).

Plate 18.1 Leaf necrosis in *Mercurialis perennis* on the floor of Hayley Wood near Cambridge in June 1966.

Plate 21.1. Effects of heat damage on fully exposed leaves of a mature bush of *Camellia sinensis* var *assamica*. For details and discussion, see p. 489.

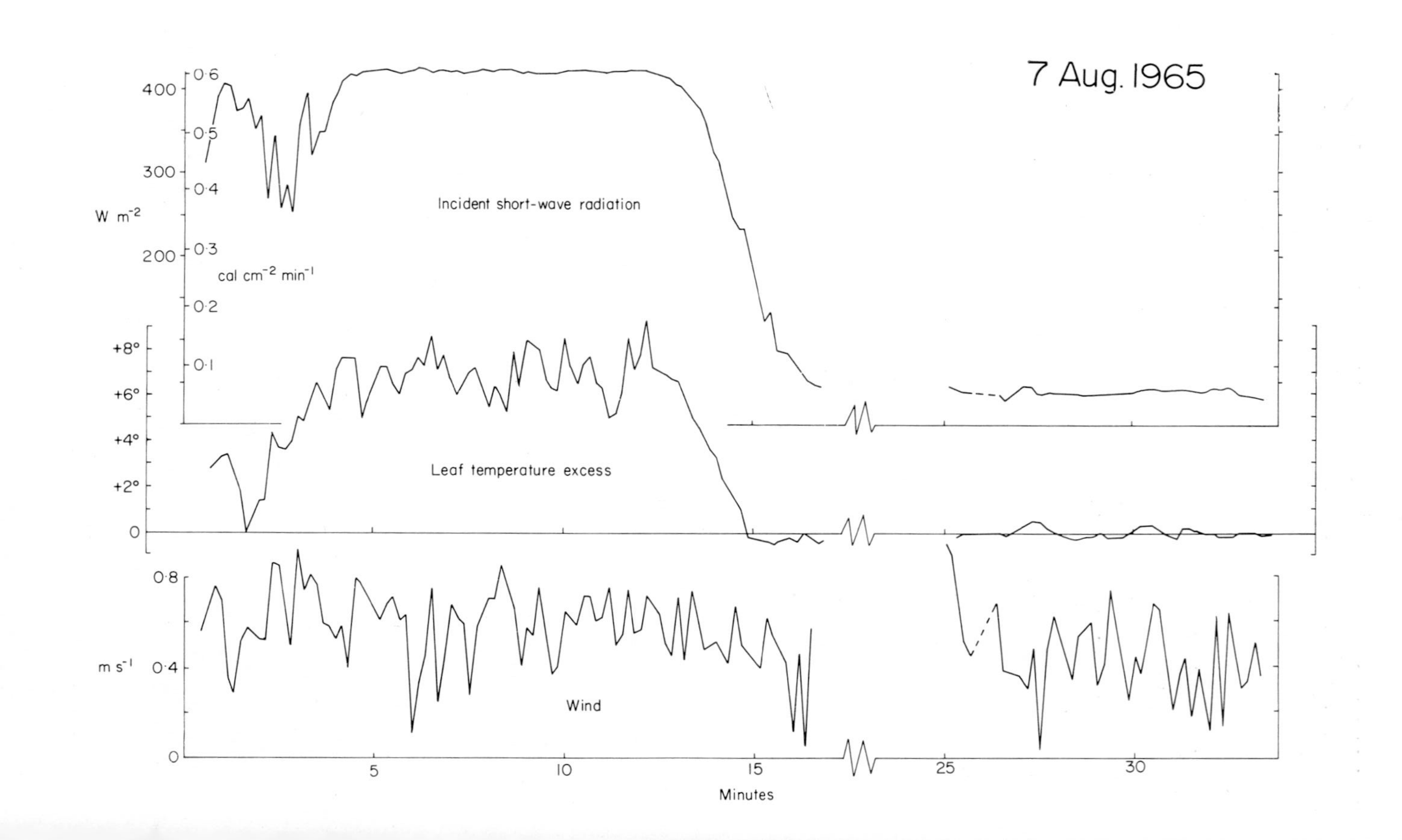
7 Aug. 1965
W m^{-2}
400
300
200
0·6
0·5
0·4
0·3
0·2
0·1
cal cm^{-2} min^{-1}
Incident short-wave radiation
+8°
+6°
+4°
+2°
0
Leaf temperature excess
m s^{-1}
0·8
0·4
0
Wind
5
10
15
25
30
Minutes

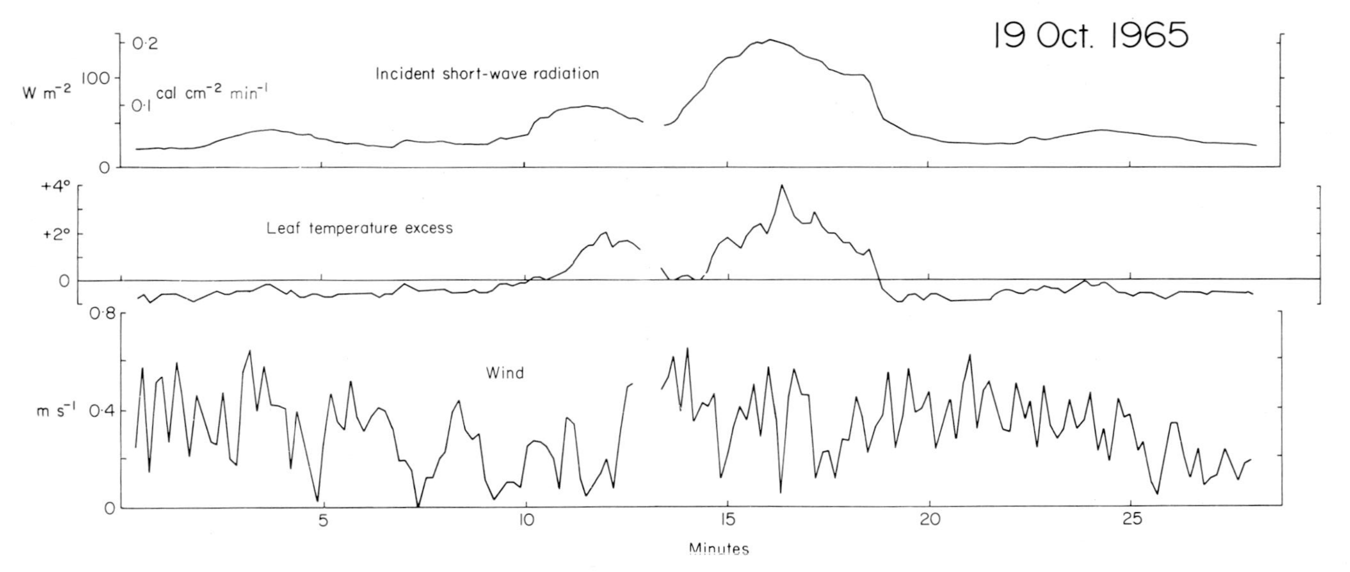

Figure 18.5. Sunflecks in Madingley Wood and their effects on the leaf temperature of *Impatiens parviflora*. These examples illustrate (7 August) an unusually intense sunfleck followed by shade, and (19 October) a period of shade broken by minor sunflecks. Short-wave radiation incident on a horizontal surface was measured by a Monteith-type solarimeter (Monteith 1959). Difference between leaf and air temperature was measured with a thermocouple needle on a typical crown leaf of a potted *Impatiens* plant (2–3 m away from the solarimeter, hence the apparent time-lag). Wind was measured with a home-made hot-wire anemometer. Recorded by hand in a 10-sec cycle.

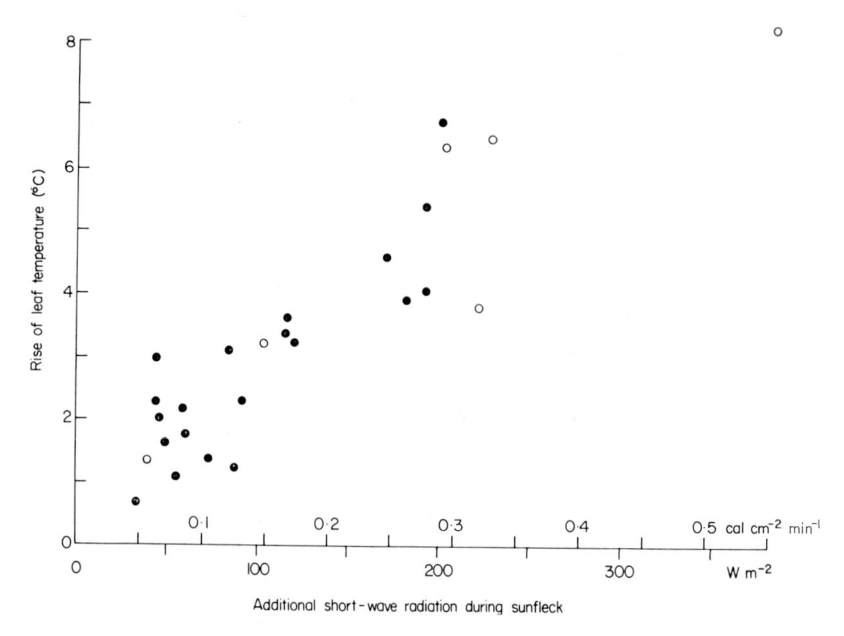

Figure 18.6. Relation between the intensity of sunflecks and the rise of leaf temperature which they produce. The rise in incident short-wave radiation during each sunfleck, as compared with the shade period immediately before or after, is plotted against the corresponding change of leaf temperature excess. These quantities are determined from curves such as those of Fig. 18.5, short-term fluctuations being smoothed out. *Impatiens* in Madingley Wood: white points refer to the site known as the 'small clearing', August 1965; black points refer to the 'large clearing' site in October 1965.

Long-wave photography in the Bradfield Woods

The Bradfield Woods Nature Reserve near Bury St. Edmund's consists of two ancient woods—Felsham Hall Wood and Monks' Park Wood—which are still actively managed. The structure (Fig. 18.4) perpetuates most features of medieval woods as known from documents and archaeological samples of their produce. Underwood is cut in panels of a few hectares on a rotation of about 11 years. There are now also areas of 'old' underwood of up to 27 years' growth. After felling, the new coppice shoots begin growth in June and grow in the first season to a height of 1 to $3\frac{1}{2}$ m depending on species and weather. They form a closed canopy in about four years and increase slowly in height for many years thereafter. At any one time the woods therefore consist of a mosaic of panels having different combinations of standard trees, middle-aged and old underwood forming a closed canopy beneath the standards, young underwood, tall herbaceous vegetation between young underwood stools, short herbaceous vegetation under closed-canopy underwood, and bryophytes. There are also other microclimatic habitats such as rides and glades. Each of these structural niches is occupied by several different plant species. Woodland in the middle stages of regrowth has a structure somewhat resembling that of the tea plantations with shade trees which are the subject of Hadfield's paper in this volume, but is more varied in species composition.

An opportunity of photographing the Bradfield Woods in the far-infra-red arose when Cambridge University Department of Aerial Photography had the use of a line-scan camera, being developed by Messrs. Hawker Siddeley, in mid-May 1973. At this time of year (determined by the requirements of other users of the camera) the ground vegetation had been in leaf for several weeks and the trees were about halfway into leaf. Variations in temperature might be produced by differences of species or structural relationship, but differences of soil moisture had not had time to develop. In general the leafing of the underwood, especially of young underwood, was appreciably more advanced than that of standard trees, but within each layer there were wide variations from species to species and in the case of some species (e.g. *Quercus robur*) also between individuals.

In Plate 18.2 a line-scan thermal photograph is compared with an ordinary oblique air photograph taken in the visible spectrum at the same time. The calibration of the line-scan instrument is not known, but from later ground observation I estimate that the range from black (cold) to white (hot) is about 10°C.

Most of the temperature variation is related to woodland structure. The following list begins with the coldest structural elements in the wood and ends with the hottest.

1. Areas in the shadow of big trees (whether in the wood or in surrounding fields).
2. Canopies of standard trees and windward edges of panels of older underwood.
3. General canopy of older underwood.
4. Canopy of middle-aged underwood.
5. Young underwood.
6. Ground vegetation of newly-felled areas and between the stools of 1–3-year-old underwood.
7. Ground vegetation of rides and of wind-sheltered edges of newly-felled panels.
8. Dead bracken (*Pteridium aquilinum*) in glades.*

These temperature differences are presumably related to the insolation and wind exposure of each structural element. In contrast, variation from

* Bracken is a perennial plant which mulches itself with its own dead fronds (cf. Watts, 1975, discussion).

(a)

(b)

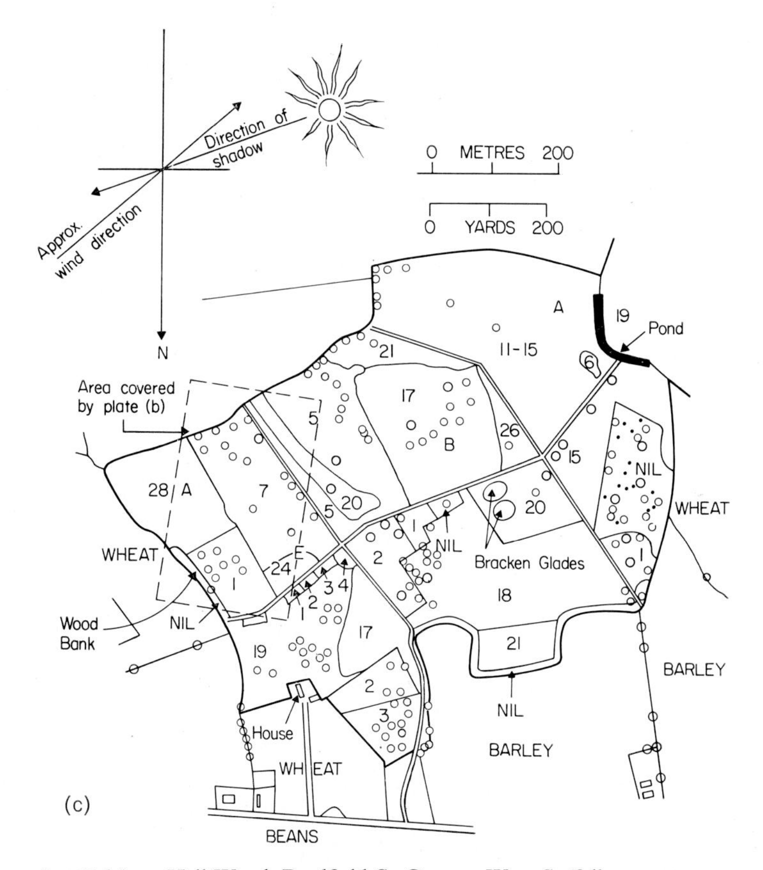

Plate 18.2. Felsham Hall Wood, Bradfield St George, West Suffolk.

(a) Aerial photograph taken with line-scanning equipment in the thermal infra-red region of the spectrum. 16 May 1973.

(b) Oblique aerial photograph taken at the same time with a conventional camera using visible light.

(c) Explanatory map showing features of the wood and of surrounding fields.

Underwood panels are shown with numbers giving the number of years' growth since the last felling. NIL indicates areas which have just been felled. A, B, E give examples of areas where there is a clear predominance of ash (*Fraxinus excelsior*), birch (*Betula pubescens*), or elm (*Ulmus procera* × *glabra*) in the underwood.

Parallel lines show rides.

Circles indicate the more prominent standard trees in the wood and hedgerow trees outside. These are mainly *Quercus robur*.

Black dots in NIL areas are bonfire sites.

The area shown in Plate (b) is outlined by broken lines.

Discrepancies of outline between Plates (a) and (c) are due to distortions in the former, caused partly by the cylindrical projection which the line-scanner generates and partly by instability in the aircraft flight.

Plates (a) and (b) are from *Cambridge University Collection; copyright reserved.*

species to species within any one structural element is relatively insignificant. Even the difference between leafy and leafless trees is barely detectable by this method. Oaks that have come into leaf are only marginally cooler than those still leafless. Underwood dominated by *Betula* (in leaf) is warmer, probably by a fraction of a degree, than that dominated by the still-leafless *Fraxinus* and *Ulmus*. No difference can be detected between areas of *Filipendula ulmaria*, *Allium ursinum*, or *Mercurialis perennis* as ground dominants.

Surface measurements in the Bradfield Woods

This is a survey of the range of surface temperatures encountered in the lower layers of the woodland structure on one particularly hot day, 14 August 1973. Many different plants were examined with the Barnes pyrometer between 11 and 18 h GMT. Air temperature reached a maximum of 27·4°C at 15 h and declined to 23° by 18 h. The relative humidity at 15 h was 41%. Wind, integrated over periods of about 1 h at 0·7 m above the ground in a relatively open part of the wood, ranged from 0·25 to 0·82 m sec^{-1}. (The Madingley observations were made in much lower air temperatures and much higher relative humidities. In Assam, Hadfield (1975, this vol. pp. 479–85) typically worked in an air temperature around 30°C with R.H. around 75% and a wind speed of about 0·6 m sec^{-1}: see his Figs. 4 & 5.)

These observations agree with the earlier evidence that surface temperatures are determined mainly by a plant's position in the woodland structure and not by its species.

The more or less horizontal foliage of 1-year-old underwood—of underwood which in August had nearly completed its first season's growth after felling, and was then about 2 m high—was on average about 1½°C *warmer* than air temperature where exposed to the sun. Variation within species was comparatively wide (Table 18.3); the highest actual temperature recorded was 31·3°. There is some evidence that *Fraxinus* and *Betula* ran about 1° cooler than *Corylus* and *Alnus*, but variation between species was small and doubtfully significant. Parallel measurements in Hayley Wood on an overcast day showed that with an air temperature of 21·5°C the foliage of 1-year-old underwood was consistently about 1° *cooler* than the air.

The element in the community that most closely resembles the *Impatiens* plants observed at Madingley is the ground vegetation under the closed canopy of old underwood of 12 years' growth or more. This was between 1° and 4° below air temperature. There was little difference between the species examined (*Filipendula*, *Mercurialis*, *Geum rivale*, *Stachys sylvatica* etc.). A small sunfleck caused by a canopy gap raised the surface temperature by only

1°–2°, so that it was still normally below air temperature. A few exceptions were found, mostly where plants had wilted, where the temperature in the sunfleck rose between 5 and 10 degrees; *Filipendula* in one instance went up to 35°C.

Table 18.3. Means and ranges of surface temperatures encountered in 1-year underwood. The figures express the difference in °C between foliage temperature (as measured pyrometrically) and air temperature.

	Mean	Range
BRADFIELD, 14 AUGUST 1973		
Corylus avellana	2·0	$\frac{1}{2}$ to 5
Alnus glutinosa	1·4	−1 to 4
Betula pubescens	0·7	$-\frac{1}{2}$ to 3
Fraxinus excelsior	0·5	−1 to $2\frac{1}{2}$
Salix cinerea and caprea	0·5	−2 to 4
HAYLEY, 17 AUGUST 1973		
Corylus avellana	−0·9	$-1\frac{1}{2}$ to $-\frac{1}{2}$
Acer campestre	−0·8	−1 to $-\frac{1}{2}$
Cornus sanguinea	−0·7	$-1\frac{1}{2}$ to $-\frac{1}{2}$

In the first year after felling, the ground vegetation among the regrowing stools is broadly similar in structure and composition to that under old underwood. (Major changes occur later in the cycle.) Temperatures in sunshine were much more variable and in general much higher than in the small sunflecks of the old canopy. The leaves of turgid plants spent much of their time over 5° above air temperature; there were several absolute readings over 35°.

The sunny side of rides and glades was intermediate between the newly-coppiced panel and the sunflecks under old canopy. Ground vegetation ran 3°–5° warmer than air temperature, often reaching 30°. Live *Pteridium* in a glade, in contrast to the dead *Pteridium* of the air survey, was only $1\frac{1}{2}$° above air temperature.

On the east side of Felsham Hall Wood the great medieval bank that encloses the wood had been exposed by felling. The air survey shows a difference of several degrees between the two sides of the bank (Plate 18.2a) and gives the false impression that it has a triangular profile. Ground measurements in August showed the *Mercurialis* carpet to be above air temperature by a mean of 6·5° on the slope facing the sun and 1·7° on the opposite slope.

Summary of the woodland observations

As far as these observations go, temperature variations, although complex, are related mainly to structural factors and depend little on species composition. Temperature distributions are highly skewed. Under shade, the ground vegetation (if turgid) is commonly slightly below air temperature and is unlikely much to exceed air temperature even during a small sunfleck. For temperatures to rise by several degrees requires more intense insolation, as in an exceptional sunfleck (Madingley) or a recently-felled area (Bradfield). The highest temperature recorded on this occasion at Bradfield (37·2° in *Mercurialis*) only just comes within the range of possible damage, but this is unlikely to be the limit, as although the weather was hot it was windy and not particularly humid and soil moisture was sufficient to prevent extensive afternoon wilting. Temperatures several degrees higher might arise under circumstances less favourable to transpiration. Moreover, temperatures as measured are an average over an area of foliage and underestimate, probably by several degrees, the temperature of its hottest parts. Thermal damage is therefore a feasible if rare phenomenon.

Higher temperatures still were recorded on the sunward sides of underwood poles, 8–15 cm diameter, exposed by felling at the edges of the panel or left standing in its interior. *Fraxinus* and *Quercus* ranged from 32° to 40° during the day. *Alnus* poles, which are dark in colour, exceeded 40° and went in one case to 46°. Because these are near-vertical objects the highest temperatures were reached late in the day; several readings over 40° were obtained after 17 h GMT. The shade side of poles seldom exceeded 28°. Although these particular trees showed no subsequent signs of distress it is clear that high trunk temperatures are a feasible cause of the 'sun-scorch' type of damage to thin-barked trees left standing after their neighbours have been felled (Peace 1962). Similarly high temperatures must be withstood by epiphytic bryophytes and lichens, particularly by those that are dark in colour.

Appendix

Thermocouple needle for measuring the temperature of leaves and other plant parts

This is a slender thermocouple that is stiff enough to be inserted into a leaf without undue handling. It is robust and cheaply made and can be used for measuring leaf temperatures in places remote from a power supply. It has a response time of a fraction of a second.

As described, the needle is about 14 cm long; one thermo-junction is in the point, the reference junction being near the other end of the needle. If the reference junction is shielded from direct radiation by a paper cap (J in Fig. 18.8) the needle gives a direct measure of leaf temperature excess. The thermocouple can if desired be made longer so that the reference junction can be kept at constant temperature.

One of the metals forming the thermo-junction has to be copper to avoid parasitic voltages elsewhere in the circuit. The needle itself is steel in order to be sufficiently stiff and sharp.

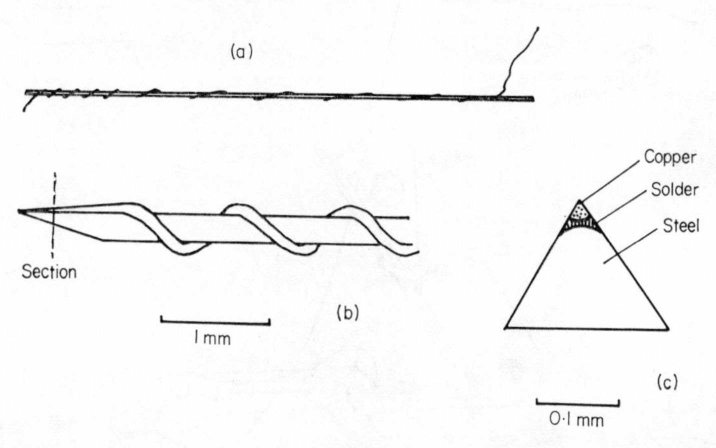

Figure 18.7. Making a thermocouple needle. (a) shows the method of winding the copper wire round the steel (not to scale); (b) shows the needle point after soldering and grinding; (c) shows a section of the type of point which the grinding process should seek to produce.

The needle is made from a length of steel harpsichord wire, no. 4 gauge (0·25 mm). A length of 44-gauge (0·081 mm) copper wire is wound in a helix round the steel wire (Fig. 18.7a), being stuck to the latter, and insulated from it, by coats of Araldite. At the tip of the needle a short length of the wires is bared and soldered together (Fig. 18.7b) (found to be more satisfactory than welding). The junction is trimmed and ground on a fine stone under a binocular microscope (Fig. 18.7c); the object is to achieve a triangular point with the copper wire running along one edge and making contact with the steel for the last 0·3 mm (Fig. 18.7).

The leads to the instrument are lengths of stranded plastic-insulated copper wire. One lead is soldered directly to the steel needle, about 2 cm from its free end, to form the air-temperature junction. The second lead is connected mechanically to the Araldite insulation of the needle, and electrically to the fine copper thermocouple wire, by means of thermo-setting silver cement.* In this way the 44-gauge thermocouple wire is protected from mechanical damage.

Following Wilson (1957), a clip is used to hold the leaf and the thermocouple and to bring them into adjustable contact. The device shown in Fig. 18.8 holds the leaf with minimum disturbance to it or its environment and

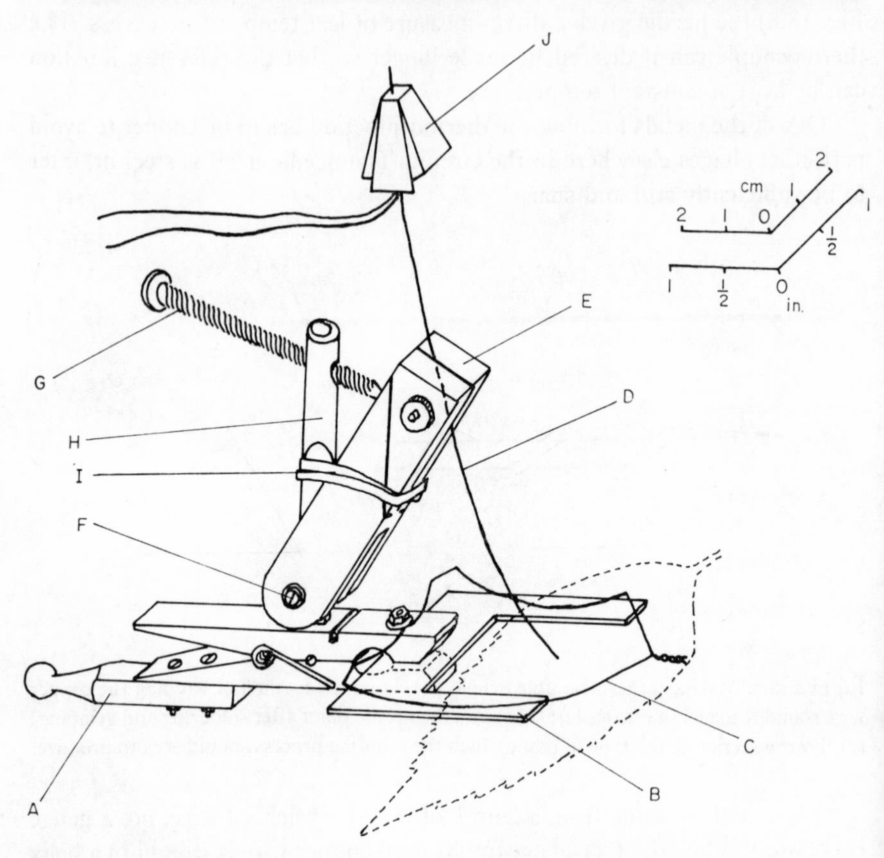

Figure 18.8. Thermocouple needle in a holder suitable for leaves like those of *Impatiens parviflora*. Explanation in text.

allows the needle point to be slowly advanced until any desired degree of contact with the leaf is obtained. The basis is a wooden spring clothes-peg attached to an articulated supporting arm, A (only the last joint is shown) by which it may be adjusted to the attitude of the leaf. The leaf is held lightly between a Perspex jaw B and a wire jaw C. The needle D is held in a wooden holder E. The latter pivots about the bolt F and is tilted forwards by advancing the screw G which works in a threaded hole in the fixed brass pillar

* Johnson, Matthey & Co. Ltd.; this is an epoxy resin which combines metallic conductivity with the Araldite-like property of sticking to non-metals.

H against the resistance of the rubber band I. J is a paper cap, carried on a small blob of solder near the blunt end of the needle, which protects the air-temperature junction from direct radiation.

Owing to its length and flexibility the needle can easily be calibrated against mercury thermometers in tubes immersed in two water baths. The calibration constant is about 5·4 μV deg C^{-1}, the standard deviation between needles in a batch being about 2% of this.

The needle is connected (with precautions against unwanted thermal voltages) to a microvoltmeter or its equivalent; a sensitivity of about 1 μV is required. The readings in this paper were taken with a spot galvanometer; on other occasions a potentiometric data-logger has been used. More recently it has become possible to buy a battery-driven amplifying microvoltmeter* which can be used either on its own or as the first stage of a recording system.

Possible sources of error include direct heating of the needle by solar radiation, electrochemical voltages, and interference with the energy balance of the leaf caused by conduction of heat along the needle. The *solar radiation* problem could in principle be largely avoided by causing the needle to enter the leaf from below, although this would create practical difficulties in manipulating the needle point and in arranging a position for the air-temperature junction. Observations of the behaviour of the needle with its point in air and exposed to the sun's rays on an intermittently sunny day failed to detect any direct response to short-wave radiation. Even with the needle point in air, it appears that heat transfer to its surroundings is so efficient as to minimize any effect of direct absorption of radiation. In use, much of the needle point is normally buried in leaf tissues. *Electrochemical voltages* (Middlehurst, Board, & Elbourne 1964) are minimized by the fact that the thermocouple wires are sealed in Araldite right up to the soldered junction. The *conduction* problem is common to all contact methods of measuring leaf temperatures, although here it is reduced by making the needle as slender as possible consistent with rigidity. It can be expected to lead to readings of leaf temperature excess which are smaller than the true values but have the correct sign.†

* Comark Ltd. This is a small, robust, and relatively cheap instrument which can be used in remote sites.

† It is difficult to measure or predict the size of this error. A rough calculation shows that, with the unusually high temperature gradient in the needle of 1 deg C cm^{-1}, the rate of conduction of heat is approximately 6 mcal min^{-1} (about half the conduction being in the copper), approximately equal to the heat load on a bright summer day on 1 mm^2 of leaf. Wilson (1957) anticipated the problem by calibrating his junctions with their tips touching the surface of the calibration bath; but this has the objection that if the bath is well stirred it transfers heat to or from the needle at a different rate from that of conduction in the leaf, while if the stirring is not perfect the temperature of the bath surface is difficult to determine.

This needle has been used in various studies on *Impatiens parviflora* in the laboratory (Rackham 1966), despite the often very thin and delicate leaves of that plant. It has given good results, under much more arduous conditions, with woody vegetation in southern Crete (Rackham 1972b).

Acknowledgements

I am indebted to Dr. G. C. Evans and Mr. D. R. Grant for much valuable discussion, in addition to the use of Mr. Grant's data which is acknowledged in the text. I am grateful for help with computing given by Mr. G. K. G. Campbell and the staff of the Plant Breeding Institute, Trumpington. Mr. C. Haynes, Mr. P. Orchard, and Mrs. S. Schumann (Trumpington) and Mrs. A. Hart and Mrs. S. Ranson (Bradfield) helped with the field-work. Parts of the work were done under a Research Fellowship of Corpus Christi College, Cambridge, and under research grants from the Nuffield Foundation (barley) and the Natural Environment Research Council (woodlands).

References

ANDERSON M.C. (1964) Studies of the woodland light climate. I. The photographic computation of light conditions. II. Seasonal variation in the light climate. *J. Ecol.* 52, 27–41, 643–63.

BLACKWELL M.J. (1966) Radiation meteorology in relation to field work. In *Light as an ecological factor* (eds R. Bainbridge, G.C. Evans and O. Rackham) pp. 17–39. Blackwell Scientific Publications, Oxford.

COOMBE D.E. (1966) The seasonal light climate and plant growth in a Cambridgeshire wood. In *Light as an ecological factor* (eds R. Bainbridge, G.C. Evans and O. Rackham) pp. 148–66. Blackwell Scientific Publications, Oxford.

EVANS G.C. & COOMBE D.E. (1959) Hemispherical and woodland canopy photography and the light climate. *J. Ecol.* 47, 103–13.

GRANT D.R. (1970) Some measurements of evaporation in a field of barley. *J. agric. Sci., Camb.* 75, 433–43.

HADFIELD W. (1975) The effect of high temperatures on some aspects of the physiology and cultivation of the tea bush, *Camellia sinensis*, in North East India. *This vol.*: 477–494.

MIDDLEHURST J., BOARD P.W. & ELBOURNE R.G.P. (1964) Electrochemical effects in thermocouples. *J. sci. Instrum.* 41, 676–8.

MONTEITH J.L. (1959) Solarimeter for field use. *J. sci. Instrum.* 36, 341–6.

MONTEITH J.L. (1973) *Principles of environmental physics.* Edward Arnold, London.

PASSIOURA J.B. (1972) The effect of root geometry on the water relations of temperate cereals (wheat, barley, oats). In *Structure and function of primary root tissues, Symp. Inst. Bot. Slovak Acad. Sci.*, Czechoslovakia.

PEACE T.R. (1962) *Pathology of trees and shrubs with special reference to Britain.* Oxford.

RACKHAM O. (1965) *Transpiration, assimilation, and the aerial environment*. Ph.D. Dissertation, Cambridge.

RACKHAM O. (1966) Radiation, transpiration, and growth in a woodland annual. In *Light as an ecological factor* (eds R. Bainbridge, G.C. Evans and O. Rackham) pp. 167–85. Blackwell Scientific Publications, Oxford.

RACKHAM O. (1972a) Responses of the barley crop to soil water stress. In *Crop processes in controlled environments* (eds A.R. Rees, K.E. Cockshull, D.W. Hand and R.G. Hurd) pp. 127–38. Academic Press.

RACKHAM O. (1972b) The vegetation of the Myrtos region. In P. Warren, *Myrtos: an Early Bronze Age settlement in Crete*, British School of Archaeol. at Athens, 283–98.

Smithsonian Meteorological Tables (1951). 6th ed. Washington.

WATTS W. (1975) Soil albedo and its consequences for soil temperature and plant growth. *This vol.*: 409–21.

WILSON J.W. (1957) Observations on the temperature of arctic plants and their environment. *J. Ecol.* 45, 500–31.

19 Energy transfer and surface temperature over plants and animals

J. A. CLARK *Department of Physiology & Environmental Studies, University of Nottingham, Sutton Bonington, Loughborough, Leicestershire*

Introduction

The energy available to a terrestrial ecosystem originates wholly from solar radiation. While the biomass and its rate of turnover are determined primarily by the photosynthetic fixation of sunlight by green plants, only a small fraction of the available energy is fixed photochemically. The major proportion of the radiation absorbed produces thermal effects which have profound influences on the activity and productivity of organisms, on their ecological range and on their behaviour and morphology. For example, the activities of poikilothermic animals are controlled by their body temperatures and hence by their thermal environments (Digby 1955, Kevan & Shorthouse 1970, Porter & Gates 1969). The importance of energy balances in determining the temperature of living organisms has motivated many studies on both animals and plants. Most of this work has either employed simplified physical or mathematical models of the organisms concerned, or, for measurements on living organisms, has been restricted to the simplified environments of the wind tunnel, phytotron or calorimeter.

The aim of this paper is to demonstrate the nature of the simplifications which are usually made in analyses of the physical energy balance of living organisms; to show examples from measurements in situations with a minimum of simplification; and to temper our enthusiasm for physical and mathematical modelling with a douche of realism.

The Energy Balance

The principle of the energy balance is the law of energy conservation expressed in the First Law of Thermodynamics. We may state this in a form appropriate to a biological system as 'The sum of inputs and outputs of energy, together with changes in stored energy, must equal zero.'

We may write the conservation equation for an organism in the most general form as:

$$S + M + C + \lambda E + L_n + P + J = 0 \qquad (1)$$

where S is the incident flux of solar radiation, M is the metabolic heat production of the organism, C is the flux of heat by convection to the surrounding air, λE is the flux of latent heat to the air, L_n is the net thermal radiation flux, P is the photosynthetic fixation of energy, and J is a storage term, all reduced to watts per square metre of surface area.

The sign convention usually adopted is for fluxes of energy towards the surface to be considered positive, together with inputs such as metabolic heat production.

Inevitably, since we are considering the effects of oversimplifications in theory and experiment, we now embark on some of our own. First, *either* M or P may be discarded, since animals do not photosynthesize and the metabolic rate of the photosynthetic organs of green plants is small compared to the other components of their energy balance. Indeed for our purposes we may even neglect photosynthesis itself, since it rarely comprises more than 5% of the incident energy flux. Second, by choosing to study only equilibrium situations we reduce J to zero. The revised energy conservation equations for plants and animals then become respectively

$$S + C + \lambda E + L_n = 0 \qquad (2a)$$

and

$$S + M + \lambda E + L_n = 0 \qquad (2b)$$

From the Stefan-Boltzman Law, L_n is expressed as:

$$L_n = \varepsilon\sigma(T_s^4 - T^4) \qquad (3)$$

where ε is the emissivity of the surface for thermal radiation

σ is Stefan's constant

T_s and T are the mean surroundings temperature and the surface temperature respectively, both expressed in degrees Kelvin.

S and L_n are often conveniently considered together, the sum being the Net Radiation, R_n:

$$R_n = S + L_n \qquad (4)$$

The value of L_n is usually negative so that for most of the time $R_n < S$ for terrestrial organisms. The net radiation is the main driving flux in the energy balance of plants in most circumstances. Even in large animals radiation sometimes exceeds metabolic heat production by an order of magnitude (Finch 1972). The energy balance equation may be further expanded to describe the balancing fluxes due to conduction processes driven by gradients of temperature and humidity. The equilibrium temperature of an organism may then be regarded as the thermodynamic potential at which its energy balance establishes a dynamic equilibrium between the input fluxes of radiation and metabolism and the processes of energy dissipation by convection and evaporation.

The balancing heat fluxes by convection and evaporation may both be related to the dynamics of the fluid flow round the organism. The relevant theoretical aspects of the transfer of heat and mass are treated fully in a number of texts (e.g. Schlichting 1968, Monteith 1973) and need not be pursued further in the proceedings of a conference whose subject is the effects of radiation.

Methods

The results presented in this paper were largely obtained by thermography. Thermography presents the distribution of surface temperature on a body as a *thermal image* of the field viewed by a camera. The indicated temperature of any point is determined by the flux of radiation in a narrow band at the high-energy end of the thermal radiation spectrum. The advantages of the technique are numerous. (1) The surface temperature distribution is seen almost instantaneously, and in much greater detail than is possible using a contact probe or even thermopile radiometers as described by Idle (1968). (2) It allows measurements of the radiative surface temperature of objects which have such a small thermal capacity that solid thermometer probes give false readings, e.g. animal coats, insects and leaves. (3) Measurement of the temperatures of otherwise inaccessible objects is often possible. (4) Surface temperatures can often be determined in a situation where the proximity of an observer would disturb the subject of measurement.

Apart from the considerable cost of present instruments, the main disadvantages of the technique are low precision in point temperature measurements, comparable with that of an ordinary mercury-in-glass thermometer, and relatively poor accuracy in the determination of absolute temperatures. Thermography has found many uses in industry, for example in the monitoring of thermal pollution of watercourses and in the detection of faulty electronic components. It is currently being widely investigated as a diagnostic tool in medicine, principally for cancer detection, but has seen little application in environmental biology.

The supplementary measurements necessary for energy balance analysis included the temperature of the air surrounding the subject, the net radiation flux incident on the surface, and the electrical input energy in the case of heated models. The methods employed in measurements on plants are described in more detail by Wigley & Clark (1974) and Clark & Wigley (1974);

those relating to measurements on animals are described by Cena & Clark (1973, 1974).

An Agavision 680 Thermal Imaging Camera was employed in the present experiments both in the measurement of surface temperatures and in the recording of temperature distributions over plant and animal surfaces. The mechanics of most thermal imaging cameras are analogous to those of the original Baird television. The Agavision 680 employs a transmission optical system, built from solid silicon lenses, to focus the image field in the plane of a stationary detector. The image field is then scanned across the detector element by a synchronized arrangement of rotating silicon prisms, and the electrical signal produced is displayed on a cathode ray tube as a visual grey-tone analogue of the temperature field (Plate 19.1). Alternatively, electronic filtering may be applied to discriminate areas of the field within a particular band of temperature, which is then displayed on the screen as a bright 'isotherm' band. Plates 19.2–5 are false-colour images constructed by changing the set temperature of such an isotherm band in conjunction with photography through an array of colour filters. The construction of each false-colour image requires about 30 seconds. Real-time colour displays are now available which would allow the recording of temperature distributions over unrestrained animal subjects, though with some loss of image quality and accuracy.

The detector of the Agavision 680 is an InSb semiconductor element, cooled to the temperature of liquid nitrogen to reduce thermal noise. The device is not ideal for biological purposes since it is sensitive principally to radiation close to 5 μm in wavelength, at the short-wavelength end of the thermal emission spectrum for 293°K, and the scale is non-linear. Moreover the detector has a significant response to the thermal band of the solar spectrum. Because of the transmission of the silicon optics beyond 1·1 μm, additional filtering was incorporated in our instrument to minimize errors due to reflected solar radiation.

Results and Discussion

Energy Balance and Surface Temperature Measurements on Mammals and Birds

Animals are most difficult subjects for the environmental physiologist. Unless unnaturally restrained, they adopt attitudes inconvenient for the simple measurement of both their metabolic heat production and their microclimate, and their posture itself can have considerable influence on energy exchange. Moreover the body of a terrestrial animal has surface elements facing in all directions, and therefore subject to varying environmental loads according to their position on the body. Experimental studies of the energy balances of mammals have necessarily been largely restricted to the measurement of metabolic rate by the monitoring of oxygen consumption. However, Finch (1972) has recently reported measurements monitoring the total energy balance of animals over periods of a day. Her surface temperature measurements were limited in number and still allow only an estimate of the mean energy balance over the whole animal. The alternative of computer simulation has made possible the assessment of the effects on the mean energy balance of features such as the stripes on Zebra Finches (Porter & Gates 1969). Even the simpler calculations of Hutchinson & Brown (1969) and Hutchinson *et al.* (1973) suffice to indicate how the suitability of breeds of cattle for hot climates depends on their coat colour. However, the measurements and calculations are usually applied to derive an energy flux average over the surface of the whole animal. This integrated flux is ultimately what the animal reacts to, but is the average of energy fluxes which vary from site to site on the surface of animals. The principal causes of variation in energy flux over the surface of a mammal are variations in the thickness of its insulation and variations in the radiant energy fluxes over its surface.

Let us consider the radiation flux first. The net radiant energy flux may comprise the largest component of the energy balance even in mammals with coats; for example Finch (1972) reports radiation loads on eland lying in the Kenyan sun of up to ten times their own basal metabolic rate. The major part of the daytime radiant load on a mammal results from the absorption of direct and diffuse solar radiation by the outer layers of the coat. The magnitude of the energy flux absorbed may vary widely from point to point on the coat because of differences in pigmentation. Plate 19.1 shows a grey-tone thermogram of a giraffe standing in sunshine. The correlation between the visible pigmentation—the familiar surface markings—and the thermal pattern is plain, and may be ascribed to differential solar heat gain between the dark and light areas of the coat. Concurrent measurements showed corresponding differences in temperature of up to 6°C in solar irradiances approaching 700 W m^{-2}, measured on a horizontal surface, even in strong wind conditions. The temperatures on shaded parts of the animals' coats were essentially uniform. Similar results were obtained with zebra and other wild species with visible patterns of pigmentation (Cena & Clark 1973, Cena 1974), and are also exhibited by domesticated animals such as Friesian dairy cows (Clark & Cena 1972). The thermal consequences of such differences in radiation absorption by the coat are not negligible; Cena (1966) measured skin temperatures up to 4°C higher under black hair than under adjacent white areas of piebald horses and cows in sunshine.

The second cause of variation in the surface energy balance due to differences in the absorbed radiation lies in the effects of angle of incidence, principally of direct solar radiation. The solar irradiance on a surface is proportional both to the beam irradiance and to the cosine of the angle between the beam and the normal to the surface. When the receiving surface is rough or curved the flux will vary between the irradiance of the full solar beam and values corresponding to diffuse radiation only. Such effects are easily visible in Plate 19.2. The sheep in Plate 19.2a had recently been sheared, and had a coat thickness of approximately 5 mm. Because the shorn coat was almost uniform in thickness and had a smooth surface, the measured surface temperature is much more uniform than that on the rough coat of an unshorn animal with a fleece length of about 50 mm (Plate 19.2b). The conditions in which the two thermograms were recorded were identical. The temperature on the outside of the rough coat varied between 30° and 51°C even within the area of the flank, and the majority of this variation may be ascribed to variations in the angle of incidence. The effects of differences in insulation on the partition of energy are evident in the measurements made concurrently (see caption). The higher insulation on the unshorn animal between the site of radiation absorption and the 'thermostated' skin of the animal results in an energy balance in which more of the incident energy is dissipated to the outside environment. The difference in the mean net radiative fluxes of 350 W m^{-2} and 220 W m^{-2} for the shorn and unshorn sheep respectively is due almost entirely to the higher coat temperature at which equilibrium is established in the case of the unshorn animal. Despite the higher coat temperature the environmental load would be smaller for the unshorn animal. Similar effects allow sheep to survive in the tropical outback of Australia where McFarlane *et al.* (1956) measured temperatures of 85°C on a sheep's back.

The other principal cause of variation in the local energy balance, and hence surface temperature, is the thickness of insulation on the body. Whole-body measurements such of those of Blaxter *et al.* (1959) allow the mean insulation on the body to be estimated. The external insulation, for instance on a sheep, varies from zero on the feet to a maximum of approximately 10 cm of fleece on the back. Poorly-insulated areas of the body, such as the face, ears and legs of mammals, may have very different surface temperatures from the major part of the trunk. Thermograms shown by Clark & Cena (1972) clearly reveal the contrast between well and poorly insulated animals. Beneath good insulation the tissue temperatures may be high, and these are extended by vasodilation into the peripheral regions to allow the dissipation of excess heat. The same areas of poorly-insulated animals have low surface temperatures and tissue insulation by vasoconstriction is necessary to conserve heat.

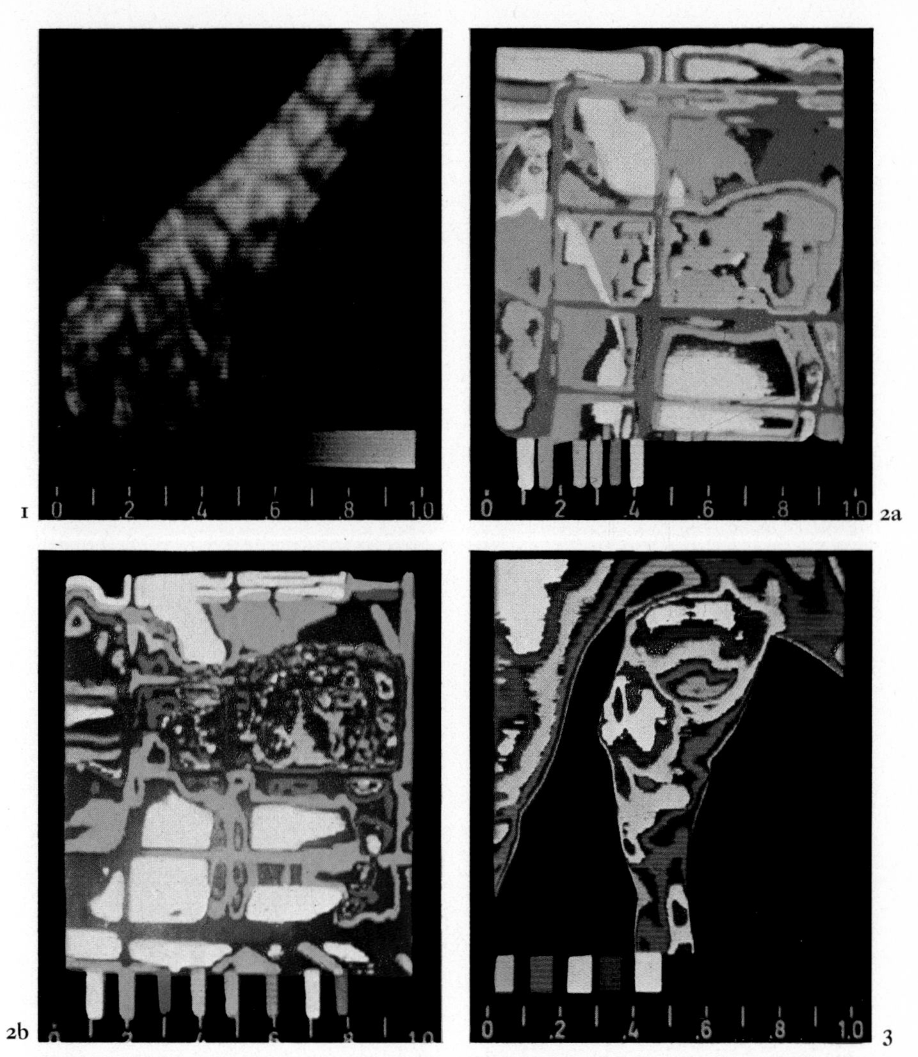

Plate 19.1. Grey-tone thermogram of a giraffe standing in the sun, showing a temperature distribution caused by differential solar heat gain between the dark and light areas of the familiar visible coat pattern. The warmest areas of the coat are shown as light areas on the thermogram print. The nominal temperature span (0–1·0) of the grey-tone scale below the thermogram is 10°C.

Plate 19.2. False-colour thermograms of (a) shorn and (b) unshorn sheep in strong sunshine. The scale below the thermograms indicates the relative temperatures of the different colour bands within the field. The scale (0–1·0) is approximately linear; in this plate it spans approximately 50°C and is the same in (a) and (b) to allow direct comparison of temperature distributions. The surface temperature of the shorn animal is comparatively uniform, while that of the unshorn animal shows a wide range of variation. The grid between the animals and the camera is the crate used to restrain the sheep during measurements. The bar across the left flank was removed to allow measurements to be made on the most uniform area of coat. Conditions: air temperature 23°C, wind speed 2·5 m s^{-1}.

	(a)	(b)
Radiative surface temperature (°C)	35·2	30 to 51
Skin temperature (°C)	34·5	35·5
Net radiation to coat in sun (W m^{-2})	350	220

continued overleaf

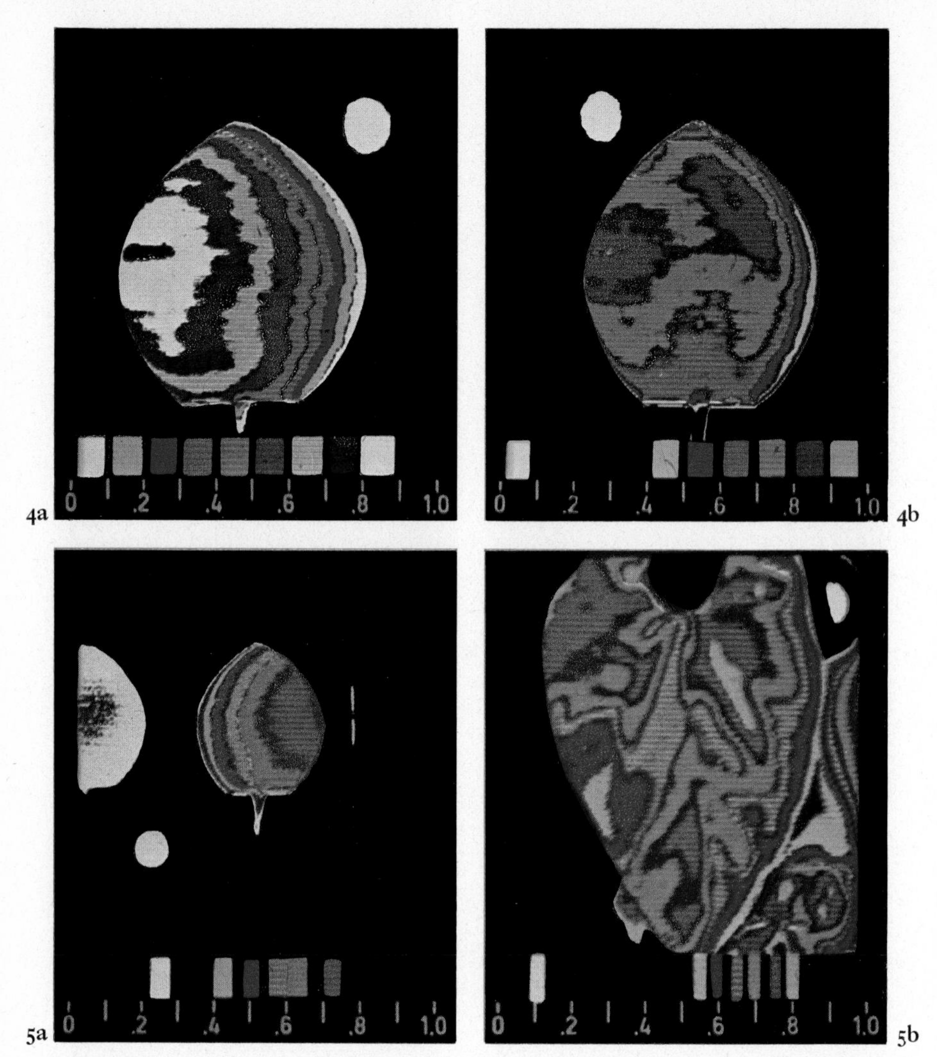

Plate 19.3. False-colour thermogram of a bull's scrotum viewed from the forward right side of the animal at approximately 60° to the body axis. This shows changes in apparent surface temperature with angle of view both on the almost hairless scrotum and on the hair-covered under surface of the animal (see text). These changes may cause errors in radiative temperature measurements. Scale span = 10°C.

Plate 19.4. False-colour thermograms showing temperature distribution over a model leaf in a laminar air flow of 0·7 m s^{-1} (a) parallel to the flow and (b) at an angle of incidence of 20° to the flow. The cold object close to the tip of the leaf, also visible in Plate 5, is a black-body cavity, used as a temperature reference for the measurement of absolute temperatures.

Plate 19.5. False-colour thermograms showing temperature distributions over (a) a model leaf in the field and (b) a *Phaseolus vulgaris* leaf in a wind tunnel. Scale span = 10°C.

Errors in Radiative Temperature Measurements

Thermal imaging cameras offer no greater precision in the measurement of temperatures than do the simple thermopile radiometers described by Idle (1968). They do, however, provide spatial information which can aid our understanding of some of the sources of error in the measurement of radiative surface temperature. For example a thermopile directional radiometer pointed at the subjects in Plates 19.1 and 19.2 would have provided a measure of the mean surface temperature which varied with the site of measurement. Moreover, the indicated temperature, and the apparent standard error of its measurement, could change with the distance of the observer from the surface. Two further potential sources of error may be deduced from Plate 19.3. The subject of the thermogram is the scrotum and nearby ventral and leg surfaces of a bull. The scrotum itself is almost hairless and may be considered as a smooth surface. Watmough *et al.* (1970) have shown that where such surfaces have an emissivity for thermal radiation even a few percent less than unity, the thermal emission at glancing angles of view is much reduced. Significant errors may occur for angles greater than 45° between the line of view and the normal to the surface. This effect causes the isotherm surrounding the scrotum in Plate 3, which is mainly pale blue, indicating an anomalous temperature. The tips of hairs are closer to air temperature than the bulk of the coat, so that, where a hair is viewed along a line close to the tangent to the body surface, a temperature is measured which is not characteristic of the total thermal radiation emitted by the coat (Cena & Clark 1973). Again, the outline of a colder, pale green, isotherm can be observed on Plate 19.3 over the hair-covered leg and belly surfaces of the bull. The scrotum of mammals is an interesting example of thermoregulation in which the input of heat to the organ by blood flow from the trunk is comparatively constant, as is the rate of heat production within the organ itself. Thermoregulation is achieved by control of the surface area. Birds use similar mechanisms for control of their whole-body energy balances, being able to increase heat dissipation by the extension of wings and necks to increase their heat-exchange area.

Temperature and Energy Balance of Insects

Thermography may be employed with advantage not only in the inspection of large intractable subjects, but also to measure temperatures of objects of small size. Insects are suitable subjects, frequently considered as poikilotherms, in the sense that they may be expected to have body temperatures close to air temperatures. Digby (1955) showed that this is in fact only true of the smallest insects, and that insects with body diameters greater than a few millimetres attain temperatures several degrees above the ambient air temperature when solar radiation is significant. The temperature excess varies in proportion to the solar irradiance and depends on colour. Kevan & Shorthouse (1970) have reported field observations of thermoregulation by basking in arctic species of butterflies, which they conclude have developed pigmentation of the wings suited to this practice. The thermal effects of pigment patterns on wings have been observed by Clark *et al.* (1973). The small size of insects has made direct observation of the body temperature of unrestrained specimens difficult. The application of thermography has therefore allowed the first temperature measurements of free flying specimens. The results presented in Fig. 19.1 (reproduced from Cena & Clark 1972)

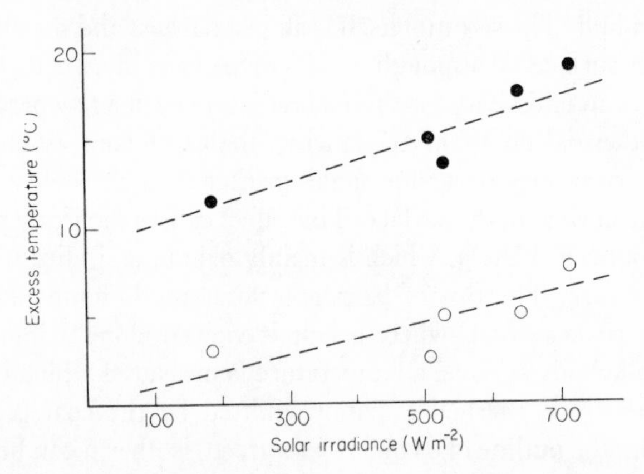

Figure 19.1. The variation with solar irradiance of the excess temperature above air temperature of the thorax (●) and abdomen (○) of bees (*Apis mellifica*) alighting at the hive entrance after foraging flights. Air temperatures in the range 15–17°C. (Reprinted from *Nature*.)

show that the surface temperature of bees in sunshine follows the level of solar irradiance. Although the effect of flight metabolism on the observed thorax temperatures is even larger, the absorption of solar radiation would be critical for an insect's activity in conditions marginal for flight.

Energy Balance and Surface Temperatures of Leaves

Most studies of the energy balance of plant leaves assume that the surfaces are isothermal and that measurements may be referred to the theoretical and empirical results of engineering studies of heat and mass transfer from flat plates. A few authors (e.g. Vogel 1970) have expressed their doubts on the validity of this procedure, while Raschke (1956), Idle (personal communication), and Parlange *et al.* (1971) have measured temperature distributions over leaves. However, Thom (1968), Parkhurst *et al.* (1968), Gates & Papian (1971), Pearman *et al.* (1972), and other authors have employed essentially isothermal models in order either to simplify the measurements or to facilitate the comparison of heat and mass transfer with that of momentum. Plates 19.4 and 19.5 illustrate the considerable variations in surface temperature which may in fact be obtained on leaves and realistic leaf models.

The reason why so much work has been done with isothermal surfaces lies principally in the availability of information from engineering sources for comparison purposes. The engineer is mainly interested in transfer from metallic bodies which are ideally isothermal, and the engineering literature for other cases is slight. The Nusselt and Sherwood numbers for an isothermal surface predict the *mean* rate of transfer for surfaces of appropriate scale dimensions. In fact the boundary layer varies in thickness with position, generally increasing in the downwind direction. The rate of transfer therefore varies with position over the surface, and if the surface is uniformly supplied with thermal energy, a considerable quantity of energy must be transferred laterally from hot parts of the body to others at a lower temperature. In effect, if the energy balance equation (2a) for an isotherm plate was re-written for a particular locality rather than for mean transfer, a conduction term would also be necessary. However, real leaves are composed principally of water and air, both of which have thermal conductivities approximately two orders of magnitude lower than metals (Vogel 1970); moreover most leaves have a small thickness compared to their width. Lateral transfer of heat is therefore likely to be negligible for broad leaves. Since the transfer resistance varies across the surface, the local energy balance must therefore result in an equilibrium temperature which varies with position, as we observe in the Plates. The partition of energy may change considerably from the leading to the trailing edge. For example, from measurements on a dry model leaf, made of Perspex, with an electrical heater embedded to simulate uniform absorption of solar radiation, the thermal radiant flux in laminar parallel flow increases from negligible values close to the leading edge to 25% of the heat input at the trailing edge (Plate 19.4).

In parallel flow, the temperature of the dry model (Plate 19.4a) increases monotonically across the surface in the downwind direction, approximately with the square root of the distance from the leading edge. In turbulent-wake flow conditions—simulating the flow in a crop canopy—the increase is less rapid, indicating higher transfer rates, but is still monotonic. The pattern of surface temperatures shown in Plate 19.4b illustrates the complications introduced by flow at oblique angles to the surface, when the rates of transfer show no simple relationship with position on the surface. Published empirical relationships between heat transfer and Reynolds numbers break down in oblique flow, presumably because there is no heat transfer analogue of bluff-body pressure-drag and the two-dimensional Reynolds number is therefore inapplicable. There is some comfort in the record of Plate 19.5, which shows that temperature distributions at low wind speeds, over the model in the field (Plate 19.5a) and a leaf of similar shape in the wind tunnel (Plate 19.5b), are similar to those observed on the model in the laboratory.

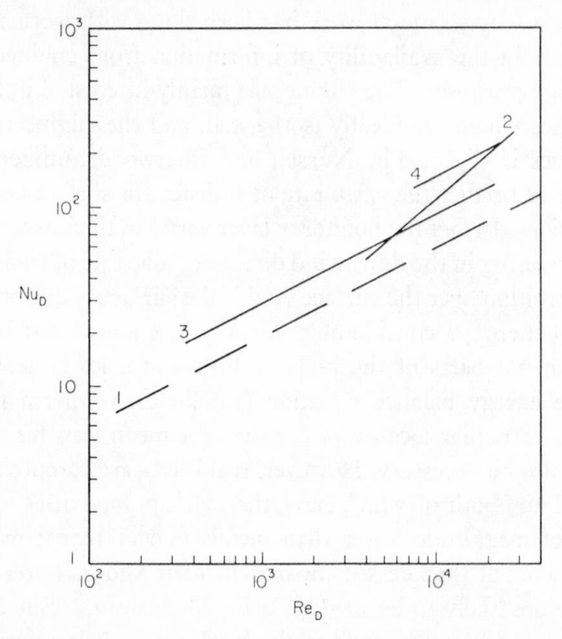

Figure 19.2. Comparison of relationships between the dimensionless heat transfer (Nusselt number) and the dimensionless flow parameter (Reynolds number) for model leaves. Line 1—the theoretical prediction for laminar flow, Line 2—measured relationship in turbulent-wake flow found by Wigley & Clark (1974), Line 3—measured in the field by Pearman *et al.* (1972), Line 4—measured in a fan wake by Parlange *et al.* (1971). (Reprinted from *Boundary-layer Meteorology.*)

In the literature of heat transfer from leaves there remains a very considerable discrepancy between the results from wind tunnel studies in laminar flow and field experiments. Reported values of heat transfer coefficients measured in field conditions have been commonly up to 2·5 times the value for laminar flow predicted by the Polhausen equation. The problem has been exacerbated by the fact that the differences have been too consistent to be ascribed to experimental error, and that the relevant wind velocities around plant canopies are generally between one and two orders of magnitude lower than those at which transition from laminar to turbulent flow might be expected over surfaces the size of plant leaves even in the presence of turbulent free-stream flow. Wigley & Clark (1974) have recently proposed one mechanism which could explain such an anomaly in heat transfer. They suggest that in turbulent flow the transition to a turbulent boundary layer takes place at lower velocities in the presence of pressure gradients. Their measurements in parallel flow (Fig. 19.2, reproduced from their paper) lie close to the predictions of the Polhausen equation in laminar conditions, but in a turbulent-wake flow the slope of the logarithmic plot of mean Nusselt number against mean Reynolds number is close to 0·8, the value appropriate to a turbulent boundary layer.

Conclusions

The experimental results discussed in this paper, in conjunction with the theory presented, emphasize that biological organisms are poorly approximated by isothermal surfaces. Simple physical and mathematical models, such as those employed by Gates & Papian (1971), can therefore give only the order of magnitude of the equilibrium state of a plant or animal energy balance. Large local variations in energy balance exist on the surfaces of both animals and plants. The leaves of plants have low thermal conductivities and approximate more closely to constant-flux surfaces than to the isothermal surfaces well documented in the engineering literature. In contrast, variations in solar radiation absorption over animals, due to pigment patterns on the coat and shape, produce an energy balance whose total magnitude is highly variable in space and time and as the animal changes its posture. It is clear that coated animals in no way approximate to the physicist's ideal of a uniform isothermal woolly ball spinning rapidly to average out the radiation flux.

Ultimately the plant or animal is the only integrator of the environment it experiences. Progress depends on better co-operation between physiology and physics to allow examination of the micro-scale of energy balances and their effects on the longer-term productivity and development of animals and plants.

Acknowledgements

The author thanks the Science Research Council for the grant which supported this work. His thanks are also due to Kris Cena and Graham Wigley who were his partners in the experimental measurements and co-authors in several of the papers cited. The measurements on giraffes and other non-native animals were made at Twycross Zoo, with the kind permission of the proprietors.

References

Blaxter K.L., Graham McC.N. & Wainman F.W. (1959) Temperature, energy metabolism and heat regulation in sheep. III. *J. Agr. Sci.* 52, 41–9.

Cena K. (1966) Investigations of absorption of solar radiation by cattle and horses of various coat colours. *Acta Agr. Silvest.* 6, 93–138.

Cena K. (1974) Radiative heat loss from animals and man. In *Heat loss from animals and man* (eds J.L. Monteith & L.E. Mount). Butterworths, London.

Cena K. & Clark J.A. (1972) Effect of solar radiation on temperature of working bees. *Nature* 236, 222–3.

Cena K. & Clark J.A. (1973) Thermal radiation from animal coats. *Phys. Med. Biol.* 18, 432–43.

Cena K. & Clark J.A. (1973) Thermographic measurements of the surface temperatures of animals. *J. Mammalogy* 54, 1003–7.

Cena K. & Clark J.A. (1974) Heat balance and thermal resistances of sheep's fleece. *Phys. Med. Biol.* 19, 51–65.

Clark J.A., Cena K. & Mills N.J. (1973) Radiative temperatures of butterfly wings. *Z. Ang. Ent.* 73, 327–32.

Clark J.A. & Cena K. (1972) The application of thermal imaging techniques to animals. *Dtsch. Tierartzl. Wschr.* 79, 292–6.

Clark J.A., Cena K. & Monteith J.L. (1974) Measurements of the local heat balance of animal coats and human clothing. *J. appl. Phys.* 35, 751–4.

Clark J.A. & Wigley G. (1974) Heat transfers from real and model leaves. In *Proc. 1974 conf. Int. Centre for Heat and Mass Transfer* (ed. D.A. Vries). Scripta, Washington.

Digby P.S.B. (1955) Factors affecting the temperature excess of insects in sunshine. *J. Exp. Biol.* 32, 279–98.

Finch V. (1972) Thermoregulation and heat balance of the East African eland and hartebeest. *Amer. J. Physiol.* 222, 1374–9.

Gates D.M. & Papian L.E. (1971) *Atlas of energy budgets of plant leaves.* Academic Press, London & New York.

Hutchinson J.C.D. & Brown G.D. (1969) Penetrance of cattle coats by radiation. *J. appl. Physiol.* 26, 454–64.

Hutchinson J.C.D., Brown G.D. & Allen T.E. (1973) Effect of the coat on heat dissipation of cattle in a radiant environment. *1st Australian Conf. on Heat and Mass Transfer.* 3, 15–24.

Idle D.B. (1968) The measurement of apparent surface temperature. In *The measurement of environmental factors in terrestrial ecology* (ed. R.N. Wadsworth). Blackwell, Oxford.

Kevan P.G. & Shorthouse J.D. (1970) Behavioural thermoregulation by High Arctic butterflies. *Arctic.* 23, 268–79.

McFarlane W.V., Morris R.J. & Howard B. (1956) Water economy of tropical Merino sheep. *Nature* 178, 304.

Monteith J.L. (1973) *Principles of environmental physics.* Edward Arnold, London.

Parkhurst D.F., Duncan P.R., Gates D.M. & Kreith F. (1968) Convection heat transfer from broad leaves of plants. *J. Heat Transfer* (*Trans ASME*) 90, 71–76.

PARLANGE J-Y., WAGGONER P.E. & HEICHEL G.H. (1971) Boundary layer resistance and temperature distribution on still and flapping leaves. *Plant Physiol.* 48, 437–42.

PEARMAN G.I., WEAVER H.L. & TANNER C.B. (1972) Boundary layer heat transfer coefficients under field conditions. *Agric. Meteorol.* **10**, 83–92.

PORTER W.P. & GATES D.M. (1969) Thermodynamic equilibria of animals with environments. *Ecol. Monogr.* **39**, 227–44.

RASCHKE K. (1956) Mikrometeorologisch gemessne Energieumsätze eines Alocasiablattes. *Arch. Met. Geophys. Bioklim.* **37**, 240–68.

SCHLICHTING H. (1968) *Boundary layer theory*. McGraw-Hill, New York.

THOM A.S. (1968) The exchange of momentum, mass and heat between an artificial leaf and the air flow in a wind tunnel. *Quart. J. roy. Met. Soc.* **94**, 44–55.

VOGEL S. (1970) Convective cooling at low airspeeds and the shapes of broad leaves. *J. Exp. Bot.* **21**, 91–101.

WATMOUGH D.J., FOWLER P.W. & OLIVER R. (1970) The thermal scanning of a curved isothermal surface. *Phys. Med. Biol.* **15**, 1–8.

WIGLEY G. & CLARK J.A. (1974) Heat transport coefficients for constant energy flux models of broad leaves. *Boundary Layer Meteorol.* **7**, 139–50.

Questions and discussion

A number of ways were discussed in which studies of leaves in the laboratory might be extended to include complications affecting the behaviour of plants and plant communities in the field. Dr. Clark considered that lobes, teeth, and hairs on leaves might increase heat transfer by promoting the transition to turbulent-wake flow. Heat exchange might also be increased by fluttering. As yet there was little quantitative information about these factors, or about the problems caused by the rapidly-changing conditions that occurred in the field.

20 The temperature variation of the canopy and trunk of Scots pine and its relation to the net input of radiation and to air temperature

W. J. SHUTTLEWORTH *Institute of Hydrology Wallingford*

Introduction

The ecology of a natural environment is influenced not only by the prevailing meteorological conditions but also by the response of the local vegetation to changes in those conditions. Part of that response is its change in temperature and it is of interest to know how the surface and internal temperatures of the vegetation are influenced by the input of radiation. In this respect the behaviour of the plant (and its component parts) is dictated by the amount of radiation it intercepts and by its ability to absorb that energy or to dissipate it into the air, either as heat or as some other form of energy. This will depend on the type and form of the vegetation and on the extent of its interaction with the air mass.

The data presented in this paper are drawn from a much larger body of information, collected in the 'Thetford Project', and undertaken, by the Institute of Hydrology, with the primary object of understanding the mechanisms controlling transpiration from a pine plantation, and the extent and rate of evaporation of intercepted rainfall. Since the application of the data to environmental temperature response is retrospective, it necessarily represents a less than perfect experimental investigation of this phenomenon: however, the particular response of a forest plantation turns out to be so simple that the data probably present an adequate picture of the more significant features.

The study area forms part of Thetford Forest, Norfolk, England. It consists of a stand of Scots and Corsican Pines (*Pinus sylvestris* and *P. nigra* ssp *laricio*) planted around 1930 and now about 16 m high. Similar plantations extend at least 2 km from the measuring site in all directions and form a sufficiently extensive and homogeneous cover for the theory of aerodynamic exchanges in field crops to be applicable.

Instrumentation

The Thetford Project involves accurate measurements of wet and dry bulb temperature at twelve heights below, through and above the canopy, together with similar multiple measurements of windspeed. Measurements are also made of the net radiation above and below the canopy and of the more important components of that radiation. (The canopy is here defined as the dense living vegetation near the top of the trees: the dead branches below this level have been removed in accordance with local forestry practice.) The experimental apparatus is controlled and recorded by a computer-controlled data acquisition system which produces five-minute averages of the meteorological data on paper-tape punches; it is these five-minute averages that are presented in this paper.

Air temperature measurements were made with aspirated quartz-crystal thermometers mounted on booms at each level; these thermometers were accurate to within a few hundredths of a degree centigrade. Measurements of net radiation were made with Funk-type net radiometers similarly mounted on booms, six above and three below the canopy, while solarimeters were used to make measurements of solar radiation.

The surface temperature of the canopy was measured with a precision radiation thermometer (Barnes Engineering Co., model PRT-5). The accuracy of this instrument was somewhat less than that of the other thermometers. Apart from the possibility of thermal drift (which, it is believed, may give rise to variations of the order of 0·1°C) the measurement is influenced by changes both in the emissivity of the canopy and in the amount of infra-red radiation reflected from the canopy. These two effects compensate to some extent and, to a first approximation, the fractional error in the radiometric canopy temperature caused by neglecting reflected radiation and by assuming an emissivity of unity is given by:

$$\frac{\Delta T}{T} = \frac{1-\varepsilon}{4} \frac{L_D - L_U}{L_U} \tag{1}$$

where L_D and L_U are the downward and upward fluxes of long-wave radiation intercepted by and emitted from the canopy, and ε is its true emissivity. L_U depends on temperature but is typically 350 W m^{-2} while $(L_U - L_D)$ varies from about 30 W m^{-2} in cloudy conditions to a maximum of 100 W m^{-2} in clear sky conditions. It is possible to estimate ΔT at least once and often twice each day by taking the difference between radiometric canopy temperature and air temperature in periods when there is no sensible heat flux into or out of the canopy, these periods being identified by zero gradient in the potential temperature profile (Fig. 20.1). Its value was typically 0·5°C, corresponding to a dry canopy emissivity of about 95%: when the canopy is wet the emissivity is greater, typically 98–99%, and ΔT is less. It is believed that changes in emissivity and thermal drift could together give errors of the order of ¼°C in the radiometric measurements of surface temperature.

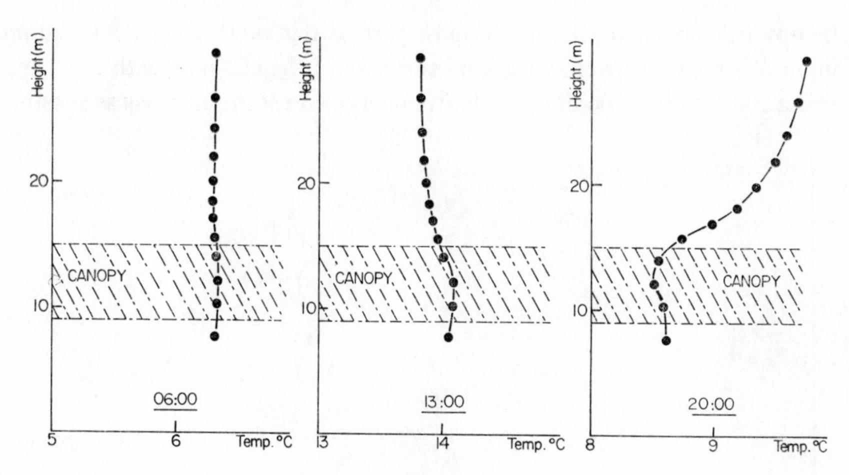

Figure 20.1. Typical air temperature profiles (15 March 1973).

Air temperature in a forest canopy and its relationship to radiation input

Figure 20.1 gives typical measurements of the air temperature profile through and above the canopy. It should be emphasized that this figure and all the subsequent data presented here are merely examples arbitrarily selected from many days of similar data. Any conclusions drawn from such days apply in general. The figure shows typical profiles of air temperature at dawn, mid-afternoon and late evening.

From the point of view of this paper it is important to notice that the relative temperature variation within each profile is small compared to the absolute temperature variation between the different profiles. It is also important to notice that the temperature variation falls off quickly with height so that, *in comparison with the diurnal temperature range*, in daylight hours, measurements of air temperature near a *forested* surface are fairly representative of the whole boundary layer air mass. To reinforce these points, the canopy air temperature (defined here as the mean temperature of the three measurements within the canopy) is plotted for a complete day in Fig. 20.2, together with the air temperature at the highest measurement level: the relative magnitude of the differences involved is then obvious.

It is necessary to understand how the temperature of the boundary layer air mass is related to radiation input: a simplified model will illustrate this. Consider an idealized day on which there is a boundary-layer air mass travelling inland from the sea at a constant speed u. It receives radiation input and is warmed; on this occasion the boundary layer happens to be capped by a stable inversion which prevents it from losing any of this heat by upward convection. Let us suppose further that on this day the radiation input R is the same over all the land surface and changes only with time, and that a constant fraction of the radiative energy enters the air mass as sensible

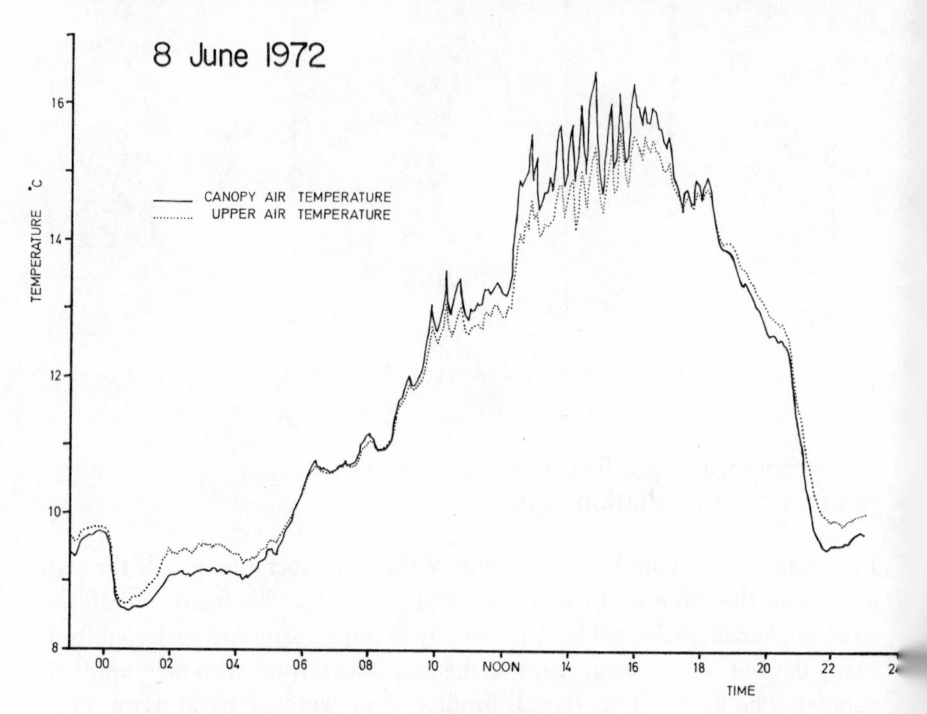

Figure 20.2. Variation of the air temperature within the forest canopy compared to that 15 m above the trees.

heat. If T_o is the temperature of the air mass as it leaves the coast, and T_l is its temperature when a distance l downwind of the coast, then:

$$T_l = T_o + K \int_{t_l - l/u}^{t_l} R(t)dt \tag{2}$$

where t_l is the time at which the air, which left the coast at time $(t_l - l/u)$, is at the point l, and K is an undefined constant.

Obviously the relationship between the situation just described (illustrated in Fig. 20.3) and reality is somewhat tenuous; nevertheless it does make the basic point that the present temperature of the boundary layer air mass is related primarily to its medium and long term history of energy input, rather than to the immediate and local radiation input. One more obvious effect of this result is a time lag between the maximum radiation input and maximum air temperature. Figure 20.4 shows this effect in practice for a day with a

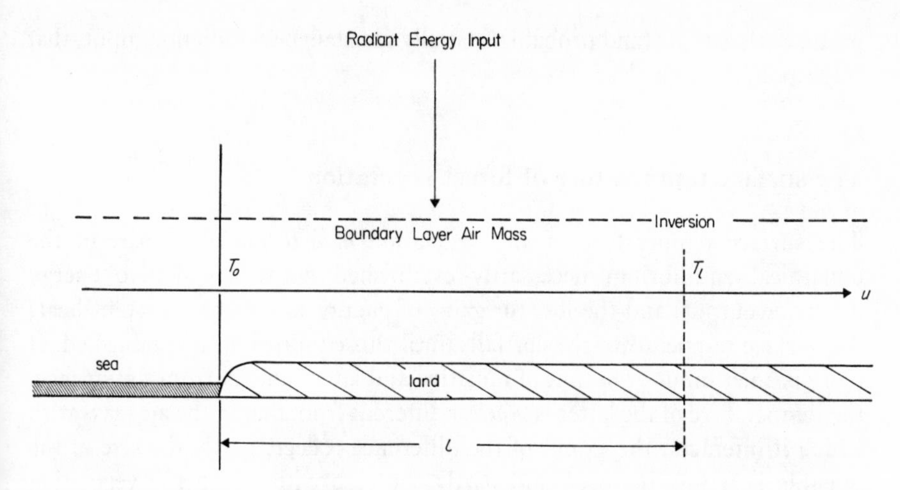

Figure 20.3. Schematic sectional view illustrating that T_l, the temperature of the boundary layer air mass at distance l from the coast, is determined partly by the initial air temperature at the coast, T_o, and partly by the integrated radiational input since the air left the coast.

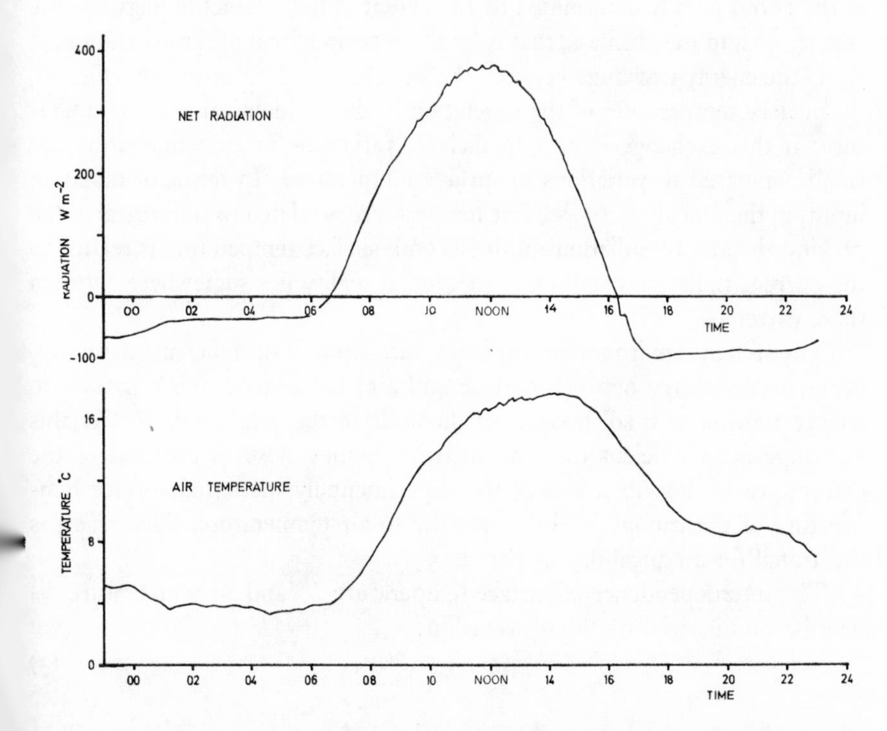

Figure 20.4. Variation of net radiation and air temperature in clear sky conditions.

particularly simple (and probably spatially constant) net radiation input, that corresponding to a clear sky.

The surface temperature of forest vegetation

The surface temperature of any vegetation is determined as part of the dynamical equilibrium necessarily established between radiative energy input (or output) and the loss (or gain) of energy as sensible or latent heat; the surface temperature rises or falls until this equilibrium is established. If there is a net input or output of radiative and latent energy from the surface, the temperature of the latter is *a priori* different from that of the air mass with which it interacts: the extent of the difference reflects partly the size of the sensible heat flux involved and partly the ease with which the vegetation interacts with the air.

It is interesting to compare the difference between surface and air temperature with the diurnal variation in air temperature. Since both have more or less direct and independent relationships to the radiation input, the ratio of the two is largely determined by the extent of the interaction between the vegetation and the air mass, that is by the type and form of the vegetation.

If the energy exchange between the surface and air is extremely efficient, the surface temperature of the vegetation is almost identical to air temperature; if this exchange is very inefficient, variations in air temperature are small compared to variations in surface temperature. In terms of radiation input, in the first situation, surface temperature is related to an integral of the previous history of radiation; in the second, surface temperature is related to the current radiation conditions. In general reality lies somewhere between these extremes.

The forest environment probably has the most efficient naturally-occurring exchange between surface and air; the *aerodynamic resistance* to energy transfer is small because of the scale of the vegetation. In fact this exchange is so efficient that the situation comes close to the first of the extreme cases described above; the experimentally measured surface temperature of the canopy is always similar to air temperature. This result is illustrated for a typical day in Fig. 20.5.

The interdependence of surface temperature T_S and air temperature T_A may be summarized by the relationship:

$$T_S = T_A + \frac{Hr_A}{\rho c_p} \qquad (3)$$

where H is the flux of sensible heat out of the surface, ρ and c_p are the density and specific heat of air, and r_A is the aerodynamic resistance between the air and the surface. In the absence of any radiative and latent energy exchange, H becomes zero and the surface temperature is identical to air temperature; it is air temperature which determines the long-term variation of both curves. Local fluctuations in radiative input give rise to fluctuations in H and related fluctuations in the surface temperature of the canopy. These

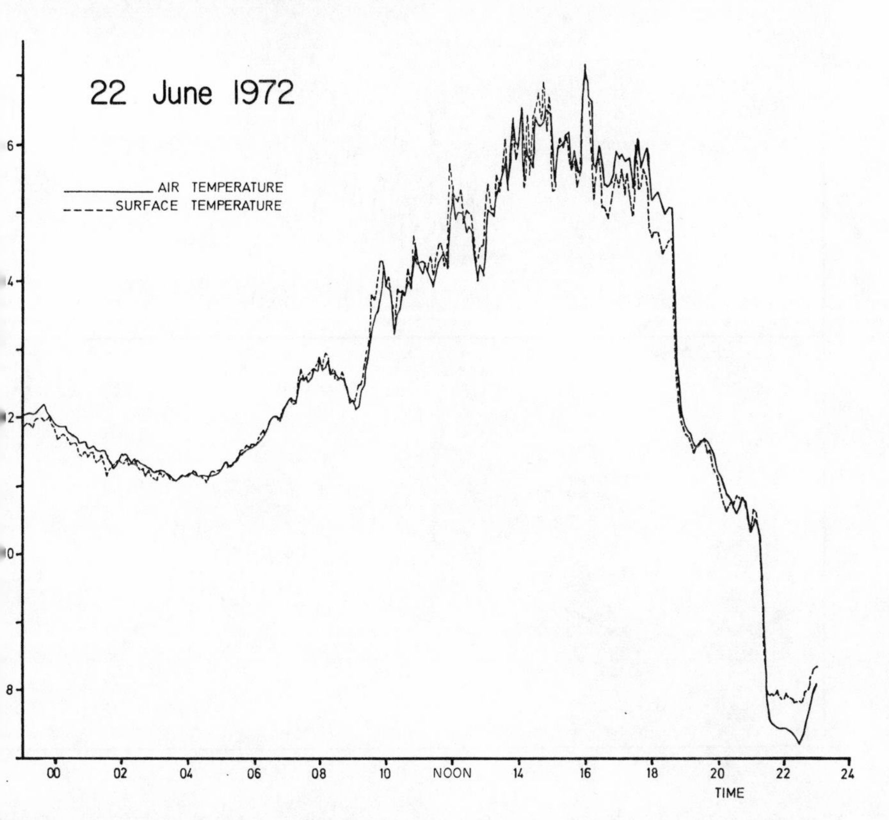

Figure 20.5. Variation of air temperature within the forest canopy compared to the radiative surface temperature of the canopy.

fluctuations not only give the varying deviations between air temperature and surface temperature expected from equation (3) but also cause similar short-term fluctuations in the observed air temperature. It is, of course, necessary that they should, since they form part of the integrated energy input which determines this temperature (Equation (2)). Since it is the short-term fluctuations in surface temperature which cause those in air temperature rather than the other way around, surface temperature fluctuations tend to precede those in air temperature and to be more extreme. The fact that they are themselves caused by fluctuations in local radiation input is illustrated in Fig. 20.6.

All the preceding comments regarding overall surface temperature can strictly be taken to apply only to the canopy, since this is the only surface temperature measured; however it is plausible that they apply equally to all the tree's exposed surfaces and in particular to its trunk. The only physical mechanism which can cause the surface temperature of the trunk to differ

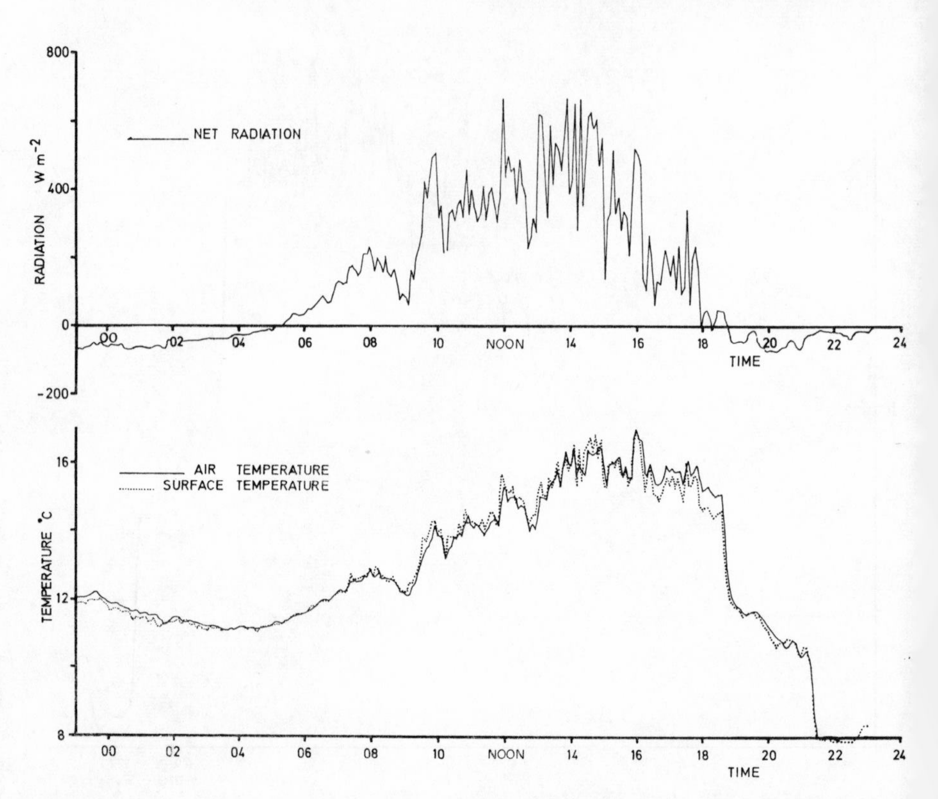

Figure 20.6. Variation in air and surface temperature compared to the local input of net radiation.

from air temperature is radiative energy exchange. The overall scale of this perturbation is less in the trunk space than in the canopy by at least an order of magnitude, which fact is illustrated in Fig. 20.7. For this reason deviations between the *average* trunk surface temperature and air temperature are likely, if anything, to be less than those of *average* canopy surface temperature. In the following section the surface temperature of the trunk and branches is equated to air temperature, but it is worth remembering that gaps in the canopy can cause local abnormalities in radiation intensity and related abnormalities in the surface temperature of branches and trunks.

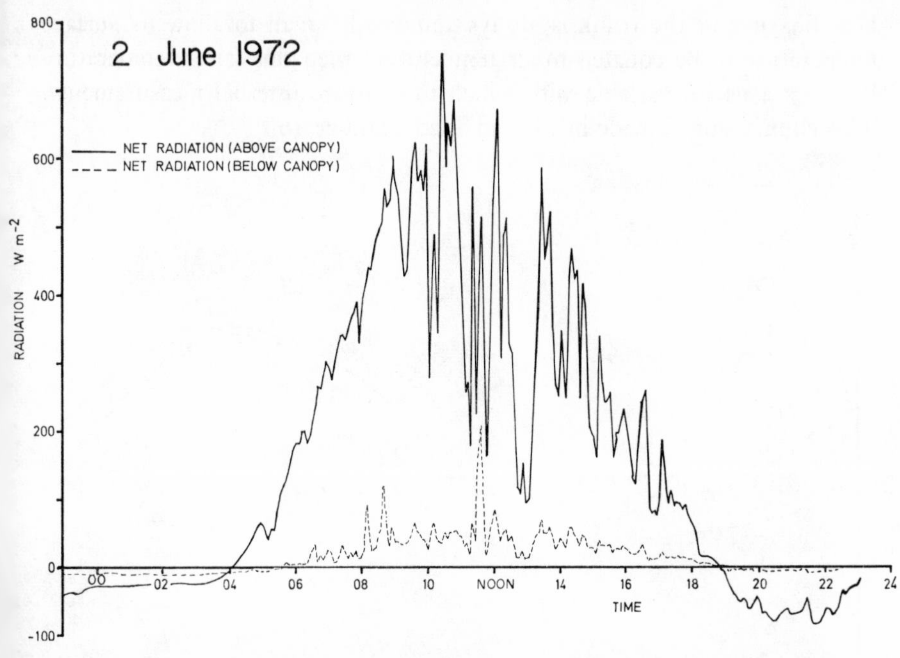

Figure 20.7. Net radiation measured above and below the forest canopy in Thetford Forest.

Trunk temperature measurements

Although a complete and precise knowledge of the heat content of every part of the biomass would be extremely useful to the energy budget measurement carried out in the Thetford Project, obtaining such knowledge is impractical. It was thought useful to have at least some knowledge of the temperature variation inside a typical tree trunk, and a tree of average dimensions (20 cm diameter at breast height) was selected for this purpose. Two quartz-crystal thermometers were inserted into the trunk at a height of 4 m and to a depth of 2 cm and 4 cm respectively, and their average temperatures taken as part of the 5-minute data collection cycle. These measurements have very little quantitative use, since they apply only to one tree and even then are prone to sensor-environment interaction errors, but they do give some qualitative insight into typical behaviour. The heat content, and therefore the temperature, at any particular location inside the tree is determined from the present and past surface temperature via the mechanisms of conductive and convective heat transfer. Part of the plant's surface is underground, and therefore controlled by soil temperature, while part is above ground with a surface temperature determined by equation (3). If it is assumed that the sensible heat flux out of the trunk is always sufficiently small to allow its surface temperature to be equated to air temperature then this latter temperature becomes a useful variable with which to compare internal measurements. This comparison is made in Figs 20.8, 20.9 and 20.10.

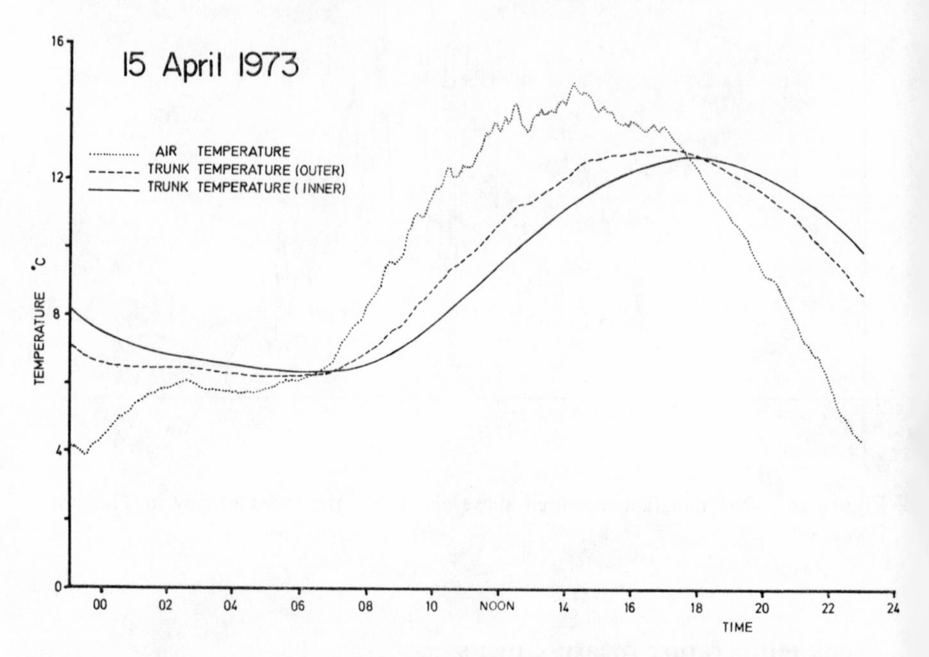

Figure 20.8. Temperature variation inside a typical tree trunk compared to air temperature. INNER temperature was taken at a depth of 4 cm, OUTER at a depth of 2 cm.

Both temperatures follow both the broad variation and to some extent the detailed variation in the air (surface) temperature. The diurnal temperature cycle in the trunk is delayed with respect to that in the air by some hours, typically 5–6 hours, while the response to short-term fluctuations is practically immediate. The outer (2 cm) temperature is in general intermediate between the inner (4 cm) temperature and air temperature: when it is higher than both, heat is conducted both inwards *and* outwards and the temperature of this part of the trunk falls quickly.

The fact that the outer temperature is intermediate, together with the observation that the interrelationship between trunk temperature T_T and air (surface) temperature appears to have the approximate form:

$$\frac{\partial T_T}{\partial t} \propto (T_A - T_T)$$

(corresponding to a primarily integral dependence of T_T on T_A), indicate that

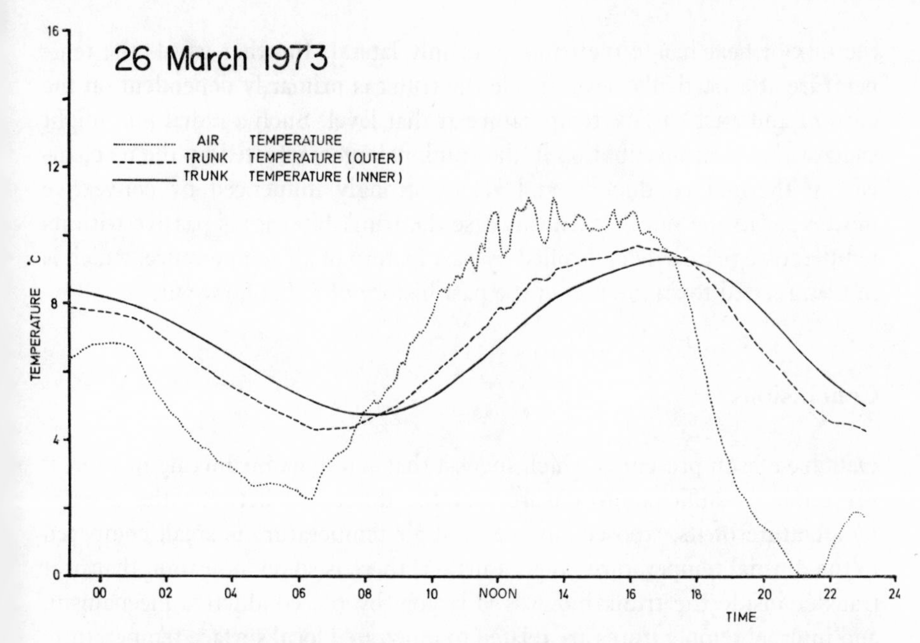

Figure 20.9. Temperature variation inside a typical tree trunk compared to air temperature. INNER temperature was taken at a depth of 4 cm, OUTER at a depth of 2 cm.

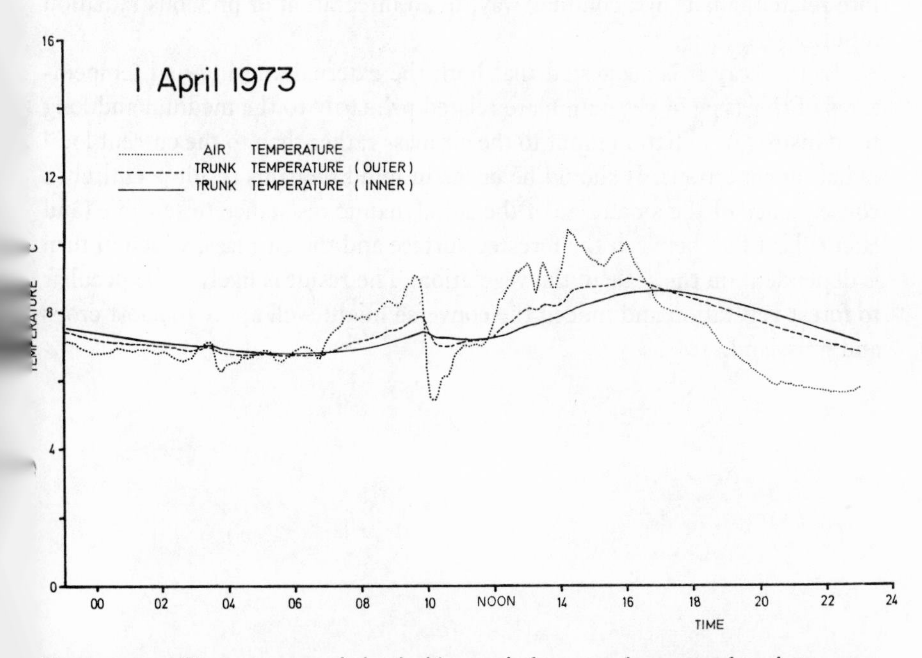

Figure 20.10. Temperature variation inside a typical tree trunk compared to air temperature. INNER temperature was taken at a depth of 4 cm, OUTER at a depth of 2 cm.

the flux of heat inside the trunk is mainly lateral. In other words the temperature at a particular level inside the trunk is primarily dependent on the current and past surface temperature at that level. Such a radial flux might suggest that heat distribution in the trunk is largely effected by the mechanism of thermal conduction and is not strongly influenced by convective movement in the sap flux. In this case the trunk biomass is passive with its temperature primarily controlled by past history of air temperature, which is in turn related to an integral of the past history of radiation input.

Conclusions

Data have been presented which suggest that aerodynamic mixing in a forest plantation is sufficient to ensure that the difference between the average temperature of its exposed surfaces and air temperature is small compared to the diurnal temperature range. Further, there is some indication that heat transfer inside the trunk biomass is largely by the conductive mechanism, and internal temperatures are related to integrated local surface temperature. It has been pointed out that the air temperature in a forest canopy is symptomatic of the whole boundary layer air mass in daytime conditions and therefore related, albeit in a complex way, to an integration of previous radiation input.

In this way it is suggested that both the external and internal temperatures of this type of vegetation are related primarily to the medium and long term history of radiation input to the air mass rather than to the current local radiation conditions. It should be borne in mind that this result is entirely a consequence of the small size of the aerodynamic resistance to sensible (and latent) heat flux between the forested surface and the air mass, which in turn is dependent on the scale of the vegetation. The result is likely to be peculiar to forest vegetation and indeed the converse might well apply to short crops and grassland.

21 The effect of high temperatures on some aspects of the physiology and cultivation of the tea bush, *Camellia sinensis*, in North East India

W. HADFIELD *Tocklai Experimental Station of the Indian Tea Research Association, Assam*[1]

Introduction

With the exception of some 18,000 hectares in the hills of Darjeeling, tea in North East India is produced on the plains, generally below 100 metres above sea level. Out of a total of rather over 300 million kilograms of tea produced from North East India, over 220 million are produced in the Brahmaputra Valley and a further 24 million in the Surma Valley which is similar in most respects to the Brahmaputra Valley. The experimental work to be described in this paper has been carried out under environmental conditions typical of those under which over 70% of North East Indian tea is grown, although the lack of detailed meteorological information for most remote tea areas makes rigorous comparisons of local differences impossible. Tea requires an acid soil of between pH 4·5–6·0 for optimal growth and the soils of N.E. Indian tea areas, though varying in other characteristics, mainly fall within this range. Soil moisture is usually sufficient for growth for most of the year and during the monsoon is such as to make drainage essential in most localities. Land drains are, in any case, necessary to cope with run-off from monsoonal downpours.

Fungus disease and insect pests generally give little trouble and are more easily controllable than with many other crops.

The tea of commerce is produced from the shoots of an extremely heterogeneous population of tea bushes mainly derived by hybridization between the extreme forms of *Camellia sinensis* L. (O. Kuntze). These forms are designated as *C. sinensis* var *sinensis* and *C. sinensis* var *assamica* by Purseglove (1968), the former having been introduced to India before the indigenous Assam plant (var *assamica*) was found in the forests near the Burma-Assam-China borders. Quality of made (black) tea is widely believed to be different for the *assamica* and *sinensis* types but the differences have never been adequately quantified. The typical forms of the two varieties are shown in

[1] Present address: Instituto Nacional de Investigaciones Agropecuarias, Quevedo, Ecuador.

Fig. 21.1 together with some hybrids and the terms 'China hybrid' and 'Assam hybrid' will be used to refer to experimental material depending on its supposed closeness to the varietal type.

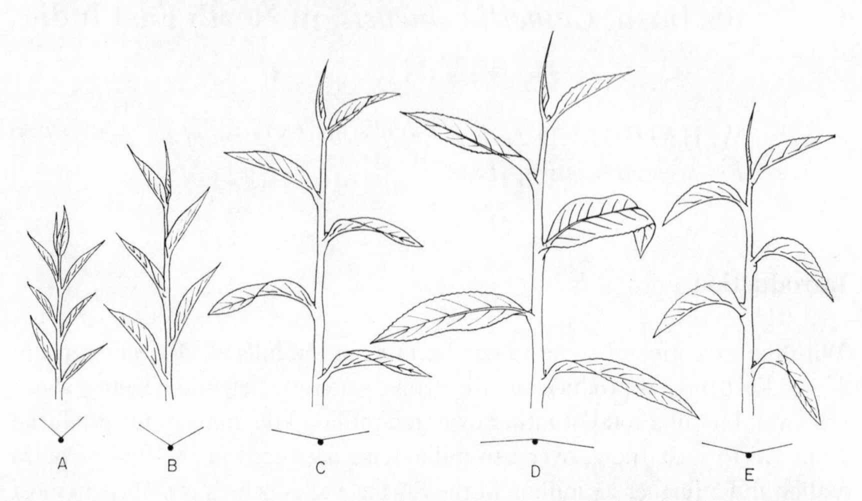

Figure 21.1. Shoots of tea bushes (*Camellia sinensis*): (A), extreme var *sinensis*; (B and C), typical hybrids between the extreme varieties; (D), extreme var *assamica*; and (E), a dependent leafed var *assamica* produced by selection and inbreeding. Approximately to scale. Cross sections of leaves are shown beneath each shoot.

Climate: Seasonality

The radiation pattern in mid-Assam is greatly influenced by the S.W. Monsoon which lasts from June to October and reduces the hours of bright sunshine to 36% of the possible value compared with 48% during the November to May period (Hadfield 1974a). The mountain ranges bounding the Brahmaputra Valley prevent the rapid dispersal of monsoon clouds by winds, the rate of air movement being extremely low throughout the year (Table 21.1).

Because of increased cloud cover during the monsoon the reduction in radiation reaching the ground during the summer months makes the radiation pattern during the cropping season remarkably constant over 10 day periods when compared with the expected values for a station at 26°N. Figure 21.2 shows the 10 day means of short wave radiation measured by a Kipp & Zonen solarimeter in the meteorological plot at Tocklai (latitude 26° 47′ N, longitude 94° 12′ E).

In terms of the climate and the tea crop we can divide the year quite conveniently into two seasons of six months each and Table 21.1 gives the major features of these seasons.

Table 21.1. Seasonality and yield of the tea bush, *Camellia sinensis* var *assamica*, in the neighbourhood of the Tocklai Experimental Station in Assam. Daylengths are calculated from sunrise to sunset. Figures are means over: 50 years, of daily observations for temperature, rainfall, humidity, sunshine and wind run; 12 years for mean soil temperature at 10 cm below bare soil; and 9 years for percentage total crops.

Period	Nov.–April	May–Oct.
% total rainfall	19·4%	80·6%
% total crop	15·1%	84·9%
Mean maximum temperature, °C	24·9°	31·0°
Mean minimum temperature, °C	14·0°	23·6°
Mean relative humidity at 1300	60%	73%
Mean hours of bright sunshine	6·3	5·0
Daily wind run (miles per day)	18·6	24·5
Daily wind run (kilometres per day)	29·9	39·3
Mean soil temperature at 10 cm below bare soil, °C	25·6°	32·3°
Mean daylength	11 hrs. 23 min.	12 hrs. 55 min.

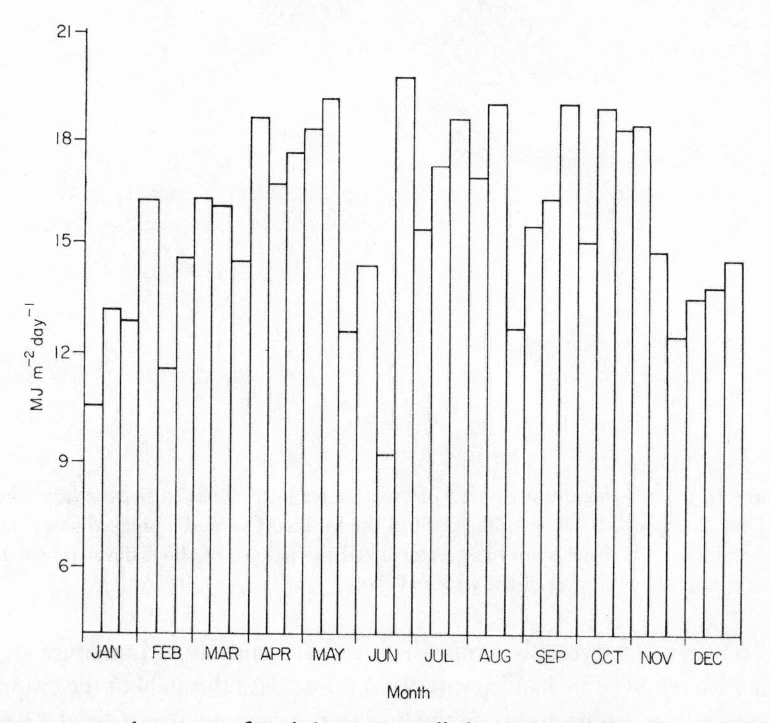

Figure 21.2. 10-day means of total short wave radiation measured by a Kipp solarimeter for Tocklai Experimental Station (Latitude 26°47′N) in Megajoules per square metre per day. The very low value for mid-June indicates the onset of the Monsoon.

During the main cropping season there is a reduction in the hours of bright sunshine of 1·3 hrs per day in spite of an increase in daylength of 1·5 hrs per day. Mean minimum temperatures are almost 10°C higher than in the November–April season, and mean soil temperatures are about 7°C higher. The markedly seasonal distribution of crop production is in sharp contrast to most other major tea growing areas which produce tea throughout the year.

Tea cultivation in Assam

In its natural state *Camellia sinensis* var *assamica* is a large shrub of up to 8 m high and 4 m diameter while var *sinensis* is much smaller, a height of 3 m being normal.

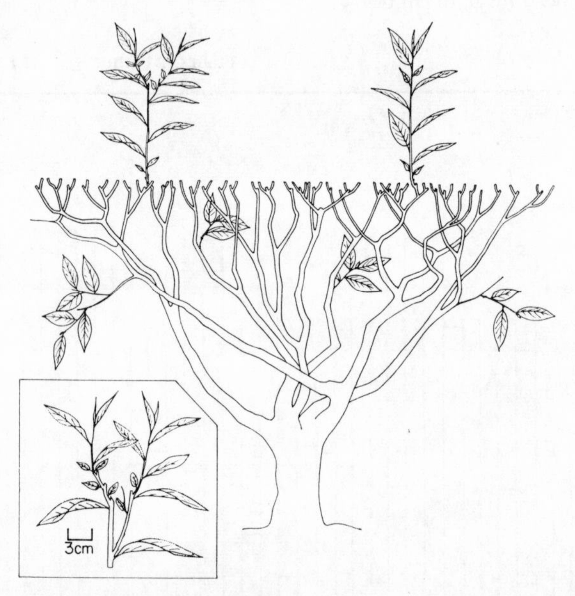

Figure 21.3. Semi-diagrammatic view of a mature tea bush, *Camellia sinensis* var *assamica*, showing an unplucked shoot on the right of the bush surface and a plucked shoot on the left-hand side. The inset shows in greater detail the pattern of shoot development from axillary buds after removal of the terminal shoot.

In order to harvest the crop which consists entirely of immature shoots which are removed by hand at intervals of 6–10 days throughout the cropping season, it is necessary to prune the bushes to a horizontal surface at a height convenient to the pluckers whose average height is about 5 feet (1·5 m). This operation drastically alters the foliage pattern and the comparatively open-crowned small tree is converted to a densely foliated shrub. After the prune, which is done in December–January, regrowth from just below the cut ends of the branches is allowed to grow unchecked until the primary shoots are between 20–25 cm long. The terminal shoot, consisting of two or three developing leaves, is then removed and from the axils of the older leaves new secondary shoots arise which, as they reach the predetermined plucking height of between 15–20 cm, are also removed. The continuation of this operation results in the production of higher order lateral shoots and after some six or seven rounds of plucking a dense mass of foliage is produced in a shallow layer. Figure 21.3 shows the above sequence of events diagrammatically.

Tea bushes are planted in a pattern and density which is a compromise between the necessity of labour having to move between rows of bushes and the achievement of complete ground cover in the shortest possible time to reduce weed growth and produce an early economic yield. With modern planting populations of about 12,000 plants per hectare, ground cover is complete within about 4–5 years and at full foliation there is a very high interception of incident radiation in comparison to most perennial plantation crops. The Leaf Area Index of such populations varies roughly from 4 to 7 depending on the variety of bush, the *assamica* types having the lower values.

Leaf temperature

For most of the harvesting season we have a densely planted bush population at a mean maximum air temperature of over 30°C during the day and one would expect that fully-exposed leaves would reach temperatures above this. Factors affecting leaf temperature include insolation, leaf size and structure, leaf pose, wind speed, relative humidity and the rate of transpiration (itself dependent on some of these factors). Even if the effect of all these factors and their many possible combinations on leaf temperature were known, a mere knowledge of leaf temperatures is of little value unless their effects on physiological processes are also known.

The extensive investigations on the tea bush at Tocklai have included the effects of temperature on net photosynthesis, on dark respiration of leaves, on respiration of non-photosynthetic parts of the bush, and on transpiration and stomatal behaviour. Before considering these aspects, however, it is first necessary to establish the range of leaf temperatures encountered in a normal plantation of tea plants in the field.

Field observations of leaf temperature

In Fig. 21.4 the temperatures of fully exposed leaves on the surface of a var

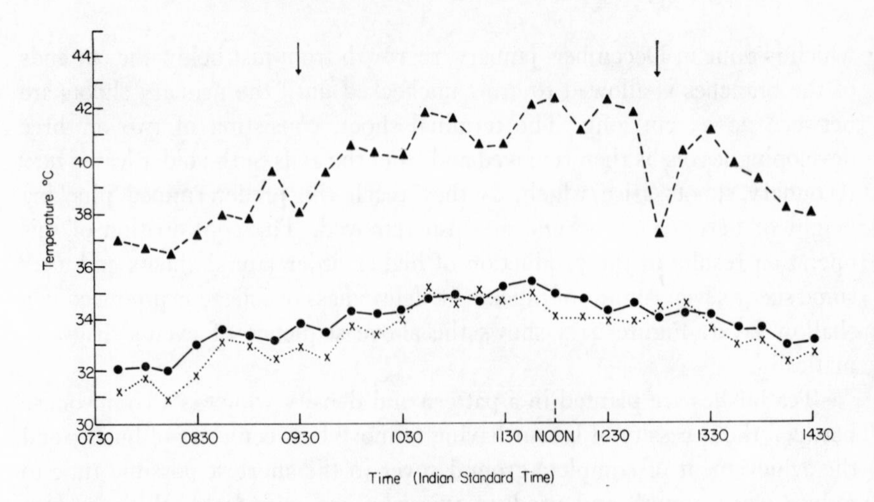

Figure 21.4. Leaf temperature of tea bushes, *Camellia sinensis* var *assamica*, over a seven hour period on a mainly clear day in late May. Air temperature ●——●, fully exposed leaves on the bush surface ▲ — — — ▲, leaves inside the bush canopy × · · · · · · ×. Values are the means of six readings for each class of leaves. Mean wind speed was 1·3 miles per hour over the period and light clouds partially obscured the sun's disc at the times indicated by arrows.

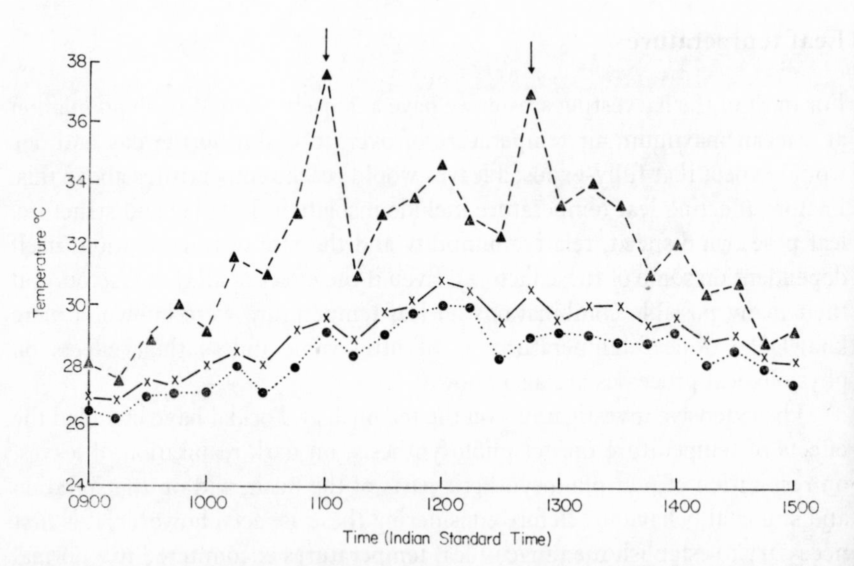

Figure 21.5. Temperatures of unshaded mature leaves of *Camellia sinensis* var *assamica* under almost wholly cloudy conditions for a six hour period in early June. Air temperature ×——×, fully exposed leaves on the bush surface ▲ — — — — ▲, leaves inside the bush canopy ● · · · · · · · ●. Values are the means of six readings for each class of leaves. Mean wind speed was 1·1 miles per hour. The sun's disc was completely obscured by cloud except for short periods indicated by arrows.

assamica tea bush are shown for a seven hour period in late May on a clear day with an average wind speed of 1·3 miles per hour (2·1 km h^{-1}) during the period of the observations. The temperature of leaves inside the canopy which did not receive any direct sunlight are also shown together with ambient air temperatures. Temperatures were measured using thermocouples as described by Rackham (1975: *this vol.*, p. 445): for full details see Hadfield (1976). This figure shows that leaves inside the canopy were either at or slightly below ambient air temperatures while fully exposed leaves were much higher than air temperatures (up to 12°C higher in Fig. 21.7).

Under dull overcast conditions the range of temperatures is greatly reduced as shown in Fig. 21.5, which gives results of a six-hour period in early June when ambient temperatures were similar to those given in Fig. 21.4, but heavy cloud was present for most of the period of observation. Wind speeds averaged 1·1 miles per hour (1·8 km h^{-1}) and the relative humidity at noon was 78% compared to 71% for the May readings.

Factors affecting leaf temperature

Under field conditions it is difficult to separate the effects of single environmental factors and therefore the effect of wind run under different radiation regimes was investigated in the laboratory. By the use of an Anglepoise lamp fitted with a G.E.C. Osram 230 V 60 Watt clear bulb and a variable speed stand fan it is possible, by placing the bulb at varying distances from the leaf surface, to achieve a wide range of radiation regimes; and due to the high proportion of infra-red radiation emitted by incandescent lamps they are very suitable for producing high leaf temperatures (Kuiper 1961). The bulb had previously been calibrated against a Kipp & Zonen solarimeter and results for four different radiation regimes are shown in Fig. 21.6, which demonstrates the marked cooling effects of wind under conditions of high irradiance.

The effect of solar radiation at constant windspeed, air temperature and relative humidity on leaf temperature is shown in Fig. 21.7. These results were obtained by abstracting from the very numerous field records those readings in which the only measured environmental factor varying was solar radiation as measured by a Kipp & Zonen solarimeter situated some 400 metres from the experimental plots. Continuous data-logging equipment would render the compilation of such figures a relatively simple process, but in the present studies the anemometer and Assman psychrometer were hand held and recorded manually, as were the leaf temperatures. It will be seen that over the range of intensities of solar radiation considered, extending from dull days to full noon sunshine, the relationship between leaf temperatures and total radiation is effectively linear.

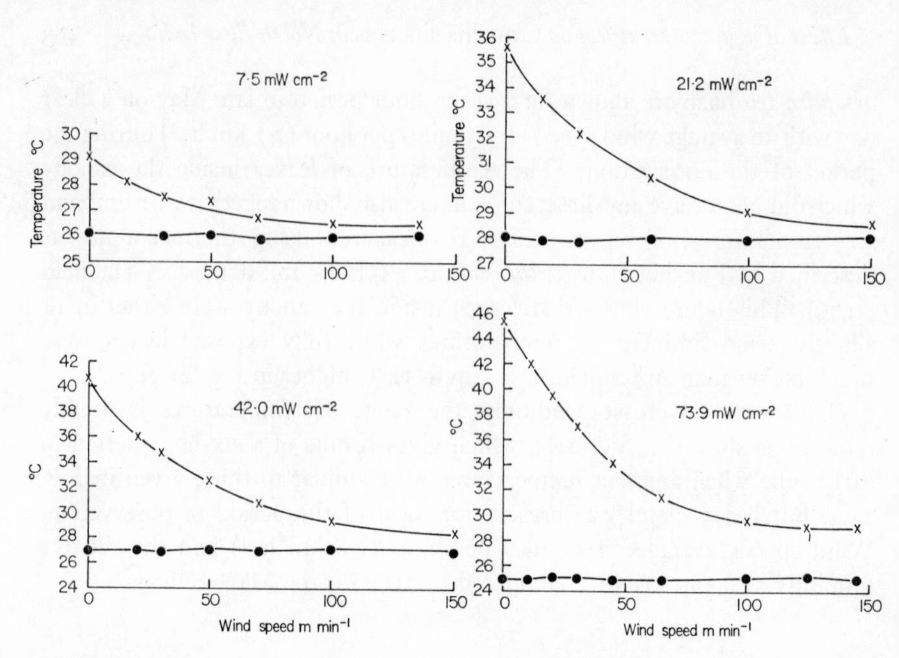

Figure 21.6. Temperatures of mature leaves of *Camellia sinensis* var *assamica* determined in the laboratory under four different intensities of total short-wave radiation, as a function of wind speed, in metres per minute. ×——× leaf temperature, ——●—— air temperature.

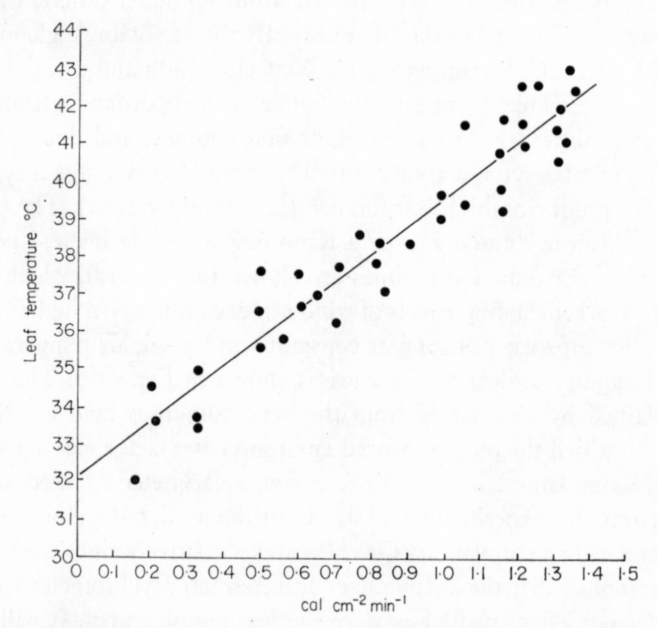

Figure 21.7. Temperatures of mature leaves of *Camellia sinensis* var *assamica* in the field as a function of total short-wave radiation under conditions of constant wind run, air temperature and relative humidity. Wind run 15–20 metres per minute, ambient air temperature 30–31°C, relative humidity 70–75%. 1 cal cm^{-2} min^{-1} $\equiv$ 69·8 mW cm^{-2}.

The cooling effect of transpiration was investigated by measuring the temperature of leaves in which transpiration had been prevented either by (a) severing the xylem by making deep cuts 1 cm apart at opposite sides of a stem or by (b) smearing the lower surface of leaves with petroleum jelly, and comparing temperatures with leaves of normal plants. Results of one such experiment carried out under high insolation are shown in Fig. 21.8 and

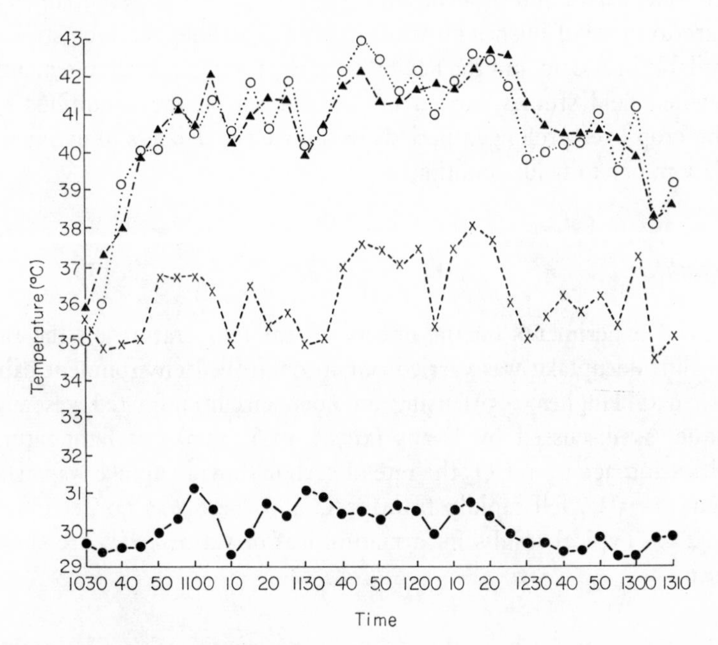

Figure 21.8. The effect on leaf temperatures of unshaded tea bushes growing in the open of reducing transpiration by (a) coating the under surface of leaves with petroleum jelly (▲ · — · ▲) and (b) severing the xylem tissue by transverse cuts on both sides of the stem (○ · · · · · ○ · · · · · ○) compared to untreated plants (×— — —×). The solid line represents air temperature. For further details see text.

under these conditions the cooling effect of transpiration is about 5°C. There is virtually no difference between the results of the two methods of reducing transpiration either in terms of leaf temperature or the time taken to produce visible symptoms of heat damage (see Plate 21.1, facing page 435).

Physiological effects of high leaf temperatures

In summary, the numerous field observations of leaf temperatures showed that under the environmental conditions of low wind speed, relatively high radiation regime and high relative humidity which are commonly found in the Brahmaputra Valley during the harvesting season, fully exposed leaves may remain 10°C above ambient temperatures for many hours a day. We can now consider the physiological effects of this situation. The results to be described have been obtained under controlled environmental conditions at Cambridge using bushes from a stock derived from Assamese seed and maintained for many years in a tropical experimental house at the Botanic Garden. The leaves and branches of these bushes had an essentially similar structure to those of bushes grown in Assam. Facilities for such studies are not available in Assam, but the flexibility of the Cambridge arrangements and the detailed field studies carried out at Tocklai on the conditions of the growing crop over prolonged periods minimize the dangers of extrapolation from laboratory to field conditions.

Photosynthesis

A series of experiments on the effects of leaf temperature on the rate of carbon dioxide uptake was carried out in controlled environment cabinets (Evans 1959, Hughes 1959) using an open circuit infra-red gas analysis technique, as discussed by Evans (1972; 12.6, 12.7). At light-saturating intensities and 300 ppm CO_2 the rate of carbon dioxide uptake was maximal between 30–35°C, fell rapidly from 37°C and there was no net uptake at about 42°C. Typical results for a mature leaf of var *assamica* are shown in Fig. 21.9.

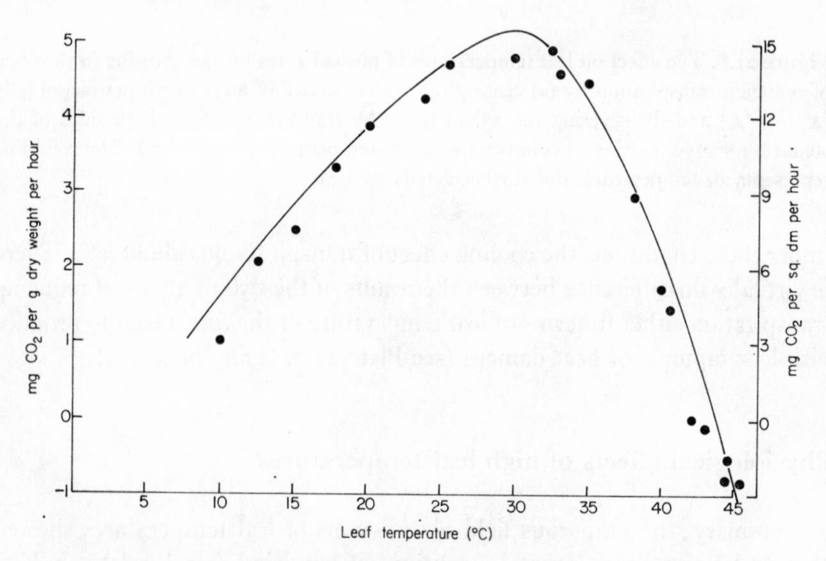

Figure 21.9. The rate of carbon dioxide uptake at saturating light intensity and increasing leaf temperature for a mature leaf of *Camellia sinensis* var *assamica*.

Respiration

In measuring respiration care was taken to avoid the disturbing effects of wounding and handling, and generally to observe the precautions discussed by Evans (1972, Ch. 12). Using attached leaves over a 12-hour period in the dark at 42°C, the mean rate of loss of carbon dioxide per gram dry weight per hour was around 8 mg for mature leaves, and 12–15 mg for leaves under 2 months old. Between 15° and 40°C the Q_{10} lies between 1·8 and 2·1.

Although only those leaves subjected to high radiation intensity reach these high temperatures, the branches and main stem may reach 32–34°C and roots will approximate to soil temperature, which may rise to 34°C in the top 9″ (23 cm) of soil where up to 80% of the young roots are found.

Respiration of whole tea bushes

An excavated root system of a 14-year-old bush is shown in Plate 21.2 where the shallow root system characteristic of var *assamica* bushes in the Brahma-

Plate 21.2. Excavated root system of a mature bush of *Camellia sinensis* var *assamica* showing the shallow nature of the system developed under central Assam conditions. The divisions on the scale are 2 inches (5 cm).

putra Valley is clearly seen. By pooling of data from the respiration studies on leaves and non-photosynthetic parts of tea bushes and determinations of fresh and dry weights of bushes in the field it has been possible to estimate total respiratory losses on an annual basis using the extensive meteorological data available at Tocklai. Such calculations are necessarily tedious as diurnal and seasonal variations in soil and air temperatures must be taken into account as well as the increasing mass of plant tissues as the season progresses.

Productivity

It is also possible to estimate net productivity with reasonable accuracy by the systematic uprooting and weighing of tea bushes of different ages from the same area (they should be from the same seed source, or, better still, from the same clone). The weight of shoots harvested as crop is always very accurately measured in order to pay labour, and pruning litter can be quickly separated into leaves and stems for fresh weight/dry weight ratio determinations. During the course of these investigations bushes of between 1 and 70 years old have been uprooted and mean annual increments to the permanent frame and root system estimated.

The balance sheet for a 20-year-old section of var *assamica* tea with a bush population of 8,000 plants per hectare is given in Table 21.2.

Table 21.2. Respiration and productivity of a 20-year-old section of tea bushes (near *Camellia sinensis* var *assamica*) in the neighbourhood of the Tocklai Experimental Station in Assam. Bush population 8,000 per hectare. Total dry weight of bushes (exclusive of leaves) 45 tonnes per hectare. The tea bushes were shaded by *Albizia odoratissima* planted 12 metres apart. Respiration of leaves in the dark is based on a mean leaf area index of 4 maintained for 250 days.

(a) Annual respiratory losses from non-photosynthetic tissues, in tonnes dry weight per hectare per annum

Organ	Dry Weight	tonne ha^{-1}	Annual Loss
Main stems	8		1·0
Large branches	16		6·2
Pruning stems	3		4·5
Large roots	16		3·0
Small roots	2		4·5
Leaves during darkness	4		3·2
		Total	22·4

(b) Annual net production, in tonnes dry weight per hectare, divided into the following elements:

Harvested tea shoots	2·8
Leaves	4·0
Pruning stems	3·4
Addition to permanent frame of stems	2·0
Addition to root system	2·5
Total	14·7

These results indicate that at this age some 60% of the gross productivity is utilised in respiration. Similar calculations for 50-year-old bushes, having a total dry weight exclusive of leaves of 86 tonne ha^{-1}, shows that respiration of the bushes accounts for 70% of the gross production.

Light profiles within the bush canopy

We therefore have a system where the upper foliage is light-saturated, the leaf temperature is well above optimum and the whole plant is at a temperature producing a large respiratory loss. The leaves lower in the canopy are within the optimal temperature range for photosynthesis, but light-limited due to the high degree of self-shading in the compact canopy. In some cases the light inside the canopy is insufficient for full stomatal opening to take place while upper leaves have their stomata closed due to the high leaf temperatures. Light profiles for extreme Assam and China varieties and an intermediate type are shown in Fig. 21.10 under cloudless conditions, at solar noon on a June day.

The horizontal leaf pose of the *assamica* variety cuts off most of the photosynthetically active part of the spectrum at the bush surface and most of the foliage is light-limited in comparison with the *sinensis* types with semi-erect leaves which allow penetration of light to a much greater depth and leaf area index.

Heat damage to leaves

As leaf temperatures rise above about 35°C the rapid transpiration causes moisture to be lost at a greater rate than it can be replaced and the relative turgor of the leaves falls. At around 74% of full turgor this results in stomatal closure and the cooling effect of transpiration, which may be as much as 5°C, is lost. In extreme cases the resultant temperature rise causes irreversible

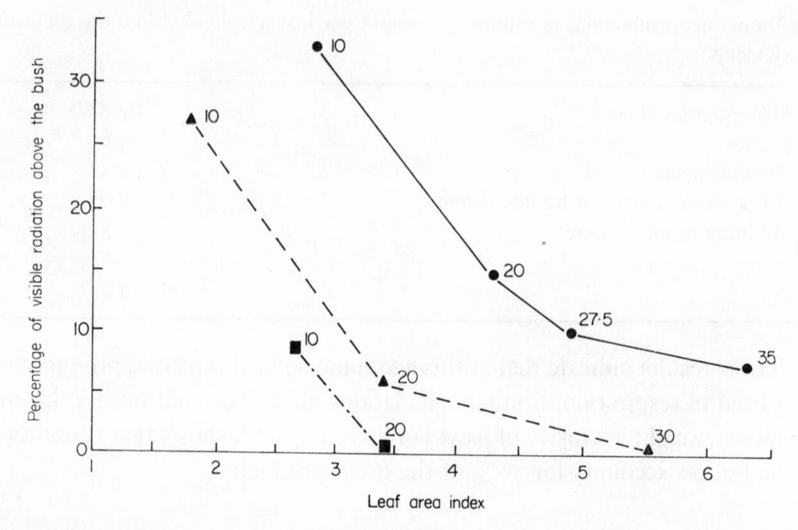

Figure 21.10. Mean values of visible radiation measured on a horizontal surface within the canopies of three types of tea bushes in Assam, as percentages of the radiation above the bushes in full sun, in relation to the leaf area index above the plane of measurement (depth in cm below the top of the canopy shown beside each point): ●——, near *Camellia sinensis* var *sinensis*; ■— · — ·, near *C. sinensis* var *assamica*; ▲ — — — — — — — —, a plant intermediate in form. Fiducial limits, all ● and ▲ 10 cm, 0·13 times each individual mean value; ▲ 20 cm, ± 1·23 per cent; 30 cm, ± 0·4 per cent; ■, 10 cm, ± 2·2 per cent; 20 cm, ± 0·2 per cent.

heat damage which becomes apparent as browning, starting at the margins and regions furthest from the main veins. These symptoms are frequently found when long dull periods are followed by hot dry spells, or when upper leaves are removed by pruning (see Plate 21.1, facing p. 435).

Even when visual symptoms of heat damage are absent, the prolonged exposure of leaves to temperatures of 40°C or more affects subsequent CO_2 uptake adversely and for at least 5 days after such exposure net photosynthesis is reduced by between 40–60% of the pre-exposure values.

Effects of shade trees

In North East India it has been traditional to interplant leguminous shade trees in sections of tea bushes, and although a detailed account of shade in tea (such as in Hadfield 1974a, 1974b) is beyond the scope of this paper, the effects of a shade tree canopy on the temperature of the bush canopy will be considered. The data already presented in Fig. 21.10 on the effects of canopy architecture of the Assam and China varieties on light penetration into the bush canopies have shown that visible radiation is greatly reduced very close to the bush surface in the Assam variety compared to the China variety.

The leaf temperature and photosynthesis studies demonstrated that under sunny, calm conditions leaves on the bush surface reach temperatures which severely restrict carbon dioxide assimilation and that horizontal leaves are generally 2–3°C hotter than erect leaves due to reduced reflection of infra-red radiation into the bush canopy. When a shade tree canopy is interposed between the bush surface and the solar beam there is a reduction in both visible and infra-red radiation, the ratios of visible to infra-red depending on the canopy characteristics of the shade tree used. In the case of the Assam variety most of the foliage is so heavily shaded that the major effect of overhead shade is on the surface leaves and as light saturation occurs at below 25 mW cm^{-2}, a considerable reduction in visible radiation on clear days has hardly any effect on the rate of photosynthesis if other conditions are favourable. There will be a reduction in the temperature of surface leaves leading to reduced moisture loss, maintenance of high leaf turgor and hence increased duration and degree of stomatal opening and reduced respiratory loss. One would expect the combination of these factors to result in increased canopy efficiency and hence increased productivity.

In the case of the China variety (or hybrids close to this) the effect of overhead shade is to bring a greater proportion of leaves to light-limited conditions and as leaf temperatures in full insolation are normally lower than the horizontal-leafed Assam types the effects on respiration and moisture stresses are correspondingly decreased. It could therefore be anticipated that shading would not necessarily be as beneficial to such foliage types (Hadfield 1974b, Hadfield 1974c).

A field trial using artificial shade

A field trial was carried out to test this hypothesis. The question is of considerable economic importance since the introduction of clones has led in recent years to the production of substantial areas of tea bushes having on average a much greater variety of leaf types, and as they all receive identical agricultural treatments, it is most unlikely that the full potential of most of them is being fully realized.

A semi-erect-leafed China hybrid and a horizontal-leafed Assam variety were planted in alternate rows in blocks of 140 bushes each (70 bushes of each clone). The China hybrid was originally selected from a mixed seed population because it possessed a combination of high quality and above-average yield under cultivation in the plains of Assam. (It would not normally have been considered as suitable planting material for Darjeeling, where made tea having a much more distinctive China character is required.) One block was unshaded and the other block was shaded by bamboo lath frames transmitting about 60% daylight. Agricultural practices were identical for both blocks and the yields for the second and seventh years after ground cover had been achieved are shown in Table 21.3. This experiment was designed to investigate the effects of shade on yield, and it was not intended as a comparison of absolute yields between the two morphologically distinct varieties used, which differed widely both in genetic make-up and in relation to selection procedures.

Table 21.3. Yields of two different forms of the tea bush at the Tocklai Experimental Station in Assam, (a) unshaded, and (b) shaded by bamboo lath frames transmitting approximately 60% of daylight. Mean yields with fiducial limits for the second and seventh years after the attainment of complete ground cover, in grams of made tea (approximately 4% moisture content) per bush per year. For further details, see text.

Bush form	Lighting	Yield and fiducial limits 2nd Year	7th Year
Near *Camellia sinensis* var *assamica*	(a) Full exposure	212±18	224±16
	(b) 60%	237±18	319±16
Near *C. sinensis* var *sinensis*	(a) Full exposure	500±42	679±33
	(b) 60%	389±42	508±33

In the second year after ground cover had been achieved the Assam variety produced a 10% increase under shade while the China hybrid produced 20% less yield under shade. By the end of the seventh year from complete ground cover the yield of the Assam variety under shade was some 40% higher than in full sun, while that of the China hybrid was some 30% less. In the case of the Assam variety the constantly increasing respiratory load of the branch and root systems, combined with the restricted and frequently overheated upper leaf canopy (in the fully exposed plots) has resulted in virtually no increase in economic yield since ground cover was achieved. This compares with an increase in yield of over 30% in the shaded plots. The decrease in variability in the 7th year is probably real, reflecting the extremely regular surfaces produced by the closely spaced bushes as they mature.

More recent studies on the 'shade dependence' of the horizontal-leafed Assam variety compared to the erect or semi-erect China hybrid types strongly suggest that the response of the former to fertilizers is inhibited by the extreme mutual shading in the canopy, resulting in a large percentage of inefficient leaves. This condition is aggravated by the necessity of providing overhead shade to maintain the surface leaves at below supra-optimal temperatures.

Conclusions

The general conclusion of the investigations carried out at Tocklai during the past decade is that under the environmental conditions of most of the Assam Valley, the indigenous tea plant and hybrids having a similar plant form require some form of overhead shade to minimize the physiological consequences of high temperatures and that this in turn limits the response to applied nutrients. With a constantly increasing respiratory load of woody tissue in the form of branches and roots and a static photosynthetic area imposed by the nature of the crop, the production of new young shoots by such plants is steadily reduced as the plants age, at the same time as mounting production costs necessitate increased crops. Plants derived from the China variety and having a similar plant form do not suffer from these disadvantages to the same extent and the economic future of the Assam tea industry must, in the long term, be dependent upon the selection and exploitation of bushes which are not shade dependent and which possess the market characters demanded from Assam teas. Such selection has already been started but, as in most perennial plants, progress will be slow.

The productivity studies indicate that the rapid achievement of complete ground cover by increasingly closer planting may result in an economically shorter life by imposing a larger respiratory load at a given age on a static photosynthetic area. If this proves to be the case a fairly drastic revision of long-term planning and land usage in the tea industry will be required as the century-old assumption of a 50-year minimum economic life of a tea bush has been based on a plant population somewhat less than half of that now being planted.

In retrospect it is ironic that the discovery, exploitation and deliberate selection for over half a century of the type of tea bush associated with Assam, with its large, horizontal leaves, should have now become the greatest single biological factor inhibiting the future progress of the industry in the lowlands of North East India.

Acknowledgments

The help and advice given by Dr. Clifford Evans throughout the preparation of this paper is gratefully acknowledged. Professor Sir Frank Engledow suggested several improvements to the original draft and the results are published with the permission of the Director of the Tea Research Association.

Bibliography

EVANS G.C. (1959) The design of equipment for producing accurate control of artificial aerial environments at low cost. *J. agric. Sci., Camb.* 53, 198–208.

EVANS G.C. (1972) *The quantitative analysis of plant growth.* Blackwell Scientific Publications, Oxford. pp. xxvi + 734.

HADFIELD W. (1974a) Shade in North-East Indian tea plantations. I. The shade pattern. *J. appl. Ecol.* 11, 151–78.

HADFIELD W. (1974b) Shade in North-East Indian tea plantations. II. Foliar illumination and canopy characteristics. *J. appl. Ecol.* 11, 179–99.

HADFIELD W. (1976) Shade in North-East Indian tea plantations. III. Leaf temperatures in the field. *J. appl. Ecol.* (in the press).

HUGHES A.P. (1959) Plant growth in controlled environments as an adjunct to field studies. Experimental application and results. *J. agric. Sci., Camb.* 53, 247–59.

KUIPER P.J.C. (1961) The effects of environmental factors on the transpiration of leaves, with special reference to stomatal light response. *Meded. Landb. Hoogesch. Wageningen* 61, (7) 1–49.

PURSEGLOVE J.W. (1968) *Tropical Crops. Dicotyledons* 2. Longmans Green (London) pp. viii, 333–719.

RACKHAM O. (1975) Temperatures of plant communities as measured by pyrometric and other methods. *This vol.*: 423–49.

Questions and discussion

Professor Willis asked how large was the effect of the shade trees on wind speed. Mr. Hadfield said that generally wind speeds in the open around Tocklai were below 1·5 km hour^{-1}, and that although the open planting of shade trees must reduce wind run, he had been unable to measure the effect with the available anemometers. Dr. Evans asked about the effect of reducing air movement yet further by placing screens around a bush. Mr Hadfield said that the result was disastrous around midday, a plant completely exposed to full sun being killed in 1–2 hours. It followed that even the reduced air movement among the shade trees of a plantation was highly beneficial.

In answer to two questions from Professor Mount, Mr. Hadfield said that there were wide differences in reflectance between the upper leaves of different types of tea bushes, var. *sinensis* having matt leaves and var. *assamica* much more shiny ones. By constraining leaves of one variety into the position characteristic of the other it could, however, be shown that differences in reflectance had much less effect on leaf temperature than differences in leaf angle.

Professor Willis asked about water deficits in the dry season. Mr. Hadfield said that in the Brahmaputra valley the problem was rather how to dispose of 2,500 mm of rain in a few months. Owing to the very shallow root system and high water table there could be rapid alternations of drought and waterlogging. The carbon dioxide uptake by plants exposed to artificial drought falls, but recovers within 48 hours of watering. On the other hand, after waterlogging there may be no sign of recovery even after two or three months.

22 Solar radiation and heat balance in animals and man

D. L. INGRAM and L. E. MOUNT *A.R.C. Institute of Animal Physiology, Babraham, Cambridge*

Solar radiation can have a considerable effect on an animal's heat balance. Of the estimated 1360 W m^{-2} irradiance of the outer atmosphere (Monteith 1973) up to 1000 W m^{-2} may reach the ground in some desert regions, and progressively less under cloudless conditions in more temperate regions. For a homeothermic animal with a resting metabolic heat production of 50 W m^{-2}, the heating effect of solar radiation is potentially overwhelming, even when the directional restriction of solar radiation is taken into account. The parts of the animal's surface which do not receive direct irradiation are exposed to diffuse and reflected short-wave radiation, and in addition receive long-wave radiation from the surroundings which is increased by solar heating. Some poikilothermic animals use solar radiation in conjunction with shade to maintain a fairly stable body temperature, but under hot conditions solar heating represents for both poikilotherms and homeotherms a heat load which is resisted or dissipated by several physiological and behavioural adaptations.

Physiological responses

Although there are some instances where metabolic heat production is important in influencing a poikilotherm's body temperature, as in the warming effect of pre-flight activity in some insects, poikilotherms tend instead to influence body temperature by using environmental diversity. Homeotherms, on the other hand, exhibit the obligatory production of a minimum amount of heat even when resting, with the result that heat dissipation must equal heat production if thermal stability is to be maintained. This is the basis of the heat balance, which, including the heat gain from solar radiation, may be written:

Metabolic heat production + heat gain by radiation = heat loss by radiation, convection, conduction and evaporation + heat storage.

The particular way in which the equation is disturbed by solar radiation depends (1) on the intensity of radiation, which may be equivalent to several times the metabolic heat, (2) the non-radiant components of the environment and (3) the animal's *critical temperature*. The critical temperature is the environmental temperature below which metabolic heat production must increase if body temperature is to be maintained; it is referred to standard conditions of free convection when air temperature and radiant temperature are equal. The variation in critical temperature between species is large, being +35°C for the new-born pig and about −30°C for some Arctic animals in their winter coat. Below the critical temperature an increase in heat gained from solar radiation can be offset by a decrease in heat production and in these circumstances the animal saves energy reserves. Above the critical temperature metabolism is already at a minimal level, and any extra heat load imposed must be absorbed as heat storage or dissipated to the environment.

Some large animals such as the camel, donkey, and cattle store heat by allowing the deep body temperature to increase during the day, and then lose heat at night (Schmidt-Nielsen 1964, Ingram & Mount 1975). Small animals cannot continue to store heat for the whole day, because as body size decreases thermal capacity decreases more rapidly than the body surface to which heat gain by radiation is related. However, the antelope ground squirrel, which lives in the desert, stores heat over a short period of time and then returns to its burrow to lose heat before again returning to the surface and allowing its body temperature to increase (Bartholomew 1964). The effective use of heat storage, nevertheless, is confined for the most part to large animals, and even so some animals do not tolerate much rise in core temperature.

An increased rate of heat loss can sometimes be achieved by an increase in peripheral blood flow, but under conditions of high solar radiation this may result in the transfer of heat into the body rather than out. As environmental temperature rises, sensible heat loss through radiation, convection and conduction declines, until evaporative loss becomes the only channel for heat dissipation; however, this uses up reserves of body water which may not be readily replaced. Animals smaller than the rabbit cannot afford to cool themselves by evaporation for more than a very short time, because, as with heat storage, as size decreases the water available decreases more rapidly than the quantity of heat to be lost. Behavioural patterns which reduce the solar heat load then become of great importance.

Behavioural responses

The poikilotherms share with the homeotherms the dangers of overheating and dehydration during exposure to heat, but in cooler environments the addition of solar radiation may represent an opportunity to increase body temperature and become active at an air temperature which would otherwise result in a low metabolic rate. Indeed some lizards are able to maintain their body temperature to within 2·8°C for 80% of the day by simply moving in and out of the sun (Bogert & Cowles 1959). In a hot desert this may mean that at mid-day and just afterwards most of the time is spent in the shade, but Schmidt-Nielsen (1964) has drawn attention to the fact that behaviour of this kind can enable some lizards to maintain a body temperature of 30°C above ambient air even when air temperature is at freezing point. The simple act of moving in and out of a patch of sunlight depends on the presence of shade; but the amount of solar radiation received can also be modified by changes in the radiation profile associated with the animal's orientation to the sun's rays. When the axis of the body is at right angles to the sun's rays a greater surface area is exposed to radiation than when it is parallel, as can be verified by looking at the size of the shadow which is cast. Lizards are known to orientate themselves to the sun (Templeton 1970), and under very hot conditions they make use of an inclined surface when lying parallel to the sun's rays and so expose an even smaller surface area (White 1973). Some insects also orientate themselves with respect to the sun, and in the case of the desert locust Stower and Griffiths (1966) have found that this sort of behaviour could make as much as 6°C difference to the body temperature.

In the laboratory, studies involving operant conditioning (Weiss & Laties 1961, Baldwin & Ingram 1968, Corbit 1970) have demonstrated that animals can adjust the amount of radiant heat which they receive and so limit the increase in metabolism demanded by a cold environment. Their responses in this situation are influenced by experimental variations in the temperatures of the hypothalamus and spinal cord in much the same way as autonomic thermoregulation (Ingram & Mount 1975), although the parallel is not exact (Carlisle & Ingram 1973). These experiments, however, all apply to studies in the laboratory where the animal is able to make only very limited responses. Under temperate conditions outdoors, pigs have been found to display relatively little attention either to operating a switch which turned an infra-red heater on for a short period, or to selecting the area of highest natural radiant temperature (Baldwin & Ingram 1968, Ingram & Legge 1972). If an animal allows thermal factors to dominate its behaviour other activities tend to suffer, and it is probable that behaviour is dominated by thermoregulation only under extreme conditions; at other times other activities take precedence. The nature of the coat and the radiation profile of the animal are therefore of considerable importance since they effectively change the solar heat load.

Thermal radiation increment

The warming effect produced by solar radiation which has been absorbed at the coat or clothing surface can be expressed as a 'thermal radiation increment' which takes the form of a temperature increment to be added to air temperature. Burton and Edholm (1955) derive this increment by postulating that in the steady state the solar radiation which is incident on a clothed man, for example, is lost from the surface of the clothing partly by reflection of short-wave radiation and partly by long-wave radiation to the environment. The temperature difference between skin and clothing surface ($T_S - T_{Cl}$) is given by HI_{Cl}, where H = heat loss per unit area of skin and I_{Cl} = insulation of clothing. The temperature difference between the clothing surface and the surrounding air ($T_{Cl} - T_A$) is similarly given by $(H+R)I_A$, where R = solar radiation per unit area and I_A = the air insulation. Then:

$$\begin{aligned} T_S - T_A &= HI_{Cl} + (H+R)I_A \\ &= H(I_{Cl}+I_A)+RI_A \end{aligned} \qquad (1)$$

RI_A then gives the thermal radiation increment, as the required temperature increment to be added to air temperature; this quantity is the product of the incident solar radiation and the air insulation. After dividing equation (1) throughout by $(I_{Cl}+I_A)$, it becomes:

$$\frac{T_S - T_A}{I_{Cl}+I_A} = H + R\frac{I_A}{I_{Cl}+I_A}$$

which indicates that the heat loss from the body surface is increased by

$$R\left\{\frac{I_A}{I_{Cl}+I_A}\right\},$$

so that the efficiency of solar radiation in adding heat to the body is given by the insulation outside the point of absorption divided by the total insulation. In this way solar radiation can be considered as an addition to metabolic heat production. This is applicable both to clothed man and to animals with coats; in both cases part of the solar radiation is lost to the environment, and part increases the heat load of the body.

Radiation exchange in animals

The general problems of radiation exchange in animals have recently been reviewed by Cena (1974) who has examined the factors which influence the amount of radiation absorbed and gives the general radiative heat balance equation as:

Net radiation $R_n = (1-p^*)(S_t+S_e)+\varepsilon(L_a+L_e)-L_b$

where S_t = direct and diffuse solar radiation

S_e = short-wave radiation reflected from surroundings

p^* = short-wave reflection coefficient of the body.

The remaining terms are the long-wave radiative fluxes from the atmosphere (L_a), environment (L_e), and the body itself (L_b); ε is the emissivity of the body. The profile intercepting radiation influences the term (S_t+S_e), and while this may be modified by orientation to the sun the basic shape of the body is also important. The term $(1-p^*)$ depends on the nature of the coat or skin of the animal, and since a black surface absorbs nearly all the solar radiation and a white surface reflects about half, coat and skin colour are of considerable importance.

The profile of the body exposed to radiation depends both on the animal's orientation to the sun's rays and the altitude of the sun. For man, Underwood and Ward (1966) have photographed the body silhouette from various angles of altitude and azimuth and produced a formula for the calculation of the area exposed to direct radiation. When the altitude is 90° (i.e. sun directly overhead) the azimuth makes no difference, as would be expected; at all other altitudes, however, the azimuth is important. A similar technique has been used by Clapperton *et al.* (1965) to determine the radiant load on sheep. They compared the sheep's profile at various solar angles with that of a cylinder of dimensions similar to those of the animal's body and found that provided the sheep orientated itself to the sun at random it could be regarded as a cylinder, although the area intercepted by the sheep's body fell as the altitude increased from 0° to 90°. With this information it was then possible to estimate the incoming solar load (a) from the area exposed and (b) directly by placing a solarimeter at various points around the cylinder and thereby also taking into account reflected solar radiation. The two methods of estimating the solar load gave results which varied somewhat with the intensity of the load and elevation, although the agreement was generally good. The discrepancies are related to the fact that since the body casts a shadow on the ground, solar radiation cannot be reflected by the whole area of ground around the animal. It was demonstrated that in Western Scotland the incident solar load on a sheep could be as high as 350 W m^{-2}, or, allowing for radiation profile, about three times the basal metabolic rate. These calculations, however, indicated only the amount of solar energy impinging on the animal; they did not indicate the amount which entered the body. In order to investigate the actual heat load which the animal must dissipate it is necessary to inquire about the reflection of short-wave irradiation from the body surface, and the penetration of the coat by transmitted radiation.

Reflection and transmission of solar energy

The reflection of radiant energy from the skins of three different types of pig has been determined by Kelly *et al.* (1954) who found considerable differences between animals with white, red and black skin. Maximum reflection occurs at about 1 μm, but while the white skin reflects nearly 80% at this wavelength, black skin reflects less than 20%, with red skin falling between the two. Beyond 3 μm there is no difference in reflectance between skin colours. The reflectivity of pig skin is similar to that for human skin (Gates 1962), so that from the point of view of thermal balance it would appear to be an advantage to have a black skin in the temperate regions and a white skin in the tropics. A black surface, however, absorbs heat in the superficial layers from which it may be lost again by convection, while solar radiation may penetrate much further into white skin (Blum 1961, Mount 1968). On theoretical grounds Kovarik (1964) predicted that the heat load due to radiant energy would be greatest for a surface which was neither black nor white. Thompson (1955) concluded from his investigation that black skin affords much greater protection from ultra-violet light, and that this more than compensates for the additional heat load. The protection afforded to black skin against solar radiation does not, however, give complete protection against tissue damage, since black people may still suffer from sun-burn if they are exposed to strong sunlight.

Most mammals are covered by a coat of hair, and in these instances it is the nature of the coat which is important with respect to the reflection of solar energy. As Monteith (1973) has pointed out the effect of an animal's coat on the amount of solar radiation received is very complex and has not yet been fully explored. A white fur scatters the solar beam both upwards away from the animal and down towards the skin, as for example in the seal (Øritsland 1971), while a black fur will absorb the radiation. The amount of solar energy reaching the skin may therefore be higher in animals with white fur than black, although the black hairs, because they get hotter, will radiate more long-wave energy to the skin. The combination of white hair and black skin would tend to increase the capture of solar energy. The presence of an air current complicates the situation because the warmer black hair loses heat faster than the cooler white hair.

A similar complication also occurs with respect to the clothing worn by desert people. Such clothing often takes the form of a large rather loose thick robe which is well suited to insulation against radiant heating, because a very large part of the incident radiation is absorbed by the robe and is lost by convective heat transfer, so that the heat load does not impinge on the body surface itself. Since the general air temperature is high, the clothing must be loose to allow the vaporization of sweat, and when the air temperature is higher than body temperature the robe serves to protect the body from scorching, which may occur from hot wind. The robe is sometimes dark in colour, with the consequence that both short-wave and long-wave radiation are absorbed and a large heat load has to be lost by convection and re-radiation. Some desert people use a light-coloured robe which allows reflection of part of the incident solar radiation, resulting in a smaller quantity of heat being absorbed by the material itself with a consequent reduction in heat to be lost by re-radiation or convection. However, a light-coloured robe may allow more short-wave radiation to penetrate than a dark-coloured robe, producing undesirable radiant heating of the skin. The effect of the radiation increment in environmental temperature is thus reduced under desert conditions by clothing, and there is a correspondingly lower rate of sweating in clothed man than in nude man. This is true for resting man, but during exercise clothing may hinder the evaporation of sweat, and above a metabolic rate of about 280 W m^{-2} the advantage of clothing disappears.

Animals show a considerable range of adaptation in meeting the radiant increment. This is achieved partly through the coat which not only prevents heat loss in the cold but also prevents heat gain under hot conditions. There are three main types of coat, which are exemplified in the Awassi sheep, the Merino sheep and *Bos indicus*; they are respectively the loose coat, the tightly-packed and the short smooth coat. Solar radiation penetrates the loose coat, and although much of the incident radiation is lost by convective heat transfer a relatively high skin temperature is produced. The tight-packed coat of the Merino sheep reaches a high temperature on its surface, up to 85°C, when exposed to a high intensity of solar radiation; a very large part of the incident radiation is lost as long-wave re-radiation and some by convection, and the skin temperature does not rise as high as in the case of the loose-coated Awassi sheep. In *Bos indicus* and the equatorial camels and gazelles the short smooth coat acts as a good reflector of incident solar radiation, and this again results in a lower skin temperature.

These examples in animals with coats indicate adaptations which involve the dissipation of solar radiation primarily by convection, re-radiation and reflection, respectively. Wind increases heat removal from coats to a very great degree; a surface temperature of 90°C in still air is reduced to 60°C within a few minutes by a wind of 5 m/sec; this wind speed is common in the desert during the hot part of the day and is important in dissipating incident solar radiation. The incident heat load which is not dissipated through the sensible channel must be lost by evaporation, mainly by panting in sheep and by sweating in cattle.

Different coat reflectances are illustrated by two kangaroos, the Red and the Euro, which live in the interior of Australia in different types of habitat.

The Red lives on the open plain, whereas the Euro uses caves as refuges from heat; the Red kangaroo is thus exposed to solar radiation, and it is found that its fur is twice as reflective as the Euro and is also more resistant to penetration by solar radiation (Dawson 1972). Finch (1972) estimated the heat absorbed by the eland and the hartebeest under a natural irradiance of 950 W m^{-2}. The larger (400 kg) eland absorbed 390 W m^{-2}, whereas the much smaller (100 kg) hartebeest with its more reflective coat absorbed 275 W m^{-2}.

The hair coat of cattle has been investigated by Bonsma (1943), Rhoad (1940) and Stewart (1953) with respect to adaptation to the environment. As in the pig there is a wide range of reflectance. Their figures indicate the amount of solar energy reflected upwards from the coat, and do not necessarily indicate the amount of radiation received by the skin, since some heat is lost from the coat by convection. The radiation received by the skin depends on the penetration of the coat, which has been studied in cattle by Hutchinson and Brown (1969) by the use of pelts mounted on a hot plate and irradiated with a xenon lamp. Penetration was estimated by determining the absorbed irradiance from the lamp which would give zero heat flow from the plate. There was considerable variation even between two samples of coat of the same colour from the same animal. The results nevertheless demonstrated the greater penetration of radiation into white than coloured coats, and the smaller penetration in a dense rather than a loose coat. The experiments also highlighted the effect of air movement in removing the heat absorbed by dark-coloured hair. Hamilton & Heppner (1967) exposed white finches to radiant heat under cool conditions and measured oxygen consumption. They found that when the feathers were dyed black the birds' metabolic rate decreased, indicating increased absorption of heat at the skin surface. This suggests that in these birds the probable reduction in penetration of the darkened feathers by solar radiation, with less radiation absorbed at the skin surface, was more than offset by decreased reflection. Under conditions of low air movement the efficiency of heating the body, as given by I_A/I_A+I_{Cl}, would be relatively high.

Among the invertebrates many desert beetles are black in colour, and it is possible that as a result radiant heat is trapped in the cuticle from where it can be lost by re-radiation and convection as in the black robes worn by man. By contrast the study of certain desert snails by Schmidt-Nielsen *et al.* (1972) has underlined the importance of the white shell. These animals survive unprotected by shade on the desert floor partly because of the properties of the shell, which reflects over 90% of the incident radiant energy, and partly because of a pocket of air in the shell which provides insulation from the hot desert floor. The importance of the surface characteristics of the shell in reflecting the solar load is emphasized by the observation that painting the shell black leads to a rapid rise in temperature and the death of the snail.

Human habitation

Heating by solar radiation can be an important factor affecting human habitation. O'Sullivan (1974) comments on the relation between the heat distribution in a building and the rate at which heat is lost to the surroundings. He derives a balance point between the gain and loss of heat, and shows how various factors, for example the area of windows, can affect the balance. The inferred importance of solar radiation in human habitation is indicated in a recent investigation into places of archaeological interest in Greece (Legge 1972). There were marked differences in radiant temperature depending on the orientation and degree of enclosure of the caves which were studied.

Observations on the pig out of doors

So far consideration has been given to the problems of reflection and absorption of solar radiation under controlled experimental conditions. In the field the heat load can be calculated from the solar altitude, the animal's orientation to the sun, the irradiance, and the animal's surface area and the penetration of the coat by radiation, if all these factors are known. A change in air movement, or in orientation to the sun, however, upsets the calculation and for practical purposes it is desirable to use some method which measures the effects of solar radiation on heat loss directly. One such method which has been explored is the use of the Hatfield heat-flow disc. These discs (Hatfield & Wilkins 1950) consist of a disc of tellurium 1·3 mm thick and 12·7 mm in diameter, with each side covered by a sheet of copper foil to which wires are attached. The discs thus form a system of two copper-tellurium thermojunctions, and if one surface is at a different temperature from the other a potential is generated in the microvolt range. When placed on a warm surface the potential difference is proportional to the heat flow through the tellurium, and the disc can be calibrated and used as a meter. Experiments in the laboratory (Ingram 1964) have shown that disc strapped to the body of a pig give readings for heat flow close to those expected. The development of a radiotransmitter which would transmit the signal over a range of 11 metres (Heal *et al.* 1974, Heal 1974) has made it possible to study heat flow from unrestrained pigs under field conditions. In this study four discs were placed symmetrically round the animal's trunk, with one disc in the middle of the back. Because the emissivity of the copper surface of the disc would not be the same as that of pig skin, surgical tape similar in colour to the skin was placed over the disc. Readings were obtained over a range of air temperature from $-3°$ to $+25°$C, mean radiant temperature from 1° to 77°C, and air movement from 35 to 377 cm sec^{-1} (Ingram *et al.* 1974). A multivariate analysis revealed that the mean radiant temperature and the air movement accounted for almost all the variation in heat loss.

The further investigation of heat loss under conditions of solar radiation involves the determination of the heat loss coefficient for convection and radiation, and as a preliminary a study was made of the effects of placing different types of cover over a heat-flow disc, and also of implanting the disc under the skin (McGinnis & Ingram 1974). The implanted disc was unsatisfactory because during vasodilatation there was a high blood flow in the skin above the disc, and the heat flow through the disc was thus by-passed. The effect of various covers was investigated by placing excised skin on a heated block, and comparing heat flow in a disc placed between the skin and the block with heat flow in a disc placed on top of the skin. The discs were exposed to a solar load in the region of 800 W m^{-2} while protected from air movement by a polyethylene tent. A bare disc gained more heat than the disc under the skin, while the addition of pink surgical tape resulted in a large increase in the amount of reflected solar radiation. When pig hairs were glued onto the surgical tape the heat loss decreased to approximately the value indicated by the disc under the excised skin. The pelts of other animals were also studied, and it was concluded that an artificial pelt made by clipping hair from the animal under investigation provided the best surface covering for a heat-flow disc used to measure heat flow under conditions of high solar radiation. This technique allows the possibility of studying the heat balance of many animals in an outdoor environment, although because the disc must hinder evaporation of moisture it is not suitable for use in a hot environment on animals which sweat.

References

BALDWIN B.A. & INGRAM D.L. (1967) Behavioural thermoregulation in pigs. *Physiol. Behav.* **2**, 15–21.

BARTHOLOMEW G.A. (1964) Homeostasis in the desert environment. In *Homeostasis and feedback mechanisms*, Soc. exp. Biol. symposium no. 18. Cambridge University Press.

BLUM H.F. (1961) Does the melanin pigment of human skin have adaptive value? *Q. Rev. Biol.* **36**, 50–63.

BOGERT C.M. (1959) How reptiles regulate their body temperature. *Scient. Am.* **200**, No. 4, 105–20.

Bonsma J.C. & Pretorius A.J. (1943) Influence of colour and coat cover on adaptability of cattle. *Farming in South Africa*, vol. 18, pp. 101–20. February issue.

Burton A.C. & Edholm O.G. (1955) *Man in a cold environment*. Edward Arnold, London.

Carlisle H.J. & Ingram D.L. (1973) The effects of heating and cooling the spinal cord and hypothalamus on thermoregulatory behaviour in the pig. *J. Physiol., Lond.* **231** 353–64.

Cena K. (1974) Radiative heat loss from animals and man. In *Heat loss from animals and man*, ed. J.L. Monteith and L.E. Mount. Butterworths, London.

Clapperton J.L., Joyce J.P. & Blaxter K.L. (1965) Estimates of the contribution of solar radiation to the thermal exchanges of sheep at a latitude of 55° north. *J. agric. Sci., Camb.* **64**, 37–49.

Corbit J.D. (1970) Behavioural regulation of body temperature. In *Physiological and behavioural temperature regulation*, eds. J.D. Hardy, A.P. Gagge and J.A.J. Stolwijk. C.C. Thomas, Springfield, Illinois.

Dawson T.J. (1972) Thermoregulation in Australian Desert Kangaroos. In *Comparative Physiology of Desert Animals*, ed. G.M.O. Maloiy. Academic Press, London.

Finch V.A. (1972) Energy exchanges with the environment of two east African Antelopes, the eland and the Hartebeest. In *Comparative Physiology of Desert Animals*, ed. G.M.O. Maloiy. Academic Press, London.

Gates D.M. (1962) *Energy exchange in the biosphere*. Harper and Row, New York.

Hamilton W.J. (1973) *Life's colour code*. McGraw.

Hamilton W.J. & Heppner F. (1967) Radiant solar energy and the function of black homeotherm pigmentation: an hypothesis. *Science N.Y.* **155**, 196–7.

Hatfield H.S. & Wilkins F.J. (1950) A new heat-flow meter. *J. Sci. Inst.* **27**, 1–3.

Heal J.W. (1974) A physiological radiotelemetry system using mark-space ratio modulation of a square wave sub-carrier. *Med. Biol. Engng.*, November, 843–9.

Heal J.W., Ingram D.L. & Legge K.F. (1970). Measurements of heat loss from the skin by radiotelemetry. *J. Physiol., Lond.* **210**, 123P.

Hutchinson J.C.D. & Brown G.D. (1969) Penetrance of cattle coats by radiation. *J. appl. Physiol.* **26**, 454–64.

Ingram D.L. (1964) The effect of environmental temperature on body temperature, respiratory frequency and pulse rate in the young pig. *Res. vet. Sci.* **5**, 348–56.

Ingram D.L., Heal J.W. & Legge K.F. (1974) Heat loss from unrestrained pigs in an outdoor environment. *Comp. Biochem. Physiol.* **50**, 71–6.

Ingram D.L. & Legge K.F. (1970) The thermoregulatory behaviour of young pigs in a natural environment. *Physiol. Behav.* **5**, 981–7.

Ingram D.L. & Mount L.E. (1975) *Man and animals in hot environments*. Springer Verlag, N.Y.

Kelly C.F., Bond T.E. & Heitman H., Jr. (1954) The role of thermal radiation in animal ecology. *Ecology* **35**, 562–9.

Kovarik M. (1964) Flow of heat in an irradiated protective cover. *Nature, Lond.* **201**, 1085–7.

Legge A.J. (1972) Cave climates. In *Papers in Economic Prehistory*, ed. E.S. Higgs, pp. 97–103. Cambridge University Press.

Mount L.E. (1968) *The Climatic Physiology of the Pig*. Edward Arnold, London.

Monteith J.L. (1973) *The Principles of Environmental Physics*. Edward Arnold, London.

McGinnis S.M. & Ingram D.L. (1974) The use of heat flow meters to estimate the rate of heat loss from animals. *J. appl. Physiol.* (in press).

O'Sullivan P.E. (1974) Criteria for the thermal control of buildings. In *Heat loss from animals and man*, ed. J.L. Monteith and L.E. Mount. Butterworths, London.

ØRITSLAND N.A. (1971) Wavelength-dependent solar heating of harp seals (*Pagophilus groenlandicus*). *Comp. Biochem. Physiol.* 40A, 359–61.

RHOAD O.A. (1940) Absorption and reflection of solar radiation in relation to coat colour in cattle. *Proc. Amer. Soc. Anim. Prod.* 33, 291–3.

STOWER W.J. & GRIFFITHS J.F. (1966) The body temperature of the desert locust (*Schistocera gregaria*). *Entomol. Exptl. Appl.* 9, 127–78.

SCHMIDT-NIELSEN K. (1964) *Desert Animals.* Oxford University Press.

SCHMIDT-NIELSEN K., TAYLOR C.R. & SHKOLNIK A. (1972) Desert snails: problems of survival. In *Comparative Physiology of Desert Animals*, ed. G.M.O. Maloiy. Academic Press, London.

STEWART R.E. (1953) Absorption of solar radiation by the hair of cattle. *Agric. Engng. St. Joseph, Mich.* 34, 235–8.

TEMPLETON J.R. (1970) Reptiles, in *Comparative Physiology of Thermoregulation*, ed. G.C. Whittow. Vol. I. Academic Press, New York.

THOMPSON M.L. (1955) Relative efficiency of pigment and horny layer thickness in protecting the skin of Europeans and Africans against solar ultra-violet radiation. *J. Physiol., Lond.* 127, 236–46.

UNDERWOOD C.R. & WARD E.J. (1966) The solar radiation area of man. *Ergonomics* 9, 155–68.

WHITE F.N. (1973) Temperature and the Galapagos marine iguana: insights into reptilian thermoregulation. *Comp. Biochem. Physiol.* 45A, 503–13.

WEISS B. & LATIES V.G. (1961) Behavioral thermoregulation. *Science, N.Y.* 133, 1338–44.

Questions and discussion

Adaptive significance of colour

Professor W. N. McFarland said that the importance of energy balance should not be overemphasized: it was only one among many adaptive influences to which the colour of animals might be subject. Sometimes these influences might compete with one another. The topic was dealt with by Hamilton (1973).

Dr. J. A. Clark argued that it was dangerous to extrapolate to wild animals from observations on domestic animals, on the grounds that the heat balance was likely to be less important relative to other influences on the colour of the former.

McFarland mentioned the wide range of behavioural responses shown by lizards, some of which changed colour at particular times of day. He pointed out the scope for behavioural mechanisms provided by habitats with highly heterogeneous radiation environments.

Dr. Ingram said that in man black skin pigment interfered with the formation of Vitamin D in temperate latitudes; this was probably an adaptive disadvantage.

Size of butterflies in relation to the radiation environment

Dr. P. S. Lloyd observed that in some European butterflies there were series of subspecies or closely-related species that became smaller in high latitudes or at high altitudes or both. He suggested that it might be advantageous for a poikilothermic animal to be small if the radiation load was small and fluctuating. Ingram replied that the difference in thermal capacity between a large butterfly and a small one was unlikely to be sufficient to make much difference to the rate of temperature change.

Differences in colour between hair and skin, or between base and tip of hair

Professor J. L. Monteith asked whether anything was known of a correlation between the pigmentation of hair and that of the underlying skin. An animal with white fur and black skin might suffer a greater heat load than one with black fur and black skin because in the former case radiation striking the fur would be scattered forward towards the skin and absorbed by it. Ingram replied that the polar bear, which had white fur and pigmented skin, might in consequence benefit from increased radiant heating by the sun. In animals with black fur, such as the seals he had himself studied, the only radiation emitted by the fur would be long-wave and in consequence the skin colour would not matter.

In further discussion, Ingram suggested that loose hairs with white tips and black shanks should be efficient in reducing the heat load, some radiant heat being reflected from the white tips and the remainder absorbed by the shank and lost by convection. But Monteith considered that hair which was unpigmented at the tip would have little advantage over an all-black hair because most of the radiation would still be scattered downwards rather than reflected.

ADDENDA

Six of the contributors to *Light as an Ecological Factor* (1966) prepared the following addenda which cover some aspects of work that has been carried out since the publication of their original chapters.

23 Field measurements of energy in the 0·4–0·7 micron range II

G. SZEICZ *Geography Department, University of Toronto, Canada*

Introduction

Recent estimates of the proportion of photosynthetically active radiation in the solar spectrum confirm that the earlier measurements reported by Szeicz (1966) were in error. The error arose because of unnoticed difficulties in the calibration of the filtered solarimeter. After the sensors were rebuilt and carefully recalibrated, extended measurements were made, and the results and their comparison with calculation from theoretical models have been described in 'Solar radiation for plant growth' (Szeicz 1974). This addition summarizes the salient points of these recent results, and is intended to bring the question of the dependence of the photosynthetically active fraction of solar radiation on weather up to date.

For our present purpose the quality of the solar spectrum is described by the ratio-symbol η—of the active part to the total received:

$$\eta = \int_{0\cdot3}^{0\cdot7} E_\lambda d\lambda \Big/ \int_{0\cdot3}^{3\cdot0} E_\lambda d\lambda,$$

with the component values of the ratio distinguished as η_T for the total, η_I for the direct beam and η_D for the diffuse irradiance from the sky and clouds. For brevity, in what follows, it is sufficient to restrict the presentation to summarizing the three main sections of the analysis.

Theoretical calculations

The analysis of the calculations by Avaste *et al.* (1962) and Avaste (1967) are given in Table 23.1, where the active fraction is shown as function of turbidity T_0, atmospheric precipitable water w, and solar elevation β. For scale, and as extra information, the value of total irradiance on a horizontal surface, T, and the calculated value of the ratio of diffuse to total radiation D/T is also given. The first combination represents exceptionally clean and dry air; the second is for the very clean and dry polar air masses; and the third is representative of the average summer cloudless days.

Table 23.1. Estimated values of η_T and D/T (in per cent), and total irradiance T (in W m^{-2}) for clear skies, as a function of atmospheric conditions and solar elevation.

					T_o				
		0·2			0·3			0·5	
					W (mm)				
		5			21			30	
β	η_T	T	D/T	η_T	T	D/T	η_T	T	D/T
10	45	130	37	49	100	50	51	80	74
15	46	220	29	48	170	40	49	150	60
20	48	320	24	50	270	36	50	230	50
30	49	520	18	52	460	27	51	410	40
42	50	730	15	53	660	22	52	610	34
90	51	1170	12	53	1090	18	53	1030	27

The range of η_T is small and the values lie between 0·45 and 0·53 for a very wide range of turbidity and solar elevation, β, because diffuse radiation makes up for almost all the short-wave radiation lost by scattering and absorption from the direct beam. This compensating effect is very noticeable when the effect of β alone is considered. With no change of air-mass the model calculations indicate that the variation of η_T should be remarkably small with time of day.

Measurements at Locarno-Monti in Switzerland

Composition in the direct beam, and the combined estimate

Observations by the Swiss Meteorological Observatory at Locarno-Monti (580 m above M.S.L.) on the southern side of the Alps (Ambrosetti, Schram & Thams, 1968) began in 1944, first measuring only the total spectrum of the direct beam; but from 1958 readings were made with the RG8 (now known as RG 695) filter as well. Readings were taken only when the atmosphere appeared very clean and dust-free, in the absence of clouds, and these occasions occurred mostly in the descending dry northerly air-streams (föhn conditions). There were measurements on 1,988 days over an 8-year period 1958–66. The details are given by Szeicz (1974) but Fig. 23.1 shows values of η_I plotted against solar elevation β in two groups of months covering the whole year. The lower curve is for the very clear and dry winter and springtime air, whereas the upper one is for the moister and more turbid air masses of June, July, August and September. When solar elevation is greater than 20°, η_I in dry air lies between 0·41 and 0·47 whereas in the moister summer air it is between 0·45 and 0·49. Comparing the upper with the lower curve, the effect of enhanced scattering by aerosol and infra-red attenuation by water vapour is obviously more important when the sun is low; at $\beta = 10°$, the change is from 0·35 to 0·41, a 15% increase, whereas at $\beta = 60°$ the change is from 0·465 to 0·49 an increase of only 5%.

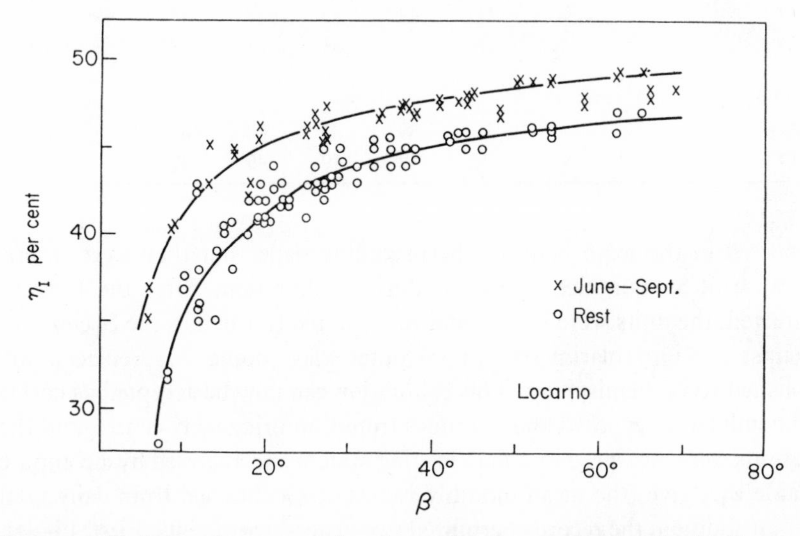

Figure 23.1. Variation of η_I on very clear days as function of solar elevation β at Locarno-Monti in Switzerland. Curves fitted by eye, ×, June to September; O rest of the year.

These measurements give a very good estimate of η_I as a function of β, and an acceptable measured value of η_D can be obtained from measurements of Bodmann & Jentzen (1964) and Henderson & Hodgkiss (1963). The analysis of their spectrophotometric measurements gives $\eta_D = 0·70 \pm 0·05$, and the two can be combined by weighting according to the measured ratios of diffuse to total radiation given by Dognieaux (Monteith 1962) for very clear days in Belgium. Table 23.2 gives the details. The values again emphasize the constancy of η_T with β and the comparison with the purely theoretical estimates of Table 23.1, especially for $\beta > 20°$, shows very good agreement.

The Cambridge measurements

The previous estimates give good values for clear days, but in ordinary weather hemispherical measurements from the full sky are needed. Attempts were already made in 1963 to record the active fraction, and these are

Table 23.2. Combined estimate of the active fraction for clear skies using measured values of η_I, η_D and D/T (values in per cent).

	β					
	10	15	20	30	40	60
	D/T					
	42	32	25	20	17	15
Summer						
η_I	41	44	45	47	48	49
η_T	53	52	51	51	52	52
Winter, Spring and autumn						
η_I	35	39	42	44	45	
η_T	50	49	49	49	49	

reported in the main body of the preceding paper but they were in error because of an unnoticed problem during calibration. After the fault was detected, the units were rebuilt and recalibrated (for details see Szeicz 1974) against a Kipp solarimeter, whose outer glass dome was replaced by a polished RG 8 hemisphere. This calibration can now be accepted as correct. The units were exposed about 3 miles from Cambridge city centre, and their outputs were measured by a data logging system and analysed by a computer. Table 23.3 gives the mean monthly values of η_T obtained from daily totals.

In addition, the records permitted two general treatments. First, Fig. 23.2

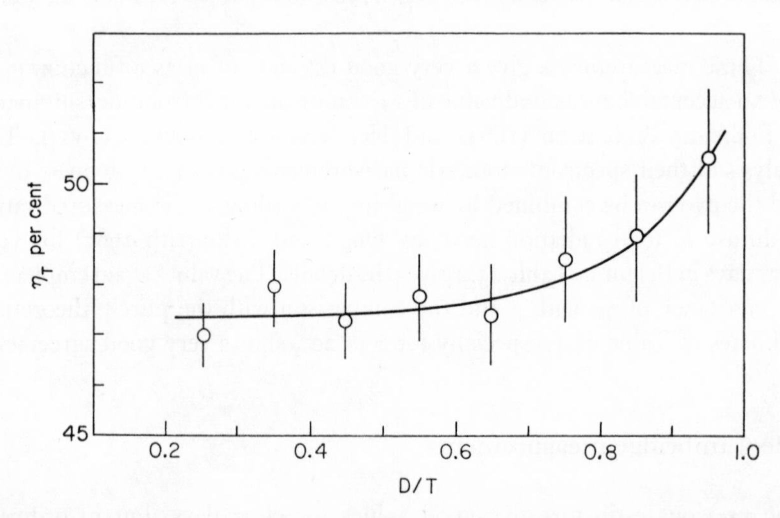

Figure 23.2. Variation of η_T with the ratio of diffuse to total solar radiation at Cambridge, England. Values calculated from daily means. Bars indicate standard deviation for each group.

Table 23.3. Cambridge results in average weather. Values in per cent. The last row gives the number of days on which measurements were made in each month.

	Jan.	Feb.	Mar.	Apr.	May	June	July	Aug.	Sept.	Oct.	Nov.	Dec.
η_T	50	51	51	48	47	48	49	49	49	48	50	50
D/T	70	71	53	63	64	62	80	58	56	54	64	55
No. of days	7	8	6	13	8	7	10	9	14	8	5	7

shows the variation of the daily values of η_T with cloudiness represented here by the fraction of diffuse radiation in the total. The plot shows that moderate cloudiness does not affect bulk spectral composition because solar radiation reflected from the sides of the smaller cumulus clouds is spectrally unmodified, and when this is combined with the still intense direct beam, η_T remains unchanged at 0·48 up to about D/T = 0·7. Beyond that, however, η_T increases and in full overcast it reaches 0·51–0·52.

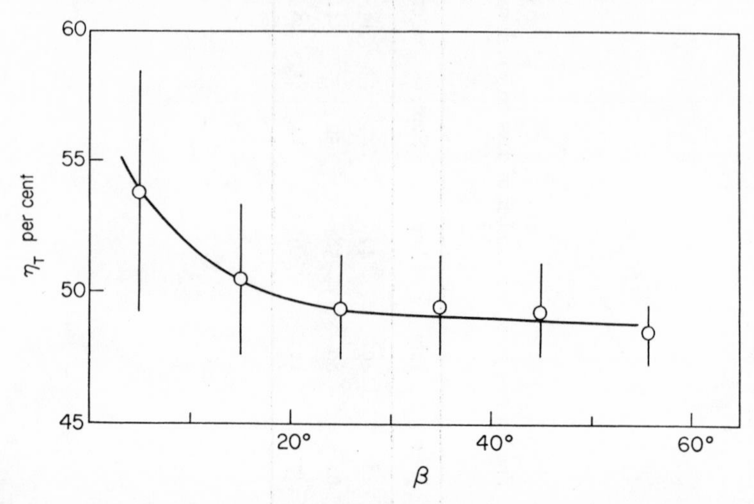

Figure 23.3. Variation of η_T with solar elevation β at Cambridge, England. Group means, calculated from hourly means, are for all conditions of the sky. Bars indicate standard deviation for each group.

The second way of treating the data is shown in Fig. 23.3, where the unselected hourly values of η_T are plotted against solar elevation. As the standard deviations show, the scatter is large, but the class means do not change significantly above $\beta = 10°$, and in this region $\eta_T \approx 0{\cdot}49 \pm 0{\cdot}02$. When $\beta < 10°$, $\eta_T \approx 0{\cdot}53 \pm 0{\cdot}04$, but because of cosine error the values here are probably less reliable. Overall, however, the analyses of the Cambridge measurements strongly support the evidence from the theoretical calculations and from the combined estimates that when the mixture of diffuse and direct solar radiation is considered, bulk spectral composition remains very close to that of the extraterrestrial solar spectrum (η for this = 0·47), and changes little with weather.

References

Ambrosetti F., Schram K. & Thams J.C. (1968) Die Intensität der directen Sonnenstrahlung in verschiedenen Spektralbereichen in Locarno-Monti. *Publ. Swiss met. Inst. No. 7.*

AVASTE O. (1967) Radiation flux divergence in the atmosphere and the flux of total radiation above the surface of the sea. *Akad. Nauk. Est. SSR. Inst. Phys. Issl. Rad. Rez. Atm.* pp. 5–42

AVASTE O., Moldau H. & Shiffrin H. (1962) Spectral distribution of direct and diffuse radiation. *Akad. Nauk. Est. SSR. Inst. Phys. Astron. No. 3.*

BODMANN H.W. & JENTZEN R. (1964) Ein registrierendes Spectralradiometer für Lichtquellen. *Lichttechnik*, 16, 20.

HENDERSON S.T. & HODGKISS D. (1963) The spectral energy distribution of daylight. *Br. J. appl. Phys.* 14, 125.

MONTEITH J.L. (1962) Attenuation of solar radiation: a climatological study. *Q. Jl. R. Met. Soc.* 88, 508–21.

SZEICZ G. (1966) Field measurements of energy in the 0·4–0·7 micron range. In *Light as an ecological factor* (eds R. Bainbridge, G.C. Evans and O. Rackham) pp. 41–52. Blackwell Scientific Publications, Oxford.

SZEICZ G. (1974) Solar radiation for plant growth. *J. appl. Ecol.* 11, p. 617–36.

24 Light measurements in the sea in terms of quanta

N. G. JERLOV *Institute of Physical Oceanography, Copenhagen University, Denmark*

This paper will consider the development of topics dealt with in an earlier article (Jerlov 1966). As mentioned, Working Group 15 (IAPSO, SCOR) has recommended that light measurements in photosynthetic studies be made in terms of quanta within the range 350–700 nm. A great deal of work has been concentrated on designing suitable irradiance meters for recording underwater quanta. The first quantum meter devised by Jerlov & Nygård (1969) is a fairly simple instrument which has the advantage that it can easily be adapted to measuring energy instead of quanta.

The direct measurement of quanta involves some difficulties. With the meter in the uppermost layers wave action causes large fluctuations in the signal due to the strong absorption of longwave light (600–700 nm). In consequence, accurate irradiance values especially just below the surface are not always secured. The wave disturbance can be practically eliminated by the use of an integrating voltmeter; on the other hand this considerably increases the cost of the equipment.

The attenuation of monochromatic light in the ocean is well illustrated (Jerlov 1966). The behaviour of quanta attenuation is somewhat different. Some examples of depth profiles of quanta irradiance for different water masses are given in Fig. 24.1. These experiments were conducted at a solar elevation of 35° off Iceland, and at elevations of 60–70° at the other stations. A common feature in the log-curves is the fast decrease in the surface stratum due to the absorption effect mentioned. For the most clear water (south of Sardinia) another marked curvature appears around 50 m, identical with that characteristic for blue light. This gradually disappears when the water becomes more turbid, and for Gibraltar Strait the curve below the surface layer is a straight line.

Attention is called to the large span between the two extreme curves in Fig. 24.1. The 10% level of surface irradiance is at 8 m off West Africa, and at 40 m south of Sardinia; in the clearest ocean water it is found as deep as

50 m. This family of curves bears witness to the large variation in the thickness of the photic zone in the ocean.

A somewhat controversial subject has been the influence of solar elevation on the attenuation of irradiance in the sea. In the upper layers maximum light is in the apparent direction of refracted sunrays. It follows that quanta have a longer path to traverse and are correspondingly more attenuated with a low than with a high sun. With increasing depth the light distribution is gradually transformed into a less directed character on account of the continuous creation of scattered light, and the irradiance attenuation tends to a value independent of solar elevation.

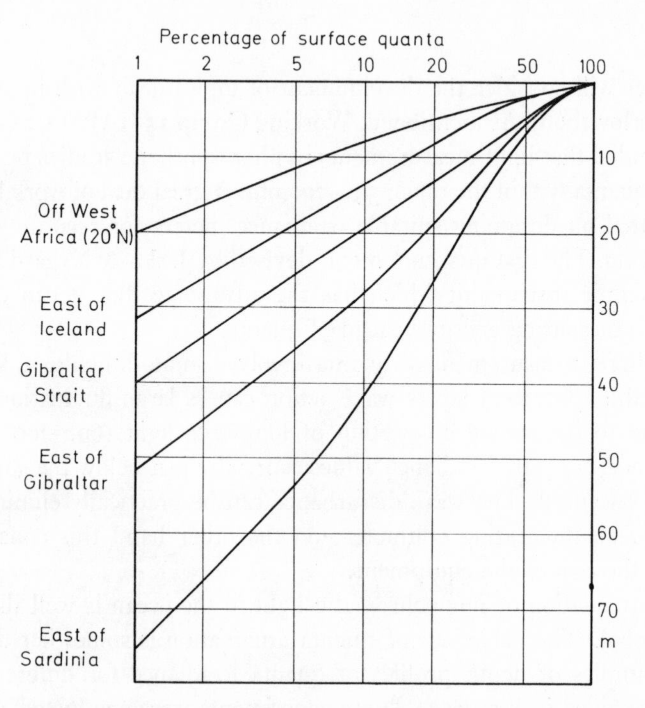

Figure 24.1. Depth profiles of quanta irradiance (350–700 nm) in percent of surface value.

It is recalled that the vertical attenuation coefficient K is generally described by:

$$E = E_0 e^{-Kz}$$

where E_0 is the irradiance just below the surface.

As a first approximation it can be assumed that total quanta are transmitted to a depth z in the surface stratum through the refraction angle j for sunlight defined by the equation

$$\frac{\sin(90-h_s)}{\sin j} = \frac{4}{3}$$

where h_s is the solar elevation.
Hence, if K_o is the coefficient for zenith sun

$$E = E_o e^{-K_o z/\cos j}$$

(Jerlov & Nygård 1969) which approximately describes the influence of solar elevation in surface water. It may be added that $\cos j = 1$ for $h_s = 90°$ and $\cos j = 0{\cdot}76$ for $h_s = 30°$.

On the other hand, Højerslev (1974) has found from direct observations in the Mediterranean that the 10% level of surface quanta changes from 38 to 30 m when the elevation goes from 72° to 30°; below 30° it is little affected. In view of these results the worker should consider whether due regard is being paid to the solar elevation effect in order to secure high accuracy in his measurement.

It is possible indirectly to arrive at a determination of underwater quanta by measuring blue light only. This simple method rests on the assumption that for ocean water of optical types I–III the ratio of quanta to the blue (465 nm) is the same function of depth. For details the reader is referred to Jerlov (1974a).

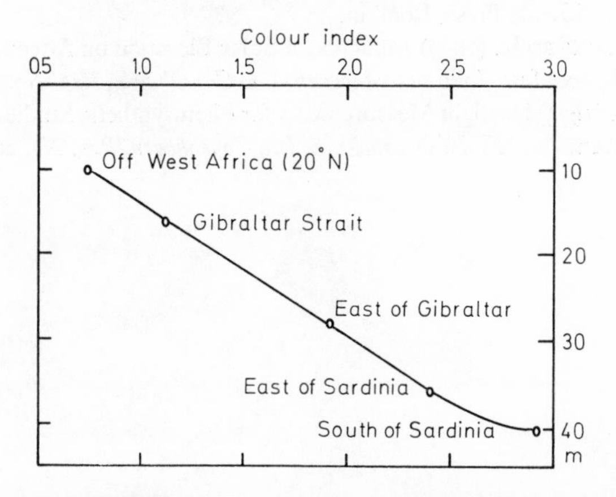

Figure 24.2. Relationship between colour index (just below the surface) and depths at which the percentage of surface irradiance of quanta (350–700 nm) is 10%.

Finally an aspect of light measurement will be considered which may lead to advantageous practical applications. The attenuation of number of quanta is determined by processes of scattering and absorption. The down-welling light is to some degree scattered upwards and the water imposes on this upwelling light a colour which is characteristic of the water mass. Jerlov (1974b) has introduced a colour index defined as the ratio of blue (450 nm) to green (520 nm) light upwelling through a cone of 20°. This index which is independent of solar elevation if above 15° can be observed in a few minutes from shipboard.

It seems logical to hypothesize that the index is closely related to the trend of the depth profiles in Fig. 24.1. The assumption is supported by the relationship exhibited in Fig. 24.2 between the colour index (just below the surface) and the depth at which the percentage of surface quanta is 10%: the plots represent mean values. The clear association between these two quantities demonstrates that a simple colour meter can be used as a tool for indirect evaluation of photosynthetic light parameters.

References

JERLOV N.G. (1966) Aspects of Light Measurement in the Sea. In *Light as an Ecological Factor*, editors R. Bainbridge, G.C. Evans and O. Rackham, pp. 91–8. Brit. Ecol. Soc. Symp. No. 6.

JERLOV N.G. (1974a) A Simple Method for Measuring Quanta Irradiance in the Ocean. *Inst. Fysisk Oceanografi, Un. Copenhagen, Rep. No. 24*, 10 pp.

JERLOV N.G. (1974b) Significant Relationships between Optical Properties of the Sea. In *Optical Aspects of Oceanography*, editors N.G. Jerlov and E. Steemann Nielsen, pp. 77–94. Academic Press, London.

JERLOV N.G. & NYGÅRD K. (1969) Influence of Solar Elevation on Attenuation of Underwater Irradiance. *Inst. Fysisk Oceanografi, Un. Copenhagen, Rep. No.* 4, 9 pp.

HØJERSLEV N.K. (1974) Daylight Measurements for Photosynthetic Studies in the Western Mediterranean. *Inst. Fysisk Oceanografi, Un. Copenhagen, Rep. No.* 26, 38 pp.

25 Shade avoidance and shade tolerance in flowering plants II. Effects of light on the germination of species of contrasted ecology

J. P. GRIME[1] and B. C. JARVIS
Department of Botany, University of Sheffield

Introduction

In the paper which appeared under this title in Volume I an attempt was made to interpret the influence of light on plant distribution by reference to measurements of the growth-rate and morphogenesis under shaded conditions of species of woodland, grassland and open habitats. It is clear, however, that the effects of light on the ecology of flowering plants are mediated through a variety of physiological responses and in this short addendum attention is turned to the influence of light on the germination of seeds.

In natural vegetation, germination frequently occurs in circumstances in which the radiation impinging on the seed has been filtered by a leaf canopy or by soil constituents such as litter, humus or mineral particles. These filters both reduce the intensity of radiation and modify its spectral composition. It seems necessary therefore, in the attempt to understand the influence of shade upon the germination of seeds in the field, to carry out investigations in which both intensity and spectral composition are experimental variables. This paper describes some preliminary experiments carried out using seeds of species of common occurrence in the Sheffield area.

Effects of light intensity on germination

Procedure

In order to make a broad comparison between species of contrasted ecology with regard to the effects of light intensity on germination, seeds of a large number of species collected from a wide range of habitats were subjected to a simple screening programme in which germination was recorded under

[1] Unit of Comparative Plant Ecology (NERC).

three standardized light conditions which henceforth are referred to as 'light', 'shade' and 'dark'. In the 'light' treatment, five replicate transparent dishes, each containing 20 seeds, were placed in a growth-room in which a radiation intensity of 4,045 μW cm^{-2} was provided by warm-white fluorescent tubes supplemented by tungsten bulbs over a daylength of 16 h. The temperature was 20°C during the day, 15°C at night. Seeds were sown on Whatman No. 1 filter-paper moistened with distilled water and germinated seeds were counted and removed daily. Conditions and procedure in the 'shade' treatment were identical to that in 'light' except that the intensity of radiation was reduced to 96·8 μW cm^{-2} (2·4% of 'light' treatment) by means of a neutral filter. In the 'dark' treatment the seeds were placed in a light-proof box which was opened briefly each day under a low intensity of green light in order to record germination. This brief illumination of the 'dark' seeds by green light causes a promotion of germination in certain species. However, experiments with seeds subjected to continuous darkness have shown that, with few exceptions, these effects are small in magnitude.

Results

A detailed account of the results, involving statistical comparisons of germination percentage and rate under the three treatments is in preparation. Here attention is confined to some of the commoner species, broadly classified into (1) grassland plants, (2) woodland plants and (3) species colonizing bare ground. These results (Table 25.1) show that, in the majority of the species, the highest percentage germination was obtained in the 'shade' treatment. In some species an inhibition was recorded in the 'light' treatment and this effect was most pronounced in one species, *Urtica urens*.[1] In three species, *Chenopodium album*, *Dipsacus fullonum* and *Juncus effusus*, all of which are colonists of bare soil or mud, germination reached a maximum in the 'light' treatment.

One of the most striking features of the results was the ability of the grasses, with only a small number of exceptions, e.g. *Brachypodium* spp., *Melica nutans*, *Milium effusum*, to germinate in darkness. Rapid germination in the 'dark' treatment was recorded in grasses from a wide variety of habitats. It is interesting to note that, in this respect, annual grasses, e.g. *Bromus sterilis* and *Hordeum murinum*, appear to differ markedly from the annual forbs, e.g. *Chenopodium* spp. and *Tripleurospermum maritimum*, with which they occur in disturbed habitats.

[1] Prolonged illumination and high intensity are known to cause inhibition of some dark-germinating seeds through the so-called high energy reaction (HER) which overrides reversible phytochrome effects (see e.g. Toole 1973).

Table 25.1. The effects of light regime on the germination of seeds of species of grassland, woodland and open habitats. The values tabulated refer to the % of seeds germinated after 30 days. For other details, see text.

(a) Grassland species

	Dark	Shade	Light
GRASSES			
Agrostis tenuis	54	83	76
Alopecurus geniculatus	99	100	100
Anthoxanthum odoratum	82	94	86
Arrhenatherum elatius	94	94	96
Brachypodium pinnatum	16	88	90
Briza media	85	93	79
Bromus erectus	87	90	91
Cynosurus cristatus	100	100	100
Dactylis glomerata	39	84	78
Festuca arundinacea	90	88	94
Festuca ovina	90	82	86
Festuca rubra	96	98	98
Holcus lanatus	58	98	100
Lolium perenne	100	100	100
FORBS			
Achillea millefolium	68	72	62
Bellis perennis	100	100	100
Centaurea nigra	3	95	90
Chrysanthemum leucanthemum	83	100	100
Helianthemum chamaecistus	86	94	98
Hieracium pilosella	91	97	95
Hypericum pulchrum	0	65	75
Hypochaeris radicata	81	97	97
Leontodon hispidus	7	95	93
Luzula campestris	71	100	82
Plantago lanceolata	25	88	64
Poterium sanguisorba	75	96	95
Rumex acetosella	58	67	48
Rumex acetosa	91	100	99
Thymus drucei	75	98	96
Tragopogon pratensis	90	100	86
Veronica officinalis	10	100	88

(b) Species of woodland and other shaded habitats

	Dark	Shade	Light
GRASSES			
Agropyron caninum	80	96	96
Brachypodium sylvaticum	22	90	94
Holcus mollis	97	90	90

(b) Species of woodland and other shaded habitats (continued)

	Dark	Shade	Light
Hordelymus europaeus	69	80	64
Melica nutans	10	98	100
Milium effusum	34	88	100
FORBS			
Cardamine flexuosa	59	90	91
Digitalis purpurea	56	100	96
Geum urbanum	4	81	82
Hedera helix	84	84	76
Moehringia trinervia	80	98	98
Mycelis muralis	39	97	89
Myosotis sylvatica	75	91	96
Silene dioica	86	97	97

(c) Species colonizing bare ground in unshaded habitats

	Dark	Shade	Light
GRASSES			
Bromus sterilis	100	100	100
Catapodium rigidum	94	98	100
Hordeum murinum	100	100	100
Poa annua	81	100	100
Poa compressa	88	98	94
FORBS			
Arctium minus	0	94	90
Blackstonia perfoliata	15	92	91
Carduus acanthoides	14	92	97
Centaurium erythraea	1	92	97
Chenopodium album	4	48	66
Chenopodium rubrum	1	100	100
Conyza canadensis	2	100	100
Crepis capillaris	8	100	99
Dipsacus fullonum	5	49	71
Juncus articulatus	0	93	94
Juncus bufonius	0	97	98
Juncus effusus	7	63	77
Linaria vulgaris	0	68	66
Matricaria matricarioides	2	58	47
Myosotis arvensis	4	100	94
Rumex crispus	1	100	93
Rumex obtusifolius	0	98	87
Sagina procumbens	1	100	100
Scleranthus annuus	23	64	57
Senecio squalidus	45	97	70
Senecio viscosus	0	98	95

Senecio vulgaris	0	100	79
Sonchus asper	4	98	100
Sonchus oleraceus	28	68	57
Tripleurospermum maritimum	0	89	89
Tussilago farfara	100	100	100
Urtica urens	29	89	3
Veronica persica	67	100	100

Effects of spectral composition on germination

Procedure

In a second screening programme germination was examined in a variety of species placed in darkness and in four spectral regions (far-red, red, green and blue) each provided at two intensities (6 μW cm^{-2} and 28 μW cm^{-2}) over a 16 h day at 20°C, with an 8 h night at 15°C. Seeds remained undisturbed until the end of the experiment, the length of which was adjusted to exceed that found to be sufficient for germination in previous experiments with unfiltered white light.

Results

The results indicate that, regardless of habitat, each of the species examined was inhibited in its germination by far-red radiation. In the majority of species there was clear evidence of a promotion of germination by the other light treatments, the order of efficiency being red > green > blue, a result consistent with phytochrome-mediated responses (Butler, Hendricks & Siegelman 1964). In Table 25.2, the four colonists of bare ground in unshaded habitats are conspicuous in terms of the comparatively high intensity of radiation which was required to induce germination. The two species of *Juncus* appear to be particularly light-demanding in that germination was virtually confined to the high intensity red-light treatment. In marked contrast, the species of grassland and woodland tended to germinate at the low intensity of red and green light and in most treatments became light-saturated at the higher intensity.

Discussion

On the basis of the limited amount of data presented here or available from other published sources it may be unwise to attempt to predict the part

Table 25.2. The effect of light of different wavebands upon the germination of seeds of species of grassland, woodland and open habitats. The values tabulated refer to the % of seeds germinated. The duration of each experiment in days is indicated by the number following the name of each species. Filters were supplied by Rank Strand Electronics Ltd. and were Deep Blue (No. 20), Primary Green (No. 39), Ruby Red (No. 14). A far-red filter was obtained by combining Nos. 14 and 20. The light source was warm-white fluorescent tubes supplemented by tungsten in the far-red treatment. Equal intensities in the various experimental containers were obtained by means of various thicknesses of muslin. For other details, see text.

	SPECIES		Low Intensity					High Intensity				
			Dark	Far-red	Red	Green	Blue	Dark	Far-red	Red	Green	Blue
	Lolium perenne	(3)	52	50	44	42	54	48	0	50	52	36
A	*Bromus erectus*	(8)	58	60	52	76	58	67	0	71	69	54
	Luzula campestris	(14)	1	1	86	31	0	1	0	86	50	19
	Rumex acetosella	(8)	14	24	50	44	48	16	3	56	59	38
	Moehringia trinervia	(8)	50	14	54	50	36	51	0	59	53	29
B	*Mycelis muralis*	(8)	0	4	40	12	0	0	0	45	47	9
	Epilobium montanum	(8)	0	0	64	68	0	0	0	72	64	0
	Silene dioica	(5)	40	—	91	57	51	42	1	85	69	50
	Juncus effusus	(14)	0	0	2	0	0	0	0	46	0	0
C	*Juncus articulatus*	(14)	0	0	2	0	0	0	0	42	0	0
	Chenopodium rubrum	(14)	0	1	0	0	0	0	1	62	12	0
	Senecio viscosus	(6)	0	0	33	0	0	0	1	74	63	0

A—Grassland species.
B—Species occurring in shaded habitats.
C—Species colonizing bare ground in unshaded habitats.

which responses to light play in determining the fate of seeds dispersed naturally into grassland, woodland or open habitats. Particular difficulties arise from the fact that in the screening experiments described in this paper it is uncertain to what extent the single seed population sampled is representative of each of the species investigated. Moreover, in natural environments, factors other than light intensity and spectral composition frequently intervene as major determinants of the location, timing and success of germination.

However daunting these complications they do not obscure the fact that Tables 25.1 and 25.2 contain evidence of differences in response between species of open and closed vegetation. In Table 25.1 it is apparent that a very high proportion of the species colonizing bare ground did not germinate effectively in darkness, a feature which is consistent with the persistence of seeds of many ruderals in an ungerminated condition for long periods following burial in the soil (see Barton 1961). It is interesting to note that, in contrast, the seeds of many of the grasses tended to germinate readily in darkness and do not appear to be incorporated into the buried seed flora of grassland soils (Spray & Grime, in preparation).

The higher light requirement for the germination of species of open habitats (Table 25.2) is again consistent with their ecology and suggests a mechanism whereby the seeds of these species might be prevented from germination below a closed canopy. However, in order to test this hypothesis, it will be necessary to carry out experiments in light regimes more nearly approximating to those beneath natural vegetation and to investigate the physiological basis for the differences in light requirement.

The control of seed germination by light has been the subject of countless investigations, particularly since Flint and McAlister (1935, 1937) reported the action spectrum for light influences on the germination of lettuce seeds. Promotion of germination occurs over a broad region of the spectrum with maximum effectiveness around 660 nm (red), while inhibition of germination results from exposure to far-red light with maximum effectiveness about 730 nm. These two effects are mediated through a single pigment, phytochrome (P), which exists in two, photoconvertible forms. One form, designated P_r, absorbs red light and is thus transformed into the second form P_{fr}. Irradiation with far-red light converts P_{fr} to P_r, although P_{fr} also reverts thermally to P_r in darkness.

A physiological explanation for the lower light requirement of seeds of grassland species and those from woodland habitats cannot be advanced on the basis of the simple experiments described here. However, Toole (1973) points out that light-stimulated seeds require a certain number of P_{fr} molecules in order to germinate and since seeds vary in their amount of total P they also differ in that proportion of P which must be converted into P_{fr}

to achieve maximum germination. If this view applies to all the species investigated, then clearly species such as *Moehringia trinervia* and *Lolium perenne* (Table 25.2) have sufficient P_{fr} initially since they germinate readily in total darkness. On the other hand, species such as *Mycelis muralis* and *Epilobium montanum* appear to have insufficient P_{fr} initially and hence show a light requirement, although this is fulfilled by the low intensity treatment. Such seeds may have a high total P content, since a seed rich in total P, at a given photoequilibrium, will produce more P_{fr} molecules than a seed low in total P. Similarly the relatively high light requirement shown by seed of the species colonizing bare ground may reflect a relatively lower total P content. However, it must be noted that differences in the light-filtering characteristics of the seed coat could account, at least in part, for some of the differences between the various species. Alternatively, species of grassland and shaded habitats may simply have a lower P_{fr} requirement to initiate germination processes.

Acknowledgements

We are indebted to Mr. A. M. Neal, Miss J. Rodman and Miss D. Smith for their assistance in the investigation, part of which was carried out with the support of the Natural Environment Research Council.

References

BARTON L.V. (1961) *Seed preservation and longevity.* Leonard Hill, London. 216 pp.

BUTLER W.L., HENDRICKS S.B. & SIEGELMAN H.W. (1964) Action spectra of phytochrome *in vitro. Photochem. Photobiol.* 3, 521–8.

FLINT L.H. & MCALISTER E.D. (1935) Wavelengths of radiation in the visible spectrum inhibiting the germination of light-sensitive lettuce seed. *Smithsonian Misc. Collection* 94, 1–11.

FLINT L.H. & MCALISTER E.D. (1937) Wavelengths of radiation in the visible spectrum promoting the germination of light-sensitive lettuce seed. *Smithsonian Misc. Collection* 96(2), 1–8.

TOOLE V.K. (1973) Effects of light temperature and their interactions on the germination of seeds. *Seed Sci. & Technol.* 1, 339–96.

26 Light, shade and growth in some tropical plants II

D. B. MURRAY *Cocoa Research Unit, University of the West Indies, Trinidad*

Since the original paper on this subject was published (Murray 1966), abundant evidence from other cocoa growing countries has confirmed the significance of the shade/mineral nutrition interaction in the growth and cropping of cocoa, *Theobroma cacao*, L. These countries include Brazil (Rosand *et al.* 1971), Ecuador (Lainez 1963) and New Guinea (Anon 1969). It is not proposed to deal any further with cropping details but attention is drawn to two aspects of light relations, the effect of shade or no shade on the soil under the cocoa and methods of measuring light intensity.

Soil deterioration

The Ghana cocoa experiment (Cunningham *et al.* 1961) was continued for 17 years of continuous cropping under the four treatments (1) shade, no fertilizer; (2) shade, fertilizer; (3) no shade, no fertilizer; and (4) no shade, fertilizer.

Soil samples had been taken at the start of the experiment in 1957 and again in 1972. In 1972 samples were also taken from an adjoining field which had been left under bush fallow for ten years. The results of the analyses have been reported by Ahenkorah *et al.* (1974) and a summary is given in Table 26.1.

Irrespective of treatment, there was a general decrease in pH over the period and this is probably associated with depletion of exchangeable bases by cropping and leaching. The Ca levels appear to show anomalies as they appear slightly higher at the end of the period as compared with the start. They are however only about one third of those found in the bush fallow. The K levels at the start are also only a half those in the bush fallow but show a fall over the period. This drop is not found in shaded plots which had been fertilized but, without shade, fertilizers did not maintain the K level.

The most startling effect however was on the levels of P which in the absence of fertilizers fell from over 20 p.p.m. to less than 1 p.p.m.

Table 26.1. Soil analyses, 0–5 cm, at the start and end of a fertilizer-shade experiment (from Ahenkorah *et al* 1974).

Treatment	pH		Exchangeable bases meq/100 g soil: Ca		Mg		K		P, p.p.m.		%C		C/N	
	1957	1972	1957	1972	1957	1972	1957	1972	1957	1972	1957	1972	1957	1972
Shade, no fertilizer	7·5	6·7	9·0	13·1	2·9	2·2	0·54	0·34	24·1	0·7	4·4	2·3	12·8	11·3
Shade, & fertilizer	7·9	6·5	7·5	10·0	2·6	1·2	0·52	0·53	25·5	21·2	4·0	1·8	12·0	9·7
No shade, no fertilizer	7·3	6·8	9·3	9·9	2·2	1·6	0·51	0·48	27·3	0·9	3·8	1·8	10·7	11·0
No shade, & fertilizer	7·2	6·7	6·7	14·0	2·3	2·4	0·42	0·34	22·5	19·3	3·6	1·9	11·9	10·5
Bush fallow	N.D.	7·6	N.D.	33·8	N.D.	4·8	N.D.	1·14	N.D.	3·4	N.D.	3·2	N.D.	7·8

N.D. Not determined.

A rather surprising result is shown in the C/N ratios. These in general show a decline but this is no greater in the absence of shade. It may be assumed that once the cocoa canopy had closed it compensated for the absence of the shade trees and that the soil was adequately shaded. The actual declines in organic carbon and nitrogen were both severe, averaging 50% for carbon and 44% for nitrogen.

Further work is needed on the complex changes that occur in the soil when tropical forest is felled, possibly burnt and the exposed land then used for man's crops. Replacement by tree crops which again shade the soil would be expected to result in reduced soil deterioration. Fertilizer regimes must be carefully planned so that imbalance does not occur after years of cropping.

Light measurements

Early work on light measurements with cocoa was usually confined to spot comparisons of light intensity under the shade trees and in the open using photographic light meters and results were reported as percentages of full sunlight. The nearest research station would usually have a Campbell Stokes recorder. Clearly for most purposes it is more useful to have measurements of total short-wave radiation, made, say, with a Kipp solarimeter, but such measurements are still available only for a limited number of sites.

In Table 26.2 are shown the monthly readings in the cocoa growing areas of Trinidad and Ecuador for both a Campbell Stokes recorder and a Kipp solarimeter. The mean number of hours of sunlight recorded by the Campbell Stokes recorder in Ecuador is only 35% of that recorded in Trinidad. However on the Kipp solarimeter Ecuador receives 75% of the monthly solar radiation received in Trinidad.

Since the last edition of the Smithsonian Meteorological Tables was published in 1951, it has been shown that the linear relationship there established between hours of bright sunshine, as measured by a Campbell Stokes recorder, and total short-wave radiation often holds well in a particular climatic region (the more limited in area, the better), during that part of the annual cycle when climatic change is not large. Outside these limits the scatter increases, until finally no relationship can be demonstrated. The data given in Table 26.2 show this point with great clarity. The fact that a reduction in hours of bright sunshine to a third or less between Trinidad and Ecuador (taking the year as a whole) can be accompanied by a reduction in total short-wave radiation of only around 25% shows that in Ecuador there must be substantial periods of light haze, reducing the intensity of direct sunlight sufficiently to prevent a mark on the Campbell Stokes record, while not making any large reduction in total short-wave radiation, the deficiency in

Table 26.2. Light measurements in Trinidad and Ecuador. Mean daily solar radiation, langleys (Kipp Solarimeter).

	Jan.	Feb.	March	April	May	June	July	August	Sept.	October	Nov.	Dec.	Mean
Trinidad	385	418	478	446	470	275	388	396	394	374	308	346	390
Ecuador	325	356	343	364	329	305	280	258	253	238	219	261	294
Hours of sunshine (Campbell Stokes)													
Trinidad	7·6	8·2	8·0	8·1	7·8	6·7	7·2	6·9	6·5	7·0	7·0	7·1	7·3
Ecuador	2·3	4·0	3·9	4·1	4·0	3·8	2·6	2·1	1·5	0·9	0·8	1·7	2·6

direct solar radiation being largely made up by the large bright aureole around the sun. Thus while under suitable restrictions the readings of Campbell Stokes recorders can give much useful information, the comparison is valuable in demonstrating the care which is needed in interpreting data derived from them. Among other consequences would be that sunflecks as such would be relatively unimportant in Ecuadorian forests, although their place would probably be taken by more diffuse radiation, having much the same spectral composition, penetrating through holes in the canopy from the bright aureole around the sun. However, this diffuse radiation would be much more difficult to characterize in quantitative terms, using methods similar to those used for measuring sunflecks (Evans 1956, Evans, Whitmore & Wong 1960).

All this must be allowed for in adjusting the tree shade in the two countries, the general conclusion being that cocoa should need less shade in Ecuador than in Trinidad. The need for adequate instrumentation for light recording in tropical countries must be emphasized.

References

ANON (1969) Shade–fertilizer trials. *Ann. Rep. 1966-67. Dept. of Agric.* Papua and New Guinea.

AHENKORAH Y., AKROFI G.S. & ADRI A.K. (1974) The end of the first cocoa shade and manurial experiment at the Cocoa Research Institute of Ghana. *J. hort. Sci.* 49, 43–51.

CUNNINGHAM R.K., SMITH R.W. & HURD R.G. (1961) A cocoa shade and manurial experiment at the West African Cocoa Research Institute, Ghana. II. Second and third years. *J. hort. Sci.* 36, 116–25.

EVANS G.C. (1956) An area survey method of investigating the distribution of light intensity in woodlands, with particular reference to sunflecks. *J. Ecol.* 44, 391–428.

EVANS G.C., WHITMORE T.C. & WONG Y.K. (1960) The distribution of light reaching the ground vegetation in a tropical rain forest. *J. Ecol.* 48, 193–204.

LAINEZ J. (1963) Fertilizacion de cacao bajo differentes condiciones ecologicas. *Ann. Rept.* 1963, I.N.I.A.P. Ecuador.

MURRAY D.B. (1966) Light shade and growth in some tropical plants. In *Light as an ecological factor* (eds R. Bainbridge, G.C. Evans and O. Rackham) Blackwell Scientific Publications, Oxford.

MURRAY D.B. (1968) Shade and light intensity. *Ann. Rep. Cacao Res.* 1967. University of the West Indies, Trinidad.

ROSAND P.C., DE MIRANDA E.R. & DO PRADO E.P. (1971) Effect of shading and fertilizer application on yield of mature cocoa in Bahia, Brazil. *Proc. 3rd Internat. Cacao Rez. Conf.* Accra, 1969.

Smithsonian Meteorological Tables (1951) Sixth revised edition. Smithsonian Miscellaneous Collections, Vol. 114. Washington D.C.

27 The enigmatic echinoids in retrospect

N. MILLOTT *Director, The University Marine Biological Station Millport, Isle of Cumbrae, Scotland*

Since publication of the original account, the problems outlined in it have received further attention, but its title, if apt in 1966, remains so.

Accounts of phototaxis remain disparate and in the realm of physiological colour change, the fundamental problems concerning the precise mechanisms that activate and effect pigment movement and control its diurnal rhythm in *Diadema*, remain unresolved. However, Dambach (1969) and Weber & Dambach (1972) have taken a fresh look at the phenomenon of colour change in *Centrostephanus longispinus* in conjunction with their light- and electron-microscope study of chromatophore structure. They echo the views of the writer concerning the singular nature of echinoid chromatophores, but they regard them as free cells which effect their overt changes by amoeboid movement. If attempts were made to interpret colour change in *Diadema* on this basis, though some features could be explained, for example, how living chromatophores can be displaced bodily, or survive bisection in the short term, the notion would not explain how the halves can seemingly re-integrate. Neither would it account for the pigment movement in non-cellular chromatoglyphs of *Diadema antillarum*.

The covering reaction of several echinoids still excites controversy. In most instances opinion continues in favour of the habit being related to light, but new aspects of its possible significance in some urchins have been introduced by Péquignat (1966) and Dix (1970) who relate the habit to feeding. Dambach and Hentschel (1970) lay emphasis on the similarity of the effectors concerned and their co-ordination, in covering and locomotion. Recognition of the similarity is not new (see Millott 1966) and in instances where light induces covering as well as taxic locomotory movements, it can be misleading. Similarity in such cases does not necessarily imply identity and is only to be expected when the same stimuli (light, contact and mechanical stress) are involved in co-ordination of the same effectors. In *Lytechinus* the two activities can be separated since artificial 'white' light stimulates positive phototaxis but little covering, whereas ultra-violet radiation stimulates negative phototaxis and intensive covering (Sharp & Gray 1962).

Analysis of the responses of spines to shading, from which much has been learned concerning the photosensitivity of echinoids and, incidentally, concerning their nervous organization, has been extended. Millott & Coleman (1969) examined the fine structure of the skin of *Diadema* in its most photosensitive areas (podial organs). Earlier contentions concerning the lack of structural and ultra-structural differentiation of photoreceptors were fully substantiated, but it was revealed that the felt of cutaneous nerves was so superficial as to show bundles of fine nerve fibrils in the spaces between the cells of the epidermis. The possible significance of such elements in relation to the neural photosensitivity already demonstrated is obvious.

The search for the photoreceptive pigments involved has continued, but has not yet been crowned with success. However, one possibility has been eliminated by microspectrophotometry: Millott & Okumura (1968a) have shown that the echinochrome in neurons of the photosensitive radial nerve is unlikely to be responsible, in that the pigment *in situ* absorbs maximally in the visible between 540 and 560 nm, which is far removed from the maximum of the action spectrum of the shadow response which falls between 455 and 460 nm.

Nervous interaction and especially inhibition, had already been shown to be important factors in the generation of the shadow response, but difficulties in recording discrete action potentials from echinoid nerve continue to form major obstacles in investigating the phenomena electrophysiologically. The demonstration by Takahashi of photically excited action potentials, including both 'on' and 'off' discharges, previously reported, has not been repeated. A step in this direction was taken by Millott & Okumura (1968b) who showed the existence in the radial nerves of *Diadema*, of a slowly propagated potential capable of inhibiting the shadow response.

More attention has been directed to the rhythm of reproductive activities in several species, leading Pearse (1972) to suggest that they are not influenced by monthly changes in moonlight but that the rhythms may be more closely related to tidal factors.

References

DIX T.G. (1970) Covering response of the echinoid *Evechinus chloroticus* (Val.). *Pacific Sci.* 24, 187–94.

DAMBACH M. (1969) Die Reaktion der Chromatophoren des Seeigels *Centrostephanus longispinus* auf Licht. *Z. vergl. Physiol.* 64, 400–6.

DAMBACH M. & HENTSCHEL G. (1970) Die Bedeckungs reaktion von Seeigeln. Neue Versuche und Deutungen. *Mar. Biol.* 6, 135–41.

MILLOTT N. (1966) Co-ordination of spine movement in echinoids. In *Physiology of Echinodermata*: 465–85. ed. R.A. Boolootian, Interscience, New York.

MILLOTT N. & OKUMURA H. (1968a) Pigmentation in the radial nerve of *Diadema antillarum*. *Nature, Lond.* **217**, 92–3.

MILLOTT N. & OKUMURA H. (1968b) The electrical activity of the radial nerve in *Diadema antillarum* Philippi and certain other echinoids. *J. exp. Biol.* 48, 279–87.

MILLOTT N. & COLEMAN R. (1969) The podial pit—a new structure in the echinoid *Diadema antillarum* Philippi. *Z. Zellforsch.* 95, 187–97.

PEARSE J.S. (1972) A monthly reproductive rhythm in the diadematid sea urchin *Centrostephanus coronatus* Verrill. *J. exp. Mar. Biol. Ecol.* 8, 167–86.

PÈQUIGNAT E. (1966) 'Skin digestion' and epidermal absorption in irregular and regular urchins and their probable relation to the outflow of spherule-coelomocytes. *Nature, Lond.* **210**, 397–9.

SHARP D.T. & GRAY I.E. (1962) Studies on factors affecting the local distribution of two sea urchins *Arbacia punctulata* and *Lytechinus variegatus*. *Ecology*, 43 (2), 309–13.

28 The visibility of objects underwater

C. C. HEMMINGS *Marine Laboratory, Department of Agriculture and Fisheries for Scotland, Victoria Road, Aberdeen, Scotland*

The review by Jerlov & Nielsen (1974) shows the two aspects of optical oceanography currently receiving most attention to be firstly, the basic physics of light and radiation in the sea, and secondly radiation in relation to primary productivity. Some advances both in the theory of image transmission through water, and in practical ways of improving it, are the result of commercial interest in underwater technology, in particular photography and television (Anon. 1968, SPIE 1966 and 1968, Mertens 1970). There is an increasing tendency to use the modulation transfer function (MTF), originally used for lens testing, to describe the image transmission of optical systems, including viewing through water, rather than to relate image transmission directly to fundamental optical properties (Replogle 1966, Duntley 1974).

It is the physical and physiological properties of the eyes of animals including man that determine the effectiveness of their vision underwater. The relevant parameters are the contrast perception threshold, spectral sensitivities and various measures of acuity. Of these, the first is perhaps the most important as it limits the maximum visible range of an object. Lythgoe (1968) relates this dependence of range on contrast to the visual pigment of the eye, and suggests that in some circumstances the possession by an animal of a visual pigment with a maximum absorption wavelength offset from the maximum wavelength transmission of the water in which it lives could give it a visual range advantage. Experiments on underwater resolution acuity using black and white barred targets are described by Muntz *et al.* (1974), who express their results in terms of the MTF. These can then be used with the appropriate Fourier transforms to describe most types of object.

If suggestions are to be made about the visual efficiency of fish, then spectral sensitivities, acuities, MTF's and contrast perception thresholds must be known for the species concerned. Of these, only the scotopic spectral sensitivity has been determined chemically for a number of fish species. Hester (1968) describes the determination of contrast perception thresholds for the goldfish, and work is currently proceeding at the Marine Laboratory, Aberdeen to determine values for the cod. Nakamura (1968) obtained resolution acuities for two pelagic tuna species using striped targets. He relates his results to the coloration of the two species, which consists in both of a pattern of permanent and transient dark bars and stripes.

In my contribution to the first symposium on this topic, I suggested that the thresholds for the onset and cessation of behaviour patterns such as general activity, schooling and feeding which are often related simply to light 'intensity' (Woodhead 1966), may perhaps be more properly and accurately related to the combination of optical parameters which control the visibility of environmental features, schooling companions, predators or prey to the species concerned. Collette & Talbot (1972) observed the dawn and dusk change-over of diurnal and nocturnal species of coral reef fish during a 17-day period underwater. On one occasion high turbidity was associated with an altered activity time-table for many species. Precise measurement of relevant optical properties could in this and similar cases demonstrate whether the behaviour was intensity- or visibility-related. It would in general be true that the ecological usefulness of any work on the vision, coloration or behaviour of aquatic animals would be greatly increased if a limited number of relevant optical and radiation measurements were made in the water in which the animals live.

References

ANONYMOUS (1968) *Bibliography on underwater photography and photogrammetry*. Kodak Pamphlet No. P-124. Eastman Kodak, Rochester N.Y., 23p.

COLLETTE B.B. & TALBOT F.H. (1972) Activity patterns of coral reef fishes with emphasis on nocturnal-diurnal changeover. *Bull. nat. Hist. Mus. Los Angeles County* **14**, 98–124.

DUNTLEY S.Q. (1974) Underwater visibility and photography. In *Optical Aspects of Oceanography* (edited by Jerlov N.G. & Nielsen E.S.). Academic Press, London 494 p.

HESTER F.J. (1968) Visual contrast thresholds of the goldfish. *Vision Res.* 8, 1315–34.

JERLOV N.G. & NIELSEN E.S. (Eds) (1974) *Optical aspects of oceanography*. Academic Press, London 494 p.

LYTHGOE J.N. (1968) Visual pigments and visual range underwater. *Vision Res.* 8, 997–1012.

MERTENS L.E. (1970) *In-water photography*. Wiley-interscience, New York 391 p.

MUNTZ W.R.A., BADDELEY A.D. & LYTHGOE J.N. (1974) Visual resolution underwater. *Aerospace Med.* **45**, 61–6.

NAKAMURA E.L. (1968) Visual acuity of two tunas *Katsuwonus pelamis* and *Euthynnus affinis*. *Copeia* 1968 (1), 41–9.

REPLOGLE F.S. (1966) Underwater illumination and imaging measurements. In *SPIE Underwater Photo-Optics Seminar Proceedings.*

SPIE (1966) *Society of Photo-Optical Instrumentation Engineers Seminar Proceedings* **7**, Underwater Photo-Optics.

SPIE (1968) *Society of Photo-Optical Instrumentation Engineers Seminar Proceedings* **12**, Underwater Photo-Optical Instrumentation Applications.

WOODHEAD P.M.J. (1966) The behaviour of fish in relation to light in the sea. In H. Barnes (ed.) *Oceanogr. mar. Biol. ann. Rev.* **4**, 337–403.

Demonstrations

Most of the demonstrations presented to the Symposium related to papers which were read, and the substance of them has been incorporated in the printed versions. The following accounts contain other material.

29 Developments in hemispherical photography

G. C. EVANS, P. FREEMAN and O. RACKHAM
Botany School, University of Cambridge

When Evans & Coombe (1959) wrote about the application to ecological research of hemispherical photography, using so-called 'fish-eye' lenses, a descendant of the original lens computed by Hill (1924) was still in production by Messrs Beck. (The proof plates of the original had been lost in a war-time fire.) This lens, incorporated in their 'Robin Hill' camera, has since gone out of production. Its greatest advantage was its wide field of view, extending roughly 4° below the horizon, 188° in all, a very great convenience in field work. On the other hand it was fitted to a plate camera, making long series of photographs awkward, and it was not fully colour-corrected, a deep red filter being needed for best definition. This created difficulties when working on plant canopies. Photographs had to be taken with very fast panchromatic plates, on a bright day of light overcast (as insufficient red light comes from a blue sky) and the exposure had usually to be at least 0·1 s, which often caused trouble with movement due to wind.

Since then several makers have produced hemispherical lenses, one of the best known being the Nikon, produced by Nippon Kogaku of Tokyo, which fits the standard Nikon camera body taking 35 mm film. We also showed a lens by Hanimex, which is cheaper but has less good definition. It would be principally useful for recording work where great accuracy is not needed. These and similar lenses have been available for a number of years, and are beginning to appear on the second-hand market.

Depth of focus

All these lenses have very great depth of focus, the Nikon having a focal length of only 8 mm, which is convenient for work in standing crops and controlled-environment cabinets. It is easy to arrange that everything beyond, say, 0·5 m will be in focus. A minor drawback when working in the field is the tendency of raindrops, etc., on the front surface of the lens to appear in the picture.

Colour correction

The Nikon is supposed to be fully colour-corrected. Two prints were exhibited taken on the same day, one with a filter transmitting the whole visible spectrum (L1A), the other with a deep red filter (R60). Definition was very slightly better with the latter, but when photographing anything liable to be moved by wind, any advantage here would be outweighed by the effects of the longer exposure needed. Thus the Nikon assembly has two marked advantages over the Beck: ease of handling of a film as compared with a plate camera, and much shorter exposure times, which make it possible to photograph a woodland canopy in very poor light, or on a windy day, when, however, the photograph is not necessarily representative of average conditions over a period.

Accuracy

Evans & Coombe (1959) concluded that for most purposes it suffices to work to an accuracy of about 15′ of arc, roughly half the apparent diameter of the sun. This represents roughly 2 mm at a distance of 50 cm from the camera, so that there is no great difficulty, for calibratory purposes, in working to 10 times this accuracy, i.e. 1·5′ or 0·2 mm at a radius of 50 cm. Such an accuracy may also be needed when taking hemispherical photographs for special purposes, but if a solar track were to be superimposed it would then be necessary to work out the particular track for the day in question (see Evans & Coombe, above).

Limitations of field and definition

Inspection of the calibration photographs showed that at the stop used the Nikon lens had very good definition near the zenith from a distance of 50 cm outwards, but less good definition near the horizon. To stop down would improve the definition but lengthen the exposure, undesirable for vegetation photographs on any but the stillest days. The Hanimex lens has less good definition over the whole field.

The field of the Nikon lens cuts off at about 4½° above the horizon, and the Hanimex at about 9½°. Neither has therefore a truly hemispherical field of view, unlike the original lens calculated by Hill (1924) and made by Beck, whose field of view extended to rather over a hemisphere.

Effects of these properties on use in the field

In order to fit an overlay showing altitudes and azimuths it is necessary to identify two points on a hemispherical photograph: most conveniently the intersection of the N–S plane through the lens with the horizon. Using the Robin Hill lens this could easily be done by driving in a pair of levelled white marking stakes at a considerable, but not necessarily determinate, distance from the camera.

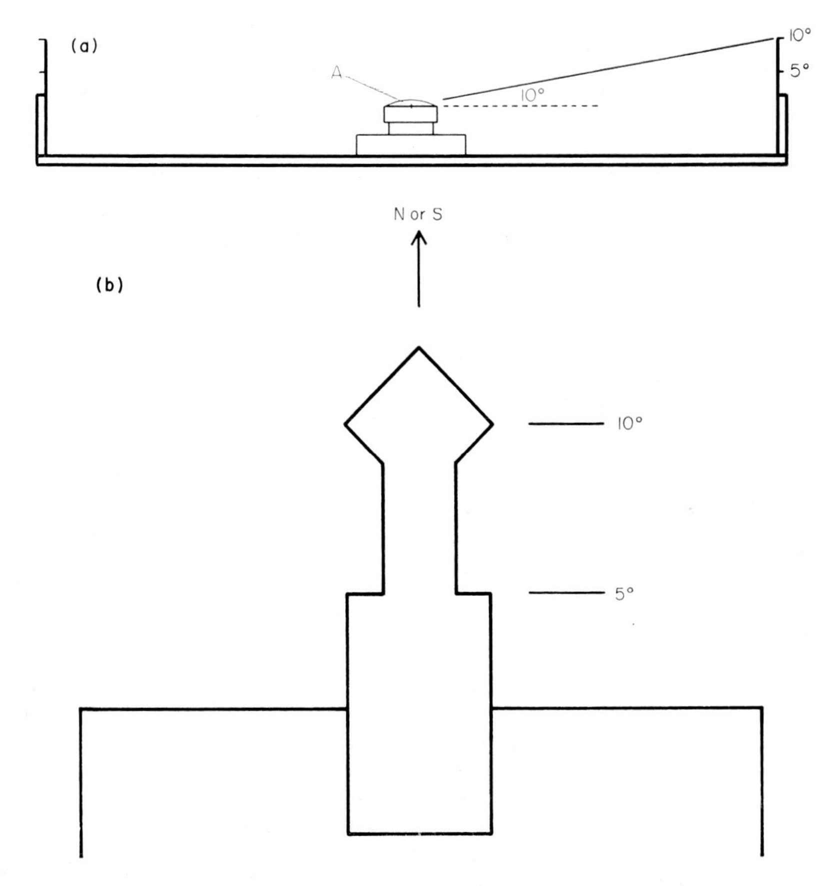

Figure 29.1. (a) Side elevation of camera carriage with altitude markers. The camera lies on its back in the centre of the baseboard, and is secured by a tripod-screw through a hole in a short length of metal angle. A, apparent optical centre of lens as determined by calibration. The altitude markers are hinged at each end of the baseboard and fold flat when not in use.

(b) End elevations of altitude marker, with shoulder at an elevation of 5° above the apparent optical centre of the (Nikon) lens, and a diamond centred on 10° (10° and 15° respectively for the Hanimex lens).

The restricted fields of view make this impracticable using either the Nikon or Hanimex lenses: it is necessary to use an altitude of at least 10° above the horizon for the Nikon and 15° for the Hanimex instead of the plane of the horizon. In turn this makes the use of distant marker stakes awkward, and it is preferable to construct a carriage for the camera, incorporating markers at a fixed elevation. However, if the carriage is not to be impossibly clumsy these markers must be close to the camera, making it essential to know the position of the apparent optical centre of the lens.

Camera mounting with altitude/azimuth markers

The carriage shown in Fig 29.1a is adapted for the Nikon lens. The white markers (Fig. 29.1b) each have a shoulder at an altitude of 5° above the apparent optical centre of the lens, and a diamond with its greatest width at 10°. For the Hanimex lens the corresponding altitudes would be 10° and 15°.

To work to an accuracy of 15′ implies levelling a board 1 m long to 5 mm. Experiment with a spirit level will show that it can readily be levelled to 1 mm.

Orientation may still most conveniently be done by providing two fixed markers defining a true North–South line, and aligning the mounting by sighting between them. This avoids any disturbance of a magnetic compass by the materials of the camera and its carriage. Allowance must of course be made for the deviation of the compass from true north, and if one intends to work in a remote part of the world it is convenient to ascertain beforehand what the magnetic declination there will be.

Calibration of fish-eye lenses

Requirements

In order to be able to construct an accurate grid overlay showing altitude, azimuth, solar tracks etc. it is necessary to know

(a) the position of the apparent optical centre of the lens
(b) the projection of the observed hemisphere on to the flat surface of the plate.

Neither of these pieces of information is readily obtainable from the makers, and in any case it is wise to calibrate a particular lens. Working to the accuracy (of 1·5′) already discussed, the time taken is not excessive in comparison with the time taken in use.

Plate 29.1. Two photographs of the same scene taken with the Nikon fish-eye lens. The scene is a recently-coppiced part of the Bradfield Woods (p. 438) with a high density of standard trees. The upper picture has been exposed so as to show details of the vegetation; it shows the north and south markers on the camera carriage (50 cm away), a disc on a cane placed so as to occlude the sun (about 1 m away), and trees and underwood at distances up to 30 m. The lower picture was given a shorter exposure in order to record more accurately the obstruction of the sky; superimposed is a grid of altitude and azimuth at 30° intervals and of the sun's track at 51°N for the 21st of each month counting from December.

Determination of apparent optical centre

For this purpose two sets of pointed markers are set out, say at elevations of 30° and 60° above a hypothetical horizon, and at two different radii, say 50 and 150 cm. Zenith markers are included to check alignment, as in Fig. 29.2. The markers at each elevation face towards each other, at right angles to the plane of the paper in Fig. 29.2.

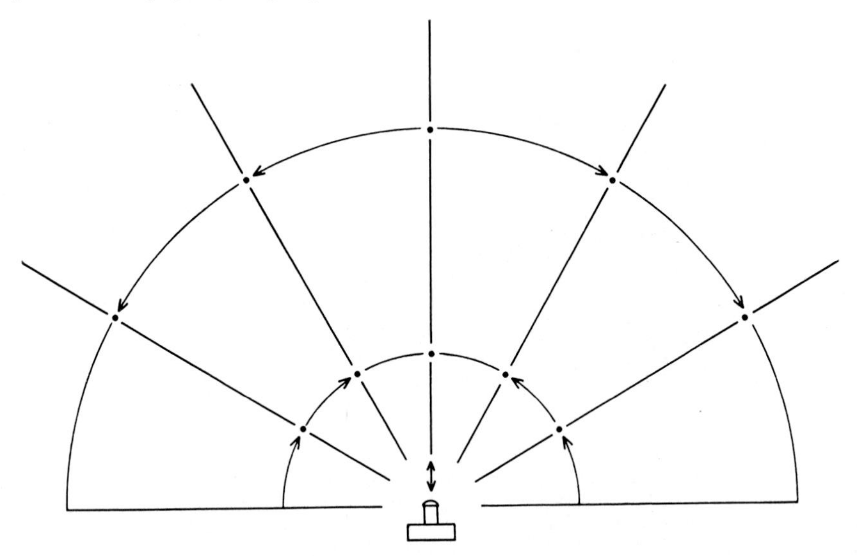

Figure 29.2. Arrangement of markers for determination of the apparent optical centre of a hemispherical lens. For details see text.

A series of photographs is then taken as the camera is moved towards or away from the zenith through the centre of alignment of the markers. When this coincides with the apparent optical centre of the lens, all the markers should appear in the correct alignment.

It turns out that for both the Nikon and Hanimex lenses which we have studied the apparent optical centre is level with the top of the metal mount of the lens.

Determination of projection

For this purpose markers are fixed at 5° intervals around a meridian semicircle of roughly 1 m diameter, and a photograph is taken with the apparent optical centre of the lens, previously determined, in the centre of the semicircle. Examples of photographs taken in this way were shown for both lenses, and it could be seen that the projection in each case is a close approximation to equiangular in the direction from horizon to zenith.

The positions of the markers on the photograph are then measured with a travelling microscope, and distances plotted against angles.

The individual deviations of each 5° step from a linear relationship with distance are then calculated, and these are shown, for the Nikon and Hanimex lenses, in Fig. 29.3 a and b. These graphs are then used to work out the corrected plot for overlays giving altitudes and solar tracks. They

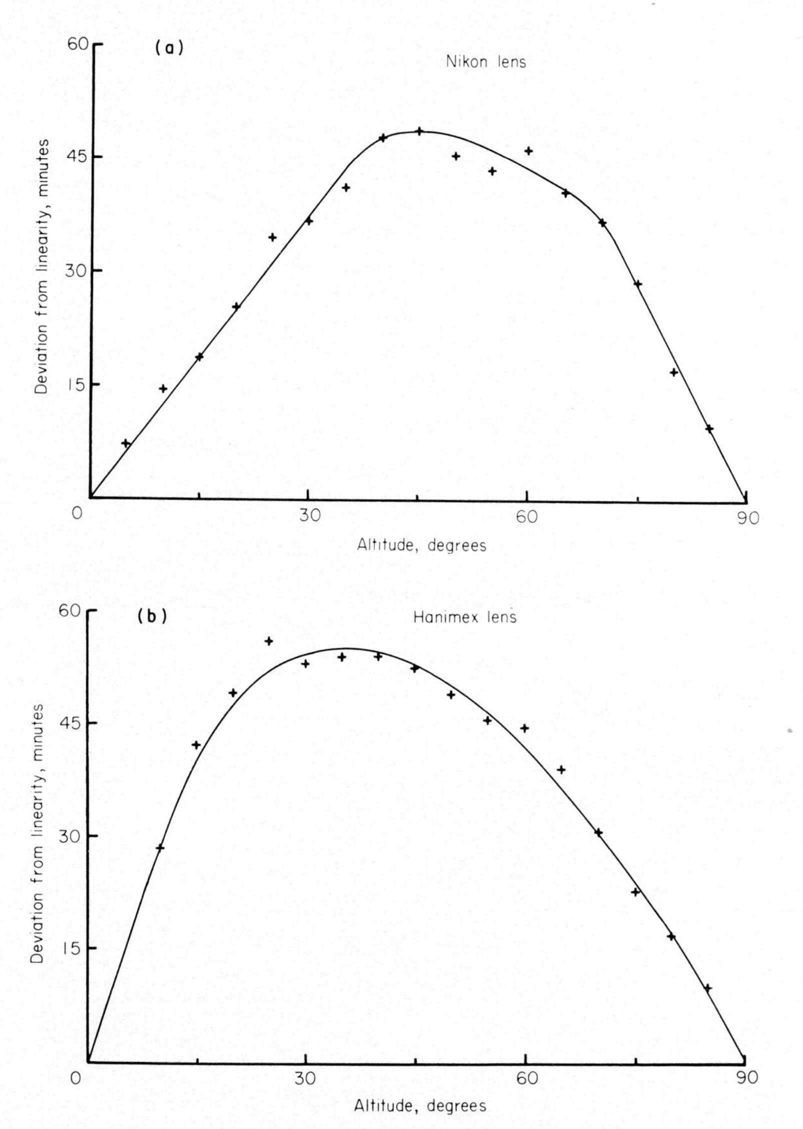

Figure 29.3. Deviations of hemispherical lenses from linear relationships between altitude of markers above a plane through the optical centre of the lens, and distances on a diameter across the photograph. (a) for the Nikon lens; (b) for the Hanimex.

demonstrate that at the worst, around an elevation of 45°, the deviation from linearity is less than 1° for both the two lenses.

Printing with overlays

The desired grid showing altitudes, azimuths, solar tracks etc. may be drawn on transparent material or reproduced photographically on a large plate (these are becoming difficult to obtain). It must be in mirror image, as it has to be used upside down. Using the same photographic enlarger as was used for printing the calibration photographs, the negative is projected through the overlay on to the printing paper. Let us suppose that we are working with a Nikon lens. Then, using an orange filter, the degree of enlargement is adjusted until the images of the 5° shoulders on the markers span across the diameter of a 5° circle on the overlay (Plate 29.1). The overlay is then adjusted to true north and south, and finally the 10° diamonds are checked against the 10° marks on the N–S azimuth line. The overlay is then ready for printing in.

References

Evans G.C. & Coombe D.E. (1959) Hemispherical and woodland canopy photography and the light climate. *J. Ecol.* 47, 103–13.

Hill R. (1924) A lens for whole sky photographs. *Quart. J. roy. met. Soc.* 50, 227–35.

30 Light and the growth of ash

GEOFFREY GARDNER[1] *Department of Botany, University of Sheffield*

Many of the deeply incised valleys (dales) of the Carboniferous Limestone in Derbyshire carry woodland having a large percentage of Ash, *Fraxinus excelsior* L. These woods have been shown generally to be young and to contain extensive even-aged stands (Pigott 1969, Merton 1970).

This work was carried out in the Lathkilldale National Nature Reserve, in Meadow Place Wood (National Grid SK 199661) on the north-facing slope of the dale. Here *F. excelsior* contributes about 95% of the total canopy trees [Merton (1970) Area B], the small remainder comprising *Ulmus glabra* and *Acer pseudoplatanus* with occasional specimens of *A. campestre*, *Sorbus aucuparia* and *Crataegus monogyna*. The shrub layer includes *Viburnum opulus*, *Corylus avellana* and *Prunus padus*.

In places the herb layer is species-rich, containing *Galeobdolon luteum*, *Endymion non-scriptus*, *Convallaria majalis*, *Arum maculatum*, *Anemone nemorosa*, *Primula vulgaris*, *Oxalis acetosella*, *Brachypodium sylvaticum*, *Deschampsia caespitosa* and *Melica uniflora*. In other places it is dominated by *Mercurialis perennis*. The frequent luxuriance of the herb layer beneath ash has always been ascribed to the fact that the shade which ash casts is not very dense.

Measurements of levels of radiation within the wood were made using tube solarimeters (Szeicz *et al.* 1964), with a specially constructed portable amplifier (Gardner & Sutton 1970). Almost simultaneous readings were made within the wood and at a point outside it. Spot readings were taken at nine sites at weekly intervals, during the period from two hours before to two hours after solar noon. The tubes within the wood were supported above the herb layer on special stands. The order of making the readings was altered systematically on each occasion. These readings were grouped according to the sky conditions outside the wood under which they were made. Three distinct groups of conditions, cloudy-dull, below 0·57 cal cm^{-2} min^{-1},

[1] Present address: Biological Sciences Department, Goldsmith's College, University of London, New Cross, London SE14 6NW.

cloudy-bright, 0·57–1·13 cal cm^{-2} min^{-1}, and sunny-bright, above 1·13 cal cm^{-2} min^{-1}, could be recognized.

For subsequent experiments, the growing season was divided into two parts of approximately equal length. The first (Period I) covers the time from germination in May to the production of the terminal resting bud, an important stage in the growth of the seedling; the second (Period II) is from this stage to just before leaf-fall. It was found that resting buds were normally laid down as soon as the tree canopy was fully expanded.

Two series of readings are given in Table 30.1. The first series, taken on

Table 30.1. Spot readings of total radiation received at nine woodland sites and outside the wood under various weather conditions (values in cal cm^{-2} min^{-1}).

		Early Period I			Period II	
Site	Sky	Outside wood	Inside wood	Sky	Outside wood	Inside wood
D	C.D.	—	—	C.D.	0·23	0·007
	S.B.	1·41	0·125	C.B.	0·85	0·027
6	C.D.	0·43	0·073	C.D.	0·18	0·003
	S.B.	1·08	0·153	C.B.	0·57	0·017
1	C.D.	0·50	0·107	C.D.	0·23	0·011
	S.B.	1·33	0·301	C.B.	1·08	0·033
7	C.D.	0·45	0·115	C.D.	0·14	0·003
	S.B.	1·19	0·361	C.B.	0·57	0·025
3	C.D.	0·40	0·167	C.D.	0·16	0·019
	S.B.	1·25	0·371	C.B.	1·08	0·065
5	C.D.	0·42	0·173	C.D.	0·14	0·019
	S.B.	1·13	0·401	C.B.	0·63	0·063
4	C.D.	0·49	0·173	C.D.	0·20	0·022
	S.B.	1·32	0·461	C.B.	0·80	0·035
2	C.D.	—	—	C.D.	0·24	0·073
	S.B.	1·20	0·461	C.B.	0·91	0·105
L	C.D.	0·38	0·141	C.D.	0·21	0·061
	S.B.	1·16	0·361	C.B.	0·94	0·097
		C.D. 21 May 1969		C.D. 18 August 1969		
		S.B. 21 May 1969		C.B. 5 September 1969		

C.D. Cloudy-dull
C.B. Cloudy-bright
S.B. Sunny-bright

D Darkest site
L Lightest site

21 May 1969, is for early Period I. At this time ash trees are just beginning to expand their leaves, so comparatively little shade is cast within the wood, though the other species there are already in full leaf.

The second series, taken on 18 August 1969 and 5 September 1969, is during Period II when the canopy is fully expanded and the shade cast by it is at a maximum.

It is difficult to obtain an absolute comparison of the sites from this type of reading because of the changing sky conditions, but the values in Table 30.1 give a good indication of the effect of the canopy on the radiation penetrating it.

The luxuriance of the vegetation of the herb layer, at all sites except D, 6 and 1, clearly indicates that the levels of radiation below the canopy are quite sufficient for the growth of a wide range of woodland herb species. At sites D, 6 and 1, on the other hand, the situation is very different. The herb layer is so sparse as to be almost non-existent. However, these sites are all under tree species other than ash, D under *Acer pseudoplatanus*, 6 under *Ulmus glabra* and 1 under *Crataegus monogyna*.

The dense herb layer below the ash trees presents considerable difficulties for the ash seedlings, since they must begin to grow below rather than within it (see Wardle 1959). In order to examine this situation more closely, it is necessary to look at the shading effect of the herb layer and in particular that under *Mercurialis perennis*. The tube solarimeters used are ideal for slipping into such vegetation without disturbance and being 0·86 m in length allow for the inclusion of sunflecks when these occur.

Spot readings were taken at regular intervals on a 16 × 16 m grid. For readings above the herb layer the tubes were supported on special stands; they were placed carefully on the ground for readings below the herb layer. The readings were obtained under constant sky conditions during the period from two hours before to two hours after solar noon, on 14 June 1968 under sunny-bright conditions and on 11 June 1968 under cloudy-dull conditions. There is a considerable further reduction in the levels of radiation owing to the herb layer (Table 30.2). In energy terms the visible part of the spectrum (400–750 nm) shows a reduction to about one-third below the herb layer. The great similarity in the levels of radiation between cloudy-dull and sunny-bright conditions below the herb layer suggests that the *Mercurialis perennis*-dominated herb layer is an effective barrier to sunflecks.

The fate of natural seedlings has been followed in a series of permanent quadrats, observed regularly over several years, and some typical results are shown in Fig. 30.1. During the first summer deaths occurred owing to establishment difficulties. After that, with one exception (a single seedling in May 1968), death took place only during the winter. Figure 30.1 follows the fate of seedlings germinating in a single year. A similar pattern has been obtained for seedlings germinating in subsequent years. In this sample the half-life was 7–8 months and no plants survived for longer than three years.

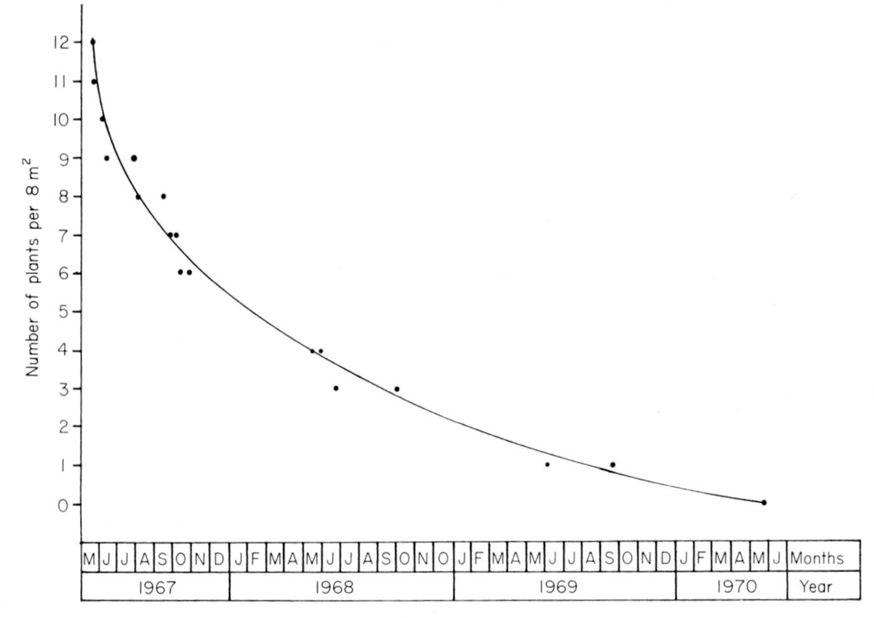

Figure 30.1. Mortality curve of ash seedlings which germinated in permanent quadrats in Lathkilldale in 1967.

On one occasion, owing to lack of seed production in the area, no germinating seedlings were present. Newly-germinated seedlings were therefore planted in the same area and their fate followed. The same pattern was again closely repeated; in fact the number of plants remaining in the third year was predicted a year in advance to within one plant.

Table 30.2. Levels of visible radiation (400–750 nm) above and below a herb layer dominated by *Mercurialis perennis* under various weather conditions.

	Mean (cal cm^{-2} min^{-1})
Above Herb Layer	
Sunny-bright (14 June 1968)	0·040
Cloudy-dull (11 June 1968)	0·033
Below Herb Layer	
Sunny-bright (14 June 1968)	0·014
Cloudy-dull (11 June 1968)	0·014

No significant difference between means for weather conditions either above or below the herb layer.

Ash seedlings grown in pots sunk into the ground at the nine sites listed in Table 30.1 all survived their first summer. At the lightest site (site L), growth was good, with four pairs of true leaves being produced per seedling. At the darkest site (site D), growth was poor and the seedlings made little increase on their original seed weight. At this site, under a sycamore tree, the herb layer of *Mercurialis perennis* was very sparse (40 stems/m²). The level of radiation throughout most of the year here is similar to that below *Mercurialis perennis* over a wide area of the wood, where the density is 320 stems/m². At site D, several seedlings exhibited the 'wirestem' effect and others disappeared without trace soon after losing their leaves in autumn. 'Wirestem' is a condition where the epidermis and cortex of the stem rot away leaving only the stele to present a 'wiry' stem. Long (1966) showed this to be the result of damping-off fungi. Plants attacked in this way decay very quickly. Subsequent mortality of the ash plants occurred only during winter, when the plants carried no leaves.

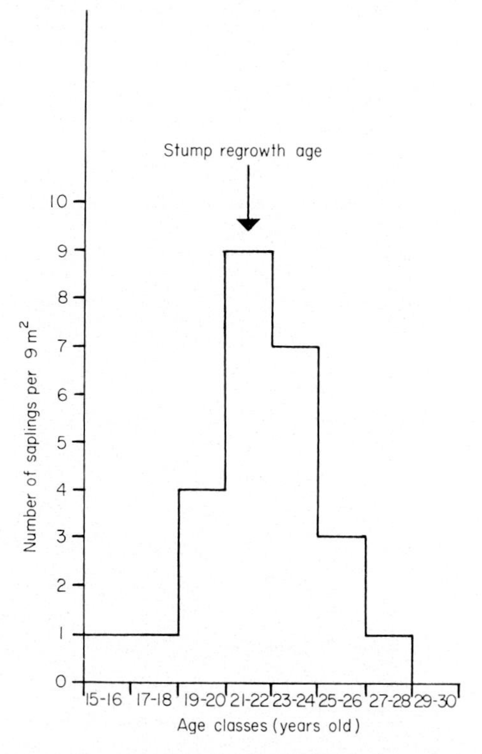

Figure 30.2. Age-distribution of a sample of individuals from a stand of sapling ash in Lathkilldale.

Susceptibility to fungal attack has been shown by many workers to be increased by a low carbohydrate content in the tissues of the plant (Read 1968). Chemical analysis of pot-grown ash seedlings shows lower levels of carbohydrate at the dark sites, where the 'wirestem' effect was observed, than at the other lighter sites.

Mortality is one end-point of seedling and sapling growth within the wood. At the other end of the scale there are stands of saplings which have grown well and reached a height of several metres. This situation is usually ascribed to improved light conditions. To examine this more closely, such a stand was investigated by harvesting the saplings in it along a belt transect comprising contiguous 1-m^2 quadrats. The saplings were measured and aged by growth-ring and node counts. The age distribution results are given in Fig. 30.2. The mean height of the saplings was 1·79 m.

In the middle of this stand a tree stump was found from which regrowth had occurred. One of these shoots was aged at 21 years, corresponding with the peak of the graph.

Discussion

The performance of ash seedlings and saplings indicates that the levels of radiation received above the herb layer at sites in Table 30.1 below ash trees are sufficient to allow them to continue growth. Much older ash trees are still being eliminated from the canopy in this area, owing to intraspecific competition for light within it, so new individuals cannot be expected to succeed where these have failed. Nevertheless, in this zone between the herb layer and the canopy, saplings can survive for many years (at least 28 in Fig. 30.2) and form a large reserve of potential replacements for the canopy, should gaps occur.

The other barrier to growth is the herb layer. It has been seen from Table 30.2 that this causes a considerable further reduction in the amount of radiation available and, though not demonstrated here, probably changes in light quality also. Levels of radiation reduced by the herb layer are similar to those of areas under trees with a denser canopy than ash (e.g. *Acer pseudoplatanus*). This means that seedlings less than 20 cm in height are growing in conditions equivalent to a sycamore wood, where the herb layer is almost non-existent. Figure 30.1 shows the result of this situation.

During the first summer, seedlings experience difficulties in establishing themselves and die, owing to chance factors which lead to the placement of seeds in situations where the seedlings cannot grow. By the end of the summer, as leaf-fall approaches, little if any increase in weight has been made on the original seed weight, root growth is poor and the seedlings possess only a pair of cotyledons and one pair of true leaves. Soon after these leaves are lost, deaths occur, probably mostly of the smaller seedlings originating from smaller seeds. Some seedlings disappear without trace within a few days. In others, there are visible symptoms indicating fungal attack via the 'wirestem' effect, probably due to the depletion of the carbohydrate content of the tissues.

These observations suggest that these seedlings have been living throughout the year close to their compensation point and that, once new photosynthate ceases to be produced, respiration quickly depletes what little reserves remain. This pattern is repeated in subsequent years with further deaths occurring only during the winter. Most plants die within a few years.

This short life-span appears to have been general for a considerable time. There is evidence of it in the results in Fig. 30.2. At the point at which the tree was felled, one plant had survived for 5 or 6 years, three had survived for 3 or 4 years and six had survived for 1 or 2 years. The improved conditions caused by the felling of the tree allowed these to survive longer than would otherwise have been the case, together with nine newly-germinated seedlings. In subsequent years, a few more plants were added until all this extra growth made the area too dark for further plants to survive. This happened 22 years before the sample was taken and there is no evidence that these saplings will not survive for many more years.

One can only guess at the changes in the herb layer during this time, but three possibilities exist which would account for the resulting sapling growth. The felling of the tree may have damaged the herb layer sufficiently to prevent it from forming an efficient barrier to radiation; or the increase in radiation following gap formation may have caused a deterioration in the herb layer; or the level of radiation below the herb layer after removal of a canopy tree may have been sufficient to allow increased sapling growth.

The evidence presented here points to the herb layer in ash woods as a very effective barrier to seedling and sapling growth.

References

GARDNER G. & SUTTON F. (1970) A portable linear amplifier for use with solarimeters. *J. appl. Ecol.* 7, 653–56.

LONG P.G. (1966) Mycotic factors and density-induced mortality in pure and mixed stands of closely related species. Ph.D. thesis, University of Sheffield.

MERTON L.F.H. (1970) The history and status of the woodlands of the Derbyshire limestone. *J. Ecol.* 58, 723–44.

PIGOTT C.D. (1969) The status of *Tilia cordata* and *T. platyphyllos* on the Derbyshire limestone. *J. Ecol.* 57, 491–504.

READ D.J. (1968) Some aspects of the relationship between shade and fungal pathogenicity in an epidemic disease of pines. *New Phytol.* 67, 39–48.

SZEICZ G., MONTEITH J.L. & DOS SANTOS J.M. (1964) Tube solarimeter to measure radiation among plants. *J. appl. Ecol.* 1, 169–74.

WARDLE P. (1959) The regeneration of *Fraxinus excelsior* in woods with a field layer of *Mercurialis perennis*. *J. Ecol.* 47, 483–97.

31 A spectroradiometer for measuring the spectral distribution of radiation in plant canopies

H. A. McCARTNEY and M. H. UNSWORTH,
University of Nottingham School of Agriculture, Sutton Bonington, Loughborough, Leics.

The instrument shown was a commercial system made by Gamma Scientific, Inc., San Diego, California. Radiation enters a 3 m long fibre-optic probe through a cosine corrected diffusing head of diameter 2·5 mm. The fibre optics transmit the light to a monochromator containing an automatically scanning Bausch and Lomb diffraction grating (1350 lines/mm). The output from the monochromator falls on a photomultiplier and the photomultiplier current is displayed on a digital electrometer. Electrical outputs proportional

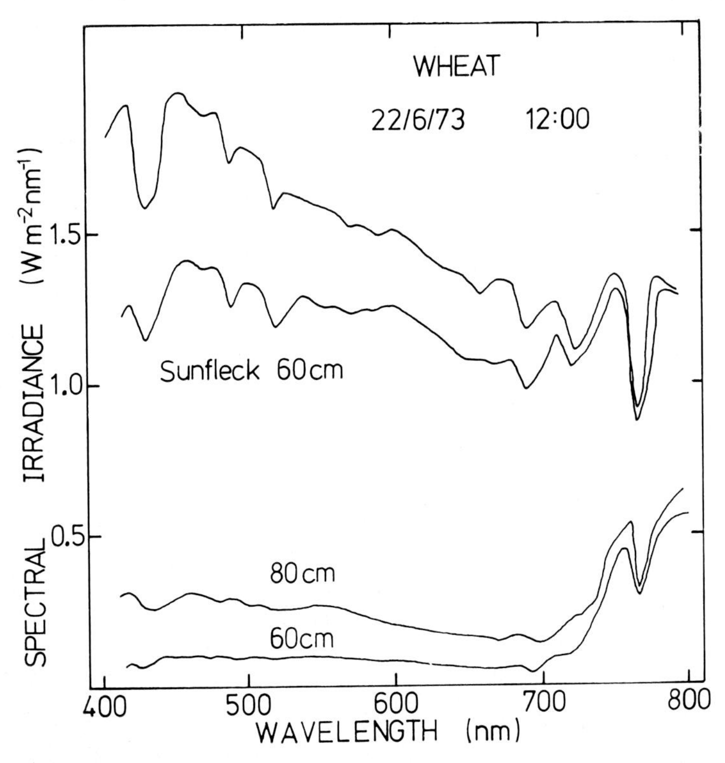

Figure 31.1. Spectral irradiance above and within a growing wheat crop at 1200 GMT on a cloudless day, 22 June 1973.

to wavelength and to photomultiplier current are taken to an X–Y recorder. The entire system is mounted on a small trolley for field use.

The system responds to wavelengths from 400 to 800 nm, the restricting factors being the transmission of the fibre optics and the photo-multiplier response respectively. The instrument is calibrated in the laboratory against a standard of spectral irradiance supplied by the National Physical Laboratory. For field use, the entrance and exit slits of the monochromator are set to give a bandwidth of about 8 nm, and a spectrum from 400 to 800 nm may be scanned in about 30 secs.

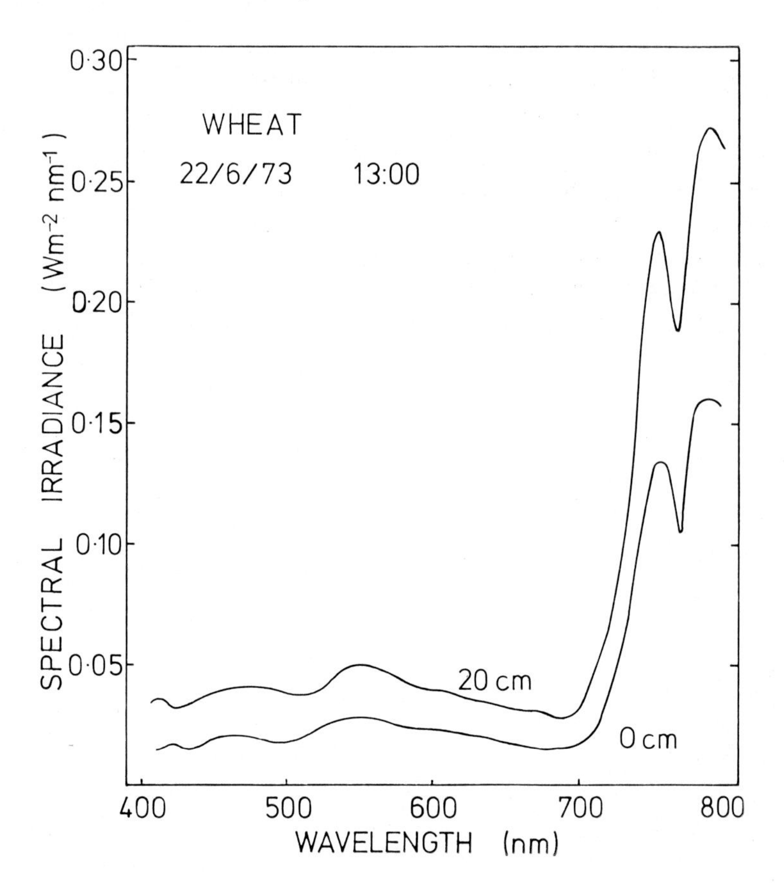

Figure 31.2. Spectral irradiance within a growing wheat crop at 1300 GMT on a cloudless day, 22 June 1973.

Figures 31.1 to 31.3 show results obtained with the spectroradiometer. Figures 31.1 and 31.2 show measurements at 1200 and 1300 on 22 June 1973, in and above an actively growing wheat crop which was about 1 m high. The height within the crop where each spectrum was measured is shown on the figures. The day was almost cloudless so that radiation above the crop varied little between 1200 and 1300. Photosynthetically active radiation between 400 and 700 nm was severely attenuated by the crop. At wavelengths greater than 700 nm the low absorbance of the leaves led to relatively large values of near infra-red irradiance within the crop. Note the contrast between radiation in a sunfleck and in shade at 60 cm (all other spectra within the crop were in shaded regions).

As a contrast Fig. 31.3 shows spectra at 1330 on 24 July 1973, an overcast day when the crop was yellow and senescent. The crop was comparatively transparent at all wavelengths and there was no strong contrast between visible and near infra-red irradiance at the bottom of the crop.

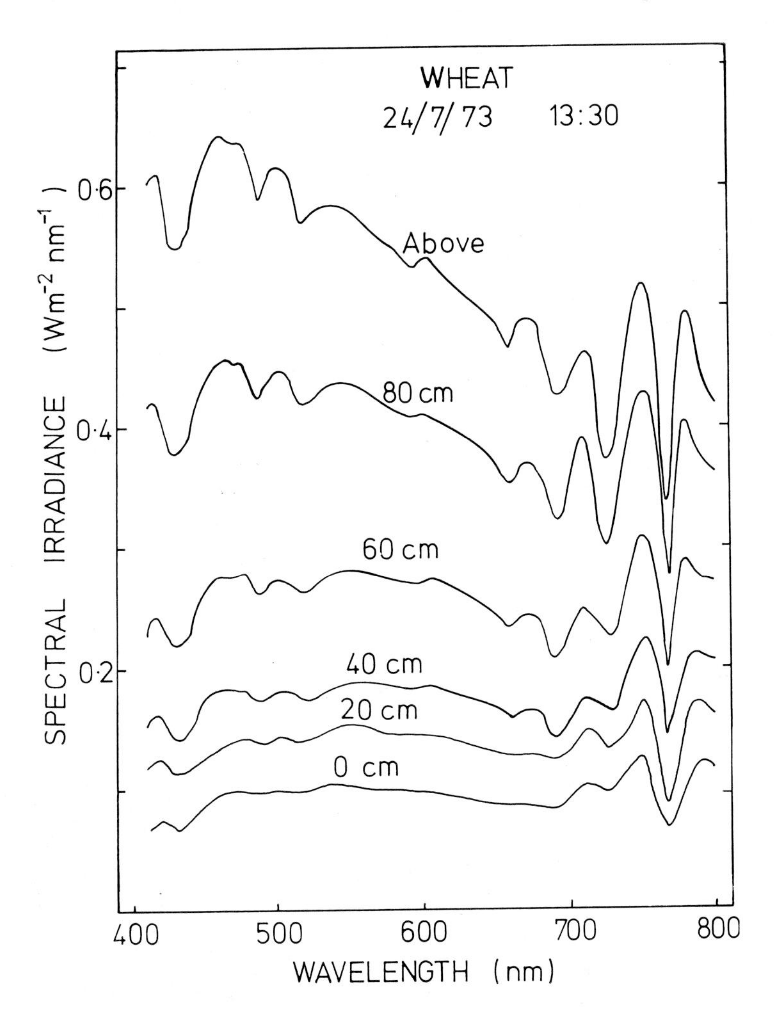

Figure 31.3. Spectral irradiance above and within a senescent wheat crop at 1330 GMT on an overcast day, 24 July 1973.

The photosynthetically active irradiance (400–700 nm) for each spectrum obtained by numerical integration from data in Figs. 31.1 to 31.3 is shown in Table 31.1. The table also shows the corresponding photosynthetically active photon flux densities obtained from the figures using the identity $P(\lambda) = I(\lambda) \times \lambda/hc$ and numerically integrating from 400 to 700 nm. ($P(\lambda)$ is the spectral photon flux density (photons m^{-2} nm^{-1}), $I(\lambda)$ is the spectral irradiance (W m^{-2} nm^{-1}), h is Planck's constant ($6{\cdot}63 \times 10^{-34}$ J s^{-1}) and c is the velocity of light in free space (300×10^{6} m s^{-1}).)

This research formed part of a project supported by the Agricultural Research Council and the Natural Environment Research Council.

Table 31.1. Photosynthetically active irradiance and photon flux density (400–700 nm) above and within a wheat crop on a cloudless day (22/6/73) when the crop was growing and on an overcast day (24/7/73) when the crop was senescent.

Date	Time	Position	Irradiance (400–700 nm) W m^{-2}	Photon flux density (400–700 nm) 10^{21} photons m^{-2} s^{-1}
28 June	1200	Above crop	471	1·28
		Sunfleck		
		0·6 m	358	0·987
		0·8 m	69	0·185
		0·6 m	23	0·063
	1300	0·2 m	16	0·042
		Ground level	7	0·018
24 July	1330	Above crop	166	0·452
		0·8 m	122	0·332
		0·6 m	77	0·212
		0·4 m	51	0·140
		0·2 m	43	0·120
		Ground level	28	0·077

32 Tube solarimeter and automatic porometer

DELTA-T DEVICES, *128 Low Road, Burwell, Cambridge CB5 0EJ.*

The instruments displayed are commercially available from Delta-T Devices. They are manufactured by agreement with Nottingham University and in collaboration with Professor J. L. Monteith.

Tube solarimeter

This instrument for measuring the average irradiance of sunlight amongst plants is based on the design of Szeicz *et al.* (1964). A sixty-junction copper-constantan thermopile wound on a former 858 mm long and 22 mm wide produces an output of about 25 mV per kW m^{-2} with an internal resistance of about 50 ohm. The borosilicate glass tube (26 mm dia) limits the response to light wavelengths from 0·35 to 2·5 μm. The response time for 63% change is 40 s and for 99% is 3 min. A similar unit fitted with an integral Kodak Wratten 88A filter restricts the waveband to 0·7–2·5 μm. This allows the visible component of solar radiation to be calculated from the difference between unfiltered and filtered outputs of units mounted side by side.

Automatic porometer

Stomatal resistance measurements are important in the analysis of photosynthesis and transpiration rates and the need for a portable field instrument for taking a large number of readings quickly and with good repeatability has strongly influenced the development of this automatic porometer based on the design of Stiles (1970) and Monteith & Bull (1970).

A small cup (22 mm dia, 5 mm deep) is clamped to the leaf surface. The relative humidity (RH) at the end of the cup (i.e. in a plane parallel to the leaf surface and 5 mm above it) is measured with a sulphonated polystyrene RH sensor. As the leaf transpires the RH in the cup rises and the time for the RH to rise past two chosen values is measured. This 'transit time' is related to the stomatal resistance to the diffusion of water vapour. By flushing the cup with dry air the measurement may be repeated.

In practice the hysteresis and response time of the RH sensor make consistent readings difficult to obtain unless all the RH limits over which the cup is operated are fixed. Whilst careful manual operation of the dry-air pump can achieve this, considerable skill and concentration are required. To relieve the operator from this effort an electric pump triggered at the appropriate RH values by electronic circuitry has been used. In addition the timing is performed by a constant-rate electronic counter which is also triggered automatically at preset RH levels. Besides reducing the strain on the operator who now has only to note the counter reading between each cycle the use of electronic techniques confers some important advantages. Fast transit times (less than 1 s) can be measured accurately to a resolution of 0·04 s. Consequently the sensor can be positioned close to the leaf surface—improving temperature equilibration between leaf and sensor and reducing the resistance of the cup itself. The RH interval for the transit can be narrowed and the whole wetting and drying cycle can take place at RH values not much different from ambient. Figure 32.1 shows the variation in

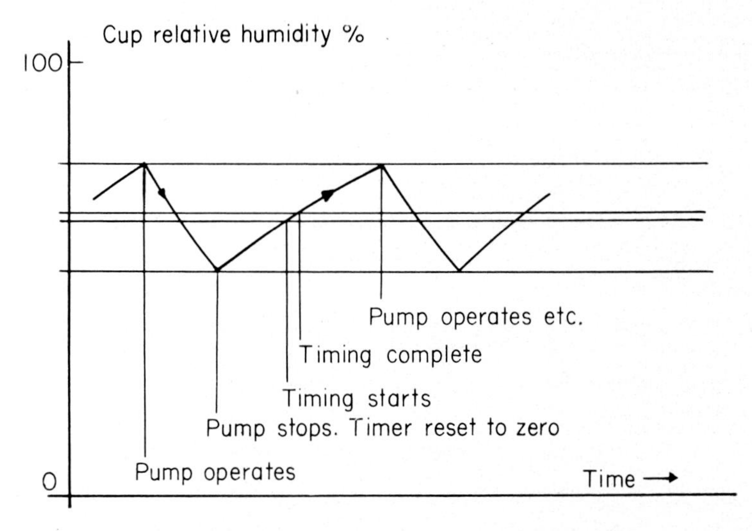

Figure 32.1. Variation of cup relative humidity with the progress of the automatic cycle.

RH at the sensor during the various phases of the automatic cycle. The progress of each cycle is displayed on an ungraduated RH scale on a meter. Normally the whole cycle takes place in the region of 60% RH inside the cup but for operation in very low ambient RH conditions the whole set of levels can be switched to produce cycling in the region of 30% RH inside the cup.

Resistance measurements in the range from about 0·5 s/cm to about 25 s/cm are practicable.

The meter is graduated also with a 0–40°C scale on which leaf and RH sensor temperature can be shown. These temperatures are important because porometer calibration is dependent on both the leaf temperature and the differential between leaf and sensor. For this reason a field calibration kit is under development which consists of a set of perforated plates backed by saturated filter paper. These plates simulate leaf surfaces each with a particular known value of resistance and should allow verification of the porometer calibration under field conditions. Care has been taken to construct the cup and calibration plates from materials with very low water absorption properties since serious errors can otherwise be introduced.

This is a portable instrument: the overall size is 400 × 200 × 150 mm and the weight including a rechargeable 12 V nickel cadmium battery is 5 kg.

References

MONTEITH J.L. & BULL T.A. (1970) A diffusion resistance porometer for field use. II. Theory, calibration and performance. *J. appl. Ecol.* 7, 623–38.

STILES W. (1970) A diffusive resistance porometer for field use. I. Construction. *J. appl. Ecol.* 7, 617–22.

SZEICZ G., MONTEITH J.L. & DOS SANTOS J. (1964) A tube solarimeter to measure radiation among plants. *J. appl. Ecol.* 1, 169–74.

33 Standards of spectral power distribution for measuring spectral irradiance

G. P. ARNOLD *Fisheries Laboratory, Lowestoft, Suffolk*

Standard lamps

A tungsten lamp of calibrated intensity and known colour temperature can be used to produce a known absolute spectral irradiance on a selected area at a given distance. Figure 33.1a shows three suitable types of photometric lamp. The G.E.C. lamp (General Electric Co. Ltd., Hirst Research Centre, Wembley, Middx.) is a straight-wire, uniplanar tungsten filament vacuum lamp rated at 100 V 75 W, with an intensity of approximately 62 cd at 2400 K. The Wotan Wi41/G (AEG (Great Britain) Ltd., Chancery Lane, London) is also a straight-wire uniplanar tungsten filament lamp but is gas-filled. This lamp is rated at 30 V 6 A and at 2856 K has an intensity of about 250 cd. A black opaque coating covers half the surface of the bulb and a window is left open in the coating opposite the filament; this prevents light reflected by the filament support from travelling in the measuring direction. The NPL miner's lamp (National Physical Laboratory, Teddington, Middx.) has a pearl bulb and a transverse tungsten filament. It is rated at 4 V 1 A and has an intensity of about 4 cd at 2856 K. The fourth lamp is a Wotan mercury-cadmium discharge lamp (Hg Cd/10 miniature spectral lamp (AEG (Great Britain) Ltd.) producing a discontinuous line spectrum, which is used to calibrate the wavelength setting of a spectroradiometer.

Figure 33.1b shows the GEC tungsten lamp mounted on a 2 m photometric bench with a pair of plumb lines for setting the position of the filament on the bench scale, which is graduated in millimetres. The lamp is connected to a stabilized D.C. power supply unit (Model 50/10 SC, Farnell Instruments Ltd., Wetherby, Yorks.) and a digital voltmeter (LM 1240 digital multimeter, Solartron Electronic Group Ltd., Farnborough, Hants.) to monitor the voltage applied to it. The power supply box contains a second stabilized D.C. supply unit (Model SCV 31E-1. A.P.T. Electronic Industries Ltd., Byfleet, Surrey) for the Wotan Wi41/G lamp. To prevent stray light reaching the photometer, the photometric bench carries two opaque black screens and the walls of the laboratory are hung with matt blackout curtain; the bench is covered with synthetic velvet.

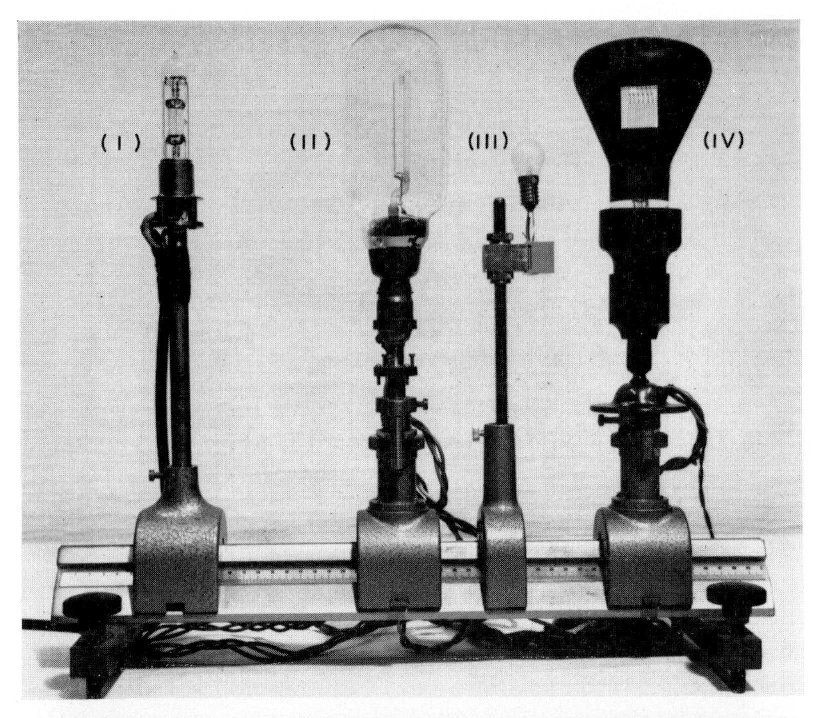

(a)

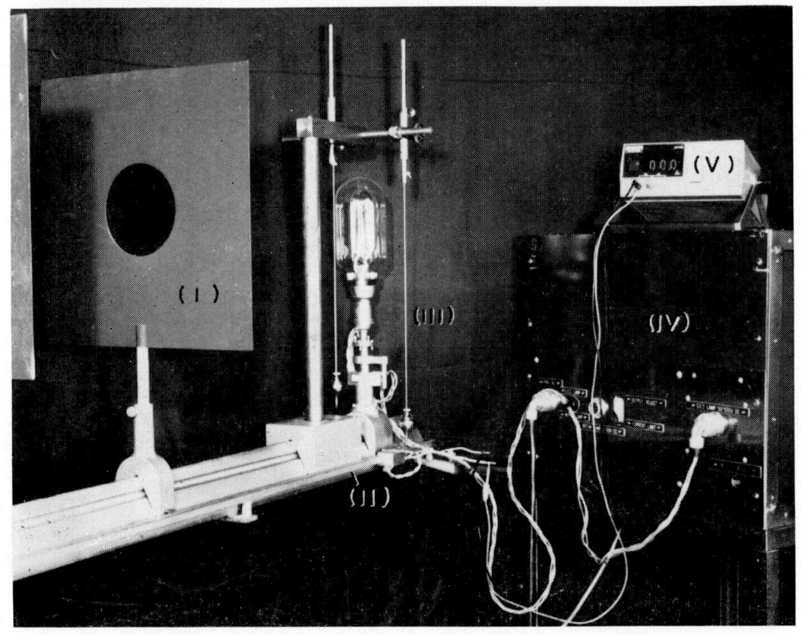

(b)

Figure 33.1. Standard photometric lamps. (**a**, upper) (i) Wotan HgCd/10 miniature spectral lamp, (ii) GEC tungsten filament lamp, (iii) NPL miner's lamp, (iv) Wotan Wi41/G lamp. (**b**, lower) GEC lamp mounted on photometric bench. (i) opaque black screen, (ii) bench scale, (iii) plumb line, (iv) power supply units, (v) digital voltmeter.

Figure 33.2 shows the National Physical Laboratory's calibration certificate for the Wotan Wi41/G lamp 7KA70.

Figure 33.3a shows the apparatus used for low-level photometric and radiometric calibrations. Light from the NPL miner's lamp is reflected by an aluminium plate, smoked with magnesium oxide (not shown), onto a disc with a range of small circular apertures. The aperture is used as a secondary source of known intensity and the photometer is set up at a measured distance from it. The whole of the inside of the box, including the lid (not shown), is painted matt black.

Spectroradiometers

Figure 33.3b shows a model 2020 SR laboratory photometer/spectroradiometer (Gamma Scientific Inc., San Diego, California). The photometer has a type PM 102 photomultiplier tube, with an S11 spectral response curve, filtered to match the V_λ function. The spectroradiometer has a Bausch & Lomb type 33–86–02 grating monochromator, with 1350 grooves per millimetre and a reciprocal linear dispersion of 6·4 nm mm^{-1}. It has a six-channel push-button attachment with independent sensitivity and dark-current controls on each channel. The monochromator covers the range 350–800 nm and the wavelength selection dial is graduated in 5 nm steps; the photomultiplier has a modified S20 spectral response curve. There are three pairs of entrance (height 20 mm) and exit slits (height 11·2 mm) with widths respectively of 5·36 and 3·0 mm (slits A), 2·68 and 1·5 mm (slits B) and 1·34 and 0·75 mm (slits C). The dimensions of the exit slit in each case are 0·56 times those of the entrance slit, equal to the magnification of the instrument. The theoretical half-power bandwidth of the three pairs of slits is 19·2, 9·6 and 4·8 nm, respectively. The performance of slits C has been checked by scanning the spectral lines of a Wotan HgCd/10 miniature spectral lamp and the actual bandwidth is approximately 5·5 nm. A second-order rejection filter eliminates overlapping ultraviolet radiation. The instrument has a large cosine receptor or alternatively a miniature cosine receptor mounted on a fibre-optic probe.

Figure 33.4a shows a QSM 2400 underwater spectroradiometer (UME-Instrument AB, Umeå, Sweden) and a deck-unit coupled to a Servoscribe 2 (Smith's Industries Ltd., Wembley, Middx.) potentiometric chart recorder. The underwater sensor, which is shown in detail in Fig. 33.4b, fits into a separate underwater case (not shown) with a maximum working depth of 80 m. The instrument scans the visible spectrum from 400 to 700 nm in 72 s and is calibrated in quanta s^{-1} cm^{-2} nm^{-1} v^{-1}. The travelling interference filter mounted under a wedge-shaped acrylic filter filled with copper sulphate

NATIONAL PHYSICAL LABORATORY

REPORT

Test of Incandescent Electric Lamps

SUBMITTED BY Department of Agriculture and Fisheries, Lowestoft, Suffolk

DESCRIPTION Wotan, W1 41/G with mask

ALIGNMENT Cap down, filament vertical. The plane of the filament was set normal to a line joining the centres of the filament and photometer screen. The mask on the lamp was towards the photometer and the filament was at the bench zero.

Laboratory mark	Consignor's mark	Rating	Results			
			Potential difference	Luminous intensity	Colour temperature	Current
			volts D.C.	candelas	°K	amperes
7 KA 70	1897	30V 6A	32.17	281.0	2856	6.010

Accuracy of candela values ± 1.0 per cent.

The results were obtained after the lamp(s) had been burning on a direct current supply at the test voltage for 10 minute(s) and do not necessarily apply to the lamp(s) after a different, especially a shorter, period of burning. The lamp(s) must not be operated at a voltage greater than the test voltage immediately prior to the stabilization period.

Preliminary burning (ageing) at this Laboratory } 25 hours at 30.0 volts A.C.

DATE 21st December 1970

MC 16/174

REFERENCE X.Phot.

MC 3129

for Director

A Laboratory Certificate or Report may not be published, except in full, unless permission for the publication of an approved abstract has been obtained in writing from the Director.

T. 222

(7086) Dd.180361 500 10/65 G.W.B.Ltd. Gp.863.

Figure 33.2. NPL calibration certificate for Wotan Wi41/G lamp.

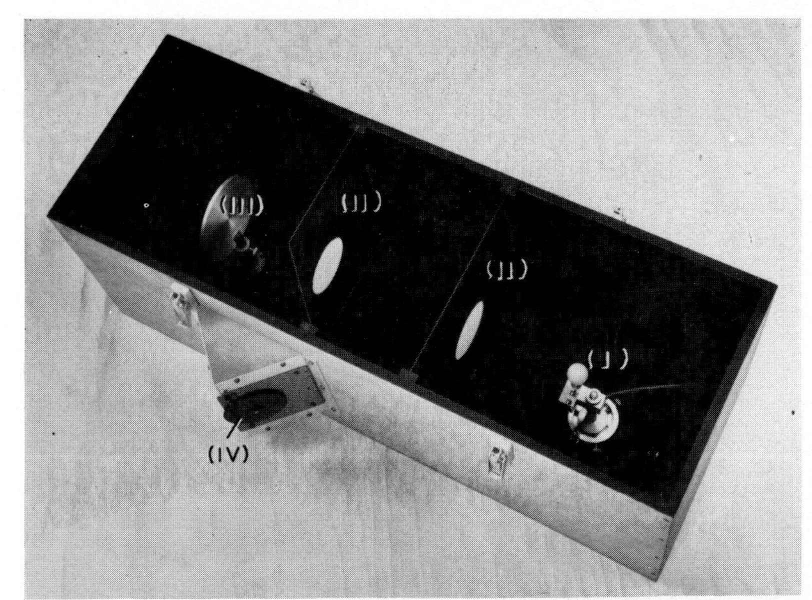

(a)

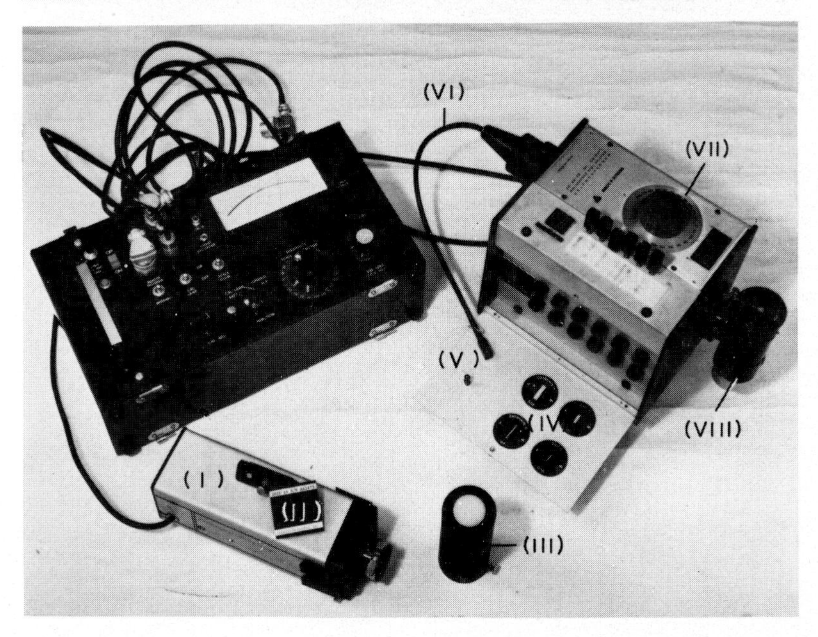

(b)

Figure 33.3. (**a**, upper) Box for low-level photometric and radiometric calibrations. (i) NPL miner's lamp, (ii) screens, (iii) aluminium plate, (iv) disc with apertures. (**b**, lower) Model 2020SR photometer/spectroradiometer. (i) photometer, (ii) V_λ filter, (iii) large cosine receptor, (iv) paired entrance and exit slits (A and C) for monochromator, (v) side- and end-on miniature cosine receptors, (vi) fibre-optic probe, (vii) wavelength dial, (viii) photomultiplier housing.

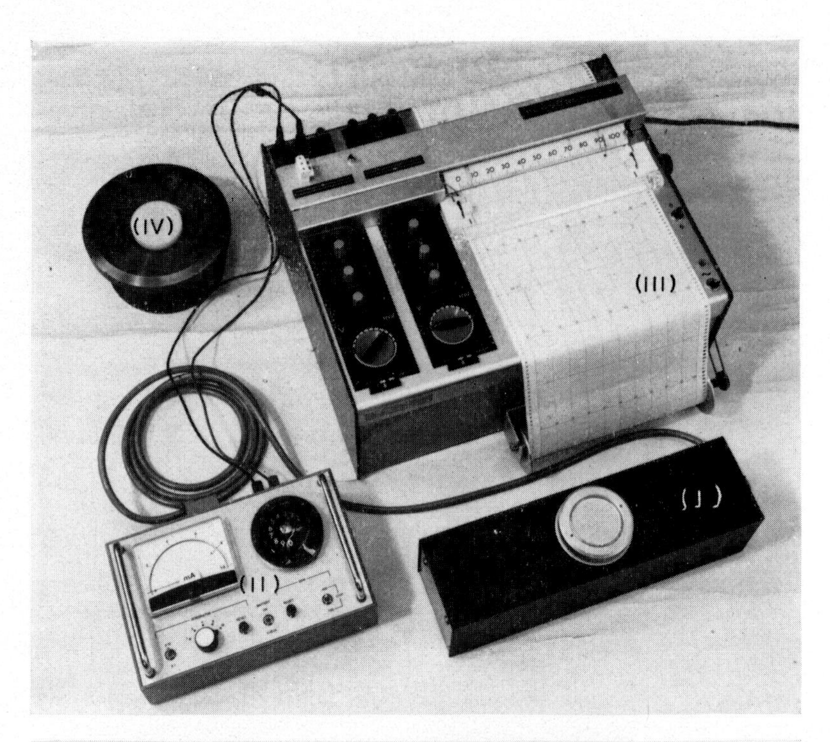

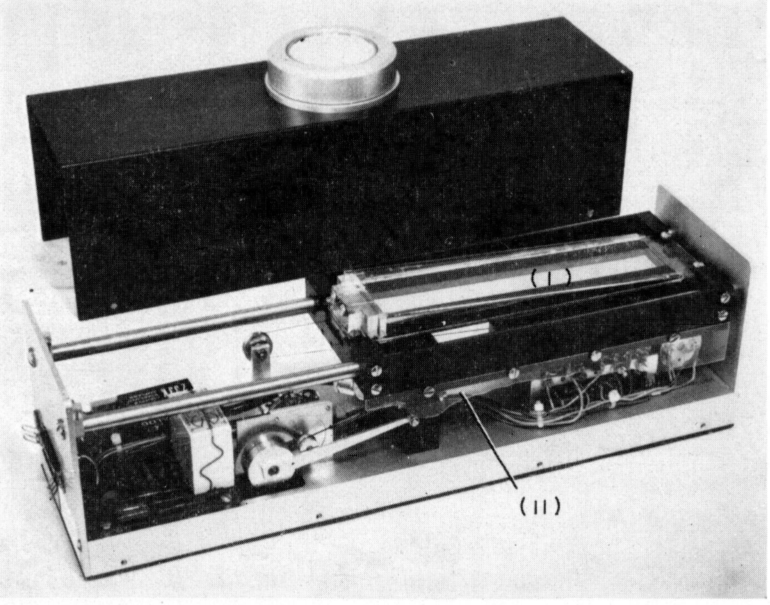

Figure 33.4. QSM 2400 quantaspectrometer. (*a*, upper) (i) underwater unit, (ii) deck unit, (iii) Servoscribe 2 potentiometric chart recorder, (iv) underwater cosine receptor (Smith, 1969). (*b*, lower) Contents of underwater unit. (i) traversing filter, (ii) cam.

solution can be seen in Fig. 33.4b. The irregular-shaped cam under the filter corrects for the spectral sensitivity of the photodiode. The amplification of the instrument is set by the telephone dial on the deck unit, which also incorporates an integrating unit. The trace on the chart recorder is a spectral irradiance curve for a warm white 65/80 W fluorescent tube. The underwater cosine collector in Fig. 33.4b is made to the dimensions given by Smith (1969).

Calibration

The spectral irradiance $E(\lambda)$ at the cosine receptor of a spectroradiometer set up at a measured distance D from the filament of a standard lamp of calibrated intensity I (cd) is given by

$$E(\lambda) = P(\lambda).\frac{\cos\theta}{D^2} \qquad \text{in W m}^{-2}\,\Delta\lambda^{-1},$$

where $P(\lambda)$ is the spectral radiant intensity of the source at any wavelength (λ) and

$$P(\lambda) = I.Q(\lambda) \qquad \text{in W sr}^{-1}\,\Delta\lambda^{-1}.$$

Values of $Q(\lambda)$ in μW sr^{-1} cd^{-1} $\Delta\lambda^{-1}$ for a range of colour temperatures are given in the NPL Tables of Spectral Power Distribution (Jones 1970). Figure 33.5 shows a data sheet for the 2020SR spectroradiometer, which incorporates values of $Q(\lambda)$ for $T = 2.400$ K; there are equivalent sheets for $T_c = 2{,}000$ K and $T_c = 2{,}856$ K. Program WP10P for the Fisheries Laboratory's Wang programmable calculating machine (370 system) calculates the spectral irradiance $E'(\lambda)$ equivalent to an experimental reading taken with the 2020SR spectroradiometer from

$$E'(\lambda) = \frac{x}{\Delta\lambda}\cdot E(\lambda)\cdot\frac{(ER)}{(CR)} \qquad \mu\text{Wcm}^{-2},$$

where x is the halfpower bandwidth of the monochromator slits in nm, $\Delta\lambda = 10$ nm and (ER) and (CR) are the experimental and equivalent calibration readings respectively. In addition to $E'(\lambda)$ the program prints out the original data, the calculated illuminance E ($E = [I\cos\theta]/D^2$) and spectral irradiance $E(\lambda)$ at the cosine receptor during calibration.

2020SR SPECTRORADIOMETER CALIBRATION & EXPERIMENTAL DATA — COLOUR TEMPERATURE 2400°K

Date	λ = ... nm	Job reference	Sheet number
Lamp	I = ... cd		
Voltage V	Cos θ = ...		
	D = ... cm		

Wavelength nm	Q (Tungsten)	Calibration readings	Experimental readings 1	2	3	4	5	6	Experimental readings 7	8	9	10
350	2·8											
360	2·946											
370	4·065											
380	5·493											
390	7·271											
400	9·45											
410	12·08											
420	15·21											
430	18·91											
440	23·21											
450	28·15											
460	33·76											
470	40·11											
480	47·23											
490	55·12											
500	63·82											
510	73·28											
520	83·53											
530	94·55											
540	106·3											
550	118·3											
560	131·9											
570	145·7											
580	159·9											
590	174·7											
			Change second order rejection filter									
600	190·2											
610	206·1											
620	222·7											
630	239·6											
640	257											
650	274·7											
660	292·6											
670	310·6											
680	328·6											
690	346·7											
700	364·9											
710	383·1											
720	401·1											
730	418·6											
740	436											
750	452·9											
760	469·9											
770	486·4											
780	502·6											
790	518·2											
800	533·3											

721483

Figure 33.5. Data sheet for 2020SR spectroradiometer with values of $Q(\lambda)$ (Jones 1970), for T_c 2400 K.

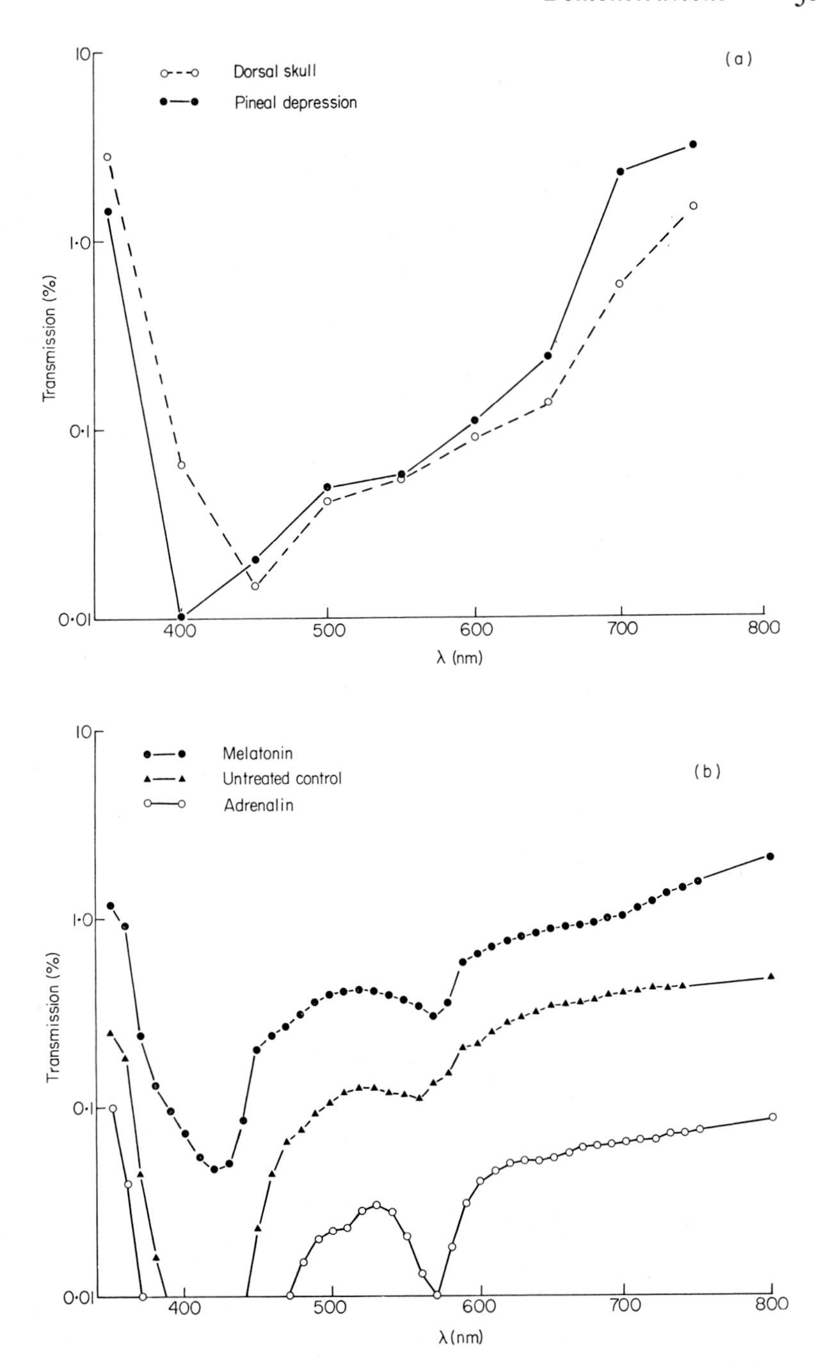

Figure 33.6. Spectral transmission of light through head of cod *Gadus morhua* L. (a) through pineal depression and adjacent area of skull, (b) through pineal depression after treatment with adrenalin and melatonin compared with control fish (V. J. Bye, Fisheries Laboratory, Lowestoft).

Transmission of light through cod head

Measurements of the transmission of light through the head of the cod *Gadus morhua* L. made by my colleague Mr. V. J. Bye demonstrate one use of the Gamma 2020SR spectroradiometer. The upper part of the head of a freshly-killed cod was mounted vertically on an optical bench and its dorsal surface illuminated by a tungsten lamp. The fibre-optic probe, without a cosine receptor, was used to measure the light penetrating the pineal depression and an adjacent area of the skull. The spectral transmission curves for a 55 cm long dark-adapted cod are shown in Fig. 33.6a. Spectral transmission curves for the pineal depression of 50–52 cm cod treated with melatonin, which contracts the melanophores of the skin, and adrenalin, which expands them, are compared with the curve for an untreated cod in Fig. 33.6b.

References

JONES O.C. (1970) Tables of spectral power distribution. *National Physical Laboratory, Division of Quantum Metrology, Report Qu. 14.*

SMITH R.C. (1969) An underwater spectral irradiance collector. *Journal of Marine Research* 27, 341–51.

The reference to proprietary products in this report should not be construed as an official endorsement of these products, nor is any criticism implied of similar products which have not been mentioned.

Addendum

Recent international comparisons have shown that, although the Wotan lamp is very stable and repeatable when handled carefully in the laboratory, significant changes can occur after travelling from one location to another. Two new types of photometric standard lamp have therefore been developed at the NPL, both to overcome this problem and to operate at colour temperatures well above that of the CIE illuminant A (2855.54 K). One is a colour temperature standard, the other a luminous intensity standard. The final prototype of the Clarke-Berry colour temperature lamp is shown in Fig. 33.7. The lamp has a conventional gas-filled glass envelope and is rated at 17 V 26 A at 3000 K. The filament comprises a pair of short, heavy-gauge self-supporting single-coil helixes with offset supports. Loose tungsten granules are enclosed in the lamp and they are periodically swirled round it to remove the heavy blackening that builds up inside the envelope. The luminous intensity standard is a more refined version of the same lamp with two side tubes each carrying an optical window (Clarke 1971).

Figure 33.7. Final prototype of the Clarke-Berry colour temperature standard lamp developed at the NPL. Both lamps have been operated for 650 h at 3000 K. The envelope of the left-hand lamp has been cleaned by swirling the loose tungsten granules enclosed in the lamp. (Photograph by courtesy of National Physical Laboratory, Teddington, Middx.)

Reference

CLARKE F.J.J. (1971) *Notes on a new type of photometric standard lamp.* 2 pp. (mimeo). National Physical Laboratory.

Author Index

Bold figures refer to pages where full references appear

Subject Index

A figure is indicated by the page number in bold type.
A plate is indicated by **Pl.** followed by the page number in bold type.
A table is indicated by the page number in italics.
Contractions and Roman letters used as symbols are placed alphabetically in the Index.
The few Greek letters used as symbols follow immediately below, before the main Index in Roman.